Mechanics of Deformable Solids: An Introduction

Keith D. Hjelmstad

Mechanics of Deformable Solids: An Introduction

Springer

Keith D. Hjelmstad
Ira A. Fulton Schools of Engineering
Arizona State University
Tempe, AZ, USA

ISBN 978-3-031-89872-3 ISBN 978-3-031-89873-0 (eBook)
https://doi.org/10.1007/978-3-031-89873-0

This Springer imprint is published by the registered company Springer Nature Switzerland AG
The registered company address is: Gewerbestrasse 11, 6330 Cham, Switzerland

For Kirsten and Annika

Preface

Mechanics of deformable solids is a fundamental subject for students studying engineering. With roots that go back hundreds of years, this subject forms the basis of many concepts used by civil, mechanical, and aerospace engineers to support the design of real-world systems. This book is accessible to engineering students in their sophomore year taking their first course on deformable solids. These students are the primary target audience. However, most chapters extend beyond what is typically covered in a first course, both to show that the basic ideas connect seamlessly with more advanced concepts and to serve as an invitation to further study of the topics.

While the nominal aim of learning mechanics of deformable solids is to sort out how solid bodies deform and transmit force when subjected to external loads, one might think of it also as basic training for developing and using mathematical models to formulate and solve engineering problems. Almost all engineers eventually make a living dealing with complex systems of one sort or another. The experience of taking a deformable solids problem from formulating theory through executing and interpreting computations is good preparation for dealing with more complicated problems.

The theory of deformable solids relies crucially on calculus, linear algebra, and differential equations, subjects required of all engineering students but subjects many do not yet use with full confidence. Employing these math concepts to solve real-world practical problems offers the chance to improve these math skills, many of which will be practically useful in other engineering contexts. Some topics, like tensors or integrating over an area to find properties of cross sections, may not have been covered in the prerequisite math courses. Those tools are built as needed from first principles.

A unique feature of this book is the focus on *derivation* of the equations governing deformable solids in a way that not only illuminates the theory, but also prepares the student for more advanced subjects. In a book with hundreds of equations, it may seem surprising that the key formulas can be summarized in a few pages. However, just starting with those end results fosters a plug-and-chug mentality, which is anathema to mastering the concepts. Today students need to understand *why* as much, if not more, than *how*. To understand the derivations is to understand the theories. The reader should return to the derivations often, using their experiences solving specific problems to help them through.

Throughout the book we use direct vector notation to do the math. This approach is a bit of a shift from traditional textbooks, which seem anxious to reduce equations to scalars as soon as possible. While students of mechanics at this level are often not yet fluent with vectors, the steady practice of using them will forge them into a useful set of tools. Using vectors religiously, even for problems that might be simple enough to work around them, opens up the vast advantages this notation can bring, particularly in understanding the geometry of two- and three-dimensional space.

Another unique feature of this book is the use of numerical computations to generate results. Traditional textbooks are largely silent on computational methods. There are many techniques that one must learn to do by hand, but stopping at that point is a lost opportunity to solve problems more broadly and to develop tools for exploration. Simply put, the computer is far better than the human at calculation. So, finding opportunities to offload computations to the computer should be embraced as a primary learning objective. Further, the discipline of coding up a solution strategy is a great way to fully understand the details. The logical structure of computing fosters logical thinking about solving problems.

Most of the MATLAB codes included in the book are organized so that the lines of code associated with printed or graphical output are relegated to functions that are not included in the book. Commercial codes generally have many more lines of code devoted to the graphical user interface than to the core analytical engine. The purpose of the codes in this book is to see the engine—to take a look under the hood—so that when you move on to the world of commercial codes, you have some idea of what those codes are doing with the theory. That said, the complete set of codes (including graphics) are available by contacting the author at keith.hjelmstad@asu.edu.

The book starts with an introduction to vector and tensor analysis. While most students have already encountered vectors before they get to the course on deformable solids, it is valuable to begin here because so much of the subsequent development depends not just on understanding what a vector is, but with developing fluency in the use of vectors to solve mechanics problems. This chapter also introduces the idea of tensors, a concept as vital to deformable solids as vectors are to force and motion. Most students will not have any prior knowledge of tensors. While there is a route through the book for which a modest understanding of tensors is sufficient, there is a huge payoff in understanding physical concepts like stress and strain that accrues from embracing the mathematical concept of tensors. We use direct vector notation because it provides the most compact derivations and the most geometric insight about force and deformation. This is a chapter to revisit any time a vector refresh is needed.

Chapter 2 is a brief, but reasonably complete, introduction to statics of rigid bodies. Equilibrium is a key concept in deformable solids. This chapter gives an opportunity to focus on the main ideas of statics, i.e., drawing free body diagrams and balancing forces and moments, without the complication of deformation or material behavior. Statics is often taught as a prerequisite to deformable solids. For readers with a background in statics, this chapter serves as a review and an opportunity to establish a footing with the notation used throughout the book.

In Chap. 3 we introduce the axial bar problem. This problem provides a fairly complete introduction to the basic ideas of deformable solids in the simplest possible mathematical framework. Here we introduce the concept of strain and show how it is related to displacement. We learn about stress, stress resultants, and stress formulas. We derive relationships between applied loads and displacement that form the basis of the problem solution methods that appear throughout the book, especially in the study of beam bending and torsion. As such, this chapter is a primer for the remainder of the book.

In Chaps. 4–6, we turn our attention to the problem of characterizing multiaxial states of strain and stress and their relationship through constitutive models. The aim of Chap. 4 is to define the idea of strain as a relative measure of deformation, to explain why it makes sense to represent strain as a tensor, and to explore what the implications of doing so are. To keep the math as simple as possible, we frame the ideas in the context of homogeneous deformation, addressing the general case at the end of the chapter. In Chap. 5 we introduce stress as another tensor quantity and show how the analysis of stress at a point is analogous to the analysis of strain. Cauchy's formula is introduced as the key to relating the internal state of stress to applied loads and to taking free body diagrams of three-dimensional solid bodies. We develop all the classical results regarding the *analysis* of strain and stress, i.e., principal values, principal directions, maximum shear strain, and the Mohr circle. Finally, Chap. 6 presents Hooke's law as a constitutive model relating stress and strain.

Chapter 7 introduces planar Bernoulli-Euler beam theory, starting with the fundamental kinematic hypothesis that plane cross sections remain plane after deformation and ending with a complete theory of planar beam bending. In Chap. 8 the beam equations from the previous chapter are linearized to open a route to developing classical solutions to practical problems. The solution methods, illustrated through several examples, mirror those first encountered in Chap. 3 for the axial bar problem.

Through the derivation of beam theory, it becomes evident that the geometric properties of the cross section of a beam are key elements of the theory. Quantities like area, centroid, and the first and second moments of area show up in virtually all the governing equations of these theories. Chapter 9 is entirely devoted to this topic, starting with methods based on the definitions of the properties, moving to practical calculation methods for cross sections composed of simple shapes, and culminating with a computational approach to solving the general problem of computing the properties for *any* cross section. The novel approach presented in this chapter shows both the power of calculus as well as the importance of coding. The code developed in this chapter can be used as a tool for exploring how the shape of the cross section can be used to achieve goals like efficiently resisting loads that cause bending or twisting.

Chapter 10 takes the ideas introduced in Chap. 8 and advances them to a point where we can write a code to do beam calculations automatically. Rethinking the solution methods for beams again from first principles and allowing some of the computations to be done numerically sheds new light on the computation of shear and moment diagrams, as well as the computation of deflections under different types of load and with different support conditions. The end product of this chapter is a code that can be used for exploration of beam problems. The code is extended to the case of a multispan beam to give an introduction to the basic ideas behind structural analysis and opens up the possibility of delving into concepts of structural design.

The computational strategies presented in Chaps. 3, 9, and 10 are best delivered through open-ended projects rather than through standard homework or exam problems. The codes can then serve as an engine to support inquiry, and mastery can be shown by summarizing the theory and observations in a written report. Traditional textbooks do not generally invite students to pursue these important alternate approaches to problem solving, but developing the ability to tackle open-ended problems and communicating ideas through writing are among the most important and durable skills an engineer can learn. In our own teaching, students complete six computing projects related to these subjects. The chapter end problems labeled as *computation* provide a starter set of prompts for these projects.

Chapter 11 covers the torsion of bars. The first part of the chapter concerns bars with circular cross sections. The developments are a good counterpoint to the chapter on the axial bar problem, as there are many similarities between the two problems. A kinematic hypothesis leads to equations that govern the relationship between torque and twist and also yields a formula for computing the internal stresses. Torsion of non-circular bars is an important topic because the kinematic hypothesis used for circular bars does not work well for commonly used open and closed thin-walled cross sections. The chapter ends with a brief introduction to methods appropriate for these important cases.

Chapter 12 is an introduction to the concepts of strength and stability of solid bodies. The main focus of the book is on linear elastic response of solids to applied load. Through a series of examples, the chapter gives a glimpse of what is at stake when the assumption of elasticity gives out, as it eventually does for all real materials, and how the capacity of a system can be limited by instability of equilibrium even if the material response is elastic. The book ends with an invitation to explore these important topics in greater detail through further study.

There are several concepts that have a longitudinal span across the book—concepts like solution of nonlinear algebraic equations, numerical solution of differential equations, numerical quadrature, and eigenvalue problems. These concepts show up in multiple places in the book, often arising from completely different physical contexts. Appendix A gathers these topics together and provides the background needed to understand what they are and how they work.

This book started out as a set of course notes meant to support a pedagogical framework called *The Mechanics Project*, which I founded in 2012. This project was an effort to rethink how students learn mechanics of deformable solids (as well as statics and dynamics). We flipped the classroom, added undergraduate teaching assistants to the instructional team, created a mastery-based grading system, and introduced computing projects. We created our own course materials because the traditional textbooks were not sufficiently supportive of the learning framework (e.g., eschewing computational methods, avoiding vectors and tensors, and often giving the process of derivation short shrift).

One of the main motivations for writing this book grew from introspecting on how the world has changed since giants like Newton, Cauchy, Euler, and Timoshenko walked the Earth. While the basic ideas of mechanics have not changed, the way we express them and how we use them has. Future engineers are not adequately served by knowing a few formulas and some shortcut methods for fast hand calculations. Engineering is not done on the backs of envelopes or on desert islands. A modest investment in math opens up so many productive ways of thinking about mechanics—often ways that are conceptually much simpler than some of the traditional approaches that are often cluttered with scalar algebra, applicable only in special cases, or steeped in analogies that must truly puzzle first-time learners.

I have learned a lot over the past decade about teaching and learning the concepts of deformable solids. I am indebted to my teaching partners Amie Baisely, Kristen Ward, and Efthalia Chatziefstratiou, who have taught mechanics with me these many years in the framework of *The Mechanics Project* with the material that grew to be this book. I am also indebted to all the students who passed through the course, each one providing a unique lesson on how humans learn. The most important thing I have learned from our students is that they *can* do the math, and when they fully engage theory, derivation, and computation, they retain the ideas better and they are more reliable in the execution of the associated computations.

Enjoy the book!

Tempe, USA
September 2024

Keith D. Hjelmstad

Contents

1 **Mathematical Preliminaries** 1
1.1 Vectors 1
1.1.1 Vector Operations and Vector Algebra 3
1.1.2 Unit Vectors 9
1.1.3 Base Vectors and Vector Components 9
1.1.4 Vector Projection 16
1.2 Tensors 18
1.2.1 Matrix Representation 18
1.2.2 Change of Basis 20
1.2.3 Tensor Invariants 24
1.2.4 Eigenvalues and Eigenvectors of a Tensor 26
1.2.5 The Spectral Representation 33
1.3 Summary 34
1.4 Problems 35

2 **Statics** 39
2.1 Newton's Laws of Motion 40
2.2 Static Equilibrium 40
2.3 Free Body Diagrams 42
2.4 Support Idealizations 46
2.5 Distributed Forces 51
2.6 Resultants 54
2.7 Applied Loads 58
2.8 How Many Equations Are There? 62
2.9 Structures 65
2.9.1 Trusses 66
2.9.2 Computation: Truss Analysis 72
2.9.3 Frames and Machines 81
2.10 Summary 84
2.11 Problems 85

3 The Axial Bar ... 93
3.1 Theory of Axial Bars ... 93
3.1.1 Deformation and Strain ... 94
3.1.2 Force and Stress ... 96
3.1.3 Boundary Conditions ... 98
3.1.4 Equations of Equilibrium ... 99
3.1.5 Elastic Constitutive Model ... 103
3.1.6 Axial Force-Displacement Relationship ... 104
3.1.7 Stress Formula ... 105
3.2 Solving Axial Bar Problems ... 108
3.2.1 Static Indeterminacy ... 109
3.2.2 Nonprismatic Bars ... 112
3.3 Computational Strategy ... 118
3.3.1 Formulation and Code ... 118
3.3.2 Study: What Is a Point Load? ... 127
3.4 Summary ... 129
3.5 Problems ... 130

4 Motion, Deformation, and Strain ... 137
4.1 Motion and Deformation ... 137
4.2 Homogeneous Deformation ... 139
4.2.1 Rigid Body Motion ... 142
4.2.2 Determining **F** from the Motion ... 143
4.3 Definition of Strain ... 146
4.3.1 Common Strain Measures ... 146
4.3.2 The Strain Tensor ... 149
4.3.3 Shear Strain ... 151
4.4 Plane Strain ... 155
4.4.1 Strain Invariants ... 157
4.4.2 The Principal Angle and Maximum Shear Angle ... 158
4.4.3 The Mohr Circle ... 162
4.4.4 Strain Gauges ... 165
4.5 Analysis of Strain in Three Dimensions ... 166
4.5.1 Principal Values and Principal Directions ... 167
4.5.2 The Mohr Circle in Three Dimensions ... 171
4.5.3 Plane Strain Revisited ... 172
4.5.4 Volume Change ... 174
4.6 Nonhomogeneous Deformation ... 176
4.6.1 The Deformation Gradient ... 176
4.6.2 Strain ... 179
4.6.3 Strain-Displacement Relationship ... 181
4.7 Summary ... 183
4.8 Problems ... 184

5 Force and Stress 191
5.1 Cauchy's Formula and the Stress Tensor 192
5.2 Normal and Shear Stresses 198
5.3 Plane Stress 199
5.3.1 Stress Transformation Equations 200
5.3.2 Principal Angle and Maximum Shear Angle 201
5.3.3 The Mohr Circle 202
5.4 Analysis of Stress in Three Dimensions 204
5.4.1 Principal Values and Principal Directions 205
5.4.2 The Mohr Circle in Three Dimensions 208
5.4.3 Plane Stress Revisited 210
5.5 Differential Equations of Equilibrium 211
5.6 Summary 218
5.7 Problems 218

6 Constitutive Models 221
6.1 Hooke's Law 222
6.2 Young's Modulus and Poisson's Ratio 224
6.3 Properties of Common Materials 227
6.4 Plane Stress and Plane Strain 229
6.5 Principal Directions of Stress and Strain 231
6.6 Combining Kinematics, Equilibrium, and Constitution 231
6.7 Other Models 237
6.8 Summary 238
6.9 Problems 238

7 Beam Theory 247
7.1 Beam Kinematics 248
7.1.1 Motion of a Beam Cross Section 248
7.1.2 Strain-Displacement Relationships 250
7.1.3 Normal Strain 253
7.2 Stress Resultants and Equilibrium 254
7.3 Constitutive Equations for Resultants 258
7.4 Beam Sign Conventions 259
7.5 Stress Formulas 260
7.5.1 Normal Stress Formula 260
7.5.2 Shear Stress Formula 261
7.6 Summary 264

8 Linear Theory of Beams 265
8.1 Linearization of the Beam Equations 265
8.2 Boundary Conditions 268
8.3 Continuity Conditions 270
8.4 Approaches to Solving Linear Beam Problems 272
8.4.1 Beam Statics 273
8.4.2 The Internal Hinge 277

8.4.3 Beam Deflections and Rotations 281
8.4.4 Statically Indeterminate Beams 287
8.4.5 Nonprismatic Beams 291
8.4.6 Structures 294
8.4.7 Normal and Shear Stresses 300
8.5 Summary 302
8.6 Problems 302

9 Properties of Cross Sections 315
9.1 Basic Approach 315
9.2 Computations for Composite Areas 326
9.3 Algorithm for Polygonal Cross Sections 331
9.3.1 The Divergence Theorem 336
9.3.2 Polygonal Cross Sections 339
9.3.3 Code Implementation 341
9.3.4 Computing Q 345
9.4 Summary 351
9.5 Problems 351

10 A Computational Approach to Beams 355
10.1 Integrating the Governing Differential Equations 355
10.2 The Moment-Area Equations 362
10.3 Automating the Solution Process 366
10.4 Code Implementation for Distributed Loads 368
10.5 Point Loads 373
10.6 Multispan Beam 380
10.7 Summary 390
10.8 Problems 391

11 Torsion 395
11.1 What is Torsion? 396
11.2 Torsion of Bars with Circular Cross Sections 397
11.2.1 Kinematic Hypothesis 397
11.2.2 Strain-Rotation Relationship 397
11.2.3 Resultant Torque and Equilibrium 399
11.2.4 Boundary Conditions 404
11.2.5 Constitutive Model for Torsion 405
11.2.6 Torque-Rotation Relationship 406
11.2.7 Torsion Stress Formula 409
11.3 Solving Torsion Problems 411
11.3.1 Static Indeterminacy 413
11.3.2 Nonprismatic Bars 416

11.4 Torsion of Bars with Non-circular Cross Sections 423
11.4.1 Non-circular Solid Sections 423
11.4.2 Open Thin-Walled Cross Sections 425
11.4.3 Closed Thin-Walled Cross Sections 428
11.5 Summary 434
11.6 Problems 434

12 Strength and Stability 441
12.1 Material Strength 442
12.2 Effects of Inelasticity 445
12.2.1 Torsion 445
12.2.2 Beam Bending 447
12.3 Stability of Static Equilibrium 450
12.3.1 Stability of a Simple System 451
12.3.2 Discrete Systems with Multiple Variables 456
12.3.3 Planar Beam Buckling 459
12.3.4 The Rayleigh-Ritz Method 472
12.4 Epilogue 482

Appendix A: Math Supplement 483

Index 505

Chapter 1
Mathematical Preliminaries

Starting an introductory book on the mechanics of deformable solids with a chapter on mathematical preliminaries is a compromise between the desire to get into the physical nature of the subject and the need to have a language to do it. Mechanics is a *mathematical model* of how the world works, and the mathematical objects we call *vectors* and *tensors* are a key part of the language required to translate physical concepts into mathematical statements. Vectors represent quantities like position, force, and displacement. Tensors represent quantities like stress, strain, and moments of inertia. Because these building blocks represent multiple disparate quantities, it is logical to start by describing what is true about vectors and tensors in general, independent of their use in the physical theory. We will get more specific about how they manifest in the mechanics of deformable solids in subsequent chapters.

Most readers will already be familiar with the concept of vectors and their operation. The review provided here establishes the notation used throughout the book. The concept of tensors may be new to many readers. The introduction to tensors in this chapter builds on knowledge of vectors and brings the subject just up to the point needed in subsequent chapters. It is important to gain facility with the math so that math operations are not an obstacle to understanding physical concepts. Read this chapter to get set for what follows, but plan on coming back to it as vectors and tensors appear naturally throughout the book.

1.1 Vectors

A vector is a directed line segment. It has two properties: (1) magnitude and (2) direction. We represent a vector with an arrow. The length of the arrow is the magnitude, and the orientation of the arrow is the direction. Vectors are basic *geometric* building blocks for describing force and motion in mechanics.

K. D. Hjelmstad, *Mechanics of Deformable Solids: An Introduction*,
https://doi.org/10.1007/978-3-031-89873-0_1

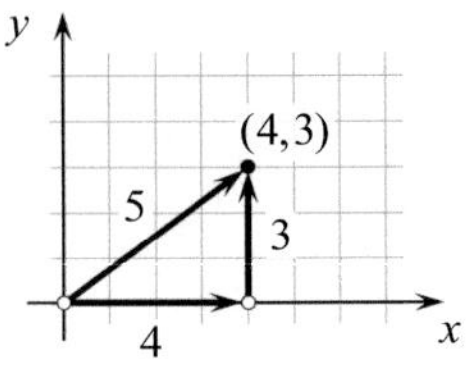

Fig. 1.1 *Coordinate geometry.* To describe the location of a point in two-dimensional space we must independently record distance along the x-axis and distance along the y-axis

One way to begin to appreciate the power and utility of vectors is to contrast them with scalars. Scalars live in the one-dimensional world of the real line wherein the notion of *direction* is limited to positive and negative, and the main property is that one scalar may be greater than, less than, or equal to another. When we endow scalars with an algebra, then we can build a mathematical system that can do useful things like write and solve equations.

What scalars cannot do is deal with *direction* in multiple dimensions. For example, to describe the location of a point in two-dimensional space, we can use *coordinate geometry*, as illustrated in Fig. 1.1. The point (4, 3) requires *two* values—the coordinates—to describe it, the first indicating how far to go in the x-direction (four units) and the second how far to go in the y-direction (three units). The distance from the origin to the point is $5 = \sqrt{4^2 + 3^2}$. But there are an infinite number of points in two-dimensional space that are a distance 5 from the origin. Hence, that single scalar value is not enough to distinguish among those points. We can always compute the distance from the coordinates, but we cannot compute the coordinates from the distance.

The preceding example is an invitation to how vectors might be useful in describing quantities that live in two- or three-dimensional space. The location of the point (4, 3) can be described with *one* arrow going from the origin to the point. It is clear that the three arrows in Fig. 1.1 point in different directions and that each is associated with a different task. It is also clear how the lengths of those arrows are related to each other, thanks to the Pythagorean theorem.

Two typical vectors, $\mathbf{u}$ and $\mathbf{v}$, are shown in Fig. 1.2. They are distinguished from one another by having different lengths and pointing in different directions. We can *see* that those vectors have different lengths and directions without laying out a coordinate system. Vectors, thought of as directed line segments, do not rely on a coordinate system, in much the same way that the concept of temperature does not depend on whether you choose to describe it with the Fahrenheit or Celsius scales. However, we will soon see that introducing a coordinate system and components will be beneficial when working with vectors.

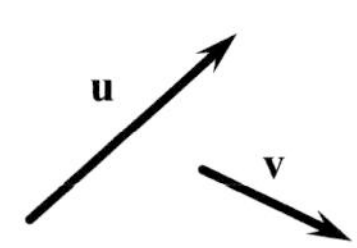

Fig. 1.2 *Vectors.* Two typical vectors, $\mathbf{u}$ and $\mathbf{v}$. Vectors are directed line segments. They have direction and magnitude

1.1.1 Vector Operations and Vector Algebra

Vectors are mathematical objects. To be useful, they must be endowed with operational characteristics: addition, subtraction, and multiplication. In this section, we define those operations.

Vector Addition Vectors are added using the *head-to-tail* rule (also known as the *parallelogram rule*), as shown in Fig. 1.3. Adding $\mathbf{u}$ to $\mathbf{v}$ results in the vector $\mathbf{u}+\mathbf{v}$. Geometrically, the resultant vector extends from the tail of $\mathbf{u}$ to the tip of $\mathbf{v}$. If we create a parallelogram by adding sides parallel to and of the same length as the vectors $\mathbf{u}$ and $\mathbf{v}$, then the vector along the diagonal of the parallelogram is the sum of the two vectors.

Vector Subtraction Vector subtraction can be viewed as a special case of vector addition if we first define what we mean by the negative of a vector. Imagine a vector $\mathbf{v}$ that when added to $\mathbf{u}$ gives zero (there and back again). Algebraically, we might write that as

$$\mathbf{u}+\mathbf{v}=\mathbf{0} \qquad \Rightarrow \qquad \mathbf{v}=-\mathbf{u} \tag{1.1}$$

Hence, the negative of a vector is that same vector just pointing in the opposite direction, as shown in Fig. 1.4.

In order for Eq. 1.1 to make sense, we must define the *zero vector*, which is designated as $\mathbf{0}$. The zero vector has zero length. Interestingly, that implies that it does not have (or need to have) a well-defined direction. The zero vector added to any other vector $\mathbf{u}$ in the space will give back that same vector, i.e., $\mathbf{u}+\mathbf{0}=\mathbf{u}$. The scalar zero is not sufficient to describe when an operation on vectors happens to produce the zero vector, because when you add two vectors together, you get another vector (not a scalar).

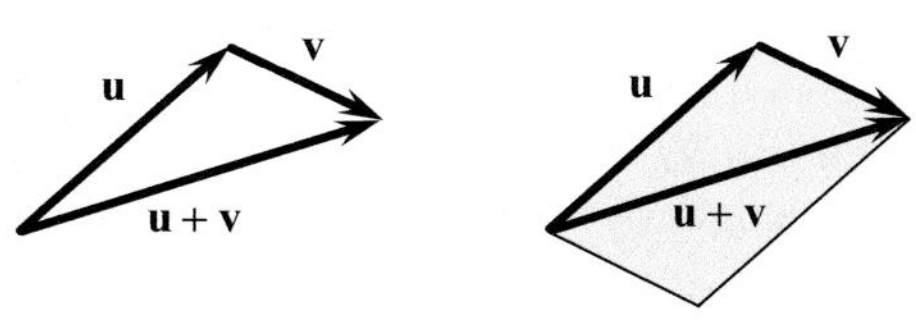

Fig. 1.3 *Vector addition.* Vectors $\mathbf{u}$ and $\mathbf{v}$ add to get the vector $\mathbf{u}+\mathbf{v}$. The operation is defined by the *head-to-tail rule* (left) or, equivalently, the *parallelogram rule* (right)

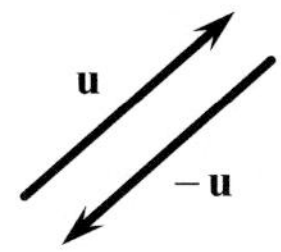

Fig. 1.4 *A vector and its negative.* The negative of the vector $\mathbf{u}$ is the vector $-\mathbf{u}$. A vector and its negative sum to the zero vector

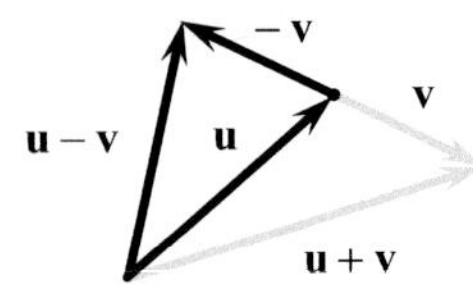

Fig. 1.5 *Vector subtraction.* The difference of two vectors **u** and **v** is the sum of the vector **u** and the negative of the vector **v**

The definition of the negative of a vector allows us to define *vector subtraction* to be vector addition with the negative of the second vector, i.e.:

$$\mathbf{u} - \mathbf{v} = \mathbf{u} + (-\mathbf{v}).$$

This operation is illustrated in Fig. 1.5, with the vector **v** both added to (as it was in Fig. 1.3) and subtracted from the vector **u** to get the sum and difference of the vectors, respectively.

Vector Multiplication There are several ways to multiply vectors, including multiplication by a scalar, dot product, cross product, and tensor product. We denote the various products with the following notation:

$a\mathbf{u}$	multiplication by a scalar
$\mathbf{u} \cdot \mathbf{v}$	dot product
$\mathbf{u} \times \mathbf{v}$	cross product
$\mathbf{u} \otimes \mathbf{v}$	tensor product

where **u** and **v** are vectors and a is a scalar. It is important to note the nature of the object that results from vector multiplication. The dot product results in a scalar. Multiplication by a scalar and the cross product result in a vector. The tensor product results in a tensor (more on that later). We can also combine these multiplication methods, as we do in the case of the *triple scalar product* of three vectors, i.e., $(\mathbf{u} \times \mathbf{v}) \cdot \mathbf{w}$, which results in a scalar.

A vector can be *multiplied by a scalar*, i.e., $\mathbf{u} \rightarrow a\mathbf{u}$, as shown in Fig. 1.6a. The result simply changes the length of the vector by multiplying the magnitude by the scalar a. If a is positive, then $\|a\mathbf{u}\| = a\|\mathbf{u}\|$. Since we have defined the negative of

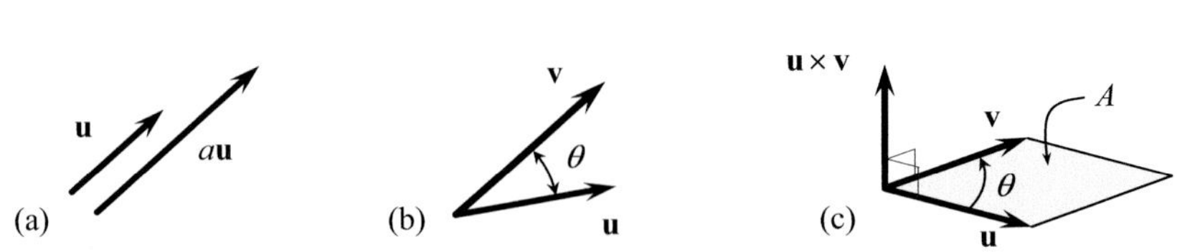

Fig. 1.6 *Vector multiplication.* **a** A vector **u** multiplied by a scalar a. **b** The dot product between two vectors **u** and **v** is equal to the cosine of the angle θ between them multiplied by the lengths of the two vectors. **c** The cross product of two vectors **u** and **v** is a vector perpendicular to both with magnitude A, which is the area of the parallelogram defined by the two vectors

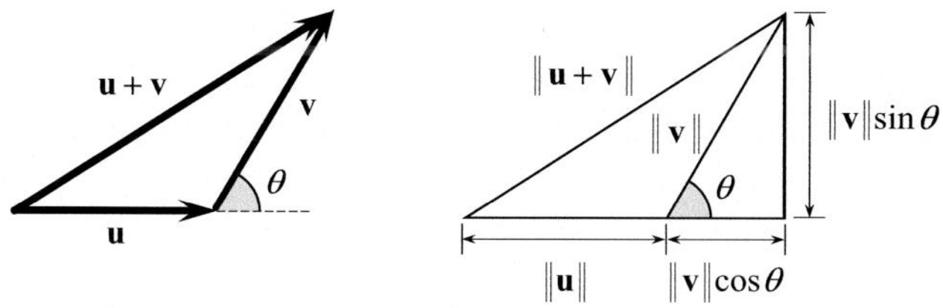

Fig. 1.7 *Proof of the dot product formula.* The dot product can be defined in terms of the cosine of the angle between vectors and their lengths

a vector to simply be the vector pointing in the opposite direction, we can interpret negative values of the scalar multiplier a as first changing the length of the vector (using the absolute value of a) and then reversing the direction.

The *dot product* can be defined through the equation

$$\mathbf{u} \cdot \mathbf{v} = \|\mathbf{u}\| \, \|\mathbf{v}\| \cos\theta \tag{1.2}$$

It should be evident from this formula that the dot product is *commutative*. In mathematical notation, that means

$$\mathbf{u} \cdot \mathbf{v} = \mathbf{v} \cdot \mathbf{u} \tag{1.3}$$

We can prove Eq. 1.2 if we assume that the magnitude of a vector is the square root of the vector dotted with itself. To wit,

$$\|\mathbf{u}\| = \sqrt{\mathbf{u} \cdot \mathbf{u}}\,.$$

Now, consider the situation shown in Fig. 1.7. The sketch on the left shows the sum of two vectors **u** and **v** and the angle θ between them. The sketch on the right shows two right triangles derived from the sketch on the left. By the Pythagorean theorem, we can write

$$\begin{aligned} \|\mathbf{u} + \mathbf{v}\|^2 &= \big(\|\mathbf{u}\| + \|\mathbf{v}\| \cos\theta\big)^2 + \big(\|\mathbf{v}\| \sin\theta\big)^2 \\ &= \|\mathbf{u}\|^2 + 2\,\|\mathbf{u}\|\,\|\mathbf{v}\| \cos\theta + \|\mathbf{v}\|^2 \cos^2\theta + \|\mathbf{v}\|^2 \sin^2\theta \\ &= \|\mathbf{u}\|^2 + 2\,\|\mathbf{u}\|\,\|\mathbf{v}\| \cos\theta + \|\mathbf{v}\|^2 \end{aligned} \tag{1.4}$$

where we have noted that $\sin^2\theta + \cos^2\theta = 1$. This result is known as the *law of cosines*. We can write the square of the length of the sum of the two vectors using the dot product and distributing the products. To wit,

$$\begin{aligned} \|\mathbf{u} + \mathbf{v}\|^2 &= \big(\mathbf{u} + \mathbf{v}\big) \cdot \big(\mathbf{u} + \mathbf{v}\big) \\ &= \mathbf{u} \cdot \mathbf{u} + \mathbf{u} \cdot \mathbf{v} + \mathbf{v} \cdot \mathbf{u} + \mathbf{v} \cdot \mathbf{v} \\ &= \|\mathbf{u}\|^2 + 2\,\mathbf{u} \cdot \mathbf{v} + \|\mathbf{v}\|^2 \end{aligned} \tag{1.5}$$

where we have made use of the fact that the dot product is commutative. It should be evident by comparing Eq. 1.4 with Eq. 1.5 that Eq. 1.2 holds. Notice, also, that Eq. 1.5 also provides a route to defining the dot product of vectors as

$$\mathbf{u} \cdot \mathbf{v} = \tfrac{1}{2}\big(\|\mathbf{u} + \mathbf{v}\|^2 - \|\mathbf{u}\|^2 - \|\mathbf{v}\|^2\big) \tag{1.6}$$

which is interesting because it defines the dot product without referring to the angle θ, as in Eq. 1.2. The subtle implication is that the dot product actually gives a route to *defining* the cosine of the angle θ as the dot product between two unit vectors with angle θ between them. Of course, both equations are true, and the form with θ is more commonly used.

It may have seemed odd to simply declare the length of a vector as the square root of the vector dotted with itself. We can use Eq. 1.2 to verify that this assertion is consistent with our definition of the dot product. If we let $\mathbf{v} = \mathbf{u}$ and note that $\cos(0) = 1$, then Eq. 1.2 gives $\mathbf{u} \cdot \mathbf{u} = \|\mathbf{u}\|^2$. We can think of the assumption about length as the primary requirement of the dot product.

The *cross product* of vectors at first seems like an odd way to multiply vectors. It turns out to be very useful in mechanics for tasks like defining moments caused by forces. The cross product $\mathbf{u} \times \mathbf{v}$ is defined to be a vector with magnitude A, the area of the parallelogram defined by the two vectors $\mathbf{u}$ and $\mathbf{v}$ shown in Fig. 1.6c, pointing in a direction that is perpendicular to both $\mathbf{u}$ and $\mathbf{v}$. There are two choices for that perpendicular direction. The one we adopt here, which is standard, is defined by the *right-hand rule*, which states that using the right hand and making the curl of the fingers go from the first vector $\mathbf{u}$ to the second vector $\mathbf{v}$, then the resulting vector points in the direction of the thumb. While the dot product is *commutative* (i.e., $\mathbf{u} \cdot \mathbf{v} = \mathbf{v} \cdot \mathbf{u}$), the cross product is not. In fact, the cross product is *anticommutative*, satisfying the property

$$\mathbf{u} \times \mathbf{v} = -\mathbf{v} \times \mathbf{u} \tag{1.7}$$

The magnitude of the vector $\mathbf{u} \times \mathbf{v}$ is given by the formula

$$A = \|\mathbf{u} \times \mathbf{v}\| = \|\mathbf{u}\|\,\|\mathbf{v}\| \sin\theta \tag{1.8}$$

as illustrated in Fig. 1.8. We will develop formulas for computing the dot and cross products of vectors once we introduce base vectors and vector components.

The *triple scalar product* is a combination of the dot and cross products of vectors, defined as

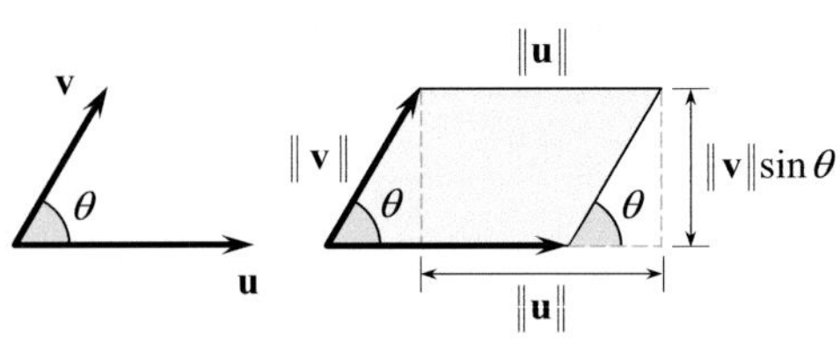

Fig. 1.8 *Magnitude of cross product.* The magnitude of the cross product $\mathbf{u} \times \mathbf{v}$ is the area of the parallelogram defined by the two vectors

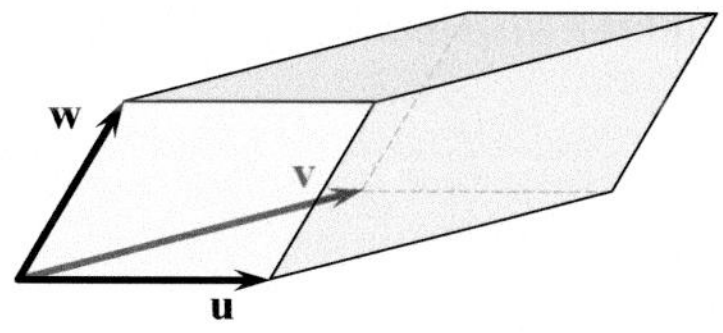

Fig. 1.9 *Triple scalar product.* The value of the triple scalar product $(\mathbf{u} \times \mathbf{v}) \cdot \mathbf{w}$ is equal to the volume of the parallelepiped defined by the three vectors

$$\big(\mathbf{u} \times \mathbf{v}\big) \cdot \mathbf{w} \tag{1.9}$$

It is important to group the cross product of vectors together because the dot product must be between two vectors and the cross product produces a vector while the dot product produces a scalar. The triple scalar product obeys the cyclic reordering of the vectors

$$\big(\mathbf{u} \times \mathbf{v}\big) \cdot \mathbf{w} = \big(\mathbf{w} \times \mathbf{u}\big) \cdot \mathbf{v} = \big(\mathbf{v} \times \mathbf{w}\big) \cdot \mathbf{u} \tag{1.10}$$

Because $\mathbf{u} \times \mathbf{v} = -\mathbf{v} \times \mathbf{u}$, any reordering of the vectors that is not cyclic changes the sign of the triple scalar product. Note also that if any vector appears twice in the triple scalar product, e.g., $(\mathbf{u} \times \mathbf{v}) \cdot \mathbf{u}$, then the result is zero. The reason is simple. The cross product produces a vector perpendicular to both vectors involved in the cross product. If we then dot that result with either one of the original vectors, we get zero because they are perpendicular to each other.

One interpretation of the physical significance of the triple scalar product is shown in Fig. 1.9. The value of the triple scalar product is equal to the volume of the parallelepiped whose edges are defined by the three vectors (and lines parallel to them).

Another way of multiplying two vectors is called the *tensor product* or *outer product* of vectors. This product is particularly important to mechanics of deformable bodies because it forms the basis of what we call a *tensor*. Symbolically, we will denote the tensor product between vectors $\mathbf{u}$ and $\mathbf{v}$ as $\mathbf{u} \otimes \mathbf{v}$. To understand what the tensor product is, we need to understand what it does. The fundamental operation of the tensor product is *defined* by

$$\big[\,\mathbf{u} \otimes \mathbf{v}\,\big]\mathbf{w} = \mathbf{u}\big(\mathbf{v} \cdot \mathbf{w}\big) \tag{1.11}$$

for *any* vector $\mathbf{w}$. Note that we can write $\mathbf{u}a = a\mathbf{u}$, i.e., we can put the scalar a either in front of the vector $\mathbf{u}$ or after it without changing the value. Hence, we can also write Eq. 1.11 as $[\,\mathbf{u} \otimes \mathbf{v}\,]\mathbf{w} = (\mathbf{v} \cdot \mathbf{w})\mathbf{u}$.

Vector algebra is similar to scalar algebra in many ways, with some additional sensitivity to order of operation and order of appearance. Rules of vector algebra include the *commutative* and *associative* laws of vector addition, respectively,

$$\begin{aligned} \mathbf{u} + \mathbf{v} &= \mathbf{v} + \mathbf{u} \\ (\mathbf{u} + \mathbf{v}) + \mathbf{w} &= \mathbf{u} + (\mathbf{v} + \mathbf{w}) \end{aligned} \tag{1.12}$$

and the *distributive* laws of vector addition with scalar multiplication:

$$
\begin{aligned}
a(\mathbf{u}+\mathbf{v}) &= a\mathbf{u}+a\mathbf{v} \\
a\mathbf{u}+b\mathbf{u} &= (a+b)\mathbf{u}
\end{aligned}
\tag{1.13}
$$

where **u**, **v**, and **w** are vectors and a and b are scalars. The commutative and associative laws imply that the order of adding vectors does not matter, a fact we will use frequently in our derivations. Vector multiplication satisfies distributive laws for vectors **u**, **v**, and **w** and scalars a, b, and c as follows:

$$
\begin{aligned}
(a\mathbf{u})\cdot\left(b\mathbf{v}+c\mathbf{w}\right) &= ab\left(\mathbf{u}\cdot\mathbf{v}\right)+ac\left(\mathbf{u}\cdot\mathbf{w}\right) \\
(a\mathbf{u})\times\left(b\mathbf{v}+c\mathbf{w}\right) &= ab\left(\mathbf{u}\times\mathbf{v}\right)+ac\left(\mathbf{u}\times\mathbf{w}\right) \\
(a\mathbf{u})\otimes\left(b\mathbf{v}+c\mathbf{w}\right) &= ab\left(\mathbf{u}\otimes\mathbf{v}\right)+ac\left(\mathbf{u}\otimes\mathbf{w}\right) \\
\left(a\mathbf{u}\times b\mathbf{v}\right)\cdot(c\mathbf{w}) &= abc\left(\mathbf{u}\times\mathbf{v}\right)\cdot\mathbf{w}
\end{aligned}
\tag{1.14}
$$

In vector multiplication, the scalar multipliers can move anywhere in the expression without changing it. Usually, all scalars are pulled out front before the vector operations are executed using the basic operations discussed previously.

Example 1.1
Use the distributive law of vector multiplication to expand the product:

$$\left(a\mathbf{u}+b\mathbf{v}\right)\cdot\left(c\mathbf{x}+d\mathbf{y}\right)$$

where a, b, c, and d are scalars and **u**, **v**, **x**, and **y** are vectors.

Solution. Expand the product as follows:

$$
\begin{aligned}
\left(a\mathbf{u}+b\mathbf{v}\right)\cdot\left(c\mathbf{x}+d\mathbf{y}\right) &= a\mathbf{u}\cdot c\mathbf{x}+a\mathbf{u}\cdot d\mathbf{y}+b\mathbf{v}\cdot c\mathbf{x}+b\mathbf{v}\cdot d\mathbf{y} \\
&= ac\left(\mathbf{u}\cdot\mathbf{x}\right)+ad\left(\mathbf{u}\cdot\mathbf{y}\right)+bc\left(\mathbf{v}\cdot\mathbf{x}\right)+bd\left(\mathbf{v}\cdot\mathbf{y}\right)
\end{aligned}
$$

While this example concerns the dot product, the same rules apply to the cross and tensor products, too.

The distributive law works the same as in scalar algebra, but the order of appearance of the vectors in the products must be respected. For example, in scalar algebra, we can write $(a+b)(a+b)=a^2+2ab+b^2$ because $ab=ba$. For the tensor product, we write

$$\left(\mathbf{u}+\mathbf{v}\right)\otimes\left(\mathbf{u}+\mathbf{v}\right)=\mathbf{u}\otimes\mathbf{u}+\mathbf{u}\otimes\mathbf{v}+\mathbf{v}\otimes\mathbf{u}+\mathbf{v}\otimes\mathbf{v}$$

which cannot be further simplified because $\mathbf{u}\otimes\mathbf{v}\neq\mathbf{v}\otimes\mathbf{u}$.

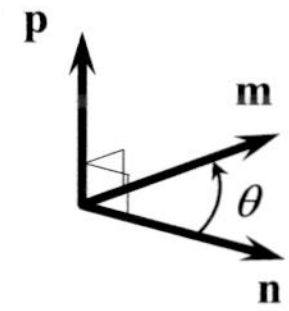

Fig. 1.10 *Dot and cross products of unit vectors.* The dot product $\mathbf{n} \cdot \mathbf{m}$ is the cosine of θ. The cross product $\mathbf{n} \times \mathbf{m}$ is the sine of θ times a perpendicular unit vector $\mathbf{p}$

1.1.2 Unit Vectors

A unit vector is simply a vector with unit length. As such, the only role it plays is in setting direction. Consider the unit vectors **m**, **n**, and **p** shown in Fig. 1.10. The dot and cross product of *unit vectors* can be expressed as

$$\mathbf{m} \cdot \mathbf{n} = \cos\theta \qquad \mathbf{m} \times \mathbf{n} = \sin\theta\, \mathbf{p} \tag{1.15}$$

where $\mathbf{p}$ is a unit vector orthogonal to both $\mathbf{m}$ and $\mathbf{n}$. From this we can see what it means if the vectors are orthogonal ($\theta = \pi/2$) or point in the same direction ($\theta = 0$):

$$\begin{array}{cc} \theta = 0 & \theta = \pi/2 \\ \mathbf{m} \cdot \mathbf{n} = 1 & \mathbf{m} \cdot \mathbf{n} = 0 \\ \mathbf{m} \times \mathbf{n} = \mathbf{0} & \mathbf{m} \times \mathbf{n} = \mathbf{p} \end{array}$$

From a geometric point of view, the dot product can quantify orthogonality, and the cross product can furnish a vector perpendicular to the plane that holds the original two vectors. We will also see that the dot product is related to the geometric concept of projection and the cross product to the computation of area.

Any vector can be used to create a unit vector by dividing the vector by its own length. In particular, the vector

$$\mathbf{n} = \frac{\mathbf{v}}{\|\mathbf{v}\|}$$

is a unit vector that points in the same direction as $\mathbf{v}$. It is a unit vector because

$$\mathbf{n} \cdot \mathbf{n} = \frac{\mathbf{v}}{\|\mathbf{v}\|} \cdot \frac{\mathbf{v}}{\|\mathbf{v}\|} = \frac{\mathbf{v} \cdot \mathbf{v}}{\|\mathbf{v}\|^2} = \frac{\|\mathbf{v}\|^2}{\|\mathbf{v}\|^2} = 1$$

1.1.3 Base Vectors and Vector Components

To carry out calculations with vectors, it is generally a good idea to first lay out a coordinate system endowed with unit *base vectors* that point along the coordinate

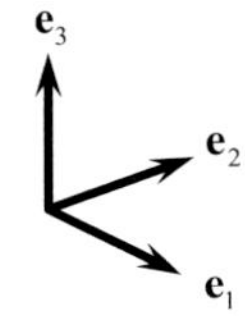

Fig. 1.11 *Unit orthogonal base vectors.* The unit orthogonal vectors $\{\mathbf{e}_1, \mathbf{e}_2, \mathbf{e}_3\}$ establish a basis for three-dimensional space

axes. These base vectors provide the building blocks for describing other vectors, and they also provide a route to devising component calculations with vectors.

Unit orthogonal base vectors To describe a problem in mechanics, we need to establish a reference frame. For that frame, we need to know what it means to be pointing in certain directions. A set of orthogonal unit base vectors provides a way to do that. As unit vectors, their primary role is to set direction. As shown in Fig. 1.11, the three *orthogonal* unit vectors $\{\mathbf{e}_1, \mathbf{e}_2, \mathbf{e}_3\}$ establish a *basis* for all vectors in the space. Orthogonal unit vectors are often called *orthonormal* vectors.

For the unit orthogonal base vectors, the dot and cross products are very simple. All possible dot products between base vectors are

$$\begin{array}{lll} \mathbf{e}_1 \cdot \mathbf{e}_1 = 1 & \mathbf{e}_1 \cdot \mathbf{e}_2 = 0 & \mathbf{e}_1 \cdot \mathbf{e}_3 = 0 \\ \mathbf{e}_2 \cdot \mathbf{e}_1 = 0 & \mathbf{e}_2 \cdot \mathbf{e}_2 = 1 & \mathbf{e}_2 \cdot \mathbf{e}_3 = 0 \\ \mathbf{e}_3 \cdot \mathbf{e}_1 = 0 & \mathbf{e}_3 \cdot \mathbf{e}_2 = 0 & \mathbf{e}_3 \cdot \mathbf{e}_3 = 1 \end{array} \tag{1.16}$$

Similarly, all possible cross products between base vectors are

$$\begin{array}{lll} \mathbf{e}_1 \times \mathbf{e}_1 = \mathbf{0} & \mathbf{e}_1 \times \mathbf{e}_2 = \mathbf{e}_3 & \mathbf{e}_1 \times \mathbf{e}_3 = -\mathbf{e}_2 \\ \mathbf{e}_2 \times \mathbf{e}_1 = -\mathbf{e}_3 & \mathbf{e}_2 \times \mathbf{e}_2 = \mathbf{0} & \mathbf{e}_2 \times \mathbf{e}_3 = \mathbf{e}_1 \\ \mathbf{e}_3 \times \mathbf{e}_1 = \mathbf{e}_2 & \mathbf{e}_3 \times \mathbf{e}_2 = -\mathbf{e}_1 & \mathbf{e}_3 \times \mathbf{e}_3 = \mathbf{0} \end{array} \tag{1.17}$$

One way to remember how the algebraic signs work for the cross product of base vectors is with the diagram shown in Fig. 1.12. If the indices in the equation are in clockwise, or cyclic, order (e.g., $1 \to 2 \to 3$), then the result is positive: $\mathbf{e}_1 \times \mathbf{e}_2 = \mathbf{e}_3$. If the indices in the equation are in counterclockwise, or acyclic, order (e.g., $3 \to 2 \to 1$), then the result is negative: $\mathbf{e}_3 \times \mathbf{e}_2 = -\mathbf{e}_1$. Of course, if the indices are the same, then the result is zero.

We will see that all formulas for the dot and cross product of general vectors are a consequence of Eqs. 1.16 and 1.17. In our work, we will usually set up the

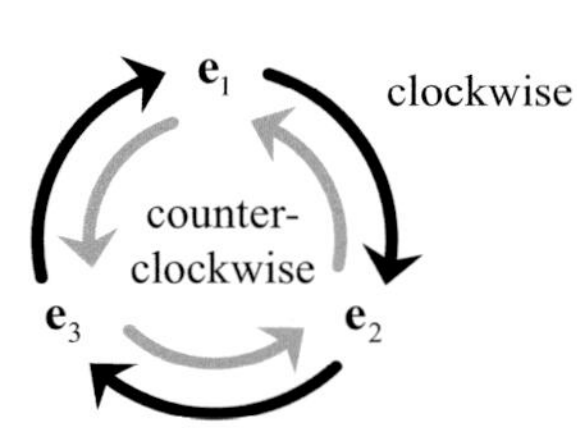

Fig. 1.12 *Cross products of base vectors.* The cross products of the unit base vectors $\{\mathbf{e}_1, \mathbf{e}_2, \mathbf{e}_3\}$ are positive if the order is clockwise and negative if the order is counterclockwise

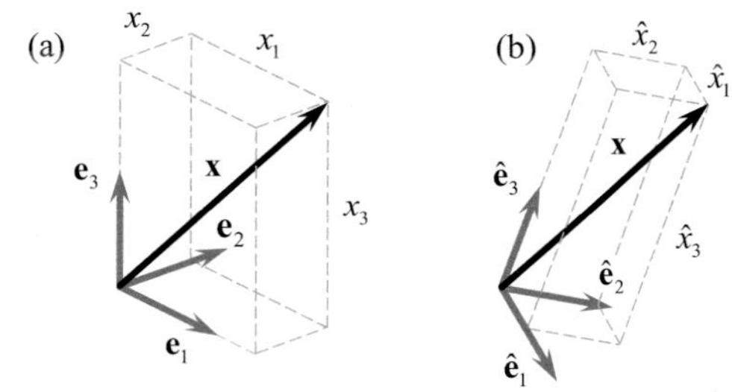

Fig. 1.13 *Vector components*. The vector $\mathbf{x}$ represented in two coordinate systems. **a** The components x_1, x_2, and x_3 relative to the unit base vectors $\{\mathbf{e}_1, \mathbf{e}_2, \mathbf{e}_3\}$. **b** The components $\hat{x}_1$, $\hat{x}_2$, and $\hat{x}_3$ relative to the unit base vectors $\{\hat{\mathbf{e}}_1, \hat{\mathbf{e}}_2, \hat{\mathbf{e}}_3\}$

calculation so that the only dot or cross products we will compute will involve the unit base vectors.

Vector Components Every vector in three-dimensional space can be expressed in terms of the unit base vectors $\{\mathbf{e}_1, \mathbf{e}_2, \mathbf{e}_3\}$. The directions of the base vectors define the axes of the coordinate system. Vectors do not require a coordinate system, but the *components* of vectors do.

Consider the vector $\mathbf{x}$ shown in Fig. 1.13a. The components of the vector (i.e., the scalar values x_1, x_2, and x_3) describe how far the vector $\mathbf{x}$ extends along each base vector. The vector $\mathbf{x}$ can then be built with the normal head-to-tail rule for vector addition. To wit,

$$\mathbf{x} = x_1\mathbf{e}_1 + x_2\mathbf{e}_2 + x_3\mathbf{e}_3 \tag{1.18}$$

The values x_1, x_2, and x_3 are the lengths of the edges of the parallelepiped shown as dotted lines.

The choice of coordinate system is not unique. For example, we could choose the system $\{\hat{\mathbf{e}}_1, \hat{\mathbf{e}}_2, \hat{\mathbf{e}}_3\}$ shown in Fig. 1.13b. The components of the vector $\mathbf{x}$ in this system are $\hat{x}_1$, $\hat{x}_2$, and $\hat{x}_3$. The values of the components of the vector in two different coordinate systems are not the same, even though it is the *same* vector. In the second coordinate system, the vector can be represented as

$$\mathbf{x} = \hat{x}_1\hat{\mathbf{e}}_1 + \hat{x}_2\hat{\mathbf{e}}_2 + \hat{x}_3\hat{\mathbf{e}}_3. \tag{1.19}$$

If we know the relationship between the two coordinate systems, then we can find the values of the components in one system given the values in the other. Let $Q_{ij} = \hat{\mathbf{e}}_i \cdot \mathbf{e}_j$ be the dot product between the ith base vector $\hat{\mathbf{e}}_i$ in the hat system and the jth base vector $\mathbf{e}_j$ in the original system. Thus, Q_{ij} is the cosine of the angle between $\hat{\mathbf{e}}_i$ and $\mathbf{e}_j$. Setting the right side of Eq. 1.19 equal to the right side of Eq. 1.18 and dotting the resulting equation with $\hat{\mathbf{e}}_i$, one can show that

$$\hat{x}_i = \sum_{j=1}^{3} Q_{ij} x_j \tag{1.20}$$

This sum can be written out in matrix notation as

$$\begin{Bmatrix} \hat{x}_1 \\ \hat{x}_2 \\ \hat{x}_3 \end{Bmatrix} = \begin{bmatrix} Q_{11} & Q_{12} & Q_{13} \\ Q_{21} & Q_{22} & Q_{23} \\ Q_{31} & Q_{32} & Q_{33} \end{bmatrix} \begin{Bmatrix} x_1 \\ x_2 \\ x_3 \end{Bmatrix} \tag{1.21}$$

We will write $\{\mathbf{Q}\}$ to stand for the matrix that has Q_{ij} in the ith row and jth column. With this notation we can write Eqs. 1.20 and 1.21 as

$$\{\hat{\mathbf{x}}\} = \{\mathbf{Q}\}\{\mathbf{x}\} \tag{1.22}$$

where $\{\hat{\mathbf{x}}\}$ is the matrix with components $\hat{x}_i$ and $\{\mathbf{x}\}$ is the matrix with components x_i. The nature of this calculation is illustrated in the following two-dimensional example.

Example 1.2
The components of a vector $\mathbf{x}$ in the system $\{\mathbf{e}_1, \mathbf{e}_2\}$ (the original system) are $x_1 = 2$ and $x_2 = 3$, as shown in Fig. 1.14a. A second coordinate system is represented by the unit base vectors $\{\hat{\mathbf{e}}_1, \hat{\mathbf{e}}_2\}$, which are rotated by 30° counterclockwise from $\{\mathbf{e}_1, \mathbf{e}_2\}$, as shown in Fig. 1.14b. Find the components of the vector $\mathbf{x}$ relative to $\{\hat{\mathbf{e}}_1, \hat{\mathbf{e}}_2\}$.

Solution. The $\mathbf{Q}$ matrix can be computed as

$$\begin{bmatrix} Q_{11} & Q_{12} \\ Q_{21} & Q_{22} \end{bmatrix} = \begin{bmatrix} \hat{\mathbf{e}}_1 \cdot \mathbf{e}_1 & \hat{\mathbf{e}}_1 \cdot \mathbf{e}_2 \\ \hat{\mathbf{e}}_2 \cdot \mathbf{e}_1 & \hat{\mathbf{e}}_2 \cdot \mathbf{e}_2 \end{bmatrix} = \begin{bmatrix} \sqrt{3}/2 & 1/2 \\ -1/2 & \sqrt{3}/2 \end{bmatrix}$$

noting that the dot product of unit vectors is the cosine of the angle between them. To wit,

$$\begin{aligned} \hat{\mathbf{e}}_1 \cdot \mathbf{e}_1 &= \cos 30^\circ = \tfrac{\sqrt{3}}{2} & \hat{\mathbf{e}}_1 \cdot \mathbf{e}_2 &= \cos 60^\circ = \tfrac{1}{2} \\ \hat{\mathbf{e}}_2 \cdot \mathbf{e}_1 &= \cos 120^\circ = -\tfrac{1}{2} & \hat{\mathbf{e}}_2 \cdot \mathbf{e}_2 &= \cos 30^\circ = \tfrac{\sqrt{3}}{2} \end{aligned}$$

Using Eq. 1.22, we can compute the components of $\mathbf{x}$ in the rotated coordinate system as

$$\begin{Bmatrix} \hat{x}_1 \\ \hat{x}_2 \end{Bmatrix} = \begin{bmatrix} \sqrt{3}/2 & 1/2 \\ -1/2 & \sqrt{3}/2 \end{bmatrix} \begin{Bmatrix} 2 \\ 3 \end{Bmatrix} = \begin{Bmatrix} \sqrt{3} + 3/2 \\ -1 + 3\sqrt{3}/2 \end{Bmatrix} = \begin{Bmatrix} 3.232 \\ 1.598 \end{Bmatrix}$$

These values are confirmed by the location of the tip of vector $\mathbf{x}$ in the picture in Fig. 1.14b.

If both $\{\mathbf{e}_1, \mathbf{e}_2, \mathbf{e}_3\}$ and $\{\hat{\mathbf{e}}_1, \hat{\mathbf{e}}_2, \hat{\mathbf{e}}_3\}$ are orthonormal sets of base vectors, then they are rotated versions of each other. In this case, the matrix $\{\mathbf{Q}\}$ must be an *orthogonal matrix*. Orthogonal matrices have the property that the inverse of the matrix is equal to its transpose, i.e., $\{\mathbf{Q}\}^{-1} = \{\mathbf{Q}\}^T$. Therefore,

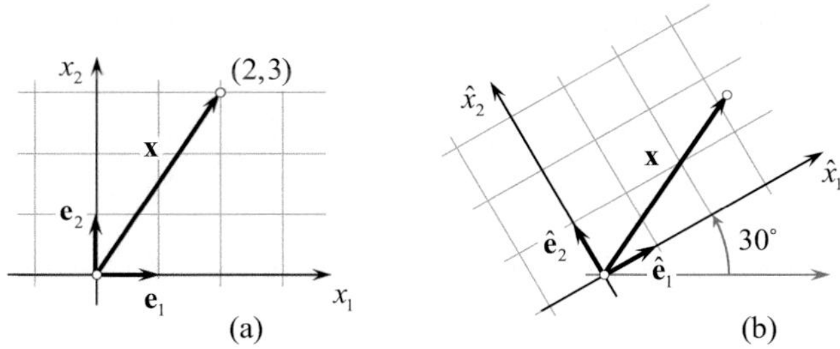

Fig. 1.14 *Example.* The vector **x** represented in two coordinate systems. **a** The system with unit base vectors $\{\mathbf{e}_1, \mathbf{e}_2\}$. **b** The system with unit base vectors $\{\hat{\mathbf{e}}_1, \hat{\mathbf{e}}_2\}$

$$\{\mathbf{Q}\}^T\{\mathbf{Q}\} = \mathbf{I}, \qquad \{\mathbf{Q}\}\{\mathbf{Q}\}^T = \mathbf{I}$$

where **I** is the identity matrix. The proof for the two-dimensional case can be carried out easily because we can write the $\{\mathbf{Q}\}$ matrix as

$$\{\mathbf{Q}\} = \begin{bmatrix} \cos\theta & \sin\theta \\ -\sin\theta & \cos\theta \end{bmatrix}$$

which is the case where the basis $\{\hat{\mathbf{e}}_1, \hat{\mathbf{e}}_2\}$ is rotated from $\{\mathbf{e}_1, \mathbf{e}_2\}$ counterclockwise by an angle θ. Now,

$$\{\mathbf{Q}\}^T\{\mathbf{Q}\} = \begin{bmatrix} \cos\theta & -\sin\theta \\ \sin\theta & \cos\theta \end{bmatrix}\begin{bmatrix} \cos\theta & \sin\theta \\ -\sin\theta & \cos\theta \end{bmatrix} = \begin{bmatrix} 1 & 0 \\ 0 & 1 \end{bmatrix}$$

In practice, when we compute with vectors, we generally do so with component calculations. The computing environment MATLAB is set up to deal with component calculations. Often in derivations we will not need to use components, but in final calculations we usually will. There are a couple of well-known formulas for computing the dot and cross products in components. They are not just equations to be remembered, they are results that emanate from doing the algebra of vectors represented with components and base vectors, as the following sections show.

Component Form of the Dot Product The vectors **u** and **v** can be written in terms of their components as

$$\mathbf{u} = u_1\mathbf{e}_1 + u_2\mathbf{e}_2 + u_3\mathbf{e}_3, \qquad \mathbf{v} = v_1\mathbf{e}_1 + v_2\mathbf{e}_2 + v_3\mathbf{e}_3$$

Compute the dot product using the distributive law to expand the dot product of the two vectors. To wit,

$$\begin{aligned}\mathbf{u}\cdot\mathbf{v} &= (u_1\mathbf{e}_1 + u_2\mathbf{e}_2 + u_3\mathbf{e}_3)\cdot(v_1\mathbf{e}_1 + v_2\mathbf{e}_2 + v_3\mathbf{e}_3)\\ &= u_1v_1(\mathbf{e}_1\cdot\mathbf{e}_1) + u_1v_2(\mathbf{e}_1\cdot\mathbf{e}_2) + u_1v_3(\mathbf{e}_1\cdot\mathbf{e}_3)\\ &\quad + u_2v_1(\mathbf{e}_2\cdot\mathbf{e}_1) + u_2v_2(\mathbf{e}_2\cdot\mathbf{e}_2) + u_2v_3(\mathbf{e}_2\cdot\mathbf{e}_3)\\ &\quad + u_3v_1(\mathbf{e}_3\cdot\mathbf{e}_1) + u_3v_2(\mathbf{e}_3\cdot\mathbf{e}_2) + u_3v_3(\mathbf{e}_3\cdot\mathbf{e}_3)\end{aligned}$$

The individual dot products of the base vectors evaluate to either 0 or 1. Thus, the final result has the form:

$$\mathbf{u}\cdot\mathbf{v} = u_1v_1 + u_2v_2 + u_3v_3 \tag{1.23}$$

The result is the standard formula for the dot product of vectors in component form.

Component Form of the Cross Product Let us carry out a similar calculation for the cross product. To wit,

$$\begin{aligned}\mathbf{u}\times\mathbf{v} &= (u_1\mathbf{e}_1 + u_2\mathbf{e}_2 + u_3\mathbf{e}_3)\times(v_1\mathbf{e}_1 + v_2\mathbf{e}_2 + v_3\mathbf{e}_3)\\ &= u_1v_1(\mathbf{e}_1\times\mathbf{e}_1) + u_1v_2(\mathbf{e}_1\times\mathbf{e}_2) + u_1v_3(\mathbf{e}_1\times\mathbf{e}_3)\\ &\quad + u_2v_1(\mathbf{e}_2\times\mathbf{e}_1) + u_2v_2(\mathbf{e}_2\times\mathbf{e}_2) + u_2v_3(\mathbf{e}_2\times\mathbf{e}_3)\\ &\quad + u_3v_1(\mathbf{e}_3\times\mathbf{e}_1) + u_3v_2(\mathbf{e}_3\times\mathbf{e}_2) + u_3v_3(\mathbf{e}_3\times\mathbf{e}_3)\end{aligned}$$

where we have used the distributive law to expand the cross products of the vectors. Each of the cross products of base vectors can be simplified in accord with Eq. 1.17 to give the final result

$$\mathbf{u}\times\mathbf{v} = \big(u_2v_3 - u_3v_2\big)\mathbf{e}_1 + \big(u_3v_1 - u_1v_3\big)\mathbf{e}_2 + \big(u_1v_2 - u_2v_1\big)\mathbf{e}_3 \tag{1.24}$$

The result is the standard formula for the cross product of vectors.

Example 1.3
Use the component formula to compute the dot and cross products between the vectors $\mathbf{u} = 2\mathbf{e}_1 + 3\mathbf{e}_2 + 4\mathbf{e}_3$ and $\mathbf{v} = 6\mathbf{e}_1 + 5\mathbf{e}_2 + 7\mathbf{e}_3$. Find the lengths of the two vectors and the angle θ between them.

Solution. The components of the vectors called out in this example can be expressed in matrix form as

$$\{\mathbf{u}\} = \begin{Bmatrix} u_1\\ u_2\\ u_3\end{Bmatrix} = \begin{Bmatrix} 2\\ 3\\ 4\end{Bmatrix}, \qquad \{\mathbf{v}\} = \begin{Bmatrix} v_1\\ v_2\\ v_3\end{Bmatrix} = \begin{Bmatrix} 6\\ 5\\ 7\end{Bmatrix}$$

Therefore, the dot product, from Eq. 1.23, is

$$\mathbf{u}\cdot\mathbf{v} = (2)(6) + (3)(5) + (4)(7) = 55$$

The lengths of the vectors are the square root of the vectors dotted with themselves. Thus,

$$\|\mathbf{u}\| = \sqrt{\mathbf{u}\cdot\mathbf{u}} = \sqrt{(2)(2)+(3)(3)+(4)(4)} = 5.3852$$

$$\|\mathbf{v}\| = \sqrt{\mathbf{v}\cdot\mathbf{v}} = \sqrt{(6)(6)+(5)(5)+(7)(7)} = 10.4881$$

From Eq. 1.2, we can compute the cosine of the angle between the vectors as

$$\cos\theta = \frac{\mathbf{u}\cdot\mathbf{v}}{\|\mathbf{u}\|\,\|\mathbf{v}\|} = \frac{55}{(5.3852)(10.4881)} = 0.9738$$

Taking the inverse cosine gives a value of θ of 0.2294 rad or 13.1458°.

Using Eq. 1.24, we can compute the cross product from the vector components as

$$\begin{aligned}\mathbf{u}\times\mathbf{v} &= \big[(3)(7)-(4)(5)\big]\mathbf{e}_1 + \big[(4)(6)-(2)(7)\big]\mathbf{e}_2 + \big[(2)(5)-(3)(6)\big]\mathbf{e}_3\\ &= 1\mathbf{e}_1 + 10\mathbf{e}_2 - 8\mathbf{e}_3\end{aligned}$$

The length of this vector is

$$\|\mathbf{u}\times\mathbf{v}\| = \sqrt{(1)(1)+(10)(10)+(-8)(-8)} = 12.8452$$

Now, we can use Eq. 1.8 to compute the sine of the angle θ as

$$\sin\theta = \frac{\|\mathbf{u}\times\mathbf{v}\|}{\|\mathbf{u}\|\,\|\mathbf{v}\|} = \frac{12.8452}{(5.3851)(10.4881)} = 0.2274$$

Taking the inverse sine gives $\theta = 0.2294$ rad or 13.1458°, which is the same angle obtained previously, as expected.

Note that these results require that the components of both vectors be expressed relative to the same basis.

The cross product of vectors is sometimes written in the form

$$\mathbf{u}\times\mathbf{v} = \det\begin{bmatrix}\mathbf{e}_1 & \mathbf{e}_2 & \mathbf{e}_3\\ u_1 & u_2 & u_3\\ v_1 & v_2 & v_3\end{bmatrix}$$

which is an abuse of mathematical notation (you cannot just stick vectors in the determinant as if they are scalars), but a good way to remember the formula for the cross product if you remember how to compute a determinant. Also, the triple scalar product can be written as a determinant:

$$(\mathbf{u} \times \mathbf{v}) \cdot \mathbf{w} = \det \begin{bmatrix} u_1 & u_2 & u_3 \\ v_1 & v_2 & v_3 \\ w_1 & w_2 & w_3 \end{bmatrix}$$

which can be written out as

$$\begin{aligned} (\mathbf{u} \times \mathbf{v}) \cdot \mathbf{w} &= u_1 v_2 w_3 + u_2 v_3 w_1 + u_3 v_1 w_2 \\ &\quad - u_1 v_3 w_2 - u_2 v_1 w_3 - u_3 v_2 w_1 \end{aligned} \tag{1.25}$$

Notice that each of the six terms is the product of three components with different indices, one from each vector. The positive contributions have the indices in clockwise order and the negative contributions have the indices in counterclockwise order (see Fig. 1.12).

In most applications in this book, we will not need the component equations for the dot and cross products of vectors because the computations will usually boil down to products of unit vectors.

1.1.4 Vector Projection

Imagine a plane that has the property that the unit vector **n** is normal to the plane. What that means geometrically is that **n** is orthogonal to every vector that lies in the plane. Thus, the dot product should have something to say about this geometry. Suppose that we have a vector **v** that is not contained in the plane, as shown in Fig. 1.15. We can find the part of **v** that lies along the vector **n** (and is, therefore, normal to the plane) and the part of **v** that is orthogonal to **n** (and is, therefore, contained in the plane).

Note that, because of the definition of the dot product

$$\mathbf{v} \cdot \mathbf{n} = \|\mathbf{v}_n\| \qquad \Rightarrow \qquad \mathbf{v}_n = (\mathbf{v} \cdot \mathbf{n})\mathbf{n} \tag{1.26}$$

The vector $\mathbf{v}_n$ is the *projection* of the vector **v** along the normal **n**. The *magnitude* of that vector is $\mathbf{v} \cdot \mathbf{n}$ and the direction is **n**. To get the component of the vector *in* the plane, we can use vector subtraction. To wit,

$$\mathbf{v}_p = \mathbf{v} - \mathbf{v}_n = \mathbf{v} - (\mathbf{v} \cdot \mathbf{n})\mathbf{n} \tag{1.27}$$

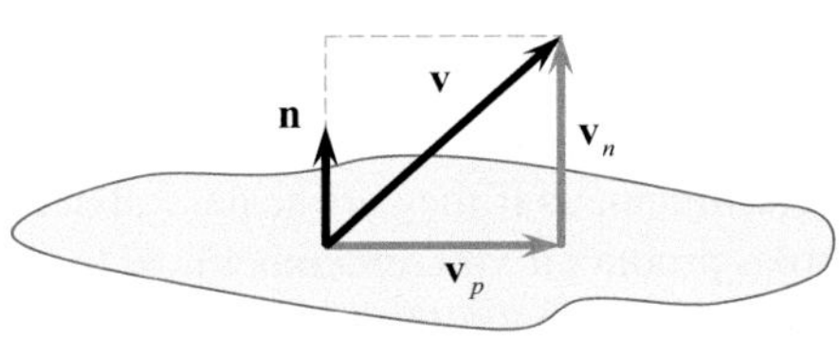

Fig. 1.15 *Projection of a vector.* The vector **v** with components projected onto and normal to a plane with unit normal vector **n**

Define the *projection* tensor $\mathbf{P}$ as follows:

$$\mathbf{P} = \mathbf{I} - \mathbf{n} \otimes \mathbf{n} \tag{1.28}$$

The purpose of $\mathbf{P}$ is to project any vector onto the plane with unit normal $\mathbf{n}$. If the sun is straight up in the sky at noon, then $\mathbf{P}$ produces the shadow of the vector on the ground. To prove that the projection tensor does what it is designed to do, compute its action on the vector $\mathbf{v}$ as follows:

$$\mathbf{P}\mathbf{v} = \big[\mathbf{I} - \mathbf{n} \otimes \mathbf{n}\big]\mathbf{v} = \mathbf{v} - \big(\mathbf{n} \cdot \mathbf{v}\big)\mathbf{n} = \mathbf{v}_p$$

To prove that this vector lies in the plane, simply dot it with $\mathbf{n}$ to get

$$\mathbf{v}_p \cdot \mathbf{n} = \big(\mathbf{v} - (\mathbf{n} \cdot \mathbf{v})\mathbf{n}\big) \cdot \mathbf{n} = \big(\mathbf{v} \cdot \mathbf{n}\big) - \big(\mathbf{n} \cdot \mathbf{v}\big)\big(\mathbf{n} \cdot \mathbf{n}\big) = 0$$

since $\mathbf{n} \cdot \mathbf{n} = 1$.

We will often use vector projection to manipulate vector equations. Rather than write equations in component form, we will write them in vector form. When we need to see the implication of, say, equilibrium in a certain direction, we will simply dot the vector equilibrium equation with a unit vector pointing in that direction.

Example 1.4
The vector equation from Newton's second law is $\mathbf{F} = m\mathbf{a}$, where $\mathbf{F}$ is the net force acting on a particle of mass m with acceleration $\mathbf{a}$. Write this equation as a set of component equations relative to the basis $\{\mathbf{e}_1, \mathbf{e}_2, \mathbf{e}_3\}$, which are base vectors pointing along the x-, y-, and z-axes, respectively.

Solution. First, write the force and acceleration vectors in component form as

$$\mathbf{F} = F_x\mathbf{e}_1 + F_y\mathbf{e}_2 + F_z\mathbf{e}_3, \qquad \mathbf{a} = a_x\mathbf{e}_1 + a_y\mathbf{e}_2 + a_z\mathbf{e}_3$$

Now, substitute the component expressions into the vector equation to give

$$F_x\mathbf{e}_1 + F_y\mathbf{e}_2 + F_z\mathbf{e}_3 = m\big(a_x\mathbf{e}_1 + a_y\mathbf{e}_2 + a_z\mathbf{e}_3\big)$$

We can dot both sides of this equation with any of the three base vectors. For example, dotting with $\mathbf{e}_1$ gives

$$\begin{aligned} \mathbf{e}_1 \cdot \big(F_x\mathbf{e}_1 + F_y\mathbf{e}_2 + F_z\mathbf{e}_3\big) &= \mathbf{e}_1 \cdot m\big(a_x\mathbf{e}_1 + a_y\mathbf{e}_2 + a_z\mathbf{e}_3\big) \\ F_x &= m a_x \end{aligned}$$

because $\mathbf{e}_1 \cdot \mathbf{e}_1 = 1$, $\mathbf{e}_1 \cdot \mathbf{e}_2 = 0$, and $\mathbf{e}_1 \cdot \mathbf{e}_3 = 0$. If we dot the equation sequentially with the base vectors $\mathbf{e}_1$, $\mathbf{e}_2$, and $\mathbf{e}_3$, then we get three scalar equations:

$$F_x = ma_x, \qquad F_y = ma_y, \qquad F_z = ma_z$$

While these equations form a perfectly acceptable interpretation of the initial vector equation, it is not always the case that the coordinate directions are the best choice for expressing the equations. We will generally leave the equations in vector form until it is evident what the most natural mathematical next steps are.

1.2 Tensors

The tensor product of vectors provides an opening to discuss another very important mathematical object, one that will be key to our description of stress and strain. A *tensor* is an object that operates on a vector and produces another vector. The definition of the tensor product of vectors is a case in point. In the relationship

$$\big[\,\mathbf{u}\otimes\mathbf{v}\,\big]\mathbf{w} = \mathbf{u}\big(\mathbf{v}\cdot\mathbf{w}\big) \tag{1.29}$$

the object $[\,\mathbf{u}\otimes\mathbf{v}\,]$ is a tensor. It is operating on the vector $\mathbf{w}$. The right side of the equation shows that the result is a vector. The resulting vector points in the same direction as $\mathbf{u}$ and its length is scaled by the amount $\mathbf{v}\cdot\mathbf{w}$. If the dot product is negative, then the resulting vector points in the $-\mathbf{u}$-direction.

1.2.1 Matrix Representation

One way to get a better handle on the tensor product is to look at it in *matrix notation* using components. Let us write the vector components in matrix form as

$$\{\mathbf{u}\} = \begin{Bmatrix} u_1 \\ u_2 \\ u_3 \end{Bmatrix}, \qquad \{\mathbf{v}\} = \begin{Bmatrix} v_1 \\ v_2 \\ v_3 \end{Bmatrix}, \qquad \{\mathbf{w}\} = \begin{Bmatrix} w_1 \\ w_2 \\ w_3 \end{Bmatrix}$$

where the notation $\{\mathbf{u}\}$ means a column matrix holding the components of the vector $\mathbf{u}$ in the coordinate system with base vectors $\{\mathbf{e}_1, \mathbf{e}_2, \mathbf{e}_3\}$. That is, the vector components from $\mathbf{u} = u_1\mathbf{e}_1 + u_2\mathbf{e}_2 + u_3\mathbf{e}_3$. One must be careful in using matrix notation because it assumes the base vectors without explicitly referring to them. We can write $\{\mathbf{v}\}^T$ to indicate a row matrix holding the components of the vector $\mathbf{v}$. With this notation, the dot product of vectors can be written as

$$\mathbf{v} \cdot \mathbf{w} = \{\mathbf{v}\}^T\{\mathbf{w}\} = \begin{Bmatrix} v_1 & v_2 & v_3 \end{Bmatrix} \begin{Bmatrix} w_1 \\ w_2 \\ w_3 \end{Bmatrix} = v_1 w_1 + v_2 w_2 + v_3 w_3$$

following ordinary rules of matrix multiplication. Now, we can write Eq. 1.29 in the form:

$$[\,\mathbf{u} \otimes \mathbf{v}\,]\mathbf{w} = \mathbf{u}\big(\mathbf{v} \cdot \mathbf{w}\big) = \{\mathbf{u}\}\big(\{\mathbf{v}\}^T\{\mathbf{w}\}\big)$$

The last expression is a 3×1 matrix multiplying a 1×3 matrix multiplying a 3×1 matrix, all of which is compatible by the rules of matrix multiplication. Matrix multiplication is associative. We can either group the last two of the three terms and compute a scalar, as we did previously, or we can group the first two terms and get a 3×3 matrix, i.e.:

$$\big[\{\mathbf{u}\}\{\mathbf{v}\}^T\big]\{\mathbf{w}\} = \begin{bmatrix} u_1 v_1 & u_1 v_2 & u_1 v_3 \\ u_2 v_1 & u_2 v_2 & u_2 v_3 \\ u_3 v_1 & u_3 v_2 & u_3 v_3 \end{bmatrix} \begin{Bmatrix} w_1 \\ w_2 \\ w_3 \end{Bmatrix}$$

Just carry out the matrix multiplication and you will see that this gives the same result as Eq. 1.11. In matrix language, we generally refer to $\{\mathbf{u}\}\{\mathbf{v}\}^T$ as the *outer product* of matrices, whereas $\{\mathbf{v}\}^T\{\mathbf{w}\}$ is called the *inner product*. As such, the inner product corresponds with the dot product of vectors and the outer product corresponds with the tensor product. Specifically,

$$\{\mathbf{u} \otimes \mathbf{v}\} = \{\mathbf{u}\}\{\mathbf{v}\}^T = \begin{bmatrix} u_1 v_1 & u_1 v_2 & u_1 v_3 \\ u_2 v_1 & u_2 v_2 & u_2 v_3 \\ u_3 v_1 & u_3 v_2 & u_3 v_3 \end{bmatrix} \tag{1.30}$$

Sometimes we will abbreviate the outer product as $\mathbf{u}\mathbf{v}^T$ (i.e., omit the curly braces) to shorten the notation. To do so is a slight abuse of vector notation and should only be used when the context makes the operation clear.

The tensor product of two vectors is not the only case of a tensor. If we look at all the combinations of the tensor products of the three base vectors, i.e., $[\,\mathbf{e}_i \otimes \mathbf{e}_j\,]$, we can see there are nine of them. These nine tensors provide a basis for expressing general tensors in component form. In fact, we can write a general tensor $\mathbf{T}$ in the form:

$$\mathbf{T} = \sum_{i=1}^{3} \sum_{j=1}^{3} T_{ij}\big[\,\mathbf{e}_i \otimes \mathbf{e}_j\,\big] \tag{1.31}$$

The scalars T_{ij} are the *components* of the tensor. We can store the components of a tensor in a matrix with i being the row index and j being the column index. To wit,

$$\{\mathbf{T}\} = \begin{bmatrix} T_{11} & T_{12} & T_{13} \\ T_{21} & T_{22} & T_{23} \\ T_{31} & T_{32} & T_{33} \end{bmatrix} \tag{1.32}$$

It is not really correct to set a tensor equal to a matrix of its components, i.e., $\mathbf{T} \neq \{\mathbf{T}\}$, because the components depend upon the basis and the matrix representation has no way to tell us that the basis is the set of tensors $[\,\mathbf{e}_i \otimes \mathbf{e}_j\,]$. The tensor itself does not depend upon the coordinate system. However, for most of the applications of tensors in this book, the basis will be understood, and sometimes we will not distinguish between tensor and matrix notation unless it is ambiguous.

With the index convention setup, the operation of a tensor on a vector is the same as matrix multiplication, i.e.:

$$\{\mathbf{Tv}\} = \begin{bmatrix} T_{11} & T_{12} & T_{13} \\ T_{21} & T_{22} & T_{23} \\ T_{31} & T_{32} & T_{33} \end{bmatrix} \begin{Bmatrix} v_1 \\ v_2 \\ v_3 \end{Bmatrix}$$

The tensor we have described here is only one of a larger family of tensors. It is what we call a *second-order* tensor. You can recognize a second-order tensor because the components have two indices (i.e., T_{ij}), and two base vectors are involved in creating the tensor basis. The only other type of tensor we will need in this book is the third-order tensor $[\,\mathbf{u} \otimes \mathbf{v} \otimes \mathbf{w}\,]$, which operates on a vector $\mathbf{a}$ according to the rule:

$$\left[\mathbf{u} \otimes \mathbf{v} \otimes \mathbf{w}\right]\mathbf{a} = \left[\mathbf{u} \otimes \mathbf{v}\right](\mathbf{w} \cdot \mathbf{a})$$

The result is a second-order tensor. The third-order tensor will come up when we create an algorithm for computing the properties of plane areas.

Many of the tensors we encounter in this book are *symmetric*. Therefore, we will restrict our attention to symmetric tensors for the remainder of the chapter. In the following sections, we will assume that the term *tensor* refers to a symmetric tensor. A symmetric tensor is one that satisfies $\mathbf{T}^T = \mathbf{T}$. In the matrix representation, the off-diagonal elements of the matrix are mirrored across the diagonal (i.e., $T_{ij} = T_{ji}$).

Tensors actually come up a lot in deformable solids. For example, characterizing stress and strain in multiple dimensions is done with tensors. Readers just getting familiar with tensors will probably be best served thinking of them through the matrix representation of the components and the operations of matrix algebra.

1.2.2 Change of Basis

Like vectors, tensors are objects that exist independent of any coordinate representation with components. A solid body subjected to force, for example, is in a state of stress. Stress, as we will show in Chap. 5, can be represented as a tensor. The body in question has no idea that we might eventually lay out a coordinate system with base vectors and then characterize the stress state with the values of the tensor components. The *components* of a tensor depend upon the coordinate system, but the tensor itself does not. Just as a vector can be represented relative to different base vectors (see Eqs. 1.18 and 1.19), so can a tensor.

The basis for vectors is a set of independent vectors that span the space. For example, the orthogonal unit vectors $\{\mathbf{e}_1, \mathbf{e}_2, \mathbf{e}_3\}$ span three-dimensional space and, therefore, form a basis for representing any vector in that space in components. Similarly, the basis for tensors is a set of independent tensors that can represent any tensor in the space. For example, the nine tensors $[\mathbf{e}_i \otimes \mathbf{e}_j]$ for $i = 1, 2, 3$ and $j = 1, 2, 3$ can accomplish that task. Since we build the base tensors from base vectors, it should be evident that we could choose a different set of base vectors, say $\{\hat{\mathbf{e}}_1, \hat{\mathbf{e}}_2, \hat{\mathbf{e}}_3\}$, and form a tensor basis with them. Thus, we could write a tensor in components relative to two different bases as

$$\mathbf{T} = \sum_{i=1}^{3}\sum_{j=1}^{3} T_{ij}\big[\,\mathbf{e}_i \otimes \mathbf{e}_j\,\big] = \sum_{i=1}^{3}\sum_{j=1}^{3} \hat{T}_{ij}\big[\,\hat{\mathbf{e}}_i \otimes \hat{\mathbf{e}}_j\,\big] \tag{1.33}$$

The components T_{ij} and $\hat{T}_{ij}$ are *not* the same numbers, just like the values of x_i and $\hat{x}_i$ were not equal in Example 1.2 in Sect. 1.1. As we did with vectors, we can find the relationship between the two sets of components. Let us compute $\hat{\mathbf{e}}_n \cdot \mathbf{T}\hat{\mathbf{e}}_m$ using both versions of the component representation of $\mathbf{T}$ in Eq. 1.33. The case on the right (the hat system) gives

$$\begin{aligned} \hat{\mathbf{e}}_n \cdot \mathbf{T}\hat{\mathbf{e}}_m &= \hat{\mathbf{e}}_n \cdot \sum_{i=1}^{3}\sum_{j=1}^{3} \hat{T}_{ij}\big[\,\hat{\mathbf{e}}_i \otimes \hat{\mathbf{e}}_j\,\big]\hat{\mathbf{e}}_m \\ &= \sum_{i=1}^{3}\sum_{j=1}^{3} \big(\hat{\mathbf{e}}_n \cdot \hat{\mathbf{e}}_i\big)\hat{T}_{ij}\big(\hat{\mathbf{e}}_j \cdot \hat{\mathbf{e}}_m\big) = \hat{T}_{nm} \end{aligned} \tag{1.34}$$

The reason it turns out this way is because $[\,\hat{\mathbf{e}}_i \otimes \hat{\mathbf{e}}_j\,]\hat{\mathbf{e}}_m = \hat{\mathbf{e}}_i\big(\hat{\mathbf{e}}_j \cdot \hat{\mathbf{e}}_m\big)$, according to Eq. 1.29. Then, the other dot product can be executed by moving $\hat{\mathbf{e}}_n$ inside the sum, which is legal because the index n is not the same as the summation indices i and j. Finally, we note that $\hat{\mathbf{e}}_n \cdot \hat{\mathbf{e}}_i$ is one if $n = i$ and zero otherwise, and $\hat{\mathbf{e}}_j \cdot \hat{\mathbf{e}}_m$ is one if $j = m$ and zero otherwise, because the vectors are orthogonal. Thus, out of the nine terms in the double sum, the only one that survives is $\hat{T}_{nm}$.

If we do the same thing for the case on the left (the original system) in Eq. 1.33, we get

$$\begin{aligned} \hat{\mathbf{e}}_n \cdot \mathbf{T}\hat{\mathbf{e}}_m &= \hat{\mathbf{e}}_n \cdot \sum_{i=1}^{3}\sum_{j=1}^{3} T_{ij}\big[\,\mathbf{e}_i \otimes \mathbf{e}_j\,\big]\hat{\mathbf{e}}_m \\ &= \sum_{i=1}^{3}\sum_{j=1}^{3} \big(\hat{\mathbf{e}}_n \cdot \mathbf{e}_i\big)T_{ij}\big(\mathbf{e}_j \cdot \hat{\mathbf{e}}_m\big) \\ &= \sum_{i=1}^{3}\sum_{j=1}^{3} Q_{ni}T_{ij}Q_{mj} \end{aligned} \tag{1.35}$$

where, as we did before for vectors, we have defined

$$Q_{ij} = \hat{\mathbf{e}}_i \cdot \mathbf{e}_j \tag{1.36}$$

Notice that the first subscript on Q_{ij} is associated with the $\hat{\mathbf{e}}_i$ (the hat system) and the second is associated with $\mathbf{e}_j$ (the original system). Now, with a suitable relabeling of the subscripts to match the terms in the equation, we get Q_{ni} and Q_{mj} in the equation. These are the direction cosines between all the combinations of vectors in the two systems. Now, equating the two versions of $\hat{\mathbf{e}}_n \cdot \mathbf{T}\hat{\mathbf{e}}_m$ gives

$$\hat{T}_{nm} = \sum_{i=1}^{3} \sum_{j=1}^{3} Q_{ni} T_{ij} Q_{mj} \tag{1.37}$$

which is nine equations, each one involving the sum of nine terms. This relationship can be written in standard matrix form if we consider the values Q_{ij} to be components of a matrix, i.e.:

$$\{\mathbf{Q}\} = \begin{bmatrix} Q_{11} & Q_{12} & Q_{13} \\ Q_{21} & Q_{22} & Q_{23} \\ Q_{31} & Q_{32} & Q_{33} \end{bmatrix} \tag{1.38}$$

Then, the matrix equation for change of basis is

$$\{\hat{\mathbf{T}}\} = \{\mathbf{Q}\}\{\mathbf{T}\}\{\mathbf{Q}\}^T \tag{1.39}$$

where $\{\hat{\mathbf{T}}\}$ represents the matrix with components $\hat{T}_{ij}$ and $\{\mathbf{T}\}$ represents a matrix with components T_{ij}. The computation is done following ordinary rules for matrix multiplication.

As an interesting side note, perhaps Eq. 1.38 has you thinking that there might be a *tensor* $\mathbf{Q}$ with components Q_{ij} relative to the standard basis. Indeed, there is. This tensor, when operating on a vector $\mathbf{v}$, produces a rotated version of the vector, $\mathbf{Qv}$, which has the same length as $\mathbf{v}$. Thus, the tensor $\mathbf{Q}$ is a pure *rotation*. When we use the matrix $\{\mathbf{Q}\}$ to compute components in a different basis, we are rotating the base vectors and leaving the vector alone (i.e., it is the *same* vector $\mathbf{v}$ in both cases, just different components). When we use $\mathbf{Q}$ as a tensor, we are leaving the basis alone (both $\mathbf{Q}$ and $\mathbf{v}$ are defined relative to the standard basis), and we are rotating the vector to produce a *new* vector. The distinction is subtle but important.

Planar Problems For *planar problems*, it is often convenient to work in two dimensions. In such cases, the vector basis will be $\{\mathbf{e}_1, \mathbf{e}_2\}$. The associated tensor basis will have four terms $\mathbf{e}_i \otimes \mathbf{e}_j$, where both i and j can take on values 1 or 2. Almost everything that is true about vectors and tensors in three dimensions reduces gracefully to two dimensions. One important exception is the cross product (and, therefore, the triple scalar product), which technically lives only in three dimensions. The appropriate space dimension to carry out vector and tensor calculations will always be clear from the context of the problem at hand.

Example 1.5
Consider the same two coordinate systems defined in Fig. 1.14, the system $\{\mathbf{e}_1, \mathbf{e}_2\}$ (the original system) and a second coordinate system represented by the unit base vectors $\{\hat{\mathbf{e}}_1, \hat{\mathbf{e}}_2\}$, which is rotated by 30° counterclockwise from $\{\mathbf{e}_1, \mathbf{e}_2\}$. The components of a tensor $\mathbf{T}$ in the original system are

$$\{\mathbf{T}\} = \begin{bmatrix} 16 & 4 \\ 4 & 8 \end{bmatrix}$$

Find the components of the tensor $\mathbf{T}$ relative to $\{\hat{\mathbf{e}}_1, \hat{\mathbf{e}}_2\}$.

Solution. The $\{\mathbf{Q}\}$ matrix can be computed as

$$\begin{bmatrix} Q_{11} & Q_{12} \\ Q_{21} & Q_{22} \end{bmatrix} = \begin{bmatrix} \hat{\mathbf{e}}_1 \cdot \mathbf{e}_1 & \hat{\mathbf{e}}_1 \cdot \mathbf{e}_2 \\ \hat{\mathbf{e}}_2 \cdot \mathbf{e}_1 & \hat{\mathbf{e}}_2 \cdot \mathbf{e}_2 \end{bmatrix} = \begin{bmatrix} \sqrt{3}/2 & 1/2 \\ -1/2 & \sqrt{3}/2 \end{bmatrix}$$

Therefore, using a two-dimensional version of Eq. 1.39, we can compute the components in the rotated coordinate system as

$$\begin{aligned} \begin{bmatrix} \hat{T}_{11} & \hat{T}_{12} \\ \hat{T}_{21} & \hat{T}_{22} \end{bmatrix} &= \begin{bmatrix} \sqrt{3}/2 & 1/2 \\ -1/2 & \sqrt{3}/2 \end{bmatrix} \begin{bmatrix} 16 & 4 \\ 4 & 8 \end{bmatrix} \begin{bmatrix} \sqrt{3}/2 & -1/2 \\ 1/2 & \sqrt{3}/2 \end{bmatrix} \\ &= \begin{bmatrix} 14 + 2\sqrt{3} & 2 - 2\sqrt{3} \\ 2 - 2\sqrt{3} & 10 - 2\sqrt{3} \end{bmatrix} = \begin{bmatrix} 17.4641 & -1.4641 \\ -1.4641 & 6.5359 \end{bmatrix} \end{aligned}$$

Obviously, the values of the tensor components are different relative in the two coordinate systems. What is less obvious is that if we sum the diagonals of the two matrices, we get the same value, i.e.:

$$\text{tr}(\mathbf{T}) = 16 + 8 = 24 = 17.4641 + 6.5359$$

We call this quantity, tr($\mathbf{T}$), the *trace* of the matrix or tensor. The trace is defined as the sum of the diagonals of the matrix. Also, the *determinants* of both matrices are the same, i.e.:

$$\det(\mathbf{T}) = (16)(8) - (4)(4) = 112 = (17.4641)(6.5359) - (-1.4641)^2$$

This correspondence is not a coincidence, but rather a property of tensors called *invariants*. We discuss invariants in the next section.

1.2.3 Tensor Invariants

In the last section, we showed that the components of a tensor in two different coordinate systems have different values. The relationship between them is given by Eq. 1.37 or Eq. 1.39. One of the most interesting features of tensors is that one can find functions of the tensor components that are the *same* in any coordinate system. These functions, called *tensor invariants*, will play an important role in the analysis of stress and strain.

To warm up to the concept of tensor invariants, let us first consider a *vector* invariant. The components of a vector $\mathbf{x}$ are different in different coordinate systems. Equations 1.20 and 1.21 give formulas for computing the values of the components $\hat{x}_i$ in the system $\{\hat{\mathbf{e}}_1, \hat{\mathbf{e}}_2, \hat{\mathbf{e}}_3\}$ given the components x_i in the system $\{\mathbf{e}_1, \mathbf{e}_2, \mathbf{e}_3\}$. There is a function

$$f(\mathbf{x}) = x_1^2 + x_2^2 + x_3^2 = \hat{x}_1^2 + \hat{x}_2^2 + \hat{x}_3^2$$

that yields the *same* value with both sets of components. Thus, the function that sums the squares of the components of a vector is invariant—the value does not change when the basis changes. Of course, we can recognize this function as the square of the length of the vector, which has a well-defined physical interpretation. This function is the only independent invariant for vectors. Of course, any function of an invariant is also an invariant. Thus, the square root of the function $f(\mathbf{x})$, which is the length of the vector, is obviously also invariant.

In three dimensions, we can identify three tensor invariants defined as

$$\mathcal{I}_1(\mathbf{T}) = \text{tr}\big(\mathbf{T}\big), \qquad \mathcal{I}_2(\mathbf{T}) = \text{tr}\big(\mathbf{T}^2\big), \qquad \mathcal{I}_3(\mathbf{T}) = \text{tr}\big(\mathbf{T}^3\big) \tag{1.40}$$

where $\mathbf{T}^2 = \mathbf{TT}$ and $\mathbf{T}^3 = \mathbf{TTT}$ are simply the tensor multiplied by itself two or three times, respectively, using ordinary rules of matrix multiplication. The notation $\text{tr}(\bullet)$ stands for the sum of the diagonal elements of $(\bullet)$, i.e., the *trace* of the matrix. We call these invariants the *primary invariants*. To economize the notation, we will sometimes write the invariants without showing the argument, i.e., write $\mathcal{I}_1$ rather than $\mathcal{I}_1(\mathbf{T})$, when it is obvious what tensor we are referring to. When it is ambiguous, e.g., when we write constitutive relationships in Chap. 6, we will include the argument.

The notation in Eq. 1.40 is subtle because the invariants are actually functions of the tensor components, T_{ij} or $\hat{T}_{ij}$, which depend upon the coordinate system. However, the result does *not* depend on the coordinate system because it is invariant. Thus, we simply put the tensor $\mathbf{T}$ as the argument of the function with the understanding that when we compute it, we must have a component representation (you cannot sum the diagonals of a matrix if you do not put the values into a matrix first).

The proof that these functions of the tensor components really are invariant is based on the fact that the matrix $\{\mathbf{Q}\}$ is an *orthogonal matrix* when the two bases $\{\hat{\mathbf{e}}_i\}$ and $\{\mathbf{e}_i\}$ are *orthonormal* (i.e., built with perpendicular unit vectors). The orthogonality relationship can be written in component form as

$$\{\mathbf{Q}\}^T\{\mathbf{Q}\} = \{\mathbf{I}\} \quad \rightarrow \quad \sum_{n=1}^{3} Q_{ni}Q_{nj} = \delta_{ij} \tag{1.41}$$

where δ_{ij} is the *Kronecker delta*, defined as

$$\delta_{ij} = \begin{cases} 1 & \text{if } i = j \\ 0 & \text{if } i \neq j \end{cases} \tag{1.42}$$

In other words, δ_{ij} is the ijth component of the identity matrix $\{\mathbf{I}\}$.

To prove the first invariant relationship, compute the trace of $\mathbf{T}$ using the formula for change of basis given in Eq. 1.37. To wit,

$$\text{tr}(\mathbf{T}) = \sum_{n=1}^{3} \hat{T}_{nn} = \sum_{n=1}^{3}\sum_{i=1}^{3}\sum_{j=1}^{3} Q_{ni}T_{ij}Q_{nj} \tag{1.43}$$

Carry out the sum on n, noting Eq. 1.41, to reduce the last expression to

$$\sum_{i=1}^{3}\sum_{j=1}^{3}\sum_{n=1}^{3} Q_{ni}Q_{nj}T_{ij} = \sum_{i=1}^{3}\sum_{j=1}^{3} \delta_{ij}T_{ij} = \sum_{i=1}^{3} T_{ii} \tag{1.44}$$

The last step converts the sum of nine terms to a sum of three terms by noting that the only three terms that survive are the ones where $i = j$ due to the definition of δ_{ij} in Eq. 1.42. Changing the dummy index i to n in the last term in Eq. 1.44 allows us to write

$$\text{tr}(\mathbf{T}) = \sum_{n=1}^{3} \hat{T}_{nn} = \sum_{n=1}^{3} T_{nn}$$

thereby proving that the trace of a tensor is invariant.

To prove that $\mathcal{I}_2$ and $\mathcal{I}_3$ are invariant, use Eq. 1.39 to write

$$\begin{aligned} \{\hat{\mathbf{T}}^2\} &= \{\hat{\mathbf{T}}\}\{\hat{\mathbf{T}}\} \\ &= \{\mathbf{Q}\}\{\mathbf{T}\}\{\mathbf{Q}\}^T\{\mathbf{Q}\}\{\mathbf{T}\}\{\mathbf{Q}\}^T \\ &= \{\mathbf{Q}\}\{\mathbf{T}\}\{\mathbf{T}\}\{\mathbf{Q}\}^T = \{\mathbf{Q}\}\{\mathbf{T}^2\}\{\mathbf{Q}\}^T \end{aligned}$$

where $\{\hat{\mathbf{T}}^2\}$ and $\{\mathbf{T}^2\}$ are the matrix representations of $\mathbf{T}^2$ in the $\{\hat{\mathbf{e}}_i\}$ and $\{\mathbf{e}_i\}$ coordinate systems, respectively. Again, the orthogonality relationship of Eq. 1.41 plays a key role. Now, following the same steps as before, we can show that

$$\text{tr}(\mathbf{T}^2) = \sum_{n=1}^{3}\sum_{i=1}^{3} \hat{T}_{ni}\hat{T}_{in} = \sum_{n=1}^{3}\sum_{i=1}^{3} T_{ni}T_{in}$$

thereby proving that the trace of the square of a tensor is invariant. The proof for the last invariant is similar. The details are left to the reader. All these calculations show that the final result does not depend on Q_{ij}. If the result does not depend on any of the components Q_{ij}, then it must be independent of the change of basis. Therefore, it does not depend on the coordinate systems chosen, which is the definition of invariance.

Any function of the invariants is also invariant. For example,

$$\begin{aligned} \mathcal{J}_1 &= \mathcal{I}_1 \\ \mathcal{J}_2 &= \tfrac{1}{2}\left(\mathcal{I}_1^2 - \mathcal{I}_2\right) \\ \mathcal{J}_3 &= \tfrac{1}{6}\left(\mathcal{I}_1^3 - 3\mathcal{I}_1\mathcal{I}_2 + 2\mathcal{I}_3\right) = \det(\mathbf{T}) \end{aligned} \tag{1.45}$$

are all invariants. These invariants are called the *principal invariants* because they play an important role in calculating the eigenvalues of a tensor (also called *principal values*), as we will show in the next section.

1.2.4 Eigenvalues and Eigenvectors of a Tensor

Tensors, like vectors, exist independent of a coordinate system. The components used to characterize both vectors and tensors depend on the base vectors associated with the coordinate system. The directed line segment, drawn as an arrow, gives a nice way to visualize a vector without referring to a coordinate system. The length of the arrow represents the magnitude of the vector, and the orientation of the arrow shows the direction of action. While the tensor invariants provide a coordinate-free way to characterize a tensor, they do not provide a direct route to visualize their action. How, then, can we visualize a tensor?

The most powerful way to think about tensors is this:

A tensor operates on a vector to produce another vector.

For example, we can write $\mathbf{w} = \mathbf{T}\mathbf{v}$ for some given tensor $\mathbf{T}$ and vector $\mathbf{v}$. The tensor $\mathbf{T}$ operates on the vector $\mathbf{v}$ to produce the vector $\mathbf{w}$. In general, the new vector $\mathbf{w}$ will have a different direction and magnitude than the original vector $\mathbf{v}$. We can represent vectors with arrows. Therefore, we can see a before ($\mathbf{v}$) and after ($\mathbf{w}$) picture that tells us what the tensor $\mathbf{T}$ does. In a sense, we know what a tensor *is* by looking at what it *does* to vectors.

To visualize the action of the tensor $\mathbf{T}$ on unit vectors $\mathbf{n}$, consider Fig. 1.16. While not directly pertinent to the discussion at hand, the tensor used to generate this figure is

$$\{\mathbf{T}\} = \begin{bmatrix} 0.80 & 0.10 & -0.30 \\ 0.10 & 0.65 & -0.40 \\ -0.30 & -0.40 & 1.20 \end{bmatrix}$$

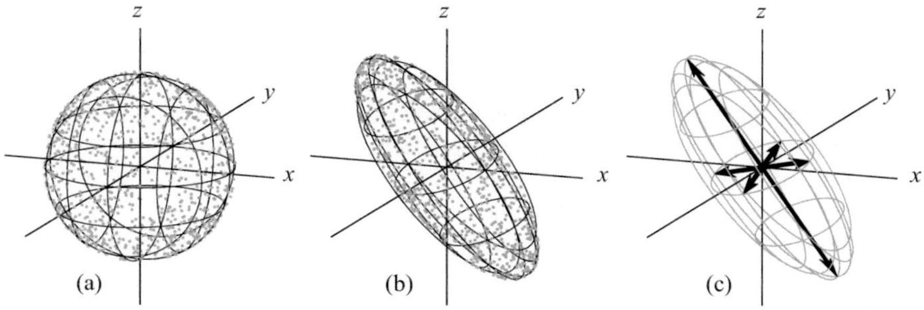

Fig. 1.16 *Tensor visualization.* The action of tensor **T**. **a** The gray dots are the tips of 1000 random unit vectors **n**. **b** The gray dots are the tips of the vectors **Tn** for the same **n** vectors. **c** The six arrows have lengths equal to the principal values and point in the principal directions

Figure 1.16a shows 1000 random unit vectors **n**. All these vectors emanate from the origin, so to keep the graphic clean, we just show the tips of the vectors as gray dots. The guidelines are there to help visualize that the dots all plot on a unit sphere. Figure 1.16b shows the 1000 vectors **Tn**. Now, the dots all fall on an ellipsoid. One way to think about what the tensor **T** does is that it takes the unit sphere and makes an ellipsoid. The *only* possible outcome of the tensor operation is a vector that lies on the ellipsoid. It cannot be inside or outside. Figure 1.16c shows the only six vectors for which the original **n** and final **Tn** point in the same direction. Notice that those vectors occur in pairs. Thus, we can say that there are three distinct directions for which this happens. We call these directions the *principal directions* of the tensor **T**. Thus, while a vector can be represented graphically with an arrow, a tensor can be represented with an ellipsoid or with a set of three vectors along the axes of the ellipsoid.

Consider a symmetric tensor **T**, and let us ask if we can find a unit vector **n** and a scalar μ that satisfy the equation:

$$\mathbf{Tn} = \mu \mathbf{n} \tag{1.46}$$

This problem is called the *eigenvalue problem*. In words, we are asking if there is a vector **n** that when operated on by **T** gives a vector that points in exactly the same direction as **n**. In such a case, the action of the tensor would be only to change the length of the original vector by a factor μ. The values μ and vectors **n** that satisfy Eq. 1.46 are called *eigenvalues* and *eigenvectors*, respectively.

The vectors **n** that satisfy Eq. 1.46 are the ones that we discovered in Fig. 1.16. The associated values μ are the semi-axis dimensions of the ellipsoid. It turns out that the semi-axes of the ellipsoid are oriented along the eigenvectors of the tensor. In fact, the six arrows shown in Fig. 1.16c are $\pm\mu_i \mathbf{n}_i$ for $i = 1, 2, 3$, where μ_i and $\mathbf{n}_i$ are the ith eigenvalue and eigenvector of **T**.

To find the eigenvalues and eigenvectors of a tensor, we must solve Eq. 1.46. We can rewrite the equation in the form

$$\left[\mathbf{T} - \mu\mathbf{I}\right]\mathbf{n} = \mathbf{0} \tag{1.47}$$

which is a linear homogeneous system of equations with coefficient matrix $[\,\mathbf{T} - \mu\mathbf{I}\,]$ and unknown vector $\mathbf{n}$. The tensor $\mathbf{I}$ is the identity tensor. It is obvious that $\mathbf{n} = \mathbf{0}$ satisfies the equation no matter what the coefficient matrix is. That solution, however, is not very interesting. We call it the *trivial solution*. What we are looking for are solutions with $\mathbf{n} \neq \mathbf{0}$.

Equation 1.47 for a tensor in three dimensions is a system of three equations with four unknowns (μ and the three components of the vectors $\mathbf{n}$). At best, we will be able to find three of those four items. It will turn out that we can find μ and the *direction* of the $\mathbf{n}$ vector, but not its magnitude. That makes sense because the vector $\mathbf{n}$ shows up on both sides of Eq. 1.46. We could scale that vector, and it will still satisfy the eigenvalue problem.

A theorem of linear algebra says that a linear homogeneous system of equations has a nontrivial solution if and only if the determinant of the coefficient matrix is equal to zero. Thus,

$$\det\left[\mathbf{T} - \mu\mathbf{I}\right] = 0 \tag{1.48}$$

which is an equation for μ that does not involve $\mathbf{n}$. Written out in components, we get (assuming that $\mathbf{T}$ is symmetric)

$$\det\begin{bmatrix} T_{11} - \mu & T_{12} & T_{13} \\ T_{12} & T_{22} - \mu & T_{23} \\ T_{13} & T_{23} & T_{33} - \mu \end{bmatrix} = 0 \tag{1.49}$$

Carrying out the computation of the determinant and collecting terms results in a cubic equation for μ in the form

$$f(\mu) = \mu^3 - \mathcal{J}_1\,\mu^2 + \mathcal{J}_2\,\mu - \mathcal{J}_3 = 0 \tag{1.50}$$

where the coefficients $\mathcal{J}_1$, $\mathcal{J}_2$, and $\mathcal{J}_3$ are the *principal invariants* of the tensor $\mathbf{T}$ given by Eq. 1.45. The simplest way to see how Eq. 1.50 comes about is to write out the determinant in Eq. 1.49, regroup terms, and compare the coefficients with the invariants of $\mathbf{T}$. Since the equation involves only invariants, its solution must also be an invariant.

In Eq. 1.50, $f(\mu)$ is called the *characteristic function* and the equation is called the *characteristic equation*. The solution of the characteristic equation yields the principal values μ_1, μ_2, and μ_3 of the strain tensor $\mathbf{T}$. We will order the principal values so that

$$\mu_1 \geq \mu_2 \geq \mu_3$$

This ordering is arbitrary but will help to clarify any comments related to the principal values. Each one of the principal values μ_i is associated with a principal direction $\mathbf{n}_i$, which we will take as a unit vector (since we will be unable to determine the

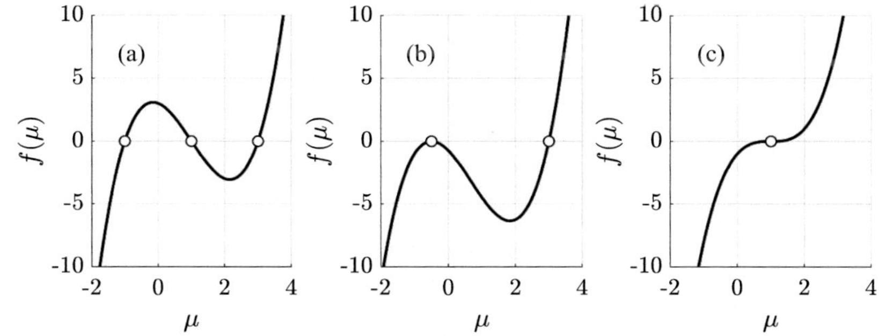

Fig. 1.17 *Characteristic equation.* The three possible cases for roots of the characteristic equation $f(\mu) = 0$ for principal values in three dimensions. **a** Three distinct roots. **b** One distinct root with two repeated roots. **c** Three repeated roots

magnitude, we might as well scale it to be one). One approach to finding the eigenvectors once the eigenvalues are known is outlined in Appendix A.5. We often call $(\mu_i, \mathbf{n}_i)$ an *eigenpair* because they go together. For example, if you plug μ_1 and $\mathbf{n}_2$ into Eq. 1.46, it will not be satisfied, i.e., $\mathbf{T}\mathbf{n}_2 \neq \mu_1\mathbf{n}_2$ (unless, of course, $\mu_1 = \mu_2$).

For the three-dimensional case, Eq. 1.49 is a cubic equation and, therefore, has three roots. There are a few cases that deserve special mention. Consider the plots of the characteristic equation for three different cases, shown in Fig. 1.17. In case (a), all three of the roots of the characteristic equation are distinct. That means the three principal values all have different values. In case (b), one root is distinct and two are repeated. For the repeated root, one of the peaks of the characteristic function just touches the axis. In case (c), all three roots are repeated. The roots of the characteristic equation are guaranteed to be real numbers if $\mathbf{T}$ is symmetric.

The eigenvectors associated with distinct roots are guaranteed to be orthogonal (see Appendix A.5 for a proof). In case (b) of Fig. 1.17, the principal directions $\mathbf{n}_2$ and $\mathbf{n}_3$ are not guaranteed to be orthogonal to each other, but both are orthogonal to $\mathbf{n}_1$ (because μ_1 is distinct). Any vector in the plane perpendicular to $\mathbf{n}_1$ is an eigenvector. We can, of course, pick two vectors from that plane that are orthogonal to each other. In case (c), any vector in three-dimensional space is an eigenvector, i.e., all directions are principal directions. Again, if needed, we can select specific vectors $\mathbf{n}_i$ such that they are orthogonal to each other.

For the two-dimensional case, the matrices are 2×2, the characteristic equation is quadratic, and there are two solutions to the eigenvalue problem. All aspects of the solution for this case are analogous to the three-dimensional case. Appendix A.5 gives more detail on how to solve an eigenvalue problem.

Example 1.6
Find the eigenvalues and eigenvectors of the eigenvalue problem $\mathbf{T}\mathbf{n} = \mu\mathbf{n}$, with the matrix

$$\mathbf{T} = \begin{bmatrix} 2 & 1 \\ 1 & 3 \end{bmatrix}$$

Solution. The eigenvalue problem can be expressed in components as

$$\begin{bmatrix} 2-\mu & 1 \\ 1 & 3-\mu \end{bmatrix} \begin{Bmatrix} n_1 \\ n_2 \end{Bmatrix} = \begin{Bmatrix} 0 \\ 0 \end{Bmatrix}$$

The characteristic equation comes from setting the determinant of the coefficient matrix equal to zero. To wit,

$$(2-\mu)(3-\mu) - (1)(1) = 0 \quad \rightarrow \quad \mu^2 - 5\mu + 5 = 0$$

Notice that $\mathcal{I}_1 = 5$ and $\mathcal{I}_2 = 5$. The characteristic equation is quadratic because the system is 2×2 and can be solved with the quadratic formula to give

$$\mu_1 = \tfrac{1}{2}(5 + \sqrt{5}) = 3.6180, \qquad \mu_2 = \tfrac{1}{2}(5 - \sqrt{5}) = 1.3820$$

Substituting μ_1 back into the eigenvaluc problem, we get

$$\begin{bmatrix} 2-3.618 & 1 \\ 1 & 3-3.618 \end{bmatrix} \begin{Bmatrix} n_1 \\ n_2 \end{Bmatrix} = \begin{bmatrix} -1.618 & 1 \\ 1 & -0.618 \end{bmatrix} \begin{Bmatrix} n_1 \\ n_2 \end{Bmatrix} = \begin{Bmatrix} 0 \\ 0 \end{Bmatrix}$$

Both equations are satisfied if $n_2 = 1.618\, n_1$. Hence, the first eigenvector (i.e., the one associated with μ_1) is

$$\mathbf{n}_1 = \begin{Bmatrix} 1 \\ 1.618 \end{Bmatrix}$$

Of course, this vector can be scaled to have length one. Now, substituting μ_2 back into the eigenvalue problem, we get

$$\begin{bmatrix} 2-1.382 & 1 \\ 1 & 3-1.382 \end{bmatrix} \begin{Bmatrix} n_1 \\ n_2 \end{Bmatrix} = \begin{bmatrix} 0.618 & 1 \\ 1 & 1.618 \end{bmatrix} \begin{Bmatrix} n_1 \\ n_2 \end{Bmatrix} = \begin{Bmatrix} 0 \\ 0 \end{Bmatrix}$$

Both equations are satisfied if $n_2 = -0.618\, n_1$. Hence, the second eigenvector (i.e., the one associated with μ_2) is

$$\mathbf{n}_2 = \begin{Bmatrix} 1 \\ -0.618 \end{Bmatrix}$$

This vector can also be scaled to have unit length. Note that the MATLAB `eig` command scales the eigenvectors so that $\mathbf{n} \cdot \mathbf{n} = 1$.

The eigenvectors for distinct eigenvalues are orthogonal, i.e.:

$$\mathbf{n}_i \cdot \mathbf{n}_j = 0, \quad \text{if } \mu_i \neq \mu_j \tag{1.51}$$

This assertion is proved in Appendix A.5. If two of the eigenvalues are equal, e.g., $\mu_2 = \mu_3$, but not equal to μ_1, then *any* vector in the plane perpendicular to $\mathbf{n}_1$ is an eigenvector associated with the repeated root $\mu_2 = \mu_3$. If all three eigenvalues are equal, i.e., $\mu_1 = \mu_2 = \mu_3$, then any vector is an eigenvector. For repeated roots, we can always select vectors that are orthogonal to form an orthogonal triad $\{\mathbf{n}_1, \mathbf{n}_2, \mathbf{n}_3\}$.

If the eigenvectors are orthogonal (i.e., $\mathbf{n}_i \cdot \mathbf{n}_j = 0$), then it is straightforward to show that $\mathbf{n}_i \cdot \mathbf{T}\mathbf{n}_j = 0$, too. Just dot both sides of the eigenvalue problem with an eigenvector. To wit,

$$\mathbf{n}_i \cdot \mathbf{T}\mathbf{n}_j = \mathbf{n}_i \cdot \mu_j \mathbf{n}_j = \mu_j \mathbf{n}_i \cdot \mathbf{n}_j = 0 \tag{1.52}$$

for $i \neq j$. The eigenpair $(\mu_j, \mathbf{n}_j)$ satisfies the eigenvalue problem. The vector $\mathbf{n}_i$ is one of the other eigenvectors. Orthogonality with respect to $\mathbf{T}$ follows from orthogonality of the vectors themselves.

Example 1.7

Show that the eigenvectors from the previous example are orthogonal.

Solution. First, note that the eigenvalues are distinct. Therefore, the vectors computed previously should be orthogonal. To make the demonstration, we need to show that $\mathbf{n}_1 \cdot \mathbf{n}_2 = 0$ and $\mathbf{n}_1 \cdot \mathbf{T}\mathbf{n}_2 = 0$. To wit,

$$\mathbf{n}_1 \cdot \mathbf{n}_2 = \{\mathbf{n}_1\}^T \{\mathbf{n}_2\} = \begin{Bmatrix} 1 & -0.618 \end{Bmatrix} \begin{Bmatrix} 1 \\ 1.618 \end{Bmatrix} = 0$$

Similarly,

$$\mathbf{n}_1 \cdot \mathbf{T}\mathbf{n}_2 = \begin{Bmatrix} 1 & -0.618 \end{Bmatrix} \begin{bmatrix} 2 & 1 \\ 1 & 3 \end{bmatrix} \begin{Bmatrix} 1 \\ 1.618 \end{Bmatrix} = \begin{Bmatrix} 1 & -0.618 \end{Bmatrix} \begin{Bmatrix} 3.618 \\ 5.854 \end{Bmatrix} = 0$$

Because the matrix is symmetric, there is no need to reverse the order of multiplication.

The *Cayley-Hamilton theorem* is related to the characteristic equation for eigenvalues. It states that for a tensor $\mathbf{T}$ with principal invariants $\mathcal{J}_1$, $\mathcal{J}_2$, and $\mathcal{J}_3$, the following relationship holds:

$$\mathbf{T}^3 - \mathcal{J}_1 \mathbf{T}^2 + \mathcal{J}_2 \mathbf{T} - \mathcal{J}_3 \mathbf{I} = \mathbf{0} \tag{1.53}$$

The proof of this theorem relies on the eigenvalues and eigenvectors of $\mathbf{T}$. If μ_i and $\mathbf{n}_i$ are the ith eigenvalue and eigenvector, respectively, then we know that $\mathbf{T}\mathbf{n}_i = \mu_i \mathbf{n}_i$. Therefore, $\mathbf{T}^n \mathbf{n}_i = \mu^n \mathbf{n}_i$, which can be shown by applying the eigenvalue relationship n times. Thus, we can write

$$\begin{aligned}
\left[\mathbf{T}^3 - \mathcal{J}_1 \mathbf{T}^2 + \mathcal{J}_2 \mathbf{T} - \mathcal{J}_3 \mathbf{I}\right] \mathbf{n}_i &= \mathbf{0} \\
\mathbf{T}^3 \mathbf{n}_i - \mathcal{J}_1 \mathbf{T}^2 \mathbf{n}_i + \mathcal{J}_2 \mathbf{T}\mathbf{n}_i - \mathcal{J}_3 \mathbf{n}_i &= \mathbf{0} \\
\mu_i^3 \mathbf{n}_i - \mathcal{J}_1 \mu_i^2 \mathbf{n}_i + \mathcal{J}_2 \mu_i \mathbf{n}_i - \mathcal{J}_3 \mathbf{n}_i &= \mathbf{0} \\
\left(\mu_i^3 - \mathcal{J}_1 \mu_i^2 + \mathcal{J}_2 \mu_i - \mathcal{J}_3\right) \mathbf{n}_i &= \mathbf{0}
\end{aligned}$$

Since μ_i satisfies Eq. 1.50, the term in parentheses in the last line is zero. This relationship is true for any of the eigenvectors. Since the eigenvectors span the space, the term in square brackets in the first line must be zero.

Example 1.8
Use the Cayley-Hamilton theorem to find the value of $\mathcal{I}_3 = \text{tr}\left(\mathbf{T}^3\right)$ for a two-dimensional tensor $\mathbf{T}$ in terms of $\mathcal{I}_1$ and $\mathcal{I}_2$. Show that the formula works by applying it to the tensor

$$\mathbf{T} = \begin{bmatrix} 5 & 2 \\ 2 & 3 \end{bmatrix}$$

and computing $\mathcal{I}_3$ both directly and from the derived formula.

Solution. The Cayley-Hamilton theorem in two dimensions has the form:

$$\mathbf{T}^2 - \mathcal{J}_1 \mathbf{T} + \mathcal{J}_2 \mathbf{I} = \mathbf{0}$$

We can multiply both sides of the equation with the tensor $\mathbf{T}$ and then take the trace of both sides of the equation. Since the trace is simply a sum, the trace of the left side is equal to the sum of the traces of the three terms. The trace of a scalar times a tensor is the scalar times the trace of the tensor. Hence, we can write

$$\begin{aligned}
\text{tr}\left(\mathbf{T}^3\right) - \text{tr}\left(\mathcal{J}_1 \mathbf{T}^2\right) + \text{tr}\left(\mathcal{J}_2 \mathbf{T}\right) &= \text{tr}(\mathbf{0}) \\
\mathcal{I}_3 - \mathcal{J}_1 \mathcal{I}_2 + \mathcal{J}_2 \mathcal{I}_1 &= 0
\end{aligned}$$

where $\mathcal{I}_1$, $\mathcal{I}_2$, and $\mathcal{I}_3$ are given by Eq. 1.40. Now, substitute the expressions for $\mathcal{J}_1$ and $\mathcal{J}_2$ from Eq. 1.45 to give

$$\mathcal{I}_3 - \left(\mathcal{I}_1\right) \mathcal{I}_2 + \tfrac{1}{2}\left(\mathcal{I}_1^2 - \mathcal{I}_2\right) \mathcal{I}_1 = 0$$

Finally, we can solve for $\mathcal{I}_3$ to get

$$\mathcal{I}_3 = \tfrac{3}{2}\mathcal{I}_1\,\mathcal{I}_2 - \tfrac{1}{2}\mathcal{I}_1^3$$

For the given tensor, we can do the matrix multiplications to compute

$$\mathbf{T}^2 = \begin{bmatrix} 5 & 2 \\ 2 & 3 \end{bmatrix}\begin{bmatrix} 5 & 2 \\ 2 & 3 \end{bmatrix} = \begin{bmatrix} 29 & 16 \\ 16 & 13 \end{bmatrix}$$

$$\mathbf{T}^3 = \begin{bmatrix} 5 & 2 \\ 2 & 3 \end{bmatrix}\begin{bmatrix} 29 & 16 \\ 16 & 13 \end{bmatrix} = \begin{bmatrix} 177 & 106 \\ 106 & 71 \end{bmatrix}$$

Now, we can directly compute $\mathcal{I}_3 = 177 + 71 = 248$. We can also compute $\mathcal{I}_1 = 5 + 3 = 8$ and $\mathcal{I}_2 = 29 + 13 = 42$. Applying the formula just derived for $\mathcal{I}_3$ for problems in two dimensions, we get

$$\mathcal{I}_3 = \tfrac{3}{2}(8)(42) - \tfrac{1}{2}(8)^3 = 248$$

which matches the value computed directly from $\mathrm{tr}(\mathbf{T}^3)$.

This example shows that there are only two independent invariants, $\mathcal{I}_1$ and $\mathcal{I}_2$, in two dimensions. The third invariant $\mathcal{I}_3$ still exists and can be computed directly as shown above, but it is not independent of the first two invariants. In fact, $\mathrm{tr}(\mathbf{T}^n)$ is invariant for any n. Only the first two of them in two dimensions and first three in three dimensions are independent.

1.2.5 The Spectral Representation

A symmetric tensor $\mathbf{T}$ can be put into *spectral* form using the eigenvalues and eigenvectors. In particular, for orthonormal eigenvectors, we can write

$$\mathbf{T} = \sum_{i=1}^{3} \mu_i \big[\, \mathbf{n}_i \otimes \mathbf{n}_i \,\big]$$

as shown in Appendix A.5. In matrix form, using the eigenvectors $\{\mathbf{n}_1, \mathbf{n}_2, \mathbf{n}_3\}$ as the basis for representing the components of the tensor, we can write

$$\{\mathbf{T}\} = \begin{bmatrix} \mu_1 & 0 & 0 \\ 0 & \mu_2 & 0 \\ 0 & 0 & \mu_3 \end{bmatrix}$$

One important ramification of this representation is that the principal invariants can be written as

$$\begin{aligned} \mathcal{J}_1 &= \mu_1 + \mu_2 + \mu_3 \\ \mathcal{J}_2 &= \mu_1\mu_2 + \mu_2\mu_3 + \mu_3\mu_1 \\ \mathcal{J}_3 &= \mu_1\mu_2\mu_3 \end{aligned} \tag{1.54}$$

These relationships can be shown either by computing the principal invariants from Eq. 1.45 or by expanding the factored form of the characteristic equation, Eq. 1.50:

$$f(\mu) = (\mu - \mu_1)(\mu - \mu_2)(\mu - \mu_3) = 0$$

The spectral representation of **T** shows that the eigenvalue and eigenvectors contain all the information that **T** does. Another important ramification of the spectral representation is that if a tensor {**T**} is diagonal, then the diagonal elements must be the principal values and the principal directions are $\mathbf{e}_1$, $\mathbf{e}_2$, and $\mathbf{e}_3$.

1.3 Summary

Vectors and tensors provide the most natural way to describe force, motion, and deformation mathematically. Many classical texts seem in a rush to express vectors in terms of their components and to proceed with scalar equations. While such an approach is not incorrect, it can limit our ability to see the geometry of the problem in the most natural light and it often makes derivation and manipulation of the equations more awkward and prone to error. Use of a direct vector notation affords us the opportunity to see the geometry while never being more than a quick vector operation away from any component expression we might need (e.g., through the operation of the dot product).

Understanding vectors and tensors starts with learning how to execute the operations faithfully. But it does not end there. The goal is to get to a level of fluency where the notation inspires a direct image of the quantity it represents. As with any language, we first learn to translate following rules of grammar, but eventually we think in that language. Attention to detail as you are learning is important. Throughout the book, scalars are represented by *italic* symbols (e.g., a, b, c), vectors with lowercase **bold** symbols (e.g., **u**, **v**), and tensors with uppercase **bold** symbols (e.g., **T**, **Q**). Computer code variables and functions are rendered in `typewriter` font.

If you are computing by hand, make sure that you distinguish vectors and tensors symbolically. One popular approach is to put an arrow or tilde above the symbol to make it a vector, or a tilde or double bar to make it a tensor, e.g.:

$$\mathbf{v} \quad \longleftrightarrow \quad \vec{v} \text{ or } \tilde{v}, \qquad \mathbf{T} \quad \longleftrightarrow \quad \tilde{T} \text{ or } \bar{\bar{T}}$$

The tilde or bar can also be positioned beneath the character. Whatever method you choose, be consistent and let the symbols help to maintain the integrity of your vector and tensor algebra.

1.4 Problems

1.1 Consider the three vectors

$$\mathbf{u} = \mathbf{e}_1 - 3\mathbf{e}_2 + 4\mathbf{e}_3, \quad \mathbf{v} = 5\mathbf{e}_1 + \mathbf{e}_2 + 2\mathbf{e}_3, \quad \mathbf{w} = 2\mathbf{e}_1 + 3\mathbf{e}_2 + \mathbf{e}_3$$

Compute $\mathbf{u} \cdot \mathbf{v}$, $\mathbf{u} \times \mathbf{v}$, and $\mathbf{u} \otimes \mathbf{v}$. Compute the triple scalar product $(\mathbf{u} \times \mathbf{v}) \cdot \mathbf{w}$ and the vector $[\mathbf{u} \otimes \mathbf{v}]\mathbf{w}$.

1.2 Consider the two vectors $\mathbf{u} = 2\mathbf{e}_1 - 3\mathbf{e}_2 + \mathbf{e}_3$ and $\mathbf{v} = \mathbf{e}_1 + \mathbf{e}_2 - 3\mathbf{e}_3$. Find the angle between the vectors using both the dot and cross products, the area of the parallelogram with edges defined by the two vectors, and a *unit* vector perpendicular to the plane containing both $\mathbf{u}$ and $\mathbf{v}$.

1.3 Consider the vector $\mathbf{v} = 5\mathbf{e}_1 + 2\mathbf{e}_2 - 7\mathbf{e}_3$. Find a unit vector that points in the same direction as the vector $\mathbf{v}$ and the projection tensor $\mathbf{P}$ (with components in matrix form) that projects vectors onto the plane perpendicular to the vector $\mathbf{v}$.

1.4 A unit vector $\mathbf{n}$ makes equal angles relative to the standard base vectors $\{\mathbf{e}_1, \mathbf{e}_2, \mathbf{e}_3\}$. Find the components of the vector $\mathbf{n}$ relative to the base vectors.

1.5 The angles between the vector $\mathbf{v}$ and the base vectors $\{\mathbf{e}_1, \mathbf{e}_2, \mathbf{e}_3\}$ are $20°$, $40°$, and $60°$, respectively. The length of the vector is 3. Find the components of the vector $\mathbf{v}$ relative to the base vectors.

1.6 Find the vector that is the projection of the vector $\mathbf{v} = 2\mathbf{e}_1 - 3\mathbf{e}_2 - 4\mathbf{e}_3$ onto the plane normal to the vector $\mathbf{u} = \mathbf{e}_1 + \mathbf{e}_2 + \mathbf{e}_3$, i.e., $\mathbf{v}_\perp$. Find the part of $\mathbf{v}$, i.e., $\mathbf{v}_\parallel$, that is perpendicular to the plane normal to $\mathbf{u}$.

1.7 The components of a vector $\mathbf{v}$ in the system $\{\mathbf{e}_1, \mathbf{e}_2\}$ (the original system) are $v_1 = 1$ and $v_2 = -3$. A second coordinate system is represented by the unit base vectors $\{\hat{\mathbf{e}}_1, \hat{\mathbf{e}}_2\}$, which are rotated by $45°$ counterclockwise from $\{\mathbf{e}_1, \mathbf{e}_2\}$. Find the components of the vector $\mathbf{v}$ relative to $\{\hat{\mathbf{e}}_1, \hat{\mathbf{e}}_2\}$. Repeat the problem rotating the second coordinate system by $45°$ in the *clockwise* direction instead.

1.8 Consider a coordinate system represented by the unit base vectors $\{\hat{\mathbf{e}}_1, \hat{\mathbf{e}}_2\}$ that is rotated by $60°$ counterclockwise from $\{\mathbf{e}_1, \mathbf{e}_2\}$ (the original system). The components of a tensor $\mathbf{T}$ in the original system are

$$\{\mathbf{T}\} = \begin{bmatrix} 5 & 2 \\ 2 & 9 \end{bmatrix}$$

Find the components of the tensor $\mathbf{T}$ relative to $\{\hat{\mathbf{e}}_1, \hat{\mathbf{e}}_2\}$.

1.9 Find the eigenvalues and eigenvectors of the tensor $\mathbf{T}$ in two dimensions with components

$$\{\mathbf{T}\} = \begin{bmatrix} 3 & 2 \\ 2 & 4 \end{bmatrix}$$

relative to the standard basis. Do the computation by hand and check the results using the `eig` function in MATLAB.

1.10 Consider a coordinate system represented by the unit base vectors $\{\hat{\mathbf{e}}_1, \hat{\mathbf{e}}_2, \hat{\mathbf{e}}_3\}$ that can be expressed in terms of $\{\mathbf{e}_1, \mathbf{e}_2, \mathbf{e}_3\}$ as

$$\hat{\mathbf{e}}_1 = \tfrac{1}{\sqrt{3}}\big(\mathbf{e}_1 + \mathbf{e}_2 + \mathbf{e}_3\big), \quad \hat{\mathbf{e}}_2 = \tfrac{1}{\sqrt{6}}\big(2\mathbf{e}_1 - \mathbf{e}_2 - \mathbf{e}_3\big), \quad \hat{\mathbf{e}}_3 = \tfrac{1}{\sqrt{2}}\big(\mathbf{e}_2 - \mathbf{e}_3\big)$$

The components of a tensor $\mathbf{T}$ in the original system are

$$\{\mathbf{T}\} = \begin{bmatrix} 5 & 2 & 1 \\ 2 & 4 & 5 \\ 1 & 5 & 7 \end{bmatrix}$$

Find the components $\{\hat{\mathbf{T}}\}$ of the tensor $\mathbf{T}$ relative to $\{\hat{\mathbf{e}}_1, \hat{\mathbf{e}}_2, \hat{\mathbf{e}}_3\}$. Compute the primary and principal invariants in both coordinate systems to demonstrate that they are equal.

1.11 Find the eigenvalues and eigenvectors of the tensor $\mathbf{T}$ with components

$$\{\mathbf{T}\} = \begin{bmatrix} 5 & 2 & 1 \\ 2 & 4 & 5 \\ 1 & 5 & 7 \end{bmatrix}$$

relative to the standard basis. Do the computation by hand and check the results using the `eig` function in MATLAB. Use the eigenvalues to compute the primary and principal invariants.

1.12 Consider the tensor $\mathbf{T}$ with components

$$\{\mathbf{T}\} = \begin{bmatrix} 7 & 1 & 1 \\ 1 & 5 & 3 \\ 1 & 3 & 5 \end{bmatrix}$$

relative to the standard basis. Show that the vectors

$$\{\mathbf{n}_1\} = \begin{Bmatrix} 1 \\ 1 \\ 1 \end{Bmatrix}, \quad \{\mathbf{n}_2\} = \begin{Bmatrix} 2 \\ -1 \\ -1 \end{Bmatrix}, \quad \{\mathbf{n}_3\} = \begin{Bmatrix} 0 \\ 1 \\ -1 \end{Bmatrix}$$

satisfy the eigenvalue problem and are, therefore, eigenvectors. Find the associated eigenvalues without directly solving the eigenvalue problem. Hint: Use the eigenvalue equation itself to show that $\mathbf{Tn}_i$ points in the same direction as $\mathbf{n}_i$. Then use the length of the resulting vector to find the eigenvalue.

Chapter 2
Statics

Mechanics is the study of force and motion. In this chapter, we present the concepts of Newtonian equilibrium, focusing on the study of rigid bodies in equilibrium, a subject often called *statics*. We introduce the concept of the *free body diagram*, establish conventions for drawing it, and show how the equations of static equilibrium emanate directly from these quantitative sketches. Along the way we introduce concepts like the moment of a force, reaction forces, and the *typical particle* for distributed effects. Throughout the chapter we focus on modeling aspects like idealization of support conditions and representation of loads, applying the machinery of statics to real-world applications like hydrostatics and structures.

Carrying out the calculations associated with statics is an exercise in vector algebra. Thus, this chapter is built on the foundations laid in the previous one. One of the best ways to master vectors is to apply them in problems of mechanics, where one can connect the abstract mathematical notions with real physical systems. In this book, we favor direct vector notation, as it often makes the geometrical aspects of a problem clearer and in many problems it helps to avoid getting lost in a sea of scalar algebra. For that reason, even a reader who already knows statics will benefit from reviewing the theory and nomenclature in this chapter and looking carefully at the examples to prepare for the main event (deformable solids).

The basic concepts of statics are useful in the analysis of deformable systems. In fact, we will consider *equilibrium* to be one of the three legs of the mechanics stool that is the foundation of the mechanics of deformable solids, the other two legs being kinematics and constitutive relationships. Equilibrium of deformable bodies is slightly different from rigid bodies in that equilibrium must hold in the *deformed* position of the body. For linear theories, in which the deformations are considered small, establishing equilibrium in the nominal position (the one and only position for a rigid body) is appropriate, and the development of equations of static equilibrium will follow those for rigid bodies exactly.

K. D. Hjelmstad, *Mechanics of Deformable Solids: An Introduction*,
https://doi.org/10.1007/978-3-031-89873-0_2

2.1 Newton's Laws of Motion

Isaac Newton published *Philosophae Naturalis Principia Mathematica* in 1686. The work, known today simply as *Principia*,[1] lays out Newton's three laws of motion:

Newton's Laws of Motion

1. A body continues in its state of rest, or of uniform motion in a right line, unless it is compelled to change that state by forces impressed upon it.
2. The change in motion is proportional to the motive force impressed and is made in the direction of the right line in which that force is impressed.
3. To every action there is always an opposed equal reaction, i.e., the mutual actions of two bodies upon each other are always equal and directed to contrary parts.

Newton's laws were created for particles and extended to rigid bodies of finite size by Leonhard Euler (1707–1783). Newton's second law is often summarized in modern notation as

$$\mathbf{F} = m\mathbf{a} \tag{2.1}$$

where $\mathbf{F}$ is the vector sum of all forces acting on the particle, m is the mass, and $\mathbf{a}$ is the acceleration vector of the particle. Newton described his laws without the benefit of vectors, which were invented about 200 years later. Vector notation is an elegant way to mathematically capture Newton's notion of *right line*, which appears both in the first law, wherein it applies to velocity (what Newton calls *motion*), and in the second law, wherein it applies to force and acceleration (what Newton calls *change in motion*). Newton's third law is the birthplace of the *free body diagram*, an indispensable tool in engineering mechanics.

2.2 Static Equilibrium

One can argue that all systems are in motion and, therefore, the equations of dynamics are the appropriate set of tools to analyze them. However, some systems find their way to a state called *static equilibrium* in which motion ceases. In static equilibrium, the acceleration is zero. Hence, Newton's second law reduces to

$$\sum_{i=1}^{N} \mathbf{F}_i = \mathbf{0} \tag{2.2}$$

[1] Sir Isaac Newton, *Principia*, Vol. 1: The Motion of Bodies. Translated by F. Cajori. University of California Press, 1934.

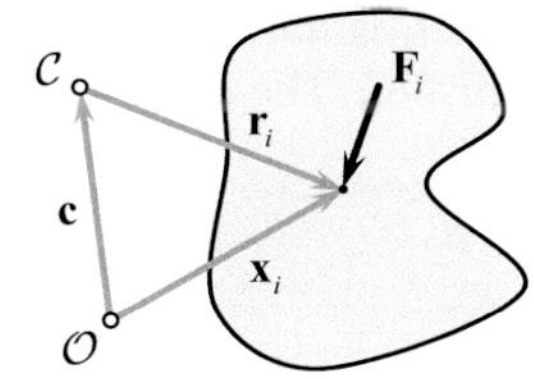

Fig. 2.1 *Forces and moments.* A body subjected to a force $\mathbf{F}_i$ located at position $\mathbf{x}_i$ relative to the origin $\mathcal{O}$. The moment arm relative to point $\mathcal{C}$ is $\mathbf{r}_i$

where the summation symbol has been added to emphasize that it is the *net* force, i.e., the vector sum of the N individual forces $\mathbf{F}_i$ acting on a particle or rigid body that must be equal to zero to have equilibrium. For a particle, all the forces act at the same point. For a rigid body, each force acts at a certain location, which can be specified by its position vector $\mathbf{x}_i$, measured relative to some arbitrarily selected *origin*, as illustrated in Fig. 2.1.

For particle mechanics, summation of forces is enough to establish equilibrium. The position vector does not affect the summation of forces. For rigid bodies, an additional concept must be added: The moment of the forces about any point in space (e.g., point $\mathcal{C}$ in Fig. 2.1) must also sum to zero for static equilibrium. The concept of the *moment of a force* is characterized mathematically as

$$\mathbf{M} = \mathbf{r} \times \mathbf{F} \tag{2.3}$$

where $\mathbf{M}$ is the moment created by the force $\mathbf{F}$. The moment requires a reference point, i.e., the point about which the moment is computed. The *moment arm* $\mathbf{r}$ is a position vector from the reference point $\mathcal{C}$ to the point of action of the force. For a rigid body, each force $\mathbf{F}_i$ is associated with its own moment arm $\mathbf{r}_i$. All moment arms must use the same reference point. The moment of a force is computed as the cross product of two vectors. Therefore, moment is a vector. Because the cross product of vectors is anticommutative, the order of the vectors in the moment calculation is important. It is customary to define moment with the moment arm first and the force second.

Euler's law of balance of moments can be stated as

$$\sum_{i=1}^{N} \big(\mathbf{r}_i \times \mathbf{F}_i\big) = \mathbf{0} \tag{2.4}$$

An imbalance of moments of forces acting on a rigid body causes the body to rotate, just as an imbalance of forces causes the body to translate. Equation 2.4 explains why balance of moments is not important for particles. Even if you have multiple forces acting on a particle, they all have the same moment arm because the point of action is the same for all forces, i.e., $\mathbf{r}_i = \mathbf{r}$ for all i. Hence, the balance of moments can be written as

$$\sum_{i=1}^{N} \big(\mathbf{r} \times \mathbf{F}_i\big) = \mathbf{r} \times \sum_{i=1}^{N} \mathbf{F}_i = \mathbf{0}$$

which is automatically satisfied if the forces are in equilibrium, i.e., if Eq. 2.2 is satisfied. The equations of static equilibrium are summarized in the box below:

Static Equilibrium

$$\sum_{i=1}^{N} \mathbf{F}_i = \mathbf{0} \qquad \sum_{i=1}^{N} (\mathbf{r}_i \times \mathbf{F}_i) = \mathbf{0}$$

For rigid bodies with adequate support, the externally applied loads will be equilibrated by the *reaction forces* associated with the supports. The equations of static equilibrium determine the amount and direction of force required at the supports. If the support reactions cannot equilibrate the applied loads, then the system will accelerate, and thus becomes a problem of dynamics, not statics.

2.3 Free Body Diagrams

A *free body diagram* is a graphical representation of the forces acting on a system. One of the key axioms of mechanics is that if a body is in equilibrium, then every part of the body must also be in equilibrium. That outcome is assured for each free body diagram by replacing the parts taken away with the forces those parts exert on what remains. From that perspective, we can view a body of finite size as a collection of particles. Each particle must be in equilibrium with the forces exerted on the particle by its neighbors. If we establish equilibrium for each particle, then equilibrium will hold for any free body diagram comprising multiple particles.

The idea of the free body diagram is illustrated in Fig. 2.2, which shows a cylindrical disk resting between two smooth surfaces, one vertical and one making an angle φ relative to horizontal. The disk has total weight W. The free body diagram removes the support provided by the surfaces and replaces that support with the reaction forces $A\mathbf{n}$ and $-B\mathbf{e}_1$. The weight of the disk is represented as a single force

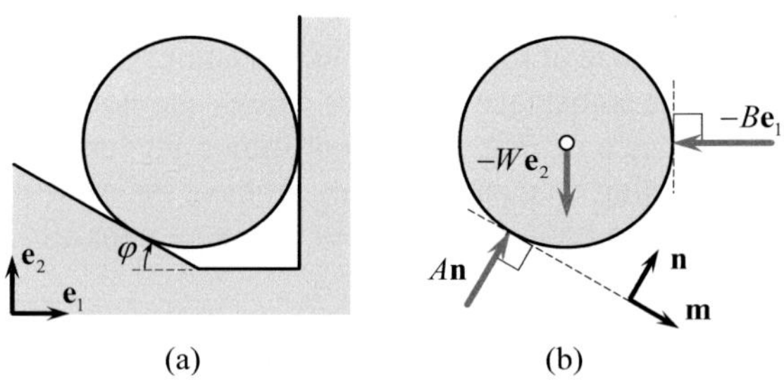

Fig. 2.2 *Free body diagram.* Circular disk of weight W resting on two smooth surfaces. **a** Problem geometry. **b** Free body diagram

$-W\mathbf{e}_2$ (we will see how to handle distributed effects in a subsequent section). The reaction forces act normal to their respective surfaces because the problem statement says that the surfaces are smooth, which is code for zero friction. Reaction forces are often governed by some sort of device like a pin or roller. In each case, the nature of the support will be described in the problem statement, either through the illustration or in words.

Observe that each force is represented as a vector on the free body diagram. The scalar magnitudes are A, B, and W, and the directions are established by the unit vectors $\mathbf{n}$, $\mathbf{e}_1$, and $\mathbf{e}_2$, respectively. The minus sign on two of the forces comes from the fact that the direction of the force is shown pointing in the opposite direction of a defined unit vector. For example, the weight force W acts downward, but the unit vector $\mathbf{e}_2$ points upward. Therefore, the force is in the $-\mathbf{e}_2$ direction.

It is important for the free body diagram to show only the forces and not the physical feature that provides those forces, as that would be redundant. For example, in Fig. 2.2b, the two surfaces in contact with the disk have been replaced by force vectors. Removing the physical feature is a good reminder that the force represents the action of the part that we took away. If that part is still there, one might wonder what the force is for. It is also important to represent the forces exactly at their point of action, especially when it comes time to sum moments about a certain point.

For this problem, we define the unit vectors $\mathbf{n}$ and $\mathbf{m}$ in terms of the angle of the incline and the standard base vectors $\{\mathbf{e}_1, \mathbf{e}_2\}$. We can think of the vectors $\mathbf{n}$ and $\mathbf{m}$ as *helper* vectors. We define these vectors to help describe the geometric features of the problem. In this case, the vector $\mathbf{n}$ defines the direction normal to the inclined surface and the vector $\mathbf{m}$ defines the plane of the incline. To carry out calculations, it is essential to relate any defined helper vectors back to the base vectors $\{\mathbf{e}_1, \mathbf{e}_2\}$. Thus, if the angle of the incline is φ, measured with respect to horizontal, then

$$\begin{aligned}\mathbf{n} &= \sin\varphi\,\mathbf{e}_1 + \cos\varphi\,\mathbf{e}_2\\ \mathbf{m} &= \cos\varphi\,\mathbf{e}_1 - \sin\varphi\,\mathbf{e}_2\end{aligned}$$

It is straightforward to show that both vectors have unit length and that these two vectors are perpendicular to each other.

The equation of balance of forces for this problem is

$$\sum_{i=1}^{N} \mathbf{F}_i = A\mathbf{n} - B\mathbf{e}_1 - W\mathbf{e}_2 = \mathbf{0} \tag{2.5}$$

Notice that we compose the equation of equilibrium directly from the free body diagram. A valid free body diagram includes all forces acting on the body and the equation of equilibrium must include all those forces.

Equation 2.5 is a vector equation. Since all forces act in the $\{\mathbf{e}_1, \mathbf{e}_2\}$ plane, there are really two scalar equations lurking in there. The solution strategy to find A and B can take many equivalent but different routes. For this problem, we can dot the equation with $\mathbf{m}$ to get

$$\begin{aligned}\mathbf{m} \cdot \big(A\mathbf{n} - B\mathbf{e}_1 - W\mathbf{e}_2\big) &= \mathbf{0}\\ -B\big(\mathbf{m} \cdot \mathbf{e}_1\big) - W\big(\mathbf{m} \cdot \mathbf{e}_2\big) &= \mathbf{0}\\ -B\cos\varphi + W\sin\varphi &= 0\end{aligned}$$

where we have noted that $\mathbf{m} \cdot \mathbf{n} = 0$. From the definition of $\mathbf{m}$, we can see that $\mathbf{m} \cdot \mathbf{e}_1 = \cos\varphi$ and $\mathbf{m} \cdot \mathbf{e}_2 = -\sin\varphi$. We can solve this equation to find the unknown reaction B. To wit,

$$B = W\tan\varphi$$

Similarly, we can dot the equation with $\mathbf{e}_2$ to give

$$\begin{aligned}\mathbf{e}_2 \cdot \big(A\mathbf{n} - B\mathbf{e}_1 - W\mathbf{e}_2\big) &= \mathbf{0}\\ A\big(\mathbf{e}_2 \cdot \mathbf{n}\big) - W\big(\mathbf{e}_2 \cdot \mathbf{e}_2\big) &= \mathbf{0}\\ A\cos\varphi - W &= 0\end{aligned}$$

We can solve for A as

$$A = \frac{W}{\cos\varphi}$$

The reactions A and B were the only unknown forces on the free body diagram. The equations of equilibrium allowed the determination of both in terms of the *given* quantities W and φ. The reactions A and B are the forces that the surfaces exert on the disk (and that the disk exerts on those surfaces).

It is always a good idea to check the results of any calculation. The expressions for A, B, and $\mathbf{n}$ can be substituted back into the original equation to give

$$\begin{aligned}\frac{W}{\cos\varphi}\big(\sin\varphi\,\mathbf{e}_1 + \cos\varphi\,\mathbf{e}_2\big) - W\tan\varphi\,\mathbf{e}_1 - W\mathbf{e}_2 &= \mathbf{0}\\ W\big(\tan\varphi\,\mathbf{e}_1 + \mathbf{e}_2\big) - W\tan\varphi\,\mathbf{e}_1 - W\mathbf{e}_2 &= \mathbf{0}\end{aligned}$$

which confirms that the equations are satisfied. The best route through the algebra of a given problem may not be evident at the start, but the vector equation generally gives some clues into the simplest steps to take to isolate and solve for the unknowns. In that sense, static equilibrium calculations are part science and part art. For this problem we did not need to balance moments because all three forces pass through the center of the disk. Therefore, moment equilibrium is satisfied.

Example 2.1
Consider the rigid bent shown in Fig. 2.3a. The bent has vertical height L and width $2L$, is supported by a pin at point $\mathcal{A}$, and sits on a roller that slides freely on the incline at point $\mathcal{C}$. The bent is subjected to forces P and $1.5P$ as shown. Assume that the self-weight of the bent is negligible. Find the reaction forces at $\mathcal{A}$ and $\mathcal{C}$ required to maintain static equilibrium.

Solution. The free body diagram of the system is shown in Fig. 2.3b. The pin support has been replaced with a force $\mathbf{A}$ of unknown magnitude and direction. The roller support has been replaced with a force $C\mathbf{n}$, where C is the unknown magnitude and $\mathbf{n}$ is a unit vector normal to the incline, given by

$$\mathbf{n} = -\tfrac{3}{5}\mathbf{e}_1 - \tfrac{4}{5}\mathbf{e}_3$$

since the slope of the incline is defined by a 3-4-5 triangle. Notice that the base vector $\mathbf{e}_3$ is downward. The equation of force equilibrium is

$$\sum_{i=1}^{N} \mathbf{F}_i = \mathbf{A} + P\mathbf{e}_1 + 1.5P\mathbf{e}_3 + C\mathbf{n} = \mathbf{0}$$

Solve this equation for $\mathbf{A}$ to give

$$\mathbf{A} = -P\mathbf{e}_1 - 1.5P\mathbf{e}_3 - C\mathbf{n}$$

Now, establish moment equilibrium by summing moments relative to point $\mathcal{A}$. To wit,

$$\sum \mathbf{M}_{\mathcal{A}} = \left(-L\mathbf{e}_3\right) \times P\mathbf{e}_1 + \left(-L\mathbf{e}_3 + L\mathbf{e}_1\right) \times 1.5P\mathbf{e}_3 + \left(-L\mathbf{e}_3 + 2L\mathbf{e}_1\right) \times C\mathbf{n} = \mathbf{0}$$

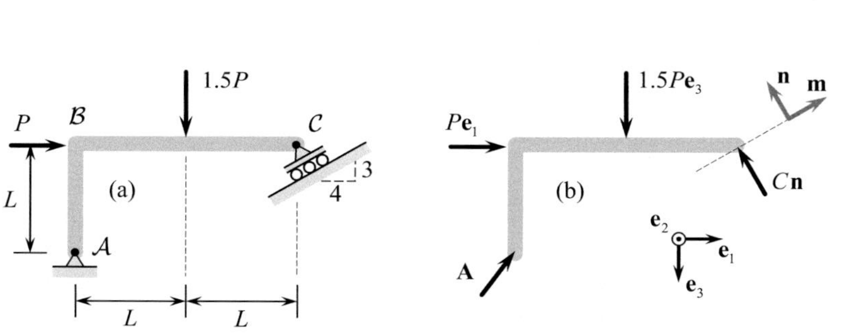

Fig. 2.3 *Example.* A rigid bent of vertical height L and width $2L$ is pinned at $\mathcal{A}$ and slides freely on a roller support at $\mathcal{C}$. Forces of magnitude P and $1.5P$ are applied. The base vectors are shown

It is a good practical strategy to sum moments about the point with the most unknowns. In this case, that point is $\mathcal{A}$ because the reaction force $\mathbf{A}$ has two unknown components. With zero-moment arm, a force at the reference point for moments contributes nothing to the equation. The quantities in parentheses are the moment arms from point $\mathcal{A}$ to the respective forces. The moment arms for the second and third terms have been created through vector addition. This equation can be simplified by noting that $\mathbf{e}_3 \times \mathbf{e}_1 = \mathbf{e}_2$, $\mathbf{e}_1 \times \mathbf{e}_3 = -\mathbf{e}_2$, that $\mathbf{e}_3$ crossed with itself is zero, and

$$\mathbf{e}_3 \times \mathbf{n} = \mathbf{e}_3 \times \left(-\tfrac{3}{5}\mathbf{e}_1 - \tfrac{4}{5}\mathbf{e}_3\right) = -\tfrac{3}{5}\mathbf{e}_2$$
$$\mathbf{e}_1 \times \mathbf{n} = \mathbf{e}_1 \times \left(-\tfrac{3}{5}\mathbf{e}_1 - \tfrac{4}{5}\mathbf{e}_3\right) = \tfrac{4}{5}\mathbf{e}_2$$

Hence, we can write the moment equilibrium equation as

$$-PL\mathbf{e}_2 - 1.5PL\mathbf{e}_2 + C\left(\tfrac{3}{5}L + \tfrac{8}{5}L\right)\mathbf{e}_2 = \mathbf{0}$$

Every term in the moment equation points in the $\mathbf{e}_2$ direction. Therefore, we can dot the equation with $\mathbf{e}_2$ and solve for C as

$$C = \frac{2.5PL}{2.2L} = 1.136P$$

With C determined, we can find $\mathbf{A}$ from the force equilibrium equation to be

$$\begin{aligned}\mathbf{A} &= -P\mathbf{e}_1 - 1.5P\mathbf{e}_3 - 1.136P\left(-\tfrac{3}{5}\mathbf{e}_1 - \tfrac{4}{5}\mathbf{e}_3\right) \\ &= -P\left(0.318\mathbf{e}_1 + 0.591\mathbf{e}_3\right)\end{aligned}$$

The vector $\mathbf{A}$ has both components negative. That means the force acts upward and to the left (in the $-\mathbf{e}_3$ and $-\mathbf{e}_1$ directions, respectively). The value of C came out positive, which means that the direction we showed on the free body diagram was correct (i.e., the force points in the positive $\mathbf{n}$ direction).

2.4 Support Idealizations

In order for a body to be in static equilibrium, it must be supported. The *support conditions* are specific locations on the body that are restrained from movement in certain ways. Generally, the restraint is created by a device that is designed for that purpose. In the real world, there is no such thing as perfect restraint—it is not possible for any device to resist force without moving at least a little. However, if the induced motions are small enough, then we are justified in idealizing the restraint to be perfect.

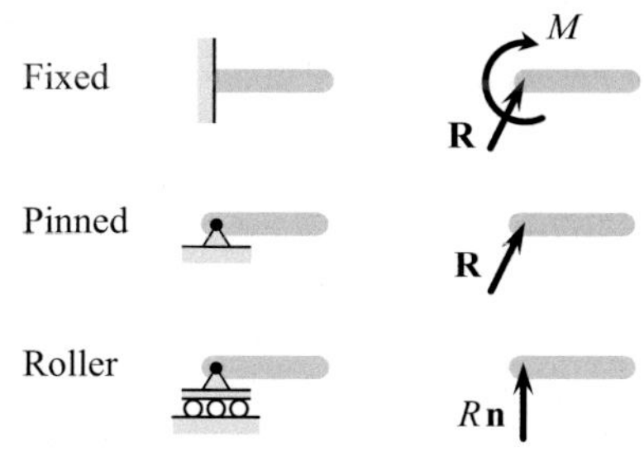

Fig. 2.4 *Support idealizations.* Three typical support conditions for planar problems, showing the name, the pictorial representation, and the associated reaction forces

Some common idealized restraint devices, and the reaction forces associated with them, are shown in Fig. 2.4. We will use these pictorial representations in problem statements throughout the book to indicate the support conditions.

A *fixed* support restrains all motion at the point of support. The point cannot translate or rotate. For a planar problem, we replace the support with a reaction force **R** and a reaction moment M. A *pinned* support completely prevents translation but does not impede rotation. In a free body diagram, we replace the pin support with a reaction force **R** (but no moment). A *roller* support allows rotation and translation along a plane. The only reaction for a roller is a force of magnitude R normal to the plane the device is rolling on. In later chapters, we will also consider the *slide* device that allows translation in one direction but impedes translation in the other direction and also does not allow rotation.

In three-dimensional problems, the support devices get more complicated. The most common cases are the fixed support that restrains all translation and rotation and the *universal ball joint* that completely prevents translation of the point of support but freely allows rotation in any direction, as illustrated in Fig. 2.5. You can think about translation and rotation of a point in three dimensions as having three components each. Thus, there are six components of motion that can be either restrained or free at a point of support. If you also include the *free* support condition, i.e., no restraint at all, then there are $2^6 = 64$ different possible support conditions for a point in three dimensions. For example, motion could be restrained in the x-direction, but not in y and z, with no rotational restraint. That would be like the end of a bar sliding on a smooth plane. Another example might be to restrain two components of translation

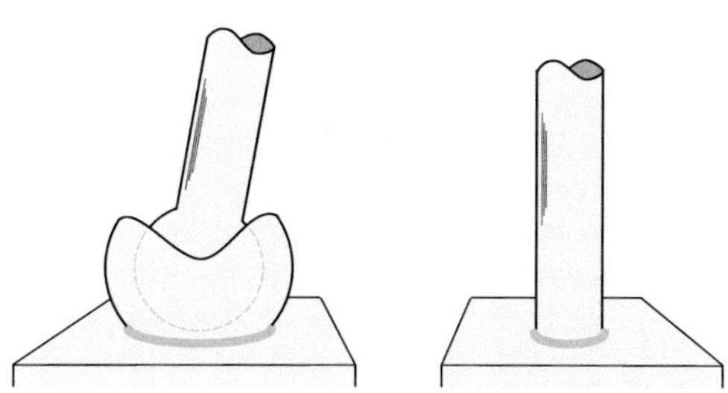

Fig. 2.5 *Three-dimensional support conditions.* The universal ball joint support is on the left. The completely fixed support is on the right

without restraining rotation, again with no rotational constraint. That would be like a bar sliding freely in a grove. There are many possibilities.

In reality, many of the support conditions are not practical because the device required to enforce the conditions would be difficult or expensive to make. A roller support is easy to build—just set the body on a round shaft. If the surface is smooth, we do not even need the shaft to enable the motion. A fixed support is also easy to make. One way to do it is to weld a bar to a massive block, as shown in the sketch on the right in Fig. 2.5. Creating a physical device that allows some translations but not others and also allows some rotation but not in all directions could be challenging to build, especially if the motions must be restrained in both directions (e.g., a roller created by placing a beam on a dowel would restrict downward motion but not upward motion normal to the plane the dowel is rolling on). Such supports might be useful in a device used in manufacturing or robotics.

It is important to remember that the support conditions we use in mechanics, like those illustrated in Fig. 2.4, are idealizations of what happens in reality. In this book, we will take those ideal conditions at face value, assuming that we have a device that will implement the conditions perfectly. Reality does not enjoy the crisp delineations of the idealized support conditions. For example, a steel I-beam might be bolted to a plate through an angle section welded to the web of the beam, as illustrated in Fig. 2.6. The actual condition is somewhere between a fixed support and a pinned support because of the dimensions and flexibility of the real connection. In practice, this joint would most likely be idealized as a pin support, even though the joint would flexibly resist rotation. However, the amount of rotational restraint is likely small compared with a case where the beam itself is completely welded to the plate. The idealization as a pin support is not exactly correct, but it might be justified, depending on the purpose of the analysis. Engineers must make these kinds of judgments when modeling real systems.

Another important device that is often used in structures is the ideal *internal pin joint*, also known as an *internal hinge*, illustrated in Fig. 2.7. The figure shows (a) the assembled pin joint, (b) an exploded view of the joint, and (c) a free body diagram of the pin. At an internal pin joint, two or more bars share the same pin. The pin acts as the arbiter of the transfer of forces. If the pin is in equilibrium, then the joint is in equilibrium. Each force shown in the free body diagram is a force coming from one of the attached members. By Newton's third law, each force exerts an equal and opposite force on the end of the member. Because the bars can rotate freely about

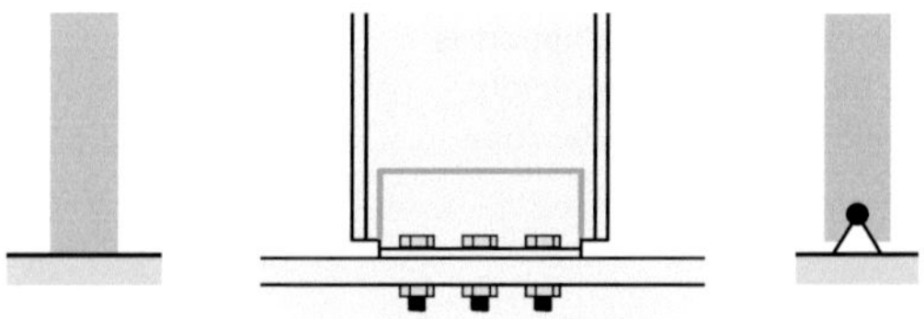

Fig. 2.6 *Idealization of support conditions.* Steel I-beam bolted to a plate through an angle section. The real condition is between the idealized conditions of the fixed support and pinned support

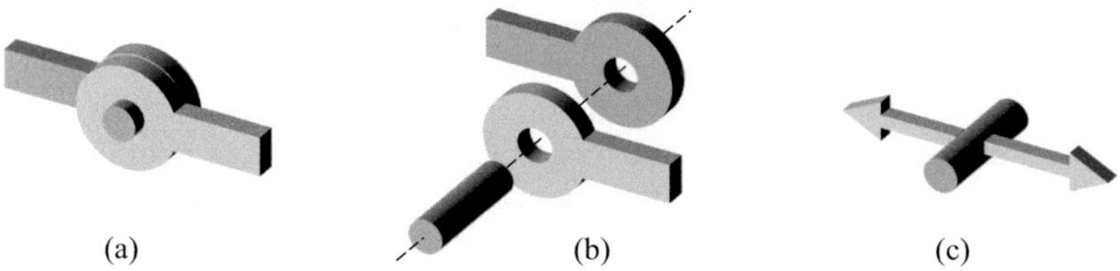

Fig. 2.7 *Internal pin joint.* An internal hinge is created by two or more bars attached to a common pin. The bars are free to rotate relative to each other and must translate together with the pin. **a** View of the assembled joint. **b** Exploded view of the joint. **c** Free body diagram of the pin

the pin, the joint cannot transmit any moment. Therefore, there can be no moment acting on the ends of the bars attached to it.

The internal hinge is *not* the same as a pin support. The basic mechanism is similar, but the members attached to a pin support are not free to translate, and therefore develop external reaction forces. The pin in an internal pin joint *can* translate and is not associated with any external reaction forces. The pin can transmit force, even though it cannot transmit moment. The only way to see the forces transmitted by the pin joint is to isolate a free body diagram with a cut through the joint that exposes the internal force.

One of the most common applications of internal pin joints is in *truss structures*, which are composed entirely of bar elements connected together with pin joints. If an internal pin joint is implemented in any other system, we generally call that a *machine* or *frame*. The absence of moment in the pin joint is the key to solving these types of systems, as we show in the next section.

A *pulley* is an important mechanical device used in lifting and other applications. The geometry of an ideal pulley of radius R is shown in Fig. 2.8a. A wheel or disk is attached to a pin at point $\mathcal{O}$ and rotates freely about it. A cable is slung around the edge of the disk. The cable does not slip on the disk as it moves because there is no resistance to rotation at the hub. This device allows the cable to be pulled around the support, changing the direction of the cable and, therefore, the force in it.

Fig. 2.8 *Pulley.* **a** An ideal frictionless pulley is a device that rotates freely on a pin support at point $\mathcal{O}$, which is the center of the disk, with a cable slung around its outer surface. **b** Free body diagram

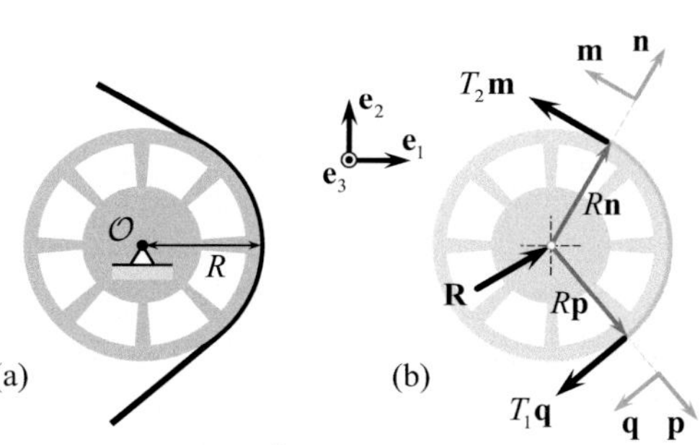

The proof that the magnitude of the force in the cable is equal on both sides as it exits contact with the disk is a simple matter of static equilibrium. Consider the free body diagram of the pulley shown in Fig. 2.8b. The pin support is removed and replaced by the reaction force $\mathbf{R}$. The part of the cable in contact with the disk is included in the free body diagram with the cables cut right at the point where the cable departs from the disk. The force in the cable can only be along the axis of the cable. Let us call the force in the cable at the top $T_2\mathbf{m}$, where the unit vector $\mathbf{m}$ is tangent to the circle at that point. Similarly, let us call the force on the bottom $T_1\mathbf{q}$, where unit vector $\mathbf{q}$ is also tangent to the disk at the point of departure from contact. The moment arm for the top force is $R\mathbf{n}$ and the moment arm for the bottom force is $R\mathbf{p}$, where $\mathbf{n}$ and $\mathbf{p}$ are unit vectors from the center of the disk to the point where the cables depart from contact. Now, moment equilibrium relative to point $\mathcal{O}$ is

$$\mathbf{M}_{\mathcal{O}} = R\mathbf{n} \times T_2\mathbf{m} + R\mathbf{p} \times T_1\mathbf{q} = \mathbf{0} \tag{2.6}$$

The force $\mathbf{R}$ has no moment arm, so it does not contribute to moment relative to $\mathcal{O}$. Unit vectors $\mathbf{n}$ and $\mathbf{m}$ are perpendicular and $\mathbf{n} \times \mathbf{m} = \mathbf{e}_3$, where $\mathbf{e}_3$ is a unit vector pointing out of the page. Similarly, $\mathbf{p} \times \mathbf{q} = -\mathbf{e}_3$. This cross product is negative due to the right-hand-rule convention for the cross product. Dividing Eq. 2.6 by R, substituting the cross-product relationships, and dotting the resulting equation with $\mathbf{e}_3$ gives

$$T_2 - T_1 = 0 \qquad \rightarrow \qquad T_1 = T_2 \tag{2.7}$$

Thus, the forces in the cable on either side of a pulley are equal. The key to this result is that both forces have moment arms of the same length. The result does not depend on the radius of the pulley or the places where the cable departs from the disk. We often use the observation that the forces are equal as a *fact* about pulleys and represent it directly on the free body diagram.

Example 2.2
Consider the pulley system shown in Fig. 2.9a. One continuous cable is attached at the left end to a support, goes under the first pulley, and over the second one. Find the force F that needs to be exerted on the right end of the cable to hold up a weight W attached to the center of the left pulley.

Solution. The free body diagrams of the two pulleys are shown in Fig. 2.9b. The force in the cable must be the same throughout because the pulleys are frictionless. Call the force in the cable T. Balance of moments at the center of the right pulley shows that $F = T$ (we could also just consider this result to be a fact about an ideal pulley, as mentioned previously). Balance of forces on the left free body diagram gives

$$T\mathbf{e}_2 + T\mathbf{e}_2 - W\mathbf{e}_2 = \mathbf{0} \qquad \rightarrow \qquad T = \tfrac{1}{2}W$$

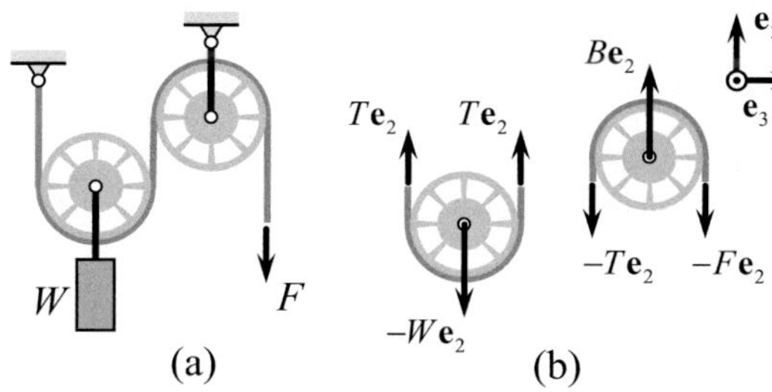

Fig. 2.9 *Example.* **a** Pulley system with two pulleys holding up a block of weight W. **b** Free body diagrams of the two pulleys

Therefore, the force needed is $F = \frac{1}{2}W$. This example illustrates the *mechanical advantage* that can be achieved with a pulley system.

2.5 Distributed Forces

Discrete forces, often called *point loads*, do not actually exist in nature. All forces are distributed, but the forces we represent with a discrete force vector are most likely the resultant of a distributed force acting over a small region. In reality, the location of the force matters in establishing equilibrium, and forces that are distributed over an area or volume need to be treated appropriately. The *typical particle* construction is a route for accomplishing that goal.

Consider the distributed load on the system shown in Fig. 2.10. The simply supported beam of length L has a pin support at the left end and a roller at the right end. A load of *intensity* q_o is distributed uniformly over the length of the beam. The load intensity has units of force per length. The distributed load is represented on the free body diagram in Fig. 2.10 using the *typical particle* construction. The length of the typical particle is $d\xi$. Hence, the total force acting on the typical particle is $q_o\mathbf{e}_3 d\xi$. The reason we call the particle *typical* is because it faithfully represents any particle along the length of the beam, the distinguishing feature being its location, measured by distance ξ from the left end of the beam.

When we establish equilibrium, we recognize that we are *summing* forces. To capture all the forces, we must add up the contributions of the forces acting on all particles. The presence of $d\xi$ in the expression for the load associated with the typical particle is also an invitation to integrate. Integration is a sum. Thus, the sum of forces on the beam shown in Fig. 2.10 is

$$\sum \mathbf{F} = \mathbf{A} - B\,\mathbf{e}_3 + \int_0^L q_o\, d\xi \mathbf{e}_3 = \mathbf{0}$$

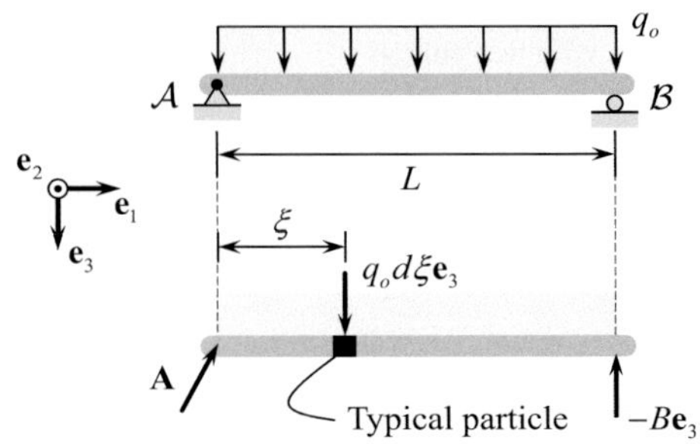

Fig. 2.10 *Typical particle construction.* Beam with uniformly distributed load acting along its length. The magnitude of the load is q_o. Top sketch shows the beam with loading. The bottom sketch shows the free body diagram with the load represented at the typical particle

Notice that the reaction at the left end is simply the vector **A**. Because it is a pin support, we do not know the magnitude or direction of that force. The reaction at the right end is represented as $-B\mathbf{e}_3$, which respects the roller condition that the force must be normal to the surface that the roller is rolling on. The direction is $-\mathbf{e}_3$ because we drew the arrow pointing up and the positive $\mathbf{e}_3$ direction is defined as down. Finally, the forces associated with the particles are represented by an integral from 0 to L. Carrying out that integral gives

$$\mathbf{A} - B\,\mathbf{e}_3 + q_o L\mathbf{e}_3 = \mathbf{0}$$

Balance of moments using the typical particle construction follows along the same lines. However, moments are summed relative to some fixed point in space, and it is essential to get the moment arm right. If we sum moments about the left end of the beam, the force **A** has no moment arm and hence does not contribute to moment equilibrium. The moment arm for the reaction force at the right end is $L\mathbf{e}_1$, and the moment arm for the force at the typical particle is $\xi\mathbf{e}_1$. Thus, moment equilibrium can be written as

$$\sum \mathbf{M}_{\mathcal{A}} = \big(L\mathbf{e}_1\big)\times\big(-B\,\mathbf{e}_3\big) + \int_0^L \big(\xi\mathbf{e}_1\big)\times\big(q_o\, d\xi\mathbf{e}_3\big) = \mathbf{0}$$

Observe that each particle contributes a different amount of moment because the moment arm is different for each one. For particles near the left end, the values of ξ are small. Therefore, those particles contribute small moments. For particles near the right end, the values of ξ are large. Those particles contribute much larger moments. To evaluate the integral that represents the sum over the particles, one must be careful to keep any quantity that depends upon ξ, the variable of integration, inside the integral. In this case, q_o is constant and $\mathbf{e}_1$ and $\mathbf{e}_3$ are constant. In fact, $\mathbf{e}_1 \times \mathbf{e}_3 = -\mathbf{e}_2$. Thus, the balance of moment equation can be processed as

$$\begin{aligned} BL\,\mathbf{e}_2 \; - q_o\mathbf{e}_2 \int_0^L \xi\, d\xi &= \mathbf{0} \\ BL\,\mathbf{e}_2 \; - \tfrac{1}{2} q_o L^2 \mathbf{e}_2 \; &= \mathbf{0} \end{aligned}$$

We can dot the equation with $\mathbf{e}_2$ and solve for B. Once we have B, we can substitute it into the expression for $\mathbf{A}$. Doing so gives

$$B = \tfrac{1}{2}q_o L, \qquad \mathbf{A} = \big(B - q_o L\big)\mathbf{e}_3 = -\tfrac{1}{2}q_o L\mathbf{e}_3$$

The minus in the expression for $\mathbf{A}$ means that the force points in the $-\mathbf{e}_3$ direction, which is upward.

For distributed loads with a more exotic spatial variation, we would replace q_o with the appropriate function $q(\xi)$ that describes the variation given in the problem statement. For example, a load that decreases linearly in intensity from q_o at the left end to zero at the right end would be described by the function:

$$q(\xi) = q_o\big(1 - \xi/L\big)$$

This function must be included in the integrals of force and moment equilibrium. The function must be written out explicitly in the integrals in order for those integrals to be evaluated, as the following example shows.

Example 2.3
Consider a simply supported beam with a distributed load given by the expression:

$$q(\xi) = q_o \sin(\pi\xi/L)$$

Compute the sum of the forces and moments about the left end for this load distribution.

Solution. The net force associated with the distributed load in the force equilibrium equation can be computed as

$$\begin{aligned}\int_0^L q(\xi)\,d\xi &= \int_0^L q_o \sin(\pi\xi/L)\,d\xi \\ &= q_o\Big[-\frac{L}{\pi}\cos(\pi\xi/L)\Big]_0^L = \frac{2q_o L}{\pi}\end{aligned}$$

The net moment relative to the location $\xi = 0$ associated with the distributed load in the moment equilibrium equation is

$$\begin{aligned}\int_0^L \xi\, q(\xi)\,d\xi &= \int_0^L \xi\, q_o \sin(\pi\xi/L)\,d\xi \\ &= q_o\Big[\frac{L^2}{\pi^2}\sin(\pi\xi/L) - \frac{L}{\pi}\xi\cos(\pi\xi/L)\Big]_0^L = \frac{q_o L^2}{\pi}\end{aligned}$$

2.6 Resultants

A *resultant* force is the net effect of a distributed force. In the example of the preceding section, we could identify the resultants as the result of integration of distributed forces. For balance of forces for the uniformly distributed load in Fig. 2.10, we computed

$$\int_0^L q_o \, d\xi = q_o L$$

The quantity $q_o L$ is the resultant of the distributed load over the length. It is the total amount of force associated with the distributed load. Observe that $q_o L$ has units of force. One common strategy to simplify statics calculations is to replace distributed forces on a free body diagram with the resultant so that only point loads are present on the free body diagram. The main question that arises with that strategy is where to place the resultant force. From the standpoint of force equilibrium, it does not matter. For moment equilibrium, it *does* matter. In our balance of moments equation, the loading term came out with magnitude

$$\int_0^L q_o \, \xi \, d\xi = \tfrac{1}{2} q_o L^2 = \left(\tfrac{1}{2} L\right)\left(q_o L\right)$$

If we placed the resultant $q_o L$ at distance $\frac{1}{2} L$ from the left end, then the moment caused by the resultant would be the same as the moment caused by the distributed load. Hence, from the standpoint of static equilibrium, there would be no difference between the distributed load and the point load.

We can generalize this idea to an arbitrary distributed load $q(\xi)$. Let us call the resultant

$$Q = \int_0^L q(\xi) \, d\xi$$

i.e., the area under the loading diagram. Taking moments about the left end gives a moment of magnitude

$$M_A = \int_0^L \xi \, q(\xi) \, d\xi$$

The proper location c of the resultant that gives the same moment as the distributed load is

$$c \, Q = \int_0^L \xi \, q(\xi) \, d\xi$$

We can solve the equation for the distance c as

$$c = \frac{1}{Q} \int_0^L \xi \, q(\xi) \, d\xi$$

The location c is, therefore, the centroid of the $q(\xi)$ diagram. Of course, the specific location of the centroid depends on the specific form of the function $q(\xi)$.

Example 2.4
Consider a simply supported beam with a trapezoidal load distributed along its length, as shown in Fig. 2.11. The magnitude of the load is q_o at the left end and $2q_o$ at the right end. Find the resultant force Q and its location c that will cause the same reaction forces. Use the resultant to find the reaction forces.

Solution. The equation of the distributed load is

$$q(\xi) = q_o\left(1 + \frac{\xi}{L}\right)$$

The resultant force is the net force over the length of the beam. Thus,

$$\begin{aligned} Q &= \int_0^L q(\xi)\, d\xi = \int_0^L q_o\left(1 + \frac{\xi}{L}\right) d\xi \\ &= q_o\left[\xi + \frac{\xi^2}{2L}\right]_0^L = \tfrac{3}{2}q_o L \end{aligned}$$

The location c where the net force must act to cause the same moment as the distributed load is

$$\begin{aligned} c &= \frac{1}{Q}\int_0^L \xi\, q(\xi)\, d\xi = \frac{1}{\frac{3}{2}q_o L}\int_0^L q_o\left(\xi + \frac{\xi^2}{L}\right) d\xi \\ &= \tfrac{2}{3}\left[\frac{\xi^2}{2L} + \frac{\xi^3}{3L^2}\right]_0^L = \tfrac{5}{9}L \end{aligned}$$

With this information, we can compute the reaction B by taking moments about the left end of the beam. To wit,

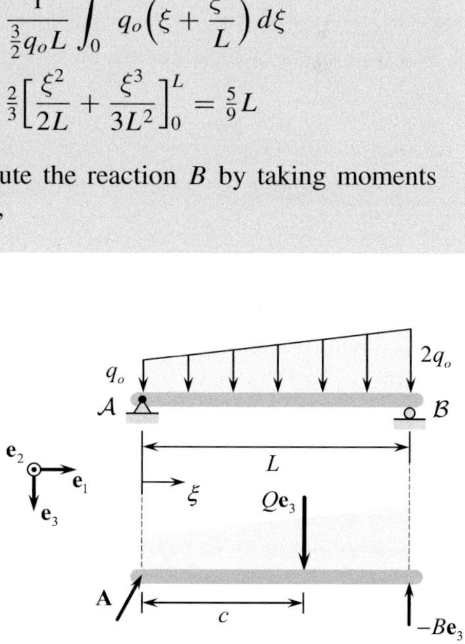

Fig. 2.11 *Example.* Beam with trapezoidal load distribution acting along its length. Top sketch shows the beam with loading. The bottom sketch shows the free body diagram with the load represented as a resultant Q

$$\left(\tfrac{5}{9}L\mathbf{e}_1\right)\times\left(\tfrac{3}{2}q_oL\mathbf{e}_3\right) + \left(L\mathbf{e}_1\right)\times\left(-B\mathbf{e}_3\right) = \mathbf{0}$$
$$-\tfrac{5}{6}q_oL^2\mathbf{e}_2 + BL\mathbf{e}_2 = \mathbf{0}$$
$$B = \tfrac{5}{6}q_oL$$

Now, we can use force equilibrium to find the reaction $\mathbf{A}$. Set the sum of forces equal to zero:

$$\mathbf{A} + Q\mathbf{e}_3 - B\mathbf{e}_3 = \mathbf{0}$$
$$\mathbf{A} + \tfrac{3}{2}q_oL\mathbf{e}_3 - \tfrac{5}{6}q_oL\mathbf{e}_3 = \mathbf{0}$$
$$\mathbf{A} = -\tfrac{2}{3}q_oL\mathbf{e}_3$$

The following example shows why we were able to lump the total weight of the disk in Fig. 2.2 at the center of the disk.

Example 2.5

Let us reconsider the uniform disk in Fig. 2.2, also shown in Fig. 2.12. The weight per unit area of the disk is w (density times acceleration of gravity times thickness of disk), and the radius of the disk is R. Show that the location of the resultant is at the center of the disk.

Solution. Let us describe the location of the typical particle in polar coordinates (r, θ). The area of the typical particle is $dA = r\,dr\,d\theta$. Define unit vectors $\mathbf{p}$ and $\mathbf{q}$ in terms of the standard base vectors as

$$\mathbf{p} = \cos\theta\mathbf{e}_1 + \sin\theta\mathbf{e}_2, \qquad \mathbf{q} = -\sin\theta\mathbf{e}_1 + \cos\theta\mathbf{e}_2$$

First, compute the net force as the sum of the weights of the particles as

$$W = \int_0^{2\pi}\int_0^R w\,r\,dr\,d\theta = w\big[\,\theta\,\big]_0^{2\pi}\big[\tfrac{1}{2}r^2\big]_0^R = w\pi R^2$$

It is not surprising that the total weight is the weight per unit area multiplied by the area of the disk.

Next, compute the moments of the weights of the typical particle about point C. The moment arm from C to the particle is $\mathbf{c} + r\mathbf{p}$, i.e., the vector from C to the center of the circle plus the vector from the center of the circle to the typical particle (which is a distance r in direction $\mathbf{p}$). Computing the moment of the typical particle weight about C gives

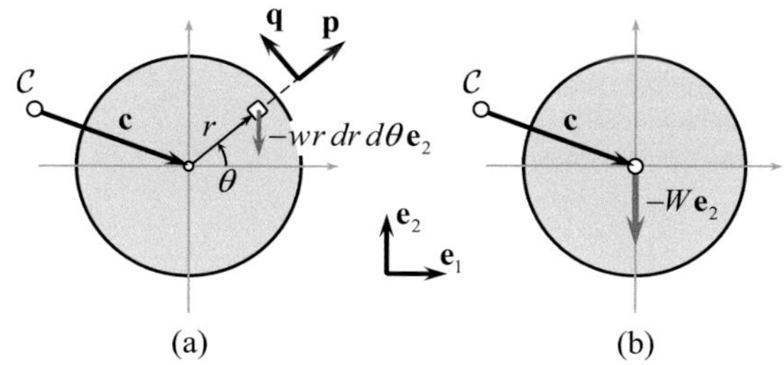

Fig. 2.12 *Example.* Uniform disk with weight per unit area w and total weight W. Compute moments relative to $\mathcal{C}$. **a** Weight of typical particle $w\,dA$ in polar coordinates. **b** Total weight W at center of disk

$$
\begin{aligned}
\mathbf{M}_{\mathcal{C}} &= \int_0^{2\pi}\int_0^R (\mathbf{c}+r\mathbf{p})\times(-w\,\mathbf{e}_2)\,r\,dr\,d\theta \\
&= \int_0^{2\pi}\int_0^R \mathbf{c}\times(-w\,\mathbf{e}_2)\,r\,dr\,d\theta + \int_0^{2\pi}\int_0^R r\mathbf{p}\times(-w\,\mathbf{e}_2)\,r\,dr\,d\theta \\
&= \mathbf{c}\times(-W\mathbf{e}_2) - w\int_0^{2\pi}\int_0^R r^2\cos\theta\,\mathbf{e}_3\,dr\,d\theta \\
&= \mathbf{c}\times(-W\mathbf{e}_2) - w\,\mathbf{e}_3\big[\tfrac{1}{3}r^3\big]_0^R\big[\sin\theta\big]_0^{2\pi} = \mathbf{c}\times(-W\mathbf{e}_2)
\end{aligned}
$$

Because $\mathbf{c}$ and $\mathbf{e}_2$ are constant, the first term integrates to W, the total weight. The cross product in the second term is $\mathbf{p}\times\mathbf{e}_2 = \cos\theta\,\mathbf{e}_3$. Finally, the second term, which represents the moment of the typical particle weight relative to the center of the disk, vanishes because the integral evaluates to zero. Thus, the moment caused by all the typical particle weights is simply $\mathbf{c}\times(-W\mathbf{e}_2)$, which is exactly the same as the moment caused by the force $-W\mathbf{e}_2$ acting at the center of the disk, as shown in Fig. 2.12b. This calculation shows that the distributed weight of the disk can be represented as a concentrated force acting at the center of the disk.

We will use the typical particle construction throughout this book rather than replacing distributed effects with point loads. One reason for doing this is to have a constant reminder that the location of force matters and for a distributed load, each bit contributes a different amount to the total. The second reason is that lumping distributed effects with this process requires knowledge of both Q and c. The computations required to get them are *exactly* the same as the computations associated with the typical particle construction. The resultant replacement approach is only an advantage for loads where the area under the curve and the centroid of the curve are known by some other means, which is the case for constant and linear load distributions.

The more important use of resultants in this book is the *stress resultant*, where we replace the tractions acting on a cross section in a free body diagram of a bar or beam, with the net effect of those tractions. In those cases, we will be able to solve

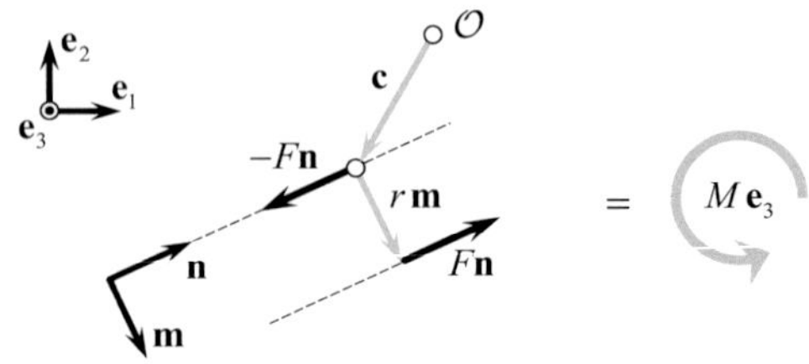

Fig. 2.13 *Couple.* Two equal and opposite forces of magnitude F separated by a distance r create a moment M but no net force

for the resultants without explicitly knowing the nature of the distributed effect. This will be a major theme throughout the book, so we leave it as a brief mention here.

A *couple* is a pair of equal and opposite forces separated by a fixed distance, as shown in Fig. 2.13. This force system causes no net force and, therefore, does not contribute anything to force equilibrium. However, the forces do cause a moment. If you sum moments about $\mathcal{O}$, you get

$$\begin{aligned}\sum \mathbf{M}_{\mathcal{O}} &= \mathbf{c} \times \big(-F\mathbf{n}\big) + \big(\mathbf{c} + r\mathbf{m}\big) \times F\mathbf{n} \\ &= \mathbf{c} \times -F\mathbf{n} + \mathbf{c} \times F\mathbf{n} + r\mathbf{m} \times F\mathbf{n} \\ &= \mathbf{c} \times \big(-F\mathbf{n} + F\mathbf{n}\big) + r\mathbf{m} \times F\mathbf{n} \\ &= rF\mathbf{e}_3\end{aligned}$$

since $\mathbf{m} \times \mathbf{n} = \mathbf{e}_3$. Notice that the result does not depend upon $\mathbf{c}$ and, therefore, it does not depend upon $\mathcal{O}$. Consequently, we can replace the two forces with a single moment $M\mathbf{e}_3$, where $M = rF$, without changing either force or moment equilibrium. When we write our equations of equilibrium, the moment $M\mathbf{e}_3$ does not contribute to the force sum and simply adds directly to the balance of moments with the other $\mathbf{r} \times \mathbf{F}$ terms. If you go back to the original representation of the moment as a couple, you can see why this is the appropriate way to handle moments. Moments do not actually exist in nature, so the concept of an applied moment in a statics problem (i.e., by putting the moment $M\mathbf{e}_3$ directly on the free body diagram) is just a shortcut for representing a couple.

2.7 Applied Loads

Often in this book, the loads acting on a system will be given as a part of the problem statement without much explanation about how those loads came about. However, the systems we analyze do not comprise the entire universe. Hence, the applied loads acting on the system come from the part of the universe that we have cut away. The forces acting on an isolated system come from some physical source, e.g., gravity or contact with another body.

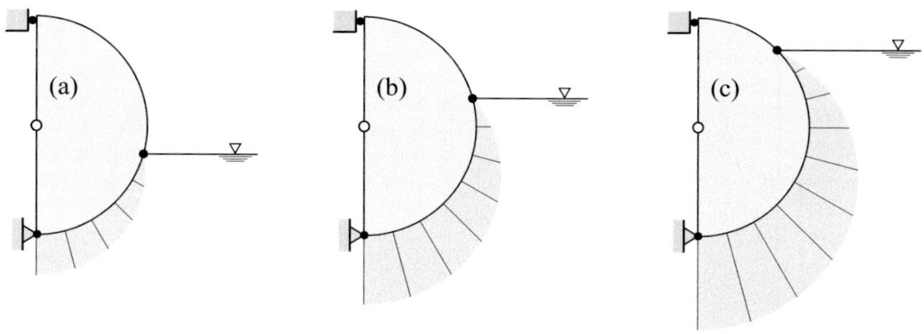

Fig. 2.14 *Hydrostatic loads.* A gate on a dam made of a half-circle cylinder holds back water on the right side. The cylinder is pinned at the bottom and is restrained by a roller at the top. **a** Low water level, **b** medium water level, and **c** high water level

A good example that connects the applied external loads to their physical source is *hydrostatics*, where the pressure from the water creates the load on the solid body. Hydrostatics is a case where we generally have to compute the loads acting on the system. As such, it provides an opportunity to contemplate where loads come from. Consider the semi-circular cylinder dam problem shown in Fig. 2.14. At any point in the water, the pressure is equal to the unit weight of water γ_w (mass density times the acceleration of gravity) multiplied by the depth from the surface. Static pressure acts normal to any surface it contacts. Figure 2.14 shows three different water levels with the associated pressure bulbs. The tractions associated with the pressure are normal to the surface and create the external loading on the dam.

Example 2.6

Consider the solid semi-circular cylinder gate used to hold back water on the right, shown in Fig. 2.15. The radius of the cylinder is R. It is pinned at $\mathcal{D}$ and rests on a roller at $\mathcal{E}$. Let the unit weight of water be designated as γ_w and the unit weight of the cylinder as γ_c. Find the reaction forces as a function of the water level (defined by θ_o). What is the minimum water level required for the reaction at $\mathcal{E}$ to be compressive?

Solution. Without loss of generality, we can assume unit width into the page. The reactions we compute then will be forces per unit width. Figure 2.15a defines the geometry of the system. We will use polar coordinates (r, θ) to describe the location of the typical particles $\mathcal{A}$ (in the cylinder) and $\mathcal{B}$ (on the surface of the cylinder). Note that the element of area is $dA = r\,dr\,d\theta$ and the element of surface length is $ds = R\,d\theta$. The water level is defined through θ_o. The sketch shows that the water depth for typical particle $\mathcal{B}$ is

$$z = R\big(\cos\theta - \cos\theta_o\big)$$

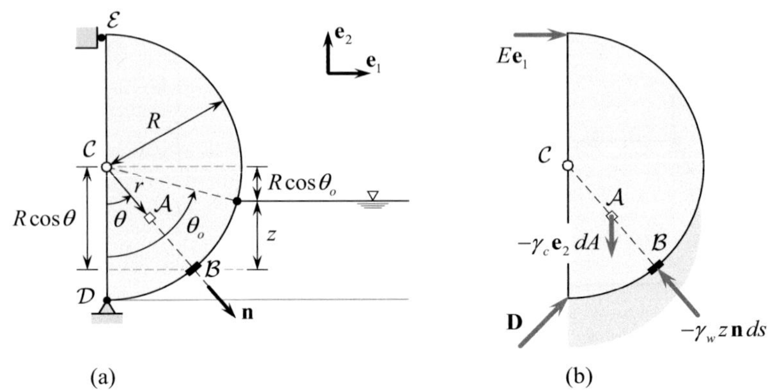

Fig. 2.15 *Hydrostatic example.* A semi-circular cylinder of radius R holds back water at the level defined by θ_o. **a** Geometry, showing polar coordinates (r, θ) of the typical particle $\mathcal{A}$ within the cylinder and the depth z to the typical particle $\mathcal{B}$ on the surface of the cylinder. **b** Free body diagram

The unit normal vector at that position is $\mathbf{n} = \sin\theta \mathbf{e}_1 - \cos\theta \mathbf{e}_2$.

The free body diagram of the cylinder is shown in Fig. 2.15b. The self-weight of the cylinder is represented at the typical particle $\mathcal{A}$, and the traction caused by the water pressure is normal to the cylinder at $\mathcal{B}$. Establish moment equilibrium relative to point $\mathcal{D}$ as follows:

$$\sum \mathbf{M}_{\mathcal{D}} = 2R\mathbf{e}_2 \times E\mathbf{e}_1 + \mathbf{M}_c + \mathbf{M}_w = \mathbf{0} \tag{2.8}$$

where

$$\mathbf{M}_c = \int_0^{\pi}\int_0^{R} \big(R\mathbf{e}_2 + r\mathbf{n}\big)\times\big(-\gamma_c \mathbf{e}_2\big)\, r\, dr\, d\theta$$

and

$$\mathbf{M}_w = \int_0^{\theta_o} \big(R\mathbf{e}_2 + R\mathbf{n}\big)\times\big(-\gamma_w z\mathbf{n}\big)\, R\, d\theta$$

are the moments caused by the weight of the cylinder and the water pressure, respectively. Noting that $\mathbf{n}\times\mathbf{e}_2 = \sin\theta\, \mathbf{e}_3$, we can evaluate these two moments as follows:

$$\begin{aligned}\mathbf{M}_c &= -\gamma_c \int_0^{\pi}\int_0^{R} r^2\big(\mathbf{n}\times\mathbf{e}_2\big)\, dr\, d\theta \\ &= -\gamma_c \mathbf{e}_3 \int_0^{\pi}\int_0^{R} r^2 \sin\theta\, dr\, d\theta = -\tfrac{2}{3}\gamma_c R^3 \mathbf{e}_3\end{aligned}$$

and

$$
\begin{aligned}
\mathbf{M}_w &= -\gamma_w R^2 \int_0^{\theta_o} z\big(\mathbf{e}_2 \times \mathbf{n}\big)\, d\theta \\
&= \gamma_w R^2 \mathbf{e}_3 \int_0^{\theta_o} R\big(\cos\theta - \cos\theta_o\big) \sin\theta \, d\theta \\
&= \gamma_w R^3 \mathbf{e}_3 \Big[-\tfrac{1}{2}\cos^2\theta - \cos\theta\cos\theta_o \Big]_0^{\theta_o} \\
&= \tfrac{1}{2}\gamma_w R^3 \big(1 - \cos\theta_o\big)^2 \mathbf{e}_3
\end{aligned}
$$

Now, we can substitute $\mathbf{M}_c$ and $\mathbf{M}_w$ into Eq. 2.8, dot with $\mathbf{e}_3$, resolve the triple scalar product in the first term, and solve for E to get

$$
E = \tfrac{1}{4}\gamma_w R^2 \big(1 - \cos\theta_o\big)^2 - \tfrac{1}{3}\gamma_c R^2 \tag{2.9}
$$

Equilibrium of forces gives

$$
\mathbf{D} + E\mathbf{e}_1 - W\mathbf{e}_2 + \mathbf{Q} = \mathbf{0} \tag{2.10}
$$

where W is the total weight of the cylinder, which can be computed as

$$
W = \int_0^{\pi} \int_0^{R} \gamma_c \, r \, dr \, d\theta = \tfrac{1}{2}\pi\gamma_c R^2
$$

and $\mathbf{Q}$ is the net force caused by the water pressure, defined as

$$
\mathbf{Q} = \int_0^{\theta_o} -\gamma_w \, z \, \mathbf{n} \, R \, d\theta
$$

To evaluate this integral, we must substitute for z and $\mathbf{n}$, as they both depend upon θ. Specifically,

$$
\begin{aligned}
z\mathbf{n} &= R\big(\cos\theta - \cos\theta_o\big)\big(\sin\theta\mathbf{e}_1 - \cos\theta\mathbf{e}_2\big) \\
&= R\big(\sin\theta\cos\theta - \sin\theta\cos\theta_o\big)\mathbf{e}_1 + \big(\cos\theta\cos\theta_o - \cos^2\theta\big)\mathbf{e}_2
\end{aligned}
$$

Substituting these expressions, integrating, and evaluating between the limits gives the net force caused by water pressure as

$$
\mathbf{Q} = -\tfrac{1}{2}\gamma_w R^2 \big[\big(1 - \cos\theta_o\big)^2 \mathbf{e}_1 - \big(\theta_o - \sin\theta_o\cos\theta_o\big)\,\mathbf{e}_2 \big]
$$

We can substitute W and $\mathbf{Q}$ back into Eq. 2.10 to find the reaction $\mathbf{D}$ to be

$$\mathbf{D} = \left[\tfrac{1}{3}\gamma_c + \tfrac{1}{4}\gamma_w\big(1-\cos\theta_o\big)^2\right]R^2\,\mathbf{e}_1 + \left[\tfrac{1}{2}\pi\gamma_c - \tfrac{1}{2}\gamma_w\big(\theta_o - \sin\theta_o\cos\theta_o\big)\right]R^2\,\mathbf{e}_2$$

It is a good idea to check that the terms associated with water pressure vanish when $\theta_o = 0$, which is the case with no water. Indeed, $\mathbf{Q} = \mathbf{0}$ for this case. It is also simple to check that the vertical component of $\mathbf{D}$ is equal to W when there is no water.

The minimum water level is the one where the reaction E just reaches zero. Setting E from Eq. 2.9 equal to zero gives

$$\tfrac{1}{4}\gamma_w R^2\big(1-\cos\theta_o\big)^2 = \tfrac{1}{3}\gamma_c R^2$$

Solving for $\cos\theta_o$, noting that only the negative value of the square root is possible because the cosine function cannot be greater than one, gives

$$\cos\theta_o = 1 - \sqrt{\frac{4}{3}\frac{\gamma_c}{\gamma_w}}$$

If the cylinder has a very small unit weight compared to water, i.e., $\gamma_c/\gamma_w \rightarrow 0$, then the reaction at the top engages with very little water. In this case, the cylinder is very buoyant and the reaction E is always compressive. The cosine function also cannot be less than -1. Therefore,

$$\sqrt{\frac{4}{3}\frac{\gamma_c}{\gamma_w}} \leq 2 \quad \rightarrow \quad \frac{\gamma_c}{\gamma_w} \leq 3$$

What this condition implies is that if the cylinder is made of a dense material, there may not be a water level that would float the cylinder and cause the reaction at $\mathcal{E}$ to be compressive.

2.8 How Many Equations Are There?

It is clear that balance of forces results in a single vector equation. However, since moments should balance relative to *any* point in space, it might seem like we could generate any number of equations (and, thus, be able to solve problems with any number of reaction forces). In this section, we show that while there are an infinite number of possible moment balance equations, they are not independent of each other.

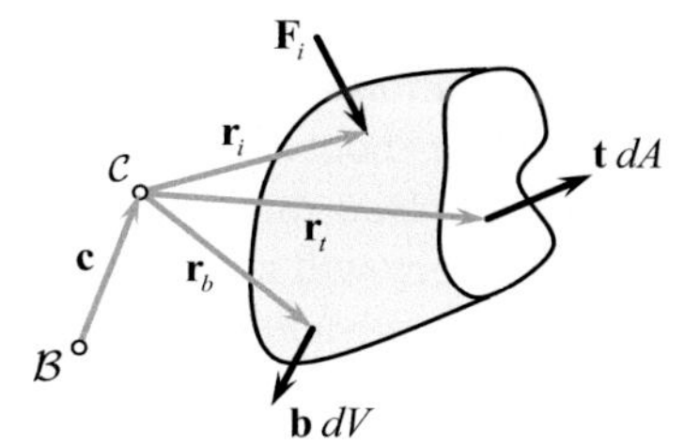

Fig. 2.16 *Equivalence of moment equations.* Solid body with applied concentrated forces $\mathbf{F}_i$, forces per unit volume $\mathbf{b}$, and forces per unit area $\mathbf{t}$. The moment arms are, respectively, $\mathbf{r}_i$, $\mathbf{r}_b$, and $\mathbf{r}_t$, relative to point $\mathcal{C}$

Consider the solid body shown in Fig. 2.16. It is subjected to concentrated forces $\mathbf{F}_i$ (only one of them is shown), forces per unit volume $\mathbf{b}$, and forces per unit area $\mathbf{t}$. The total net force can be computed by summing forces. To wit,

$$\mathbf{F} = \sum_{i=1}^{N} \mathbf{F}_i + \int_V \mathbf{b}\,dV + \int_A \mathbf{t}\,dA$$

where N is the number of concentrated forces, V is the volume over which the distributed body forces $\mathbf{b}$ act, and A is the area over which the distributed tractions $\mathbf{t}$ act. The vector $\mathbf{F}$ is the net force. If all the forces are represented on the free body diagram, then force equilibrium implies that $\mathbf{F} = \mathbf{0}$. We can take moments about any point in space. Let us first compute moments about point $\mathcal{C}$ as

$$\mathbf{M}_{\mathcal{C}} = \sum_{i=1}^{N} \mathbf{r}_i \times \mathbf{F}_i + \int_V \mathbf{r}_b \times \mathbf{b}\,dV + \int_A \mathbf{r}_t \times \mathbf{t}\,dA$$

The vector $\mathbf{M}_{\mathcal{C}}$ is the net moment of the forces about point $\mathcal{C}$. Of course, if all the forces are represented on the free body diagram, then the moments sum to zero, i.e., $\mathbf{M}_{\mathcal{C}} = \mathbf{0}$. Now, let us compute moments about point $\mathcal{B}$, noting that the vector $\mathbf{c}$ from point $\mathcal{B}$ to point $\mathcal{C}$ is constant. Thus,

$$\begin{aligned}
\mathbf{M}_{\mathcal{B}} &= \sum_{i=1}^{N} (\mathbf{c}+\mathbf{r}_i)\times\mathbf{F}_i + \int_V (\mathbf{c}+\mathbf{r}_b)\times\mathbf{b}\,dV + \int_A (\mathbf{c}+\mathbf{r}_t)\times\mathbf{t}\,dA \\
&= \mathbf{c}\times\Big(\sum_{i=1}^{N} \mathbf{F}_i + \int_V \mathbf{b}\,dV + \int_A \mathbf{t}\,dA\Big) \\
&\qquad + \sum_{i=1}^{N} \mathbf{r}_i \times \mathbf{F}_i + \int_V \mathbf{r}_b \times \mathbf{b}\,dV + \int_A \mathbf{r}_t \times \mathbf{t}\,dA
\end{aligned}$$

The term in parentheses in the second line is the net force $\mathbf{F}$ and three terms on the last line are the net moment of the forces about point $\mathcal{C}$, i.e., $\mathbf{M}_{\mathcal{C}}$. Hence,

$$\mathbf{M}_{\mathcal{B}} = \mathbf{c}\times\mathbf{F} + \mathbf{M}_{\mathcal{C}}$$

Thus, the net moment $\mathbf{M}_{\mathcal{B}}$ is *not* independent of $\mathbf{F}$ and $\mathbf{M}_{\mathcal{C}}$. The implication of this calculation is that additional moment balance equations do not generate new information about equilibrium. The moment equations taken relative to two different points will *look* different, but when taken in conjunction with the force balance equation, they will not create new equations. However, we can use multiple moment equations to advantage, as the following example shows.

Example 2.7
Consider the free body diagram of a beam subjected to a concentrated load of magnitude P applied two-thirds of the distance from the left end, as shown in Fig. 2.17. Use moment equations relative to the two ends to determine the reaction forces.

Solution. First, balance moments about the left end to find the reaction B. To wit,

$$\begin{aligned}\left(\tfrac{2}{3}L\mathbf{e}_1\right)\times\left(P\mathbf{e}_3\right)+\left(L\mathbf{e}_1\right)\times\left(-B\mathbf{e}_3\right)&=0\\ \left(BL-\tfrac{2}{3}PL\right)\mathbf{e}_2&=0\end{aligned}$$

since $\mathbf{e}_1\times\mathbf{e}_3=-\mathbf{e}_2$. We can dot the equation with $\mathbf{e}_2$ to make it a scalar equation and solve for the reaction $B=\frac{2}{3}P$. Now, balance moments about the right end:

$$\begin{aligned}\left(-L\mathbf{e}_1\right)\times\left(A_x\mathbf{e}_1+A_z\mathbf{e}_3\right)+\left(-\tfrac{1}{3}L\mathbf{e}_1\right)\times\left(P\mathbf{e}_3\right)&=\mathbf{0}\\ \left(\tfrac{1}{3}PL+A_zL\right)\mathbf{e}_2&=\mathbf{0}\end{aligned}$$

From this equation, we find that $A_z=-\frac{1}{3}P$. Finally, establish moment equilibrium about the point located at a distance a directly above point $\mathcal{A}$

$$\begin{aligned}\left(a\mathbf{e}_3\right)\times\left(A_x\mathbf{e}_1+A_z\mathbf{e}_3\right)+\left(a\mathbf{e}_3+\tfrac{2}{3}L\mathbf{e}_1\right)\times\left(P\mathbf{e}_3\right)+\left(a\mathbf{e}_3+L\mathbf{e}_1\right)\times\left(-B\mathbf{e}_3\right)&=\mathbf{0}\\ aA_x\left(\mathbf{e}_3\times\mathbf{e}_1\right)+\tfrac{2}{3}PL\left(\mathbf{e}_1\times\mathbf{e}_3\right)+BL\left(\mathbf{e}_1\times\mathbf{e}_3\right)&=\mathbf{0}\\ aA_x\mathbf{e}_2+\left(BL-\tfrac{2}{3}PL\right)\mathbf{e}_2&=\mathbf{0}\end{aligned}$$

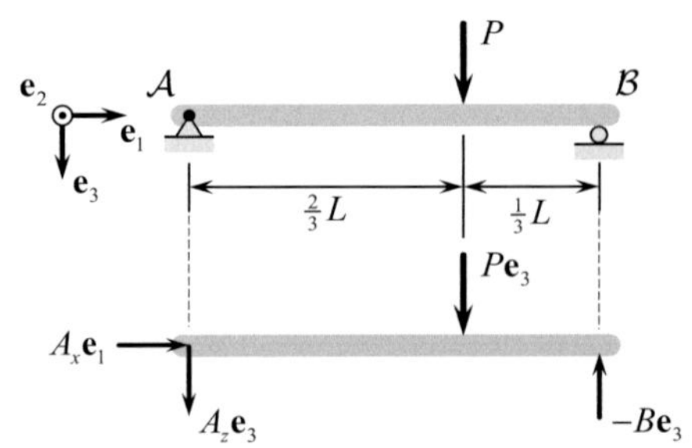

Fig. 2.17 *Example.* Simply supported beam with point load P applied two-thirds of the distance from the left end

The second line eliminates all terms with $\mathbf{e}_3 \times \mathbf{e}_3$. The third line carries out the cross products. The term in parentheses in the last line is zero from our first moment balance equation. Therefore, we get $A_x = 0$. The value of a does not matter.

We found all three reaction components using balance of moments, without using balance of forces. It is straightforward to verify that the force equilibrium is satisfied by the reactions we have computed. This strategy generates three equations that are completely equivalent to using balance of forces and one balance of moments equation.

The *science* of statics is that we have a fixed number of equations implied by the conditions of force and moment equilibrium: three in two dimensions and six in three dimensions. The *art* of statics is finding a clever way to use those equations to get to an answer with the least calculation possible. One can often avoid solving simultaneous equations simply by organizing the calculation, as we did in the previous example.

The concept of *static determinacy* simply means that all the reaction forces can be determined from the equations of equilibrium alone. In statics, all the problems you encounter are statically determinate because for rigid systems, we cannot find a unique solution for a problem that is not statically determinate. A system that is *statically indeterminate* is one in which the reaction forces cannot be determined using only the equations of static equilibrium.

In general, static indeterminacy is simply a case where the support devices create reaction forces that exceed in number the equations of equilibrium available to find them. For a planar problem, there are three equations of equilibrium. Typically, we would consider that to mean two components of force equilibrium and moment equilibrium about one point. The system is statically determinate, then, if there are three unknown reaction components. The simply supported beam of Fig. 2.10 is a case in point. The pin support has two components of reaction force and the roller has one, giving a total of three.

The good news is that for deformable solids, static indeterminacy is not a show-stopping feature like it is in rigid body statics. We will see in subsequent chapters how to use kinematic conditions associated with deformation of the system consistent with the support devices to completely solve the problem of finding all the forces, internal and external, for a system under load.

2.9 Structures

A *structure* is a system with multiple members connected together for the purpose of carrying loads from their point of application to a point of support. For example, a bridge must transmit the weight of a vehicle to the support points at the abutments. A multistory building elevates space above the ground and must carry the loads associated with what occupies the various levels (e.g., people, furniture, or books) to the ground. A *machine* is a structure designed to carry out a specific task (e.g., lifting a weight, placing a part in a manufacturing process, or bending a material to form a certain shape). Of course, all structures must carry their own weight in addition to whatever forces arise as a result of their main function. Sometimes self-weight is considered negligible compared with the other loads acting on the system, an assumption that must be verified. In this section, we describe methods of analysis for statically determinate structures.

2.9.1 Trusses

A *truss structure* is an assembly of members or elements connected together with internal pin joints (see Fig. 2.7). A typical truss is shown in Fig. 2.18. The *joints* or *nodes* of the truss can be named (e.g., joints i and j in the sketch), as can the elements (e.g., element e in the sketch). An ideal truss includes applied loads at the joints (e.g., $\mathbf{p}_i$ and $\mathbf{p}_j$ in the sketch, acting at nodes i and j, respectively) and some of the joints will have either pin or roller supports. Figure 2.18b shows that element e has length L_e and its axis aligns with the unit vector $\mathbf{n}_e$, which points from the i end to the j end of the bar. Each element e is associated with two joints. We will use the notations $i(e)$ and $j(e)$ to refer to the i and j ends of element e, respectively. The unit vector $\mathbf{n}_e$ and the length L_e can be readily computed if the coordinates of the nodes are known.

To see the implication of having pins at both ends of a bar, consider the free body diagram shown in Fig. 2.19a. The unit vector $\mathbf{n}$ points from i to j and the unit vector $\mathbf{m}$ is perpendicular to $\mathbf{n}$. The force $\mathbf{f}_i$ acts at i and force $\mathbf{f}_j$ acts at j. Since the ends are attached to internal pin joints, there are no moments acting at the ends of the bar. Moment equilibrium, with moments taken first at i and second at j gives, respectively,

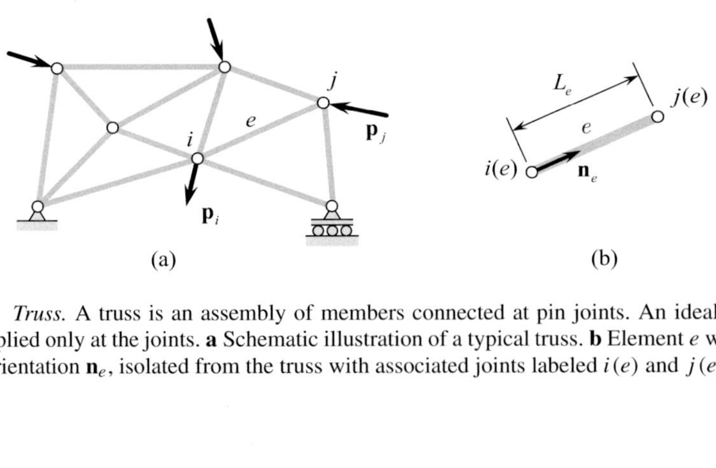

Fig. 2.18 *Truss.* A truss is an assembly of members connected at pin joints. An ideal truss has forces applied only at the joints. **a** Schematic illustration of a typical truss. **b** Element e with length L_e and orientation $\mathbf{n}_e$, isolated from the truss with associated joints labeled $i(e)$ and $j(e)$

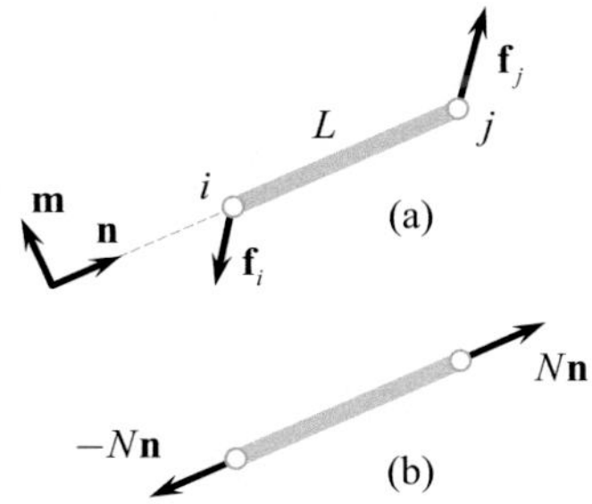

Fig. 2.19 *Equilibrium of a truss element.* Free body diagrams. **a** Element with general forces at the ends. **b** Element with equal and opposite forces oriented along the bar axis

$$L\mathbf{n} \times \mathbf{f}_j = \mathbf{0}, \qquad -L\mathbf{n} \times \mathbf{f}_i = \mathbf{0}$$

The cross product of a vector with itself is zero. Therefore, the vectors $\mathbf{f}_i$ and $\mathbf{f}_j$ must both be some scalar multiple of the vector $\mathbf{n}$. Let us write $\mathbf{f}_j = N\mathbf{n}$. Then, force equilibrium gives

$$\mathbf{f}_i + \mathbf{f}_j = \mathbf{f}_i + N\mathbf{n} = \mathbf{0} \qquad \rightarrow \qquad \mathbf{f}_i = -N\mathbf{n}$$

In other words, the forces acting on the ends of the truss bar must be as shown in Fig. 2.19b. Notice that equilibrium of the bar element does *not* determine the value of N, which we will call the member internal force. Those values must be determined from equilibrium of the joints. When the solution is obtained, N will have either a positive or negative value (or zero). A positive value of N indicates that the bar is in *tension*, with both forces pulling away from the bar in an action trying to make it longer. A negative value indicates *compression*, with both forces pointing toward the bar in an action trying to make it shorter. If we respect the orientation of the element forces when we take free body diagrams of a truss, by drawing the free body diagrams like Fig. 2.19b, then we guarantee that the elements are in equilibrium (and, therefore, we do not need free body diagrams of the members).

We can include the self-weight of the members by adding it to the free body diagram in Fig. 2.19a using the typical particle construction. Following the same derivation path, we find that if the bar has uniform weight per unit length of w that

$$\mathbf{f}_j = \tfrac{1}{2}wL\mathbf{e}_2 + N\mathbf{n}, \qquad \mathbf{f}_i = \tfrac{1}{2}wL\mathbf{e}_2 - N\mathbf{n}$$

Since the forces $-\mathbf{f}_i$ and $-\mathbf{f}_j$ are transferred to their respective joints, the self-weight of the member can simply be added as applied loads at the nodes, with half of the weight applied to node i and half to node j. The magnitude N of the axial force remains the unknown of the problem.

Static analysis of structures is a process of taking free body diagrams and writing equations of equilibrium based on them. The general aim for structures is not just to find the reaction forces, but also to find the internal forces in the members of the structure. Finding the internal forces is possible for statically determinate trusses. The term statically determinate simply means that we can find all the forces using the equations of statics alone. In what follows, we describe two methods of analysis for truss structures.

The *method of joints* takes each joint as a free body diagram and establishes equilibrium of that joint. Since the joint is a point in space, only force equilibrium is available. The following example illustrates the application of the method of joints.

Example 2.8
Consider the truss structure shown in Fig. 2.20a. Use the method of joints to determine the member forces and the reactions.

Solution. We can find the reaction forces using the free body diagram of the overall system shown in Fig. 2.20b. Taking moments about point $\mathcal{A}$ gives

$$(3a\mathbf{e}_1 + 4a\mathbf{e}_2)\times(-3F\mathbf{e}_2) + (9a\mathbf{e}_1 + 4a\mathbf{e}_2)\times(F\mathbf{e}_1) + (12a\mathbf{e}_1)\times(C\mathbf{e}_2) = \mathbf{0}$$
$$-9aF(\mathbf{e}_1 \times \mathbf{e}_2) + 4aF(\mathbf{e}_2 \times \mathbf{e}_1) + 12aC(\mathbf{e}_1 \times \mathbf{e}_2) = \mathbf{0}$$

Noting the cross products of the base vectors and dotting with $\mathbf{e}_3$, this equation reduces to

$$-9aF - 4aF + 12aC = 0$$

which can be solved to give $C = \frac{13}{12}F$. Force equilibrium requires that

$$\mathbf{A} - 3F\mathbf{e}_2 + F\mathbf{e}_1 + C\mathbf{e}_2 = \mathbf{0}$$

Substituting the value of C and solving gives $\mathbf{A} = -F\mathbf{e}_1 + \frac{23}{12}F\mathbf{e}_2$.

Figure 2.21a defines the joint and element labels, and Fig. 2.21b shows the unit vectors that define the element orientations. These vectors can be expressed in terms of the base vectors as

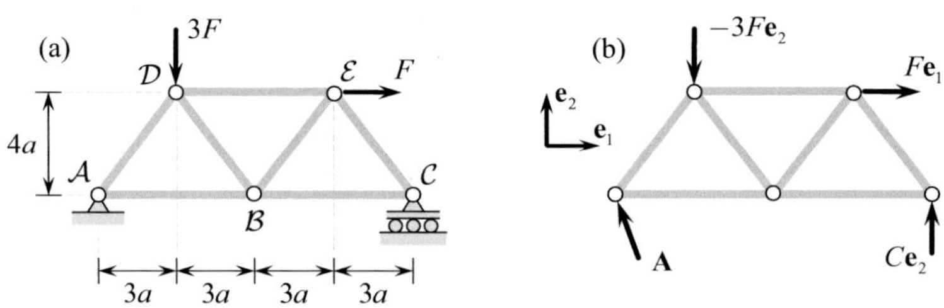

Fig. 2.20 *Method of joints.* Example truss structure. **a** Geometry of the structure in terms of the dimension a, with supports and loads in terms of the force magnitude F. The open circles are ideal pin joints. **b** Free body diagram of overall system

$$\begin{aligned}\mathbf{n}_1 &= \mathbf{n}_2 = \mathbf{n}_7 = \mathbf{e}_1\\ \mathbf{n}_3 &= \mathbf{n}_5 = \tfrac{3}{5}\mathbf{e}_1 + \tfrac{4}{5}\mathbf{e}_2\\ \mathbf{n}_4 &= \mathbf{n}_6 = -\tfrac{3}{5}\mathbf{e}_1 + \tfrac{4}{5}\mathbf{e}_2\end{aligned}$$

Figure 2.21c shows the five free body diagrams of the joints, including the applied forces at joints $\mathcal{D}$ and $\mathcal{E}$ and the reactions at joints $\mathcal{A}$ and $\mathcal{C}$. We can write equations of force equilibrium for the five joints as

$$\begin{aligned}&\text{Joint } \mathcal{A} \quad N_1\mathbf{n}_1 + N_3\mathbf{n}_3 + \mathbf{A} = \mathbf{0}\\ &\text{Joint } \mathcal{B} \quad -N_1\mathbf{n}_1 + N_2\mathbf{n}_2 + N_4\mathbf{n}_4 + N_5\mathbf{n}_5 = \mathbf{0}\\ &\text{Joint } \mathcal{C} \quad -N_2\mathbf{n}_2 + N_6\mathbf{n}_6 + C\mathbf{e}_2 = \mathbf{0}\\ &\text{Joint } \mathcal{D} \quad -N_3\mathbf{n}_3 - N_4\mathbf{n}_4 + N_7\mathbf{n}_7 - 3F\mathbf{e}_2 = \mathbf{0}\\ &\text{Joint } \mathcal{E} \quad -N_5\mathbf{n}_5 - N_6\mathbf{n}_6 - N_7\mathbf{n}_7 + F\mathbf{e}_1 = \mathbf{0}\end{aligned}$$

These are ten equations in ten unknowns (the seven element forces, the two components of the reaction $\mathbf{A}$, and the reaction C). We have already used three equations, so only seven of the joint equations are needed to complete the solution (the remaining three are redundant).

There are many ways to attack the solution of these equations. Since we have already found the reactions, a good place to start is with joint $\mathcal{A}$ because it has only two unknowns (joint $\mathcal{C}$ would work, too). Using the value of $\mathbf{A}$ already found and dotting successively with $\mathbf{e}_1$ and $\mathbf{e}_2$, we get

$$-F + N_1 + \tfrac{3}{5}N_3 = 0, \qquad \tfrac{23}{12}F + \tfrac{4}{5}N_3 = 0$$

From these equations, we can find $N_3 = -\frac{115}{48}F$ and then $N_1 = \frac{39}{16}F$. Now, look at joint $\mathcal{C}$. Using the value of reaction C already found and dotting successively with $\mathbf{e}_1$ and $\mathbf{e}_2$, we get

$$-N_2 - \tfrac{3}{5}N_6 = 0, \qquad \tfrac{13}{12}F + \tfrac{4}{5}N_6 = 0$$

From these equations, we can find $N_6 = -\frac{65}{48}F$ and then $N_2 = \frac{13}{16}F$. Now, look at joint $\mathcal{D}$. Using the value for N_3 already computed, we get

$$-\tfrac{3}{5}\left(-\tfrac{115}{48}F\right) - \tfrac{3}{5}N_4 + N_7 = 0, \qquad -\tfrac{4}{5}\left(-\tfrac{115}{48}F\right) + \tfrac{4}{5}N_4 - 3F = 0$$

from which we get $N_4 = -\frac{65}{48}F$ and then $N_7 = -\frac{5}{8}F$. Finally, look at joint $\mathcal{E}$. Dotting the equation with $\mathbf{e}_2$ and noting the value of N_6 already found, we find that $N_5 = -N_6 = \frac{65}{48}F$. Because we used three equilibrium equations on the overall free body diagram to find the reaction forces, we only needed one

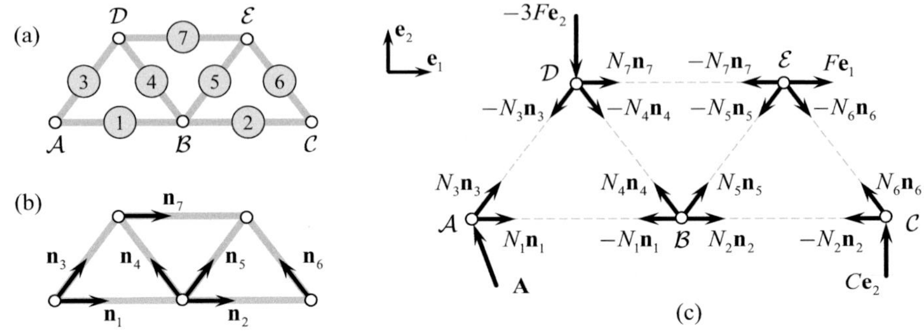

Fig. 2.21 *Example.* **a** Joint (letters) and element (circled numbers) labels. **b** Unit vectors for orientation of each element. **c** Free body diagrams of the five joints, including the loads at joints $\mathcal{D}$ and $\mathcal{E}$ and the reactions at joints $\mathcal{A}$ and $\mathcal{C}$

of the equations from joint $\mathcal{E}$ and neither of the two equations from joint $\mathcal{B}$ to determine all the bar forces. However, we can use those equations to check our calculations. Just plug the values we have into the remaining three equations to verify that equilibrium is satisfied.

To summarize the solution, we have

$$N_1 = \tfrac{39}{16}F, \quad N_2 = \tfrac{13}{16}F, \qquad N_3 = -\tfrac{115}{48}F, \quad N_4 = -\tfrac{65}{48}F$$
$$N_5 = \tfrac{65}{48}F, \quad N_6 = -\tfrac{65}{48}F, \quad N_7 = -\tfrac{5}{8}F$$

The bar forces are all proportional to F and do not depend upon the dimensions of the structure. The positive forces are tension, and the negative ones are compression.

It is interesting to note that the five joint equations can be solved simultaneously for the seven bar forces and the three reaction components, obviating the need to do the overall equilibrium calculation to first find the reaction forces. However, that strategy does not provide a direct route to solve the joints one at a time, which is more convenient for hand calculation. While the complete system of joint equations would be tedious to solve by hand, it is simple to do in MATLAB, as we show in a later section.

The *method of sections* takes a cut through multiple joints to divide the structure into two parts. One of the two parts is retained as the free body diagram with the action of the discarded part represented by the forces in the members that frame into the cut joints. There are, of course, *many* different ways to cut through a structure. All of them lead to valid statements of equilibrium, but not all of these equations are independent of each other. The following example illustrates the application of the method of sections.

Example 2.9
Consider the truss structure shown in Fig. 2.22. Use the method of sections to determine the member force in element b.

Solution. From the geometry of the structure, we can find the unit vectors pointing along elements a, b, and c to be

$$\mathbf{n}_a = \mathbf{e}_1, \qquad \mathbf{n}_b = \tfrac{6}{\sqrt{61}}\mathbf{e}_1 + \tfrac{5}{\sqrt{61}}\mathbf{e}_2, \qquad \mathbf{n}_c = \tfrac{12}{13}\mathbf{e}_1 - \tfrac{5}{13}\mathbf{e}_2$$

Figure 2.22b shows a free body diagram with a cut through joints $\mathcal{C}$ and $\mathcal{B}$, keeping the left part and discarding the right. Now, take moments about joint $\mathcal{A}$, noting that the forces in members a and c do not cause moment about that point because the moment arms point in the same direction as the forces, and any vector crossed with a vector pointing in the same direction is zero. The moment arm from $\mathcal{A}$ to $\mathcal{B}$ is $\frac{26}{3}\mathbf{n}_c$. Thus, we can write

$$\begin{aligned} 4\mathbf{e}_1 \times -F\mathbf{e}_2 + 8\mathbf{e}_1 \times -F\mathbf{e}_2 + \tfrac{26}{3}\mathbf{n}_c \times N_b\,\mathbf{n}_b &= \mathbf{0} \\ -12F\mathbf{e}_3 + \tfrac{60}{\sqrt{61}}N_b\mathbf{e}_3 &= \mathbf{0} \end{aligned}$$

where we have computed $\mathbf{n}_c \times \mathbf{n}_b$ by substituting their expressions in terms of the standard base vectors and distributing the cross products. Dotting this equation with $\mathbf{e}_3$ and solving for N_b gives

$$N_b = \tfrac{\sqrt{61}}{5}F = 1.562F$$

The positive value indicates that the force is in tension. Observe that we were able to compute the bar force without first solving for the reaction forces or any other bar force.

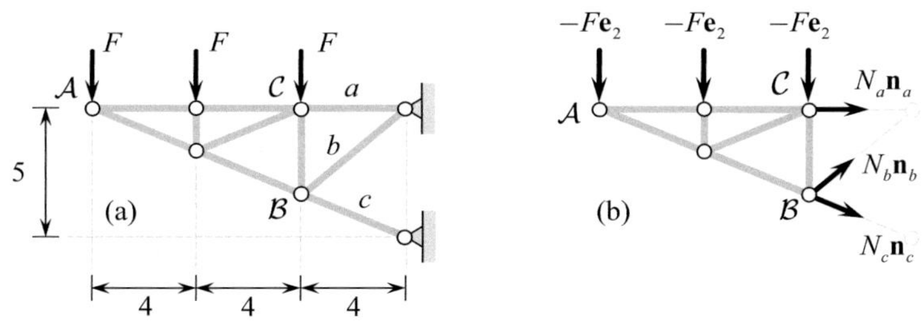

Fig. 2.22 *Method of Sections.* **a** Geometry and loading of structure with joints $\mathcal{A}$, $\mathcal{B}$, and $\mathcal{C}$ and elements a, b, and c labeled. **b** Free body diagram with cut through joints $\mathcal{C}$ and $\mathcal{B}$

This example reminds us that *any* free body diagram is valid as long as the part cut away is represented by appropriate forces on the part that remains. It is not always possible to find a way to isolate a single unknown by this method, but the equations generated are always consistent expressions of equilibrium of the forces present on the free body diagram.

2.9.2 Computation: Truss Analysis

The matrix form of the system of equations for the method of joints has a particular structure that makes it possible to write an algorithm to set up and solve the equations for general trusses in two and three dimensions. In this section, we describe that process and provide a code to carry out the computations.

To clarify the discussion, consider the specific truss structure shown in Fig. 2.23a. This truss has four joints and five elements. The joints and elements are identified by sequential numbers, each starting with one. The order is arbitrary, but once selected, it becomes the definition of the structure. The orientation of the elements is established in Fig. 2.23b. When we pick the direction of the unit vector for element e, we are deciding what is the i end and what is the j end because the element orientation vector goes *from* the i end *to* the j end. We will eventually write equations of joint equilibrium, and we need to establish an order for those equations. Figure 2.23c shows that we will number the equations sequentially starting at the first joint, going in node order, and going sequentially in x, y, and z (if it is 3D) order at each node.

When we establish joint equilibrium, every element that frames into the joint contributes a force of either $N_e\mathbf{n}_e$ or $-N_e\mathbf{n}_e$, the former being the case if the force came from the i end of the bar and the latter if the force came from the j end of the bar. If we know for each element which nodes it is attached to, then all we need to do is place either $\mathbf{n}_e$ or $-\mathbf{n}_e$ in the appropriate rows in the column of the coefficient matrix associated with the unknown member force N_e.

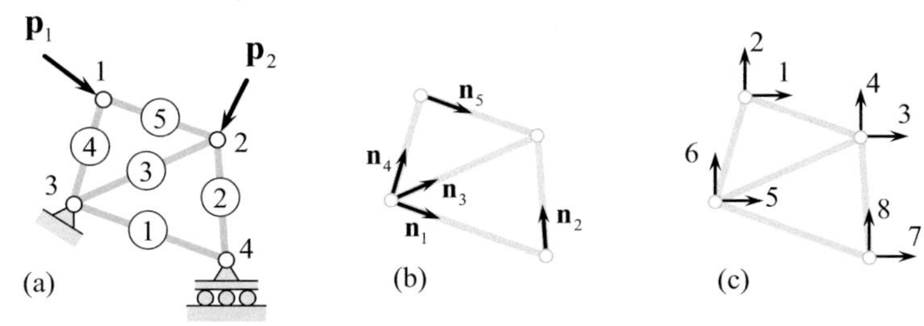

Fig. 2.23 *Truss nomenclature.* **a** Geometry and loading of structures with joints and elements labeled. **b** Element orientation unit vectors defining the i end and j end. **c** Component equation numbers for joint equilibrium

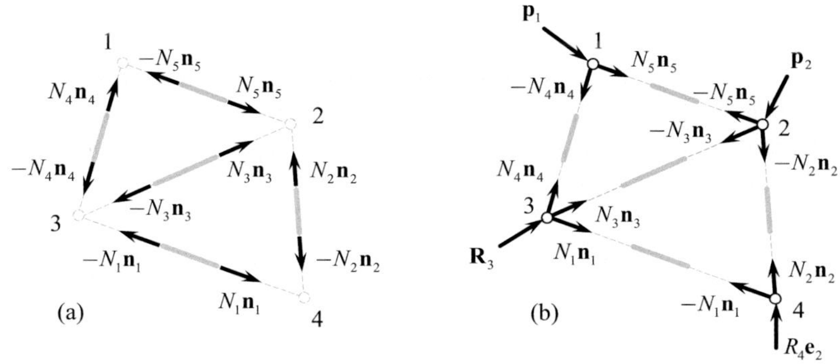

Fig. 2.24 *Free body diagrams.* **a** Free body diagrams of the five elements. **b** Free body diagrams of the four joints

Figure 2.24a shows the free body diagrams for the five elements. All forces are shown as positive (i.e., tensile). The magnitude of the force in element e is N_e. The algebraic sign of the vector force in the figure is determined by the direction of the element orientation arrow. If the force points in the $\mathbf{n}_e$ direction, then it is positive. If it points opposite $\mathbf{n}_e$, then it is negative.

The element end forces exert equal and opposite forces on the joints they are attached to. The joint forces are shown in Fig. 2.24b. The equations of joint equilibrium can be written as

$$\begin{aligned}
&\text{Joint 1} \quad -N_4\mathbf{n}_4 + N_5\mathbf{n}_5 + \mathbf{p}_1 = \mathbf{0}\\
&\text{Joint 2} \quad -N_2\mathbf{n}_2 - N_3\mathbf{n}_3 - N_5\mathbf{n}_5 + \mathbf{p}_2 = \mathbf{0}\\
&\text{Joint 3} \quad \;\; N_1\mathbf{n}_1 + N_3\mathbf{n}_3 + N_4\mathbf{n}_4 + \mathbf{R}_3 = \mathbf{0}\\
&\text{Joint 4} \quad -N_1\mathbf{n}_1 + N_2\mathbf{n}_2 + R_4\mathbf{e}_2 = \mathbf{0}
\end{aligned}$$

These equations can be put into matrix form as

$$\begin{bmatrix} \mathbf{0} & \mathbf{0} & \mathbf{0} & -\mathbf{n}_4 & \mathbf{n}_5 & \mathbf{0} & \mathbf{0} & \mathbf{0} \\ \mathbf{0} & -\mathbf{n}_2 & -\mathbf{n}_3 & \mathbf{0} & -\mathbf{n}_5 & \mathbf{0} & \mathbf{0} & \mathbf{0} \\ \mathbf{n}_1 & \mathbf{0} & \mathbf{n}_3 & \mathbf{n}_4 & \mathbf{0} & \mathbf{e}_1 & \mathbf{e}_2 & \mathbf{0} \\ -\mathbf{n}_1 & \mathbf{n}_2 & \mathbf{0} & \mathbf{0} & \mathbf{0} & \mathbf{0} & \mathbf{0} & \mathbf{e}_2 \end{bmatrix} \begin{Bmatrix} N_1 \\ \vdots \\ N_5 \\ R_{3x} \\ R_{3y} \\ R_4 \end{Bmatrix} = \begin{Bmatrix} -\mathbf{p}_1 \\ -\mathbf{p}_2 \\ \mathbf{0} \\ \mathbf{0} \end{Bmatrix}$$

where $\mathbf{0}$ is a 2×1 submatrix containing zeros and $\mathbf{n}_i$ (or $\mathbf{e}_i$) is a 2×1 submatrix containing the components of the vector $\mathbf{n}_i$ (or $\mathbf{e}_i$) relative to the base vectors $\{\mathbf{e}_1, \mathbf{e}_2\}$. As such, the coefficient matrix is 8×8. In three dimensions, the submatrices would be 3×1. We will summarize this equation as

$$\mathbf{Ca} = \mathbf{f} \tag{2.11}$$

where $\mathbf{C}$ is the $n \times m$ coefficient matrix, $\mathbf{a}$ is the $m \times 1$ vector of unknowns containing both the member forces and the reactions, and $\mathbf{f}$ is the $n \times 1$ applied load matrix. The number of rows n in the matrix $\mathbf{C}$ is the number of joint equilibrium equations available. It is equal to the number of joints times the dimension of the space. The number of columns m of $\mathbf{C}$ is the number of unknowns, which is equal to the number of elements plus the number of reaction components. In two dimensions, if we designate the number of joints as N, the number of elements as M, and the number of reactions components as R, then the matrix $\mathbf{C}$ is square if $n = m$, i.e., if

$$2N = M + R \tag{2.12}$$

In three dimensions, the matrix is square if $3N = M + R$ because $n = 3N$.

The question that remains is how to translate these results into a code. The arrays that we need to completely describe the truss geometry, support conditions, and applied loads are: (1) the coordinates of the joints, `x`; (2) the element connections, `ix`; (3) the support conditions, `id`; and (4) the applied forces, `F`. Let us define the arrays

$$\texttt{x} = \begin{bmatrix} 4, & 12 \\ 11, & 9 \\ 0, & 4 \\ 12, & 0 \end{bmatrix}, \quad \texttt{ix} = \begin{bmatrix} 3, & 4 \\ 4, & 2 \\ 3, & 2 \\ 3, & 1 \\ 1, & 2 \end{bmatrix}, \quad \texttt{id} = \begin{bmatrix} 0, & 0 \\ 0, & 0 \\ 1, & 1 \\ 0, & 1 \end{bmatrix}, \quad \texttt{F} = \begin{bmatrix} 3, & -4 \\ -2, & -4 \\ 0, & 0 \\ 0, & 0 \end{bmatrix}$$

These arrays describe the truss shown in Fig. 2.23 with applied loads

$$\mathbf{p}_1 = 3\mathbf{e}_1 - 4\mathbf{e}_2, \qquad \mathbf{p}_2 = -2\mathbf{e}_1 - 4\mathbf{e}_2$$

The origin for the joint coordinate locations has been taken directly below joint 3 and on the level with joint 4.

The number of rows of `x`, `id`, and `F` is equal to the number of joints in the truss. The number of rows in `ix` is equal to the number of elements. The number of columns in `ix` is always two because a member in a truss can only attach to two joints. The arrays `x`, `id`, and `F` have two columns in this case because it is a two-dimensional problem. For a three-dimensional problem they would each have three columns, the last representing the z-direction.

Joint coordinates The array `x` holds the coordinates of the truss joints. Each row represents a joint. The first column of `x` is the x-coordinate of the joint relative to the origin, and the second column is the y-coordinate. For example, joint 2 is located at $x = 11$ and $y = 9$.

Element connections The array `ix` describes how the elements connect the joints. Each row represents an element. The first column of the `ix` array is the node number for the i-node, and the second column is the node number for the j-node. For example, element 3 connects joint 3 to joint 2. This array determines the orientation (i.e., the $\mathbf{n}_e$ vector) of the element. It does not matter which order you put the two joint numbers in a row of `ix`, but once selected, the orientation $\mathbf{n}_e$ of the element is determined.

Applied loads and supports The arrays `F` and `id` describe the applied loads and support conditions, respectively. The columns of both arrays are associated with the x- and y-directions (just like the `x` array). For `F`, the columns give the x- and y-components of the applied force at the joint. For example, joint 1 has a force 3 in the x-direction and -4 in the y-direction. A value of 1 in the `id` array indicates that there is a support and, therefore, a reaction force. A value of zero indicates no support. In this case, joint 3 is restrained in both the x- and y-directions. Therefore, both directions get an entry of 1. At joint 4, only the y-direction is restrained. The roller has a reaction only in the direction perpendicular to the surface it is rolling on. Note that if there is an applied force at a location with support, it simply adds directly to the reaction computed.

Code 2.1 carries out the static analysis of truss structures. Code 2.2 is the input file associated with the example used in this section.

Code 2.1 MATLAB code `TrussAnaysis` to carry out the static analysis of two- and three-dimensional truss structures

```
%. Truss Static Analysis Program
   clear; clc;

%. Input problem geometry and loads
   [x,ix,id,F,nDim,nNodes,nElem] = InputExample;

%. Sort out problem dimensions
   nReac = sum(sum(id));
   nDOF = nNodes*nDim;
   nUnknowns = nElem + nReac;

%. Establish equation numbers
   dof = 1:nDim;
   DOF = zeros(nNodes,nDim);
   for i=1:nNodes
     DOF(i,:) = nDim*(i-1) + dof;
   end

%. Establish the coefficient matrix and right hand side
   C = zeros(nDOF,nElem);
   f = -reshape(F',1,[])';

%. Establish the columns of C associated with element forces
   for e=1:nElem
     iNode = ix(e,1);  jNode = ix(e,2);
     n = (x(jNode,:) - x(iNode,:))';
     n = n/norm(n);
     EqNos = [DOF(iNode,:),DOF(jNode,:)];
     C(EqNos,e) = [ n; -n ];
   end

%. Add columns of C for support reactions
   restraint = reshape(id',1,[]);
   I = eye(nDOF);
   B = I(:,restraint==1);
   C = [C,B];
```

```
%. Solve for unknowns
   a = C\f;
   N = a(1:nElem);
   R = a(nElem+1:nUnknowns);

%. Compute residual to see if equations are satisfied
   Residual = norm(C*a-f);

%. Output
   TrussOutput(nElem,nNodes,nDim,id,N,R,Residual)
```

The code starts with a call to function `InputExample`, which generates the arrays `x`, `ix`, `id`, and `f` for the target structure. It also passes back the dimension of the space (`nDim`), the number of joints (`nNodes`), and the number of elements (`nElem`). From that information, we can compute the number of reactions (`nReac`), the number of degrees of freedom (`nDOF`), and the total number of unknowns (`nUnknowns`) in the problem.

The array `DOF` is created to contain the equation numbers shown in Fig. 2.23c. First, `dof` is set to $[1, 2]$ if `nDim` is 2 and to $[1, 2, 3]$ if `nDim` is 3. Then, each row of `DOF` is created by looping over the nodes and augmenting `dof` to get the appropriate sequential values. Now, the equation numbers associated with element e are simply the rows of `DOF` associated with the i-node and j-node. The array `f` is created by putting successive rows of `F` into a single column, using the `reshape` function in MATLAB. It is easier to input the forces in joint order, but the right-hand side of the equations of equilibrium must be in a matrix with one column.

The coefficient matrix `C` is created in two steps. The first step puts in place the columns associated with the element forces. The entries in column e come from the unit orientation vector $\mathbf{n}_e$ and its negative. The unit vectors are computed from the joint coordinates of the i-node and j-node. The node numbers are stored in row e of `ix`. The i-node number for element e is stored in `ix(e,1)` and the j-node number is stored in `ix(e,2)`. The placement of the $\mathbf{n}_e$ vectors is determined by the equation numbers, which can be extracted from the array `DOF`. The second step is to create the columns associated with the reaction force unknowns. To accomplish this task, we create an identity matrix `I` of dimension equal to the number of equations. Then, we simply extract the columns of that matrix associated with the locations in `id` that have ones in them. To make that extraction simpler, we first flatten the `id` array into a single column (just like we did for `f`).

Finally, we solve the system of linear equations to get `a` and then partition it into the element forces `N` and the reaction force components `R`. The solution is checked by computing the norm of the residual force $\mathbf{Ca} - \mathbf{f}$, i.e., the amount by which the equations are not satisfied. If this value is within a tolerance of zero, then the solution is valid. These values are then printed to the command window with the function `TrussOutput` (which is not included here).

Code 2.2 MATLAB code `InputExample` develops the input arrays and problem size parameters needed for Code 2.1. This function completely describes the structure

```
function [x,ix,id,F,nDim,nNodes,nElem] = InputExample

%. Nodal coordinates
   x = [ 4, 12;
        11,  9;
         0,  4;
        12,  0];

%. Element connections
   ix = [ 3, 4;
          4, 2;
          3, 2;
          3, 1;
          1, 2];

%. Deduce problem size parameters
   nElem  = size(ix,1);                  % Number of elements
   nNodes = size(x,1);                   % Number of nodes or joints
   nDim = size(x,2);                     % Dimension of space

%. Support conditions
   id = zeros(nNodes,nDim);
   id(3,:) = [ 1, 1];
   id(4,:) = [ 0, 1];

%. Applied forces
   F = zeros(nNodes,nDim);
   F(1,:) = [ 3, -4];
   F(2,:) = [-2, -4];

   PlotTruss(nElem,nNodes,nDim,x,ix,id,F)

end
```

This function is mostly self-explanatory. Each truss structure requires an input file to generate the required arrays. The function `PlotTruss` is not included here, but its function is to render the structure, showing the elements and joints (and also their numbers), the supports, and the applied forces. This function helps with debugging the input file. If the rendered structure does not look right, then there is a problem with the input file.

Determinacy and stability As shown in Eq. 2.11, the equations of joint equilibrium for a truss can be written as the matrix equation (repeated here for convenience)

$$\mathbf{C}\mathbf{a} = \mathbf{f} \tag{2.13}$$

where **C** is the coefficient matrix, **a** is the vector of unknowns (member forces plus reactions), and **f** is the vector of applied loads. Let n be the number of rows of **C** and m the number of columns. Then n is the number of equations and m is the number of unknowns. As indicated in Eq. 2.12, if the number of equations of joint equilibrium is equal to the number of unknowns (member forces plus reactions), then the coefficient matrix **C** of the equilibrium equations is square, which fits our most basic expectation for *solvability*. However, that is only a starting point for understanding whether the equations have a unique solution or if the truss can even carry the loads.

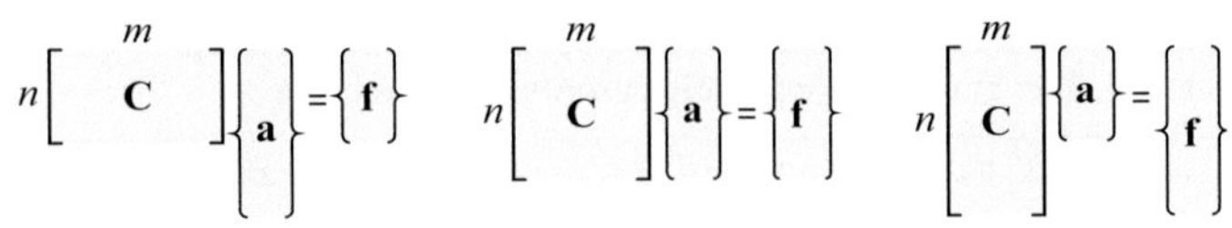

(a) $n < m$ (b) $n = m$ (c) $n > m$

Fig. 2.25 *Condition of* **C**. The number of equations is n and number of unknowns is m. **a** $n < m$. **b** $n = m$. **c** $n > m$

Figure 2.25 shows three possible cases. Case (a) with $n < m$ is an *underdetermined* system of equations. There are more unknowns than equations, so it is obvious that it will not be possible to find all the unknowns. In this case, the structure is *statically indeterminate*. It may be possible to find a solution, but the solution will not be unique. Case (b) with $n = m$ is a square system of equations, so a solution is possible. However, if **C** is *singular*, then there may be no solution. If **C** has full rank, then the structure is *statically determinate*. Finally, case (c) with $n > m$ is an *overdetermined* system of equations. There are more equations than unknowns. In the parlance of linear algebra, if **f** lives in the column space of **C**, then a solution is possible. However, that is not generally true, as the specification of the loads is independent of the connectivity of the structure. In any case, if the *rank* of **C** is less than the number of equations n, then the system is *unstable*. The rank of a matrix can be computed in MATLAB with the function `rank(C)`, or by counting the number of zero eigenvalues of the matrix $\mathbf{CC}^T$.

To illustrate what these conditions on the equations of equilibrium mean, consider the cases shown in Fig. 2.26. Case (a) is a stable and statically determinate truss with $n = m$. The rank of **C** is n. This condition allows a complete and unique solution to Eq. 2.13.

If we remove the two diagonal members, we get the structure shown in Fig. 2.26b. This structure, with $n > m$, corresponds to Fig. 2.25c. The structure is *unstable*,

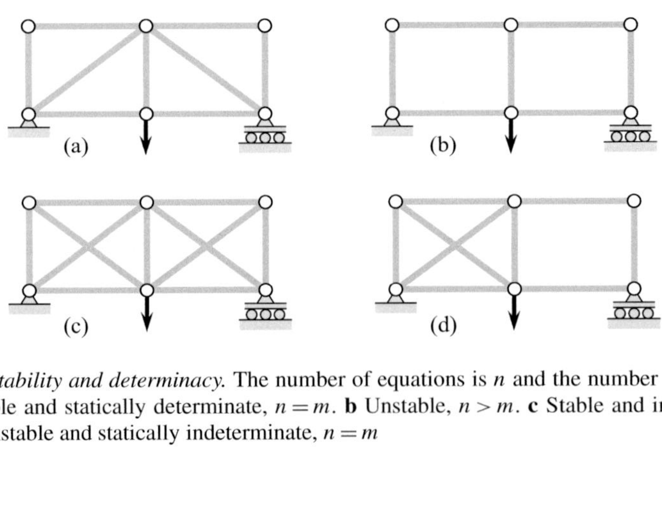

Fig. 2.26 *Stability and determinacy.* The number of equations is n and the number of unknowns is m. **a** Stable and statically determinate, $n = m$. **b** Unstable, $n > m$. **c** Stable and indeterminate, $n < m$. **d** Unstable and statically indeterminate, $n = m$

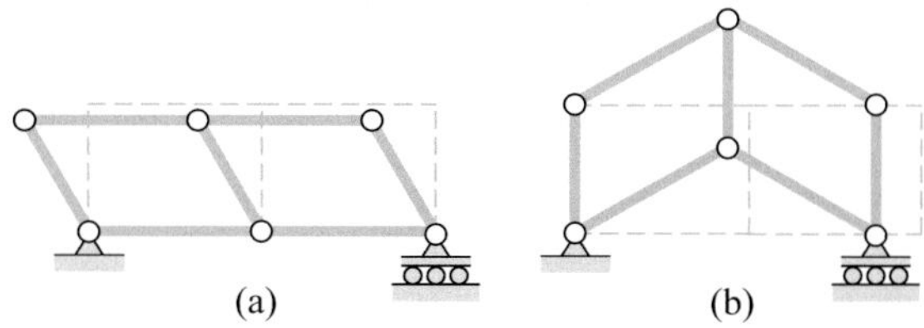

Fig. 2.27 *Rigid body modes.* Rigid body modes are motions of the structure that do not require any force. **a** Mode 1. **b** Mode 2

i.e., it is not able to carry the load shown. The difference between the number of equations and the rank of $\mathbf{C}$ is the *degree of instability* (i.e., the number of members or reactions that would need to be added to make the structure stable). The number is also the number of independent modes of motion the structure can execute without any applied forces. We call these *rigid body modes*. The two rigid body modes for case (b) in Fig. 2.26 are shown in Fig. 2.27.

If we add two additional diagonal members to the statically determinate case (a), we get the structure shown in Fig. 2.26c. This structure, with $n < m$, corresponds to Fig. 2.25a. The structure is *statically indeterminate*, i.e., the structure cannot be solved using only the equations of statics. This structure is *stable* and therefore able to carry the load. We can convert an underdetermined system of equations into a square system by multiplying both sides of the equation by $\mathbf{C}^T$ to get

$$\mathbf{C}^T\mathbf{C}\,\mathbf{a} = \mathbf{C}^T\mathbf{f} \tag{2.14}$$

The coefficient matrix $\mathbf{A} = \mathbf{C}^T\mathbf{C}$ is $m \times m$. The number of eigenvalues of $\mathbf{A}$ equal to zero is the *degree of static indeterminacy* (i.e., the number of members or reactions one would need to take away to make the structure statically determinate). Consider the eigenvalue problem

$$\mathbf{A}\boldsymbol{\varphi} = \lambda\boldsymbol{\varphi}$$

This equation has m solutions $(\lambda_i, \boldsymbol{\varphi}_i)$, where λ_i is the ith eigenvalue and $\boldsymbol{\varphi}_i$ is the ith eigenvector of $\mathbf{A}$ (see Appendix A.5). If an eigenvalue λ_i is equal to zero, then the equation reduces to $\mathbf{A}\boldsymbol{\varphi}_i = \mathbf{0}$. In the context of Eq. 2.14, a zero eigenvalue implies that there is a set of internal member forces and reactions $\mathbf{a}_i = \boldsymbol{\varphi}_i$ that are in equilibrium with no external load (i.e., $\mathbf{f} = \mathbf{0}$). We call these *self-equilibrating* forces. Thus, the solution to Eq. 2.14 is not unique because any amount of self-equilibrating force can be added without affecting the applied load. Two self-equilibrating force systems for case (c) in Fig. 2.26 are shown in Fig. 2.28. One of the main benefits of the mechanics of deformable solids is that it enables the unique solution of statically indeterminate systems.

Last, let us take a look at the case shown in Fig. 2.26d. For this truss $n = m$, but the structure is unstable *and* statically indeterminate. This case illustrates why the equation and unknown count in Eq. 2.12 is not reliable to determine static determinacy or stability. The degree of static indeterminacy is not related to the stability of the structure. A complex structure can have regions that are indeterminate and other

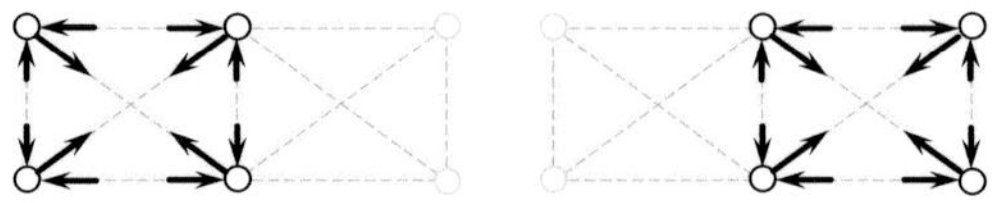

(a) (b)

Fig. 2.28 *Self-equilibrating force systems.* Force systems that are in equilibrium without external forces. **a** Force system 1. **b** Force system 2

regions that are unstable. If *any* part of a structure is unstable, then the entire structure is unstable. That is why the rank of **C** is crucial to determining the stability of the system.

Code 2.3 can be added as a call at the end of Code 2.1 to check the condition of the equations of equilibrium. It determines the stability and, if unstable, the number of rigid body modes. If the system is statically indeterminate, then the function finds the eigenvalues of $\mathbf{A} = \mathbf{C}^T\mathbf{C}$ to identify the degree of indeterminacy and computes the self-equilibrating force systems.

Code 2.3 MATLAB code `CheckCondition` to check the conditions of the equations for Code 2.1

```
function CheckCondition(C,nElem,nNodes,nDim,id)
% Check equations for static indeterminacy or instability

%. Compute dimensions of C and its rank
   [nEqns,nUnknowns] = size(C);
   RankC = rank(C);

%. Determine stability and other features if stable
   if RankC<nEqns
     fprintf('\n Structure is unstable')
     fprintf('\n Number of rigid body modes =%3i\n',nEqns-RankC)
   else
     fprintf('\n Structure is stable');

%... Look for zero eigenvalues of normal matrix C'*C
     [V,D] = eig(C'*C);
     Lambda = diag(D);

%... Establish degree of indeterminacy
     tol = 1.e-8;
     DegreeIndet = sum(Lambda<tol);
     fprintf('\n Degree of static indeterminacy = %4i\n',
         DegreeIndet)

%... Print self-equilibrating force systems
     for n=1:nUnknowns
       if Lambda(n)<tol
         fprintf('\n Self-equilibrating forces, Mode = %3i\n',n)
         a = V(:,n);
         N = a(1:nElem);
         R = a(nElem+1:nUnknowns);
         Residual = norm(C*a);
         TrussOutput(nElem,nNodes,nDim,id,N,R,Residual)
       end
     end

   end

end
```

2.9.3 Frames and Machines

If two members are pinned together with an internal pin joint, but the location of the pin is not at the ends of the members, then the structure is *not* a truss. Because the pin cannot transmit moment from one member to another, the pin has significance in the solution of the system. We usually call these kinds of structures *frames* or *machines*. To solve these kinds of problems, we must take the structure apart at the pin to expose and exploit the zero-moment condition. We will illustrate the solution of this type of system through the following example.

Example 2.10
Consider the frame structure shown in Fig. 2.29a. Two rigid bars are connected together with an ideal pin joint at $\mathcal{E}$ and subjected to the forces shown. The bars have pin supports at $\mathcal{A}$ and $\mathcal{B}$. Find the reactions at the supports and the force in the pin connecting the two elements.

Solution. Each pin support (i.e., at $\mathcal{A}$ and $\mathcal{B}$) has an unknown reaction with two components. Therefore, there are four unknown reaction components. If we take a free body diagram of the overall system, we will only have three equations of equilibrium (e.g., two force equations and one moment equation). That free body diagram will not provide enough equations to solve for all the reaction components. We must make use of the internal pin condition to solve this problem.

Divide the system into three free body diagrams, as shown in Fig. 2.30: (a) The free body diagram of bar $\mathcal{AEC}$, (b) the free body diagram of the pin at $\mathcal{E}$, and (c) the free body diagram of bar $\mathcal{BED}$. The force $\mathbf{Q}$ is the force that the pin exerts on each bar and each bar exerts on the pin. Figure 2.30b shows how that exchange happens. The pin must be in equilibrium. Therefore, the two forces must be equal and opposite. Each of those forces must be transmitted to one of the bars. By Newton's third law, those forces must be equal and opposite.

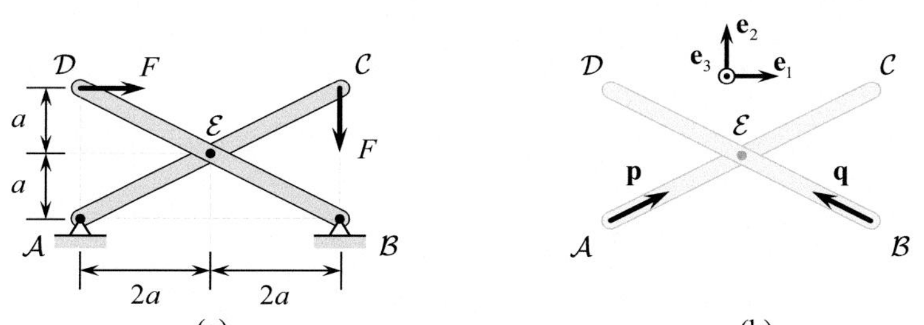

Fig. 2.29 *Frame example.* **a** Geometry, support, and applied loads on the system. **b** Definition of unit vectors **p** and **q**

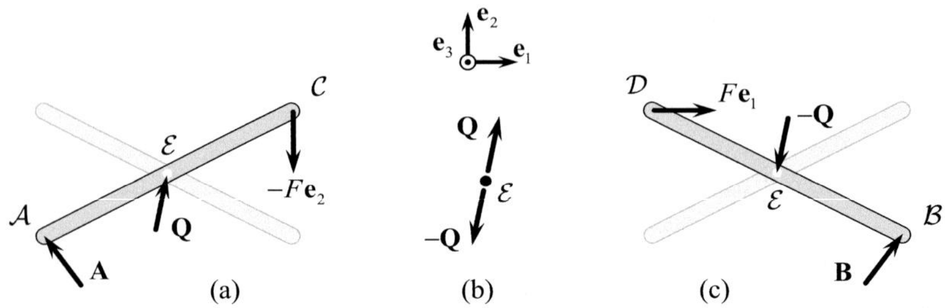

Fig. 2.30 *Free body diagrams.* **a** Free body diagram of bar $\mathcal{AEC}$. **b** Free body diagram of the pin at $\mathcal{E}$. **c** Free body diagram of bar $\mathcal{BED}$. The force $\mathbf{Q}$ is the force of the pin acting on the bars and the bars acting on the pin

The force $-\mathbf{Q}$ on the pin has its equal and opposite force $\mathbf{Q}$ on member $\mathcal{AEC}$. Likewise, the force $\mathbf{Q}$ on the pin has its equal and opposite force $-\mathbf{Q}$ on member $\mathcal{BED}$. Thus, if the force is $\mathbf{Q}$ on the free body diagram of bar $\mathcal{AEC}$, then it must be $-\mathbf{Q}$ on bar $\mathcal{BED}$, as communicated by the pin.

First, note that the unit vectors along the bars can be written in terms of the standard base vectors as

$$\mathbf{p} = \tfrac{2}{\sqrt{5}}\mathbf{e}_1 + \tfrac{1}{\sqrt{5}}\mathbf{e}_2, \qquad \mathbf{q} = -\tfrac{2}{\sqrt{5}}\mathbf{e}_1 + \tfrac{1}{\sqrt{5}}\mathbf{e}_2 \tag{2.15}$$

For convenience, let us abbreviate the distance from $\mathcal{A}$ to $\mathcal{E}$ as $L = a\sqrt{5}$. The distance from $\mathcal{A}$ to $\mathcal{C}$ is then $2L$. The distances on the other bar are similar.

Now, establish force equilibrium on both free body diagrams of the bars to get the equations

$$\mathbf{A} + \mathbf{Q} - F\mathbf{e}_2 = \mathbf{0}, \qquad \mathbf{B} - \mathbf{Q} + F\mathbf{e}_1 = \mathbf{0} \tag{2.16}$$

As an interesting side note, if we add these two equations together, the $\mathbf{Q}$ forces cancel out and we get the equation of force equilibrium of the overall system taken as a free body diagram (in which case the force in the pin would be an *internal* force and, therefore, would not appear in the free body diagram).

Next, establish moment equilibrium of the first bar about point $\mathcal{A}$ and of the second bar about point $\mathcal{B}$ to get, respectively,

$$\begin{aligned} L\mathbf{p} \times \mathbf{Q} + 2L\mathbf{p} \times -F\mathbf{e}_2 &= \mathbf{0} \\ L\mathbf{q} \times -\mathbf{Q} + 2L\mathbf{q} \times F\mathbf{e}_1 &= \mathbf{0} \end{aligned} \tag{2.17}$$

There are many ways to solve these two equations. One interesting approach is to divide out L, factor out $\mathbf{p}\,\times$ and $\mathbf{q}\,\times$, and rewrite the equations as

$$\mathbf{p} \times (\mathbf{Q} - 2F\mathbf{e}_2) = \mathbf{0}$$
$$\mathbf{q} \times (-\mathbf{Q} + 2F\mathbf{e}_1) = \mathbf{0}$$

which works because the cross product satisfies the distributive law. The first equation will be satisfied if $\mathbf{Q} = 2F\mathbf{e}_2 + x\mathbf{p}$ and the second equation will be satisfied if $\mathbf{Q} = 2F\mathbf{e}_1 + y\mathbf{q}$, where the scalars x and y are not yet known. These expressions are true because any vector crossed with itself is zero, independent of the scalar multiplier. To see this, just plug the first version of $\mathbf{Q}$ into the first moment equation. To wit,

$$\mathbf{p} \times (\mathbf{Q} - 2F\mathbf{e}_2) = \mathbf{p} \times (2F\mathbf{e}_2 + x\,\mathbf{p} - 2F\mathbf{e}_2) = \mathbf{0}$$

We can set the two versions of $\mathbf{Q}$ equal to each other and move the unknowns x and y to the left side and the known values to the right side to get the equation:

$$x\mathbf{p} - y\mathbf{q} = 2F(\mathbf{e}_1 - \mathbf{e}_2) \tag{2.18}$$

Now, we can dot this equation successively with unit vectors $\mathbf{p}$ and $\mathbf{q}$, noting that $\mathbf{p} \cdot \mathbf{q} = -\frac{3}{5}$ (just go back to the definitions of $\mathbf{p}$ and $\mathbf{q}$ and distribute the dot product), to get

$$x + \tfrac{3}{5}y = \tfrac{2}{\sqrt{5}}F$$
$$\tfrac{3}{5}x + y = \tfrac{6}{\sqrt{5}}F$$

Solving this system of equations gives $x = -\frac{\sqrt{5}}{2}F$ and $y = \frac{3\sqrt{5}}{2}F$. We can plug either of these two values back into the definition of $\mathbf{Q}$. For example, the first definition gives

$$\mathbf{Q} = 2F\mathbf{e}_2 + x\mathbf{p} = 2F\mathbf{e}_2 - \tfrac{\sqrt{5}}{2}F\big(\tfrac{2}{\sqrt{5}}\mathbf{e}_1 + \tfrac{1}{\sqrt{5}}\mathbf{e}_2\big) = F\big(-\mathbf{e}_1 + \tfrac{3}{2}\mathbf{e}_2\big)$$

The other version of $\mathbf{Q}$ gives the same value. Using this result in the balance of force equations gives

$$\mathbf{A} = F\big(\mathbf{e}_1 - \tfrac{1}{2}\mathbf{e}_2\big), \qquad \mathbf{B} = F\big(-2\mathbf{e}_1 + \tfrac{3}{2}\mathbf{e}_2\big)$$

Thus, all support reactions and the pin force are known in terms of F. The key to solvability of this problem is to note that each free body diagram gives three equations of equilibrium. When we split the system into two free bodies, we introduce the new unknown force $\mathbf{Q}$ because we have exposed the force in the pin. That adds two new unknowns (the two components of $\mathbf{Q}$), but we gain three equations. As a result, we get six equations with six unknowns.

We can generalize the notion of static determinacy for frames and machines in two dimensions by noting that each member taken as an individual free body diagram gives three equations (two components of force equilibrium and one moment equilibrium). Each pin introduces one additional vector unknown (with two components). Hence, the system is statically determinate if

$$3M = 2P + R \tag{2.19}$$

where M is the number of members, P is the number of pins connecting two members, and R is the number of reaction components. For the previous example we had $M = 2$, $P = 1$, and $R = 4$, giving $3(2) = 6 = 2(1) + 4$. Equation 2.19 does not work for cases where more than two members are connected to a pin (because then there is not just one $\mathbf{Q}$ transmitted at the pin). A system that has more than one member connecting to a pin at a joint is statically determinate if

$$3M = 2\sum_{i=1}^{P}(m_i - 1) + R \tag{2.20}$$

where P is the number of pins and m_i is the number of members framing into pin i. The modified equation will now work for trusses and frames. For example, the truss in Fig. 2.23 has five members, four joints, and three reactions. Joints 1 and 4 connect two elements and joints 2 and 3 connect three elements. The equation gives $3(5) = 2(1 + 2 + 2 + 1) + 3$, which checks out. Of course, these formulas have all the same limitations as Eq. 2.12 for trusses, as discussed in the previous section.

2.10 Summary

This chapter has outlined the consequences of applying Newton's laws to systems in static equilibrium. Newton's third law allows any system to be cut into parts, with the action of one part on the other represented by forces at the cut and noting that each part must satisfy the equations of equilibrium. Thus, one of the main challenges of solving statics problems lies in the art of isolating a *free body diagram* and faithfully representing the forces and moments acting on it. Beyond forming and manipulating the equations of equilibrium themselves, the free body diagram is the most important concept in statics. At its core, the subject of *statics* is an exploration of problems amenable to these techniques. While there are only a few basic concepts, the range of applications is vast.

The study of rigid body statics is prerequisite to the study of deformable solids, which will occupy the remainder of the book. The systems we will explore in the sequel will all be in static equilibrium. Hence, statics will form a basic building block for what follows. To that we will add the kinematics of motion, which is associated with the fact that force causes real bodies to deform, and the properties of materials, which relate how much deformation results from the applied forces. Every problem we encounter from here on will have those three features present.

2.11 Problems

2.1 Three smooth disks of diameters $3a$, $4a$, and $5a$ and thickness d (into the page) are restrained in a rectangular basin of width $7a$. The density ρ is the same for all three disks. Find the reaction forces where the disks contact the container and find the contact forces between the disks.

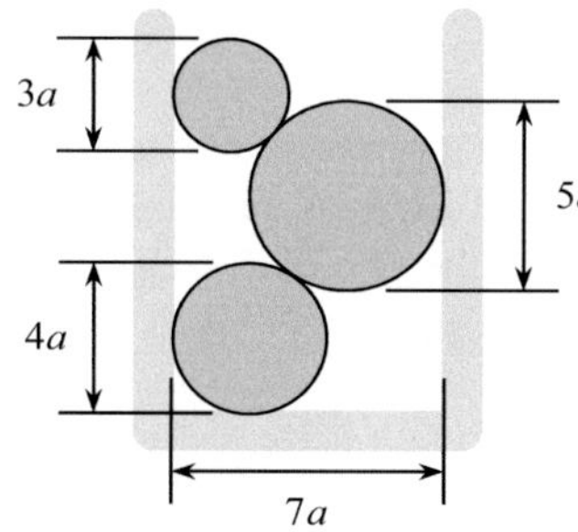

2.2 A rectangular plate of dimensions $2a \times 4a$ and uniform thickness d is pinned at the lower right corner and attached to a cable at the upper left corner. The weight per unit volume of the plate is w. The cable is attached to a pin support a distance a above the top right corner of the plate. Find the reaction forces and the force in the cable required to maintain equilibrium in the position shown. The weight of the cable is negligible.

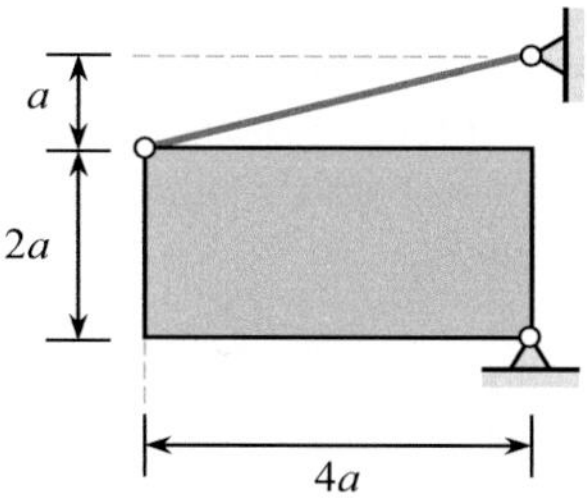

2.3 A cantilever dam holds water at a height of h above the base. The unit weight of water is γ_w. Consider a unit width of the dam (into the page). Find the reaction force and moment per unit width at $\mathcal{A}$.

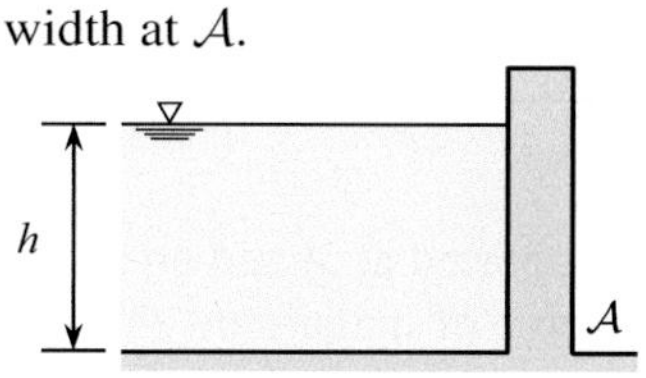

2.4 A rigid quarter-circle gate of radius R and weight per unit length of arc w is pinned at $\mathcal{A}$ and has a smooth stopper at $\mathcal{B}$. The gate holds water at a height of h above point $\mathcal{B}$. The unit weight of water is γ_w. Consider a unit width of the gate (into the page). Find the reaction force per unit width at $\mathcal{B}$ as a function of the height h of water. What is the minimum height of the water to keep the support at $\mathcal{B}$ in compression?

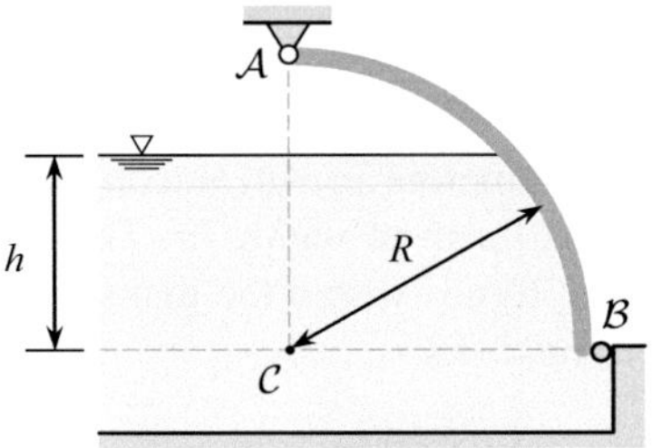

2.5 A truss has two bars with a pin at joint $\mathcal{C}$ and pin supports at $\mathcal{A}$ and $\mathcal{B}$. The truss is subjected to a downward force of magnitude P at joint $\mathcal{C}$. Find the reaction forces and the internal forces in the two bars. Neglect the weight of the members.

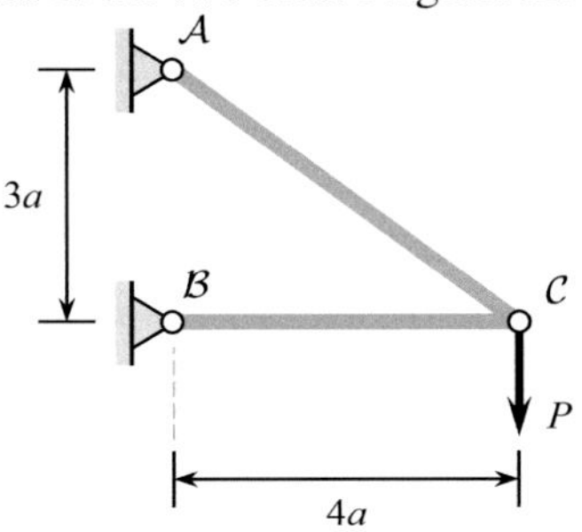

2.6 A 13-bar truss is pin-supported at $\mathcal{A}$ and on a horizontal roller at $\mathcal{B}$. The truss is subjected to downward forces of magnitudes $3P$, $4P$, and $2P$ as shown. Find the reaction forces and the internal forces in the bars using the method of joints. Verify the results by finding the forces in bars $\mathcal{CD}$ and $\mathcal{DE}$ using the method of sections. Neglect the weight of the members.

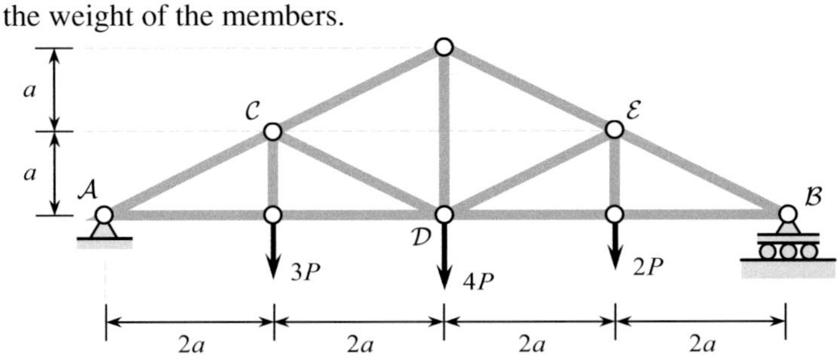

2.7 A 13-bar truss is pin-supported at $\mathcal{A}$ and on a horizontal roller at $\mathcal{E}$. The truss is subjected to a downward force of magnitude P at $\mathcal{C}$ as shown. Find the reaction forces and the internal forces in the bars using the method of joints. Verify the results by finding the forces in bars $\mathcal{BF}$ and $\mathcal{D}F$ using the method of sections. Neglect the weight of the members.

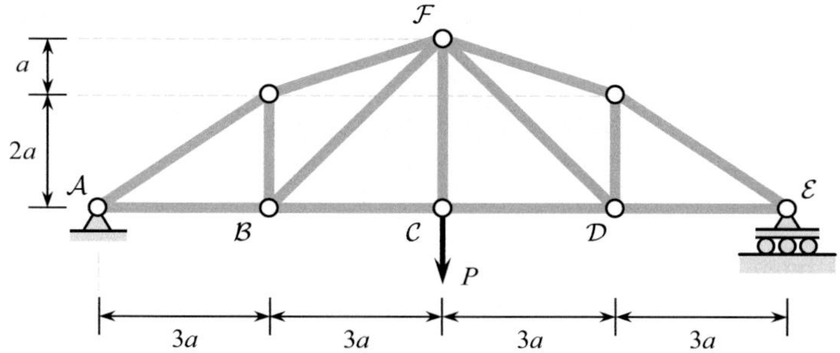

2.8 A 4-bar frame has pin supports at $\mathcal{A}$ and $\mathcal{B}$. The members are pinned together at $\mathcal{C}$, $\mathcal{D}$, $\mathcal{E}$, and $\mathcal{F}$. Member $\mathcal{CF}$ is subjected to a uniform horizontal load of intensity q_o as shown. Find the reaction forces and the forces in the pins. Neglect the weight of the members.

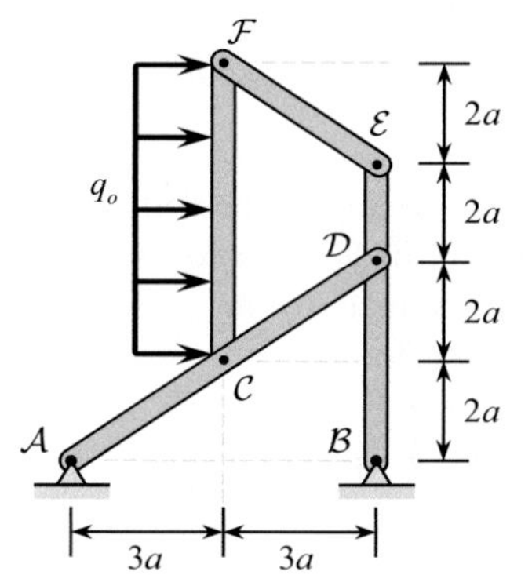

2.9 The cantilever beam $\mathcal{AB}$ is carrying a uniform load over half its length. Find the reactions at $\mathcal{A}$. Neglect the weight of the bar.

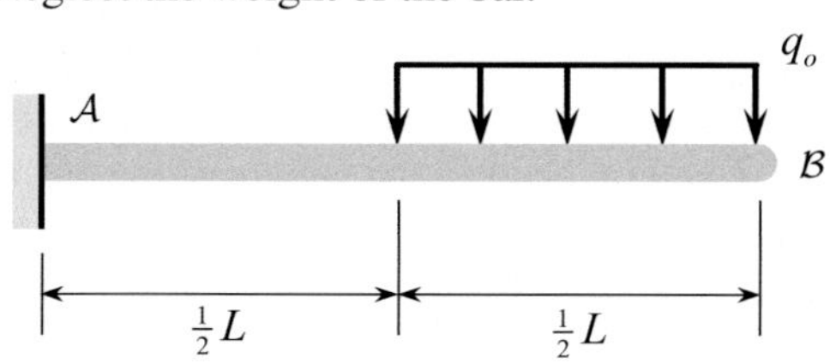

2.10 A 4-bar frame has pin supports at $\mathcal{A}$ and $\mathcal{B}$. The members are pinned together at $\mathcal{C}$, $\mathcal{D}$, $\mathcal{E}$, and $\mathcal{F}$. The structure is subjected to downward forces of magnitudes P and $2P$ as shown. Find the reaction forces and the forces in the pins. Neglect the weights of the bars.

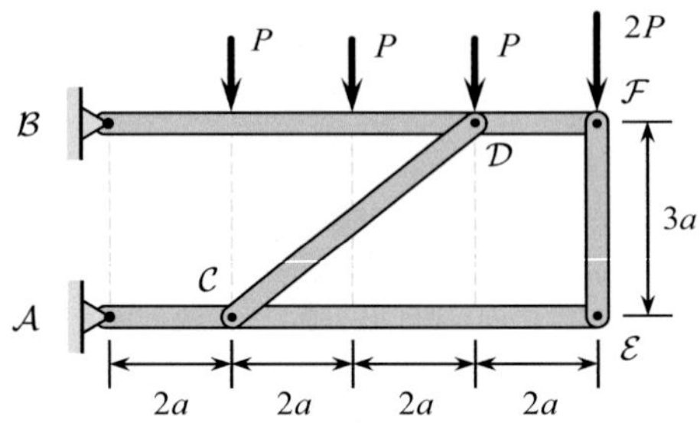

2.11 A beam of length L overhangs its supports by a distance a. The beam is subjected to a uniformly distributed load of magnitude q_o over its entire length. Find the reactions at the supports. Neglect the weight of the beam.

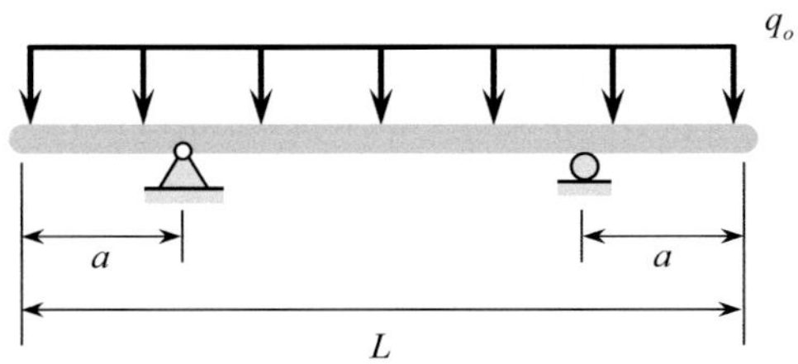

2.12 Beam $\mathcal{ABCD}$ has an overhang at one end. The beam supports a uniform load of intensity q_o over one-third of its length and a concentrated moment of magnitude M_o at $\mathcal{D}$. Find the reactions. Evaluate for $M_o = 3$ kN-m, $q_o = 12$ kN/m, and $L = 4.8$ m. Neglect the weight of the beam.

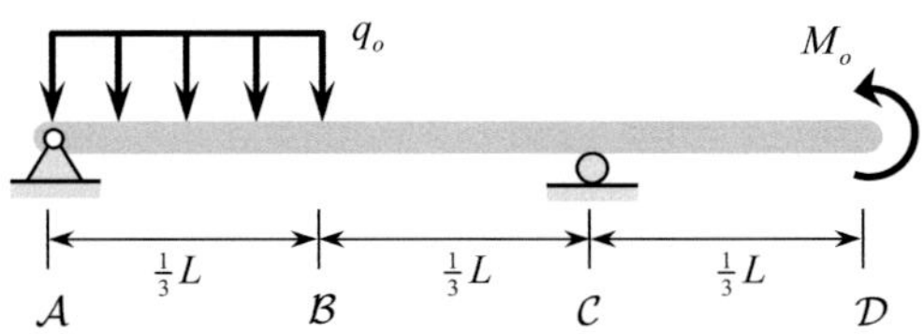

2.13 Beam $\mathcal{ABCD}$ is fixed at the right end and has a roller support at the left end. There is an internal hinge at $\mathcal{B}$. The distributed load varies linearly from zero at $\mathcal{A}$ to an intensity of q_o at $\mathcal{C}$ and is constant from $\mathcal{C}$ to $\mathcal{D}$. Find the reactions. Evaluate for $q_o = 7.5$ kN/m and $L = 6$ m. Neglect the weight of the beam.

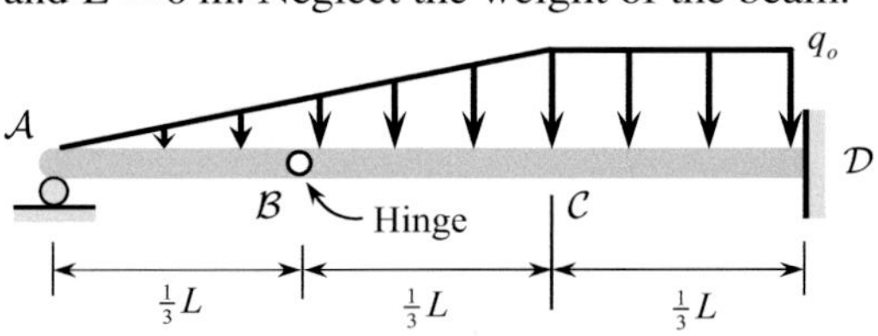

2.14 The continuous beam assembly $\mathcal{ABC}$ is made up of two straight segments connected at right angles and continuously welded together. The system is subjected to a downward distributed load of magnitude q_o over the segment $\mathcal{BC}$ and a linearly varying distributed load over segment $\mathcal{AB}$ with maximum magnitude of q_o at $\mathcal{B}$. The beam is pinned at $\mathcal{A}$ and has a roller support at $\mathcal{C}$. Find the reaction forces. Neglect the weight of the beams.

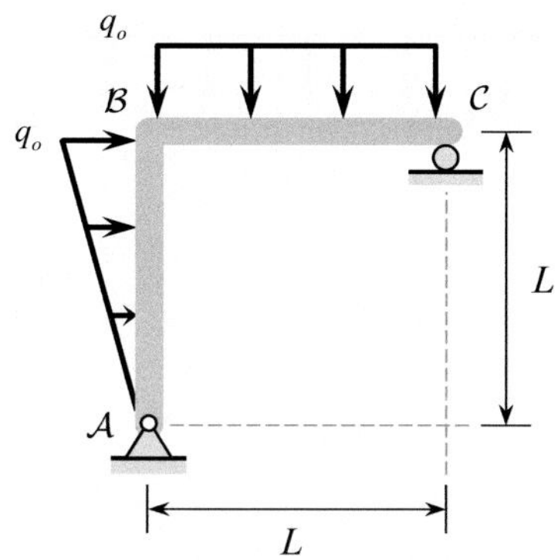

2.15 The continuous beam assembly $\mathcal{ABCD}$ is made up of three straight segments connected at right angles and continuously welded together. The system is subjected to a downward distributed load of magnitude q_o over the segment $\mathcal{CD}$ and a downward point load of magnitude P at $\mathcal{B}$. The beam is pinned at $\mathcal{A}$ and has a roller support at $\mathcal{D}$. Find the expressions for the reaction forces for the specific case where $P = q_o L$. Neglect the weight of the beam.

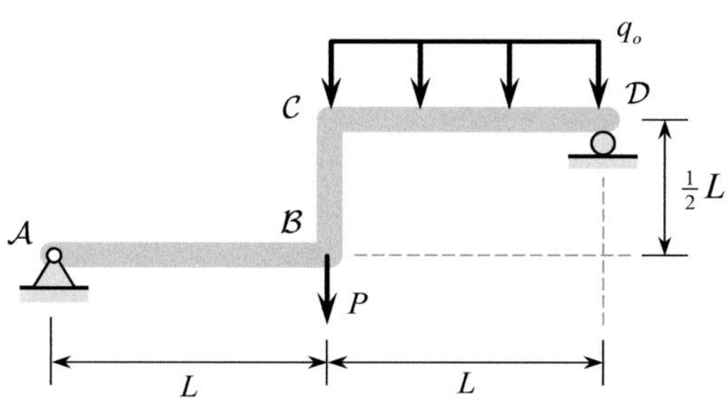

2.16 Beam $\mathcal{ABCD}$ of length L is pinned at $\mathcal{A}$ and has roller supports at $\mathcal{B}$ and $\mathcal{D}$. There is an internal hinge at $\mathcal{C}$. The beam is subjected to a constant distributed load with intensity of q_o acting downward between $\mathcal{B}$ and $\mathcal{D}$. Find the reactions and the force acting on the pin. Neglect the weight of the beam.

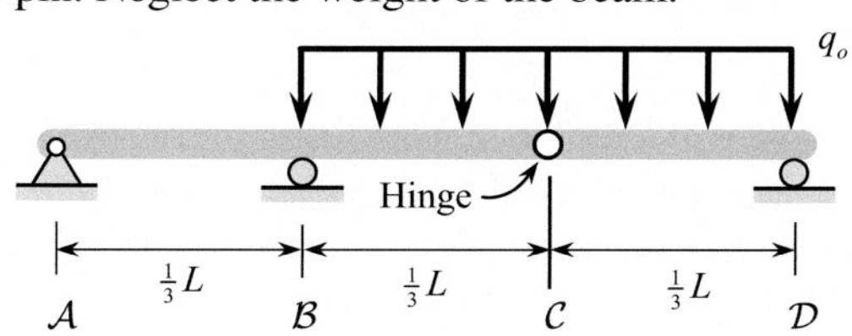

2.17 The continuous beam assembly $\mathcal{ABCD}$ is made of two straight segments connected at right angles and continuously welded together. The system is subjected to a downward distributed load of magnitude q_o over the segment $\mathcal{ABC}$. The beam is fixed at $\mathcal{D}$ and has a roller support at $\mathcal{A}$. There is an internal hinge at $\mathcal{B}$ that can transmit force, but not moment. Find the reaction forces. Neglect the weight of the beams.

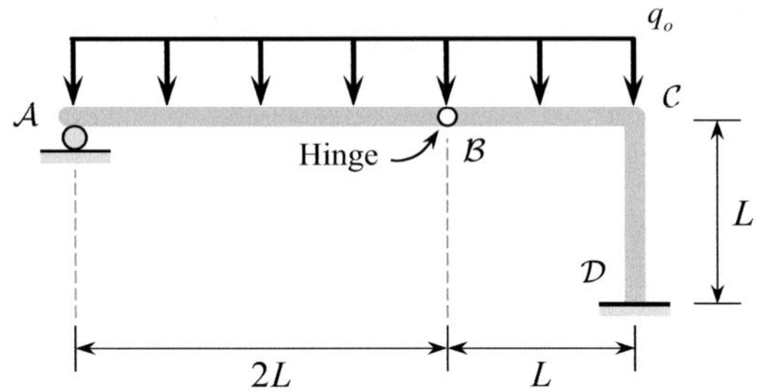

2.18 The continuous beam assembly $\mathcal{ABC}$ is made of two straight segments connected at right angles and continuously welded together. The system is subjected to a load to the right of magnitude P at $\mathcal{B}$. The beam is pinned at $\mathcal{A}$ and has a roller support at $\mathcal{C}$. Find the reaction forces. Neglect the weight of the beams.

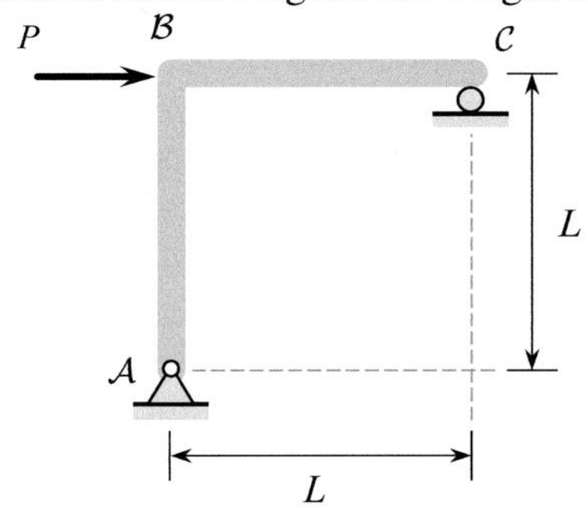

2.19 Beam $\mathcal{ABCD}$ of length $3L$ is pinned at $\mathcal{A}$, has roller supports at $\mathcal{B}$ and $\mathcal{D}$, and an internal hinge at $\mathcal{C}$. There is a triangular distributed load with intensity increasing from zero at $\mathcal{A}$ to q_o at $\mathcal{D}$, acting downward over the entire beam. Find the reactions and the force in the pin at the hinge. Neglect the weight of the beam.

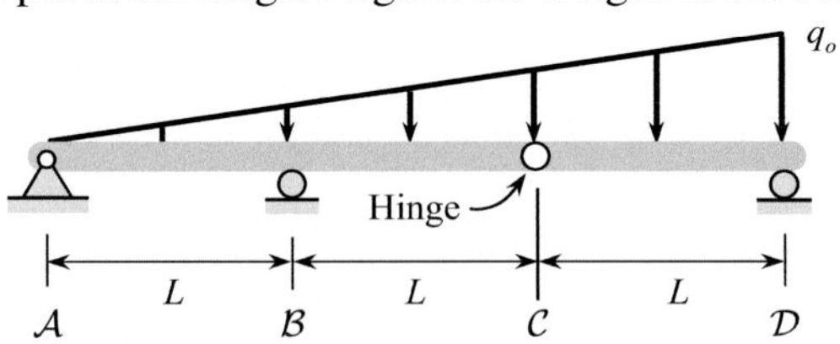

2.20 *Computation.* Verify Code 2.1, `TrussAnalysis`, by using it to solve Examples 2.8 and 2.9. Use the code to solve Problems 2.5, 2.6, and 2.7. Compare the results with hand calculations. In each case, you will need to specify numerical values for the dimensions and loads.

2.21 *Computation.* Explore what Code 2.3, `CheckCondition`, does by implementing input files for the four cases shown in Fig. 2.26. Create an interesting three-dimensional truss and check its initial stability using this code. Follow the template in Code 2.2, `InputExample`, as a guide to creating the input files.

2.22 *Computation.* Modify Code 2.1, `TrussAnalysis`, to include the self-weight of the members. Include in the input the weight per unit length, w, of the members (or sufficient information to compute it, like mass density ρ, cross-sectional area A, and the acceleration of gravity g such that $w = \rho g A$), compute the length from the nodal coordinates, and add the contributions of the element weight to the applied forces.

2.23 *Computation.* A structure is called *externally* indeterminate if there is an excess of support reactions beyond what equilibrium can determine. A structure is called *internally* indeterminate if there is an excess of member forces beyond what equilibrium can determine. Use Code 2.3 to explore the difference between external and internal static indeterminacy. Create a structure that is both internally and externally indeterminate and show the self-equilibrating force systems on a sketch of the structure.

Chapter 3
The Axial Bar

The axial bar is the simplest case of a theory of deformable solids. It has all the features of every deformable solids problem—what we call the three legs of the mechanics stool (equilibrium, kinematics, and constitution). As such, it is a good framework for introducing the basic ideas of the mechanics of deformable solids. While the axial bar has practical import on its own, the main objective of this chapter is to explore how the essential components of the theory fit together in a mathematical context that involves only a single scalar internal force function and a single scalar displacement function. The price of this mathematical simplicity is a set of restrictions on the kinematic hypothesis and the nature of forces that can be present. These restrictions will be removed in Chap. 7 when we extend the ideas to beam theory, wherein the axial bar problem will be a part of the more general theory.

3.1 Theory of Axial Bars

An *axial bar*, illustrated in Fig. 3.1, is a straight, slender solid body with one dimension, the length L, that is significantly larger than the other two. We identify a line along the long dimension of the bar as the *axis*. At any place along the bar, we can make a cut perpendicular to the axis to expose a *cross section* of the bar. To facilitate the mathematical description, let us introduce a coordinate system (x, y, z). The x-axis is coincident with the axis of the bar, and the cross sections lie in the (y, z) plane. The base vectors $\{\mathbf{e}_1, \mathbf{e}_2, \mathbf{e}_3\}$ align with the coordinate directions. In this section we develop a mathematical description of how the bar deforms under load and what internal forces develop in conjunction with the deformation.

The bar in Fig. 3.1 is illustrated with a cross section of general shape. The formulation will apply to any cross section, e.g., shapes like I-beams, angles, rectangles, circular tubes, and others. The sketch of the bar in the figure also looks like the cross sections are all the same size and shape. In general, we will consider bars that can

K. D. Hjelmstad, *Mechanics of Deformable Solids: An Introduction*,
https://doi.org/10.1007/978-3-031-89873-0_3

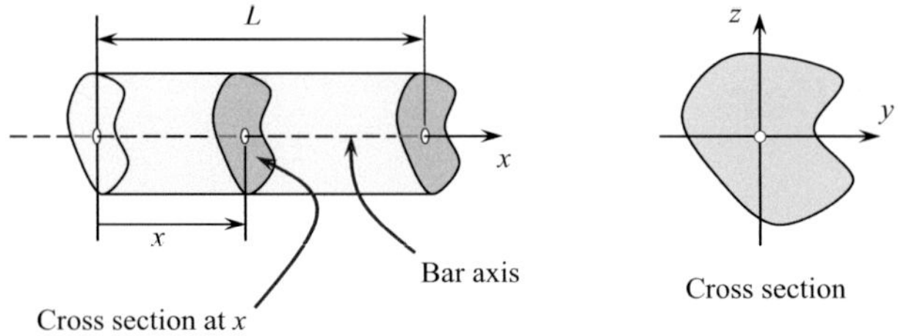

Fig. 3.1 *What is an axial bar?* An axial bar is a solid body with one dimension, along the x-axis, significantly larger than the other dimensions, i.e., the cross section in the (y, z) plane. Deformation is restricted to cross sections moving along the x-axis, and forces can only be applied along the axis of the bar

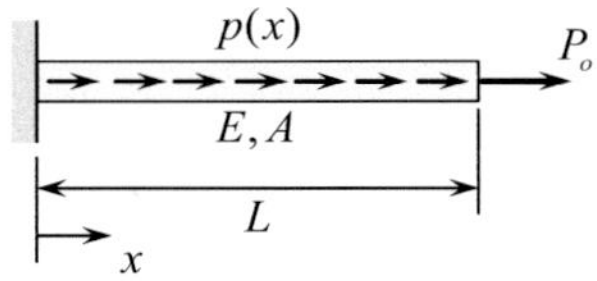

Fig. 3.2 *Axial bar problem.* Bar of area A, Young's modulus E, and length L subjected to distributed axial force $p(x)$. The bar is fixed at $x = 0$ and subjected to a concentrated force P_o at $x = L$

have a variable cross section size, making the cross-sectional area a function of x. Bars of constant cross section are called *prismatic*. Bars with cross sections that vary along the axis are called *nonprismatic*.

Figure 3.2 illustrates a typical axial bar problem. The important geometric properties are the length L and the cross-sectional area A. The independent variable describing distance along the axis of the bar is x, measured from the left end of the bar. The relevant material property E is called *Young's modulus*. It is possible to have an applied distributed load $p(x)$ as well as concentrated loads like P_o which, in this case, is acting at the end of the bar. All applied forces are constrained to act in the direction of the bar axis. The left end is fixed, meaning that the point at $x = 0$ cannot displace. All these features together comprise the *description* of the axial bar problem. The goal of the axial bar problem is to find the internal axial force $N(x)$, the reaction forces at the supports, and the displacement $u(x)$ at each point along the axis of the bar. These quantities together constitute the *response* of the bar to the applied loads.

3.1.1 Deformation and Strain

The motion of the axial bar is illustrated in Fig. 3.3. The *reference configuration* is the state of the bar before any loads are applied. This state is also known as the *undeformed configuration*. The typical cross section $\mathcal{A}$ is located a distance x from

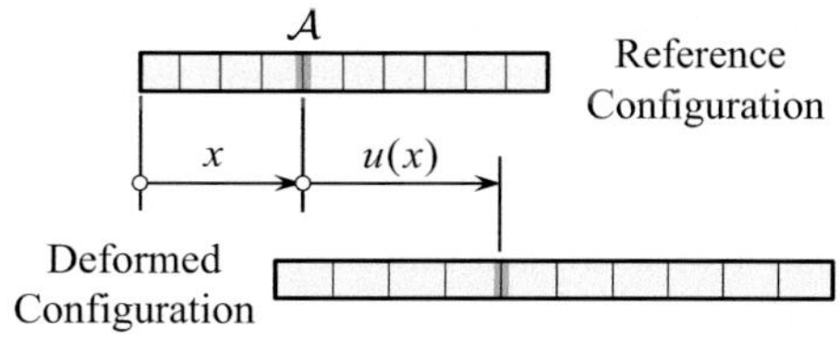

Fig. 3.3 *Motion.* Cross section $\mathcal{A}$, initially located at x in the reference configuration, has moved by an amount $u(x)$ in the axial direction in the deformed configuration

the left end of the bar in the reference configuration. In the deformed configuration, that point has moved by an amount $u(x)$ to occupy a new position. In the special context of the axial bar problem, we will restrict the motion of the bar to be only in the direction of the axis. Hence, only the single variable $u(x)$, the *axial displacement*, is required to characterize the motion. More general motions will be introduced in Chap. 7.

One way to think about the restriction on the motion is through the lens of what we call a *kinematic hypothesis*. For the axial bar, the kinematic hypothesis is that cross sections translate as rigid bodies exclusively in the direction of the x-axis (i.e., in the $\mathbf{e}_1$ direction). The cross sections do not warp or rotate from their original (y, z) plane. The consequence of the kinematic hypothesis is that the motion of the bar can be completely characterized by a single variable, the axial displacement $u(x)$. Every particle in a cross section moves by that amount. One of the primary goals of our analysis of the axial bar is to determine the function $u(x)$.

Next, consider the relative motion of two adjacent cross sections, initially separated by a distance Δx, as illustrated in Fig. 3.4. We would like to know how much the little piece of material of initial length Δx in the reference configuration deforms due to the motion $u(x)$. The left edge of the piece is initially located at x, and it displaces by an amount $u(x)$. The right edge of the piece is initially located at $x + \Delta x$, and it displaces by an amount $u(x + \Delta x)$. The length of the piece after deformation is $\Delta x + u(x + \Delta x) - u(x)$. Therefore, the change in length is $u(x + \Delta x) - u(x)$. The *engineering strain* is the change in length divided by the original length. We can take the limit as the size of the piece becomes vanishingly small to find that

$$\varepsilon(x) = \lim_{\Delta x \to 0} \frac{u(x + \Delta x) - u(x)}{\Delta x} = \frac{du}{dx} \tag{3.1}$$

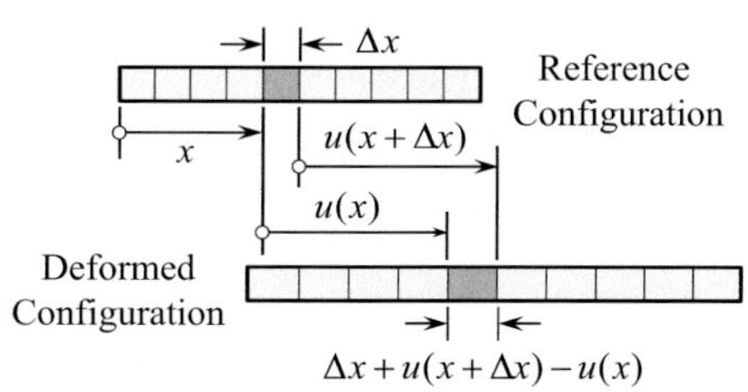

Fig. 3.4 *Deformation.* The small segment of length Δx deforms because the motions at its ends differ. The right end displaces by $u(x + \Delta x)$ and the left end displaces by $u(x)$

We call $\varepsilon(x)$ the *strain*. The strain is the rate of change (the first derivative) of the displacement function $u(x)$ with respect to the axial coordinate x. Strain is a relative measure of deformation that will be useful when it comes time to characterize the material behavior. We call Eq. 3.1 the *strain-displacement relationship* for the axial bar. The strain is a function of x, meaning that it can vary along the axis of the bar. If the displacement were to vary linearly, then the strain would be constant.

3.1.2 Force and Stress

Force, as envisioned by Newton, is an action that, if unbalanced, causes acceleration. In static equilibrium, all forces acting on a system sum to zero. We can form a free body diagram of the bar by taking a cut at some point x. This cut divides the bar into two pieces, both of which must be in static equilibrium. To respect Newton's laws, we must place forces at the cut to represent the action of the applied forces associated with the piece we took away. We call the forces placed at the cut *internal forces*, because they only show up on a free body diagram if we make a cut. The internal forces are the mechanism by which externally applied forces get transmitted through a solid body.

How do we know those internal forces exist? Because if they are *not* there, then it is *not* possible to establish equilibrium of the individual parts created by the cut. Consider the case shown in Fig. 3.5, where a bar is subjected to equal and opposite end forces of magnitude P and is, therefore, in equilibrium. If we cut the bar into two free bodies and place no force at the cut, as in Fig. 3.5b, then the sum of forces on each piece is not zero, and the two pieces will accelerate *away* from each other. For static equilibrium to hold, every piece must be in static equilibrium. For the axial bar, we need only one axially oriented force, which we call $N(x)$, to represent the part that was taken away, as shown in Fig. 3.5c. In this simple case, equilibrium holds if $N(x) = P$. The internal force is actually a resultant of what we will refer to as *stresses*, which are a normalized measure (force per area) of the internal state of force in the bar. Thus, we can think of a bar as being in a state of stress, and we can *see* the internal stresses only by taking a cut for a free body diagram.

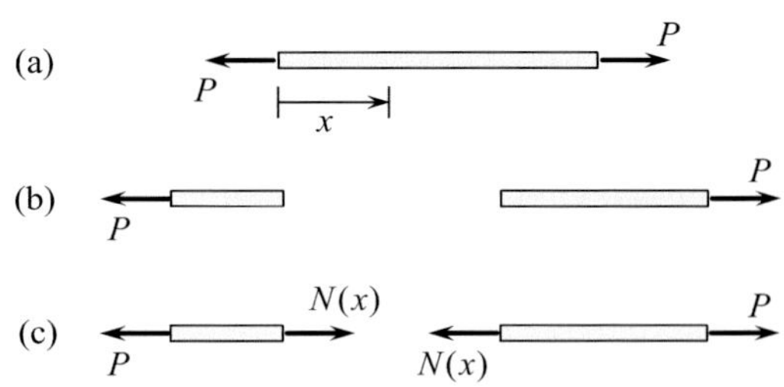

Fig. 3.5 *Internal forces.* Proof that internal forces exist. **a** Bar with end loads in equilibrium. **b** Free body diagrams with *no* forces at the cut. **c** Free body diagram with forces $N(x)$ at the cut

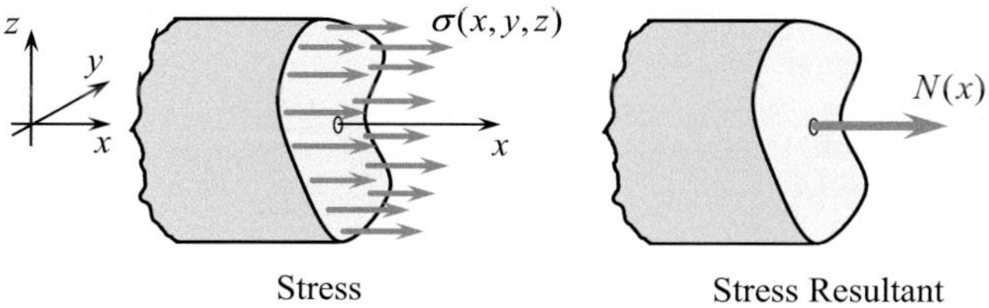

Fig. 3.6 *Axial stress resultant.* The stresses $\sigma(x, y, z)$ are distributed over the cross section as traction forces at the cut. The sum of all tractions over the cross section is the stress resultant $N(x)$

Resultant axial force Because the idea of force was first introduced in particle mechanics, we usually think of it as a discrete action represented by a single vector. In nature, point forces do not exist. We refer to force distributed over an area as *traction*, which has units of force per area. Traction multiplied by the area over which it acts gives force. To capture the net effect, we will consider a traction acting on an infinitesimal area dA and then the sum of forces becomes an integral over the region on which the tractions act.

Consider a cross section at a distance x from the left end of the bar. If we cut through the cross section, we expose the tractions acting on that face and we see a distribution of forces per unit area from the *stress* $\sigma(x, y, z)$, as shown in Fig. 3.6. Stress represents the internal state of force in a solid body. Traction is stress exposed at a cut and is a vector field, shown with the gray arrows. In general, the stress can vary from point to point in the cross section. The net effect of all the tractions acting on the cross section is called the *stress resultant*. The *resultant axial force* $N(x)$ is defined as

$$N(x) = \int_A \sigma(x, y, z)\, dA \tag{3.2}$$

where the integration is over the area of the cross section. One reason that the resultant is useful comes from realizing that we will eventually sum forces to establish equilibrium. When we compute the resultant axial force, we have done the sum over the region of the cross section and denoted it as $N(x)$. Thus, for any free body diagram with a cut at x, we can just add $N(x)$ to the other forces and we have taken care of the tractions on the area representing the cut. It is interesting to note that we can also view this issue in reverse to realize that the details of $\sigma(x, y, z)$ at the cut do not matter to the overall equilibrium beyond what the resultant captures. Hence, we can solve axial bar problems by finding $N(x)$ without knowing the details of the distribution of tractions that it comes from.

One of the main restrictions of the axial bar problem is that the tractions on a cross section have no net moment about the x-axis. Let the moment arm of a typical particle in the cross section be $\mathbf{r} = y\,\mathbf{e}_2 + z\,\mathbf{e}_3$, as shown in Fig. 3.7. The net moment caused by the tractions, relative to the axis of the bar, is

$$\mathbf{M} = \int_A \mathbf{r} \times \sigma \mathbf{e}_1\, dA = \int_A (z\,\mathbf{e}_2 - y\,\mathbf{e}_3)\sigma\, dA = \mathbf{0}$$

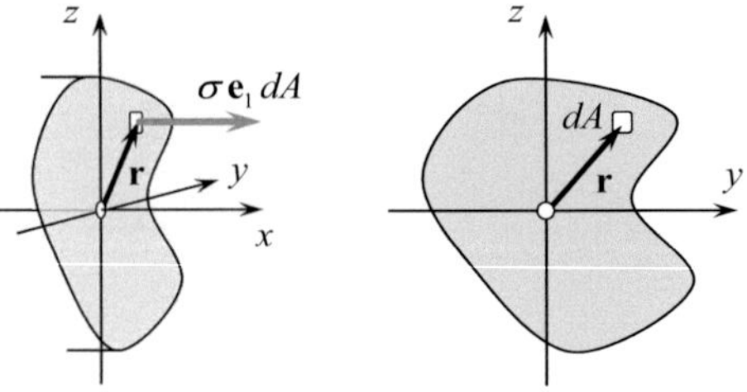

Fig. 3.7 *What about moment?* The net moment of the tractions acting on the cross section is the sum of the contributions $\mathbf{r} \times \sigma \mathbf{e}_1 dA$ of the traction on each particle in the cross section

Consider the case where σ is constant throughout the cross section. If there is no net moment, then

$$\int_A y \, dA = 0, \qquad \int_A z \, dA = 0$$

which is the definition of the *centroid* of the cross section. Thus, the requirement of having zero net moment in an axial bar boils down to the selection of the location of the x-axis. In Chap. 7 we will lift this restriction, and moments will be an important part of the response of the system.

3.1.3 Boundary Conditions

Boundary conditions are statements regarding what is known about the motion or internal force at the *ends* of the bar. For the axial bar problem, there are exactly two boundary conditions, with one at each end. The nature of the boundary conditions is generally conveyed through a symbol on the sketch, as shown in Fig. 3.8. A bar end is either *fixed* or *free* to displace. The fixed end is shown with a shaded block and the free end with the absence of the shaded block. It is possible to apply a known concentrated force at a free end, as illustrated in Fig. 3.2. The force at a fixed end is a reaction force.

At a fixed end, we know that the motion is constrained to be zero. That means $u(\bullet) = 0$, where $\bullet$ can be either 0 or L. At a free end, we know something about the internal axial force $N(x)$. If there is no concentrated applied force at the end, then the boundary condition is $N(\bullet) = 0$, where $\bullet$ can be either 0 or L. If there is a concentrated force at a free end, then the internal axial force is not zero there. The actual condition for the internal force can be derived through an equilibrium argument, as the following example shows.

(a) (c)

(b) (d)

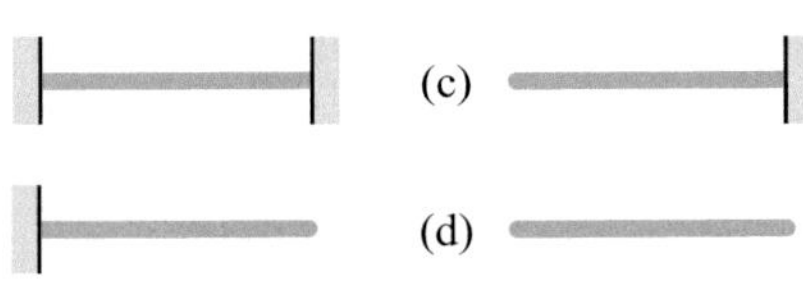

Fig. 3.8 *Boundary conditions.* Either the motion or the force must be specified at each end of the bar. The four possible cases are **a** fixed-fixed, **b** fixed-free, **c** free-fixed, and **d** free-free

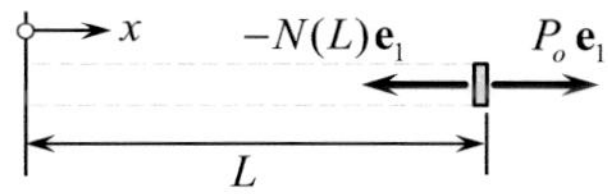

Fig. 3.9 *Example.* Free body diagram of a small piece at the right end of the bar for the problem illustrated in Fig. 3.2

Example 3.1
What are the boundary conditions associated with the problem depicted in Fig. 3.2?

Solution. The bar is *fixed* at the left end and subjected to a concentrated load P_o at the right end. The fixed end implies that $u(0) = 0$. The right end is free to move but has a concentrated applied force there. If we take a very small free body diagram of the end (to expose the internal axial force, as shown in Fig. 3.9) and establish equilibrium, we find that $N(L) = P_o$.

One of the boundary conditions is a kinematic condition because it is a constraint on the motion $u(x)$. The other is a kinetic condition because it involves the internal force and is established through equilibrium.

3.1.4 Equations of Equilibrium

To establish equilibrium equations for the axial bar, consider Fig. 3.10, which shows a piece of length Δx isolated as a free body diagram. The force acting on the left end of the piece is $-N(x)\mathbf{e}_1$ and the force acting on the right end is $N(x + \Delta x)\mathbf{e}_1$.[1] The distributed applied forces are represented as $p(\xi)\Delta x\, \mathbf{e}_1$, where, by the mean value theorem of calculus, ξ is some value of x between the two ends of the piece. For static equilibrium, the forces on the free body diagram must sum to zero. To wit,

$$N(x + \Delta x)\mathbf{e}_1 - N(x)\mathbf{e}_1 + p(\xi)\Delta x\, \mathbf{e}_1 = \mathbf{0}$$

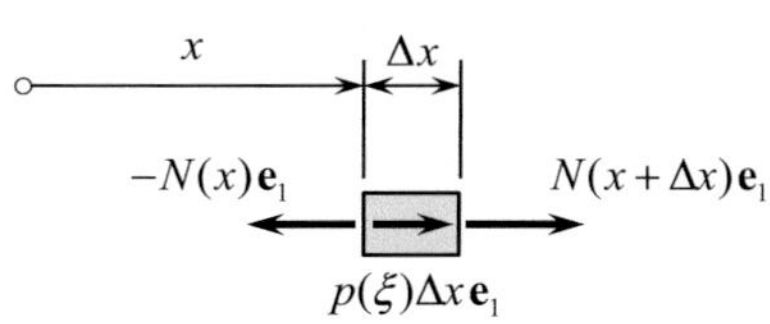

Fig. 3.10 *Free body diagram.* We can find the equation of equilibrium by taking a free body diagram of the small segment of length Δx

[1] When we take a free body diagram with a cut, we will always represent the internal forces according to their positive sign convention. For the axial bar, $N(x)$ is positive in *tension* and negative in *compression*. Hence, we represent the axial force pointing away from the cut. On the right side, the tension force is in the $\mathbf{e}_1$ direction. On the left side, it is in the $-\mathbf{e}_1$ direction.

Taking the dot product of this equation with $\mathbf{e}_1$, dividing through by Δx, and taking the limit as $\Delta x \to 0$ gives

$$\lim_{\Delta x \to 0} \frac{N(x + \Delta x) - N(x)}{\Delta x} + p(\xi) = 0$$

In the limit, the first term becomes the derivative of $N(x)$, the value of ξ in the second term approaches x, and we get the equilibrium equation

$$\frac{dN}{dx} + p(x) = 0 \tag{3.3}$$

The value of x is arbitrary, so this equation holds for all values of x. We call this a *local* equilibrium equation because it holds for a small free body diagram in the local neighborhood of x. The equilibrium equation is a differential equation. It says that the rate of change of the internal axial force $N(x)$ plus the applied distributed force $p(x)$ is equal to zero. Hence, we have an equation that relates the applied forces to the internal stress resultants. If we can find a function $N(x)$ that satisfies this equation and also satisfies any force boundary conditions at the ends of the bar, then *any* free body diagram subsequently cut from the bar will satisfy equilibrium.

Example 3.2
Consider the axial bar of length L, fixed at the left end, free at the right end, and subjected to a constant external axial force $p(x) = p_o$, as shown in Fig. 3.11. Compute the internal axial force $N(x)$ by directly integrating the equation of equilibrium in Eq. 3.3.

Solution. Since the applied force is a known function, we can integrate Eq. 3.3 as follows:

$$N'(x) = -p_o$$
$$N(x) = -p_o x + C$$

where C is the constant of integration. We can evaluate the constant of integration using the condition that the axial force vanishes at $x = L$ (because it is a free end). Using this condition gives

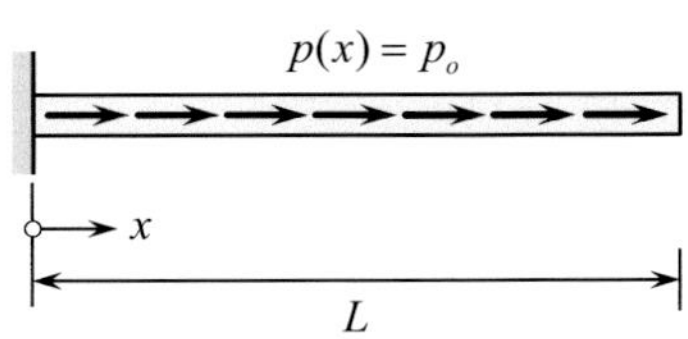

Fig. 3.11 *Example.* Axial bar subjected to a constant uniform external axial force $p(x) = p_o$

$$N(L) = -p_o L + C = 0 \quad \rightarrow \quad C = p_o L$$

Substituting this value back into the general equation gives the final expression for the internal axial force as

$$N(x) = p_o\big(L - x\big)$$

It is interesting to note that we can evaluate $N(0) = p_o L$. This value is equal to the reaction force at the left end. Note also that $N(x)$ is positive over the entire length of the bar, meaning that the bar is in *tension* under the applied loads.

We will often solve our equilibrium problems by taking finite-sized free body diagrams of the bar with a cut at x to isolate the internal axial force $N(x)$, as the following example shows.

Example 3.3
Consider the same problem as the previous example, depicted in Fig. 3.11. Find the internal axial force by taking a free body diagram with a cut at x and keeping the left piece.

Solution. Figure 3.12 shows two free body diagrams of the bar. The one on top is a free body diagram of the entire bar, showing the reaction force $\mathbf{R}$ at the left end and the applied force associated with the typical particle $p(\xi)d\xi\mathbf{e}_1$. We can use this free body diagram to find the reaction force. Establishing force equilibrium, noting that $p(\xi) = p_o$, gives

$$\mathbf{R} + \int_0^L p_o\, d\xi\, \mathbf{e}_1 = \mathbf{0}$$

$$\mathbf{R} + p_o L\, \mathbf{e}_1 = \mathbf{0}$$

Fig. 3.12 *Example.* Axial bar subjected to a constant uniform external axial force $p(x) = p_o$. Free body diagram of the whole bar (top) and free body diagram with cut at x (bottom), keeping the left piece and discarding the right piece

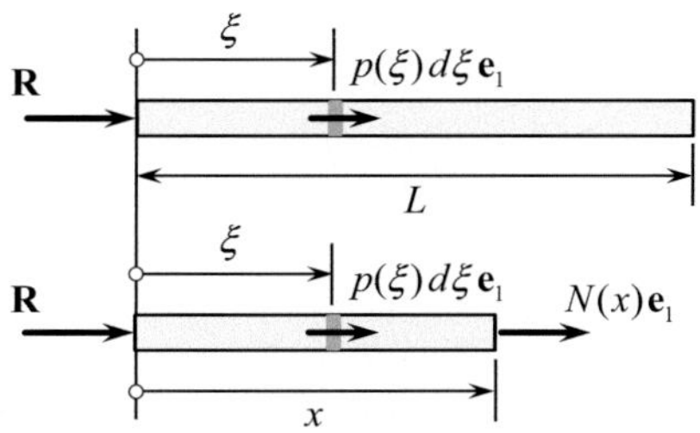

We can solve for the reaction as $\mathbf{R} = -p_o L\mathbf{e}_1$, which tells us that its magnitude is $p_o L$ and it acts in the $-\mathbf{e}_1$ direction. If all the applied forces are pulling to the right, then the reaction force must pull to the left to equilibrate them. The reaction force points along the axis of the bar because the applied forces are all in that direction.

Establishing equilibrium on the second free body diagram gives

$$\begin{aligned} \mathbf{R} + \int_0^x p_o\, d\xi\, \mathbf{e}_1 + N(x)\mathbf{e}_1 &= \mathbf{0} \\ -p_o L\mathbf{e}_1 + p_o x\, \mathbf{e}_1 + N(x)\mathbf{e}_1 &= \mathbf{0} \end{aligned}$$

where we have substituted the reaction force found from the previous calculation. Dotting the equation with $\mathbf{e}_1$ and solving for the internal axial force gives

$$N(x) = p_o\big(L - x\big)$$

which is identical to what we got in the previous example. Observe that in the second free body diagram, the range of the integration is from 0 to x, not 0 to L, as it was in the first free body diagram. When we take the cut at x and remove the right piece, we take the applied load that was on that piece away, too. The effect of the forces on the piece removed are included in the force $N(x)$ at the cut.

These two examples show the two basic ways of dealing with equilibrium. The first example solved the differential equation governing $N(x)$, while the second example determined $N(x)$ from a free body diagram, integrating the applied load over the region of the free body diagram. These two approaches are flip sides of the same coin. The second example shows that we cannot see the internal axial force until we take a cut and expose it. When we take a cut at x, we have two choices of free body diagrams. The following example shows what happens when we keep the other piece as our free body diagram.

Example 3.4
Let us take one last look at the problem depicted in Fig. 3.11. Find the internal axial force by taking a cut at x, keeping the right piece as the free body diagram.

Solution. Figure 3.13 shows a free body diagram of the bar with a cut at x. This time we will keep the right piece and throw away the left piece. The applied force associated with the typical particle is shown along with the internal axial force $N(x)$, which is acting in the $-\mathbf{e}_1$ direction (i.e., the direction of the arrow). We are using the convention that when we represent $N(x)$ on a free

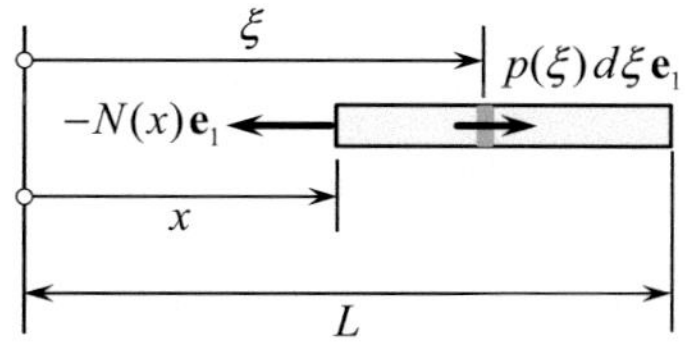

Fig. 3.13 *Example.* Axial bar problem from Fig. 3.11. Free body diagram with cut at x, keeping the right piece and discarding the left piece

body diagram, we draw it using the positive sign convention (i.e., axial force is positive in tension and, therefore, points away from the cut). Establishing force equilibrium gives

$$\int_x^L p_o\,d\xi\,\mathbf{e}_1 - N(x)\mathbf{e}_1 = \mathbf{0}$$
$$p_o\big(L - x\big)\,\mathbf{e}_1 - N(x)\mathbf{e}_1 = \mathbf{0}$$

Dotting the equation with $\mathbf{e}_1$ and solving for the internal axial force gives

$$N(x) = p_o\big(L - x\big)$$

which is identical to what we got in the previous examples. Observe that the range of the integration is from x to L because that is the region associated with the free body diagram. We do not change the definition of x, which is still measured from the left end of the bar (even when the left end is not part of the free body diagram). It is also interesting to note that the reaction force does not appear in this free body diagram. Hence, it is possible to find the internal axial force without first finding the reaction force at the left end for this problem.

3.1.5 Elastic Constitutive Model

The final piece of the axial bar puzzle is the *constitutive model*, the relationship between stress and strain. The simplest model of material behavior is Hooke's law, illustrated in Fig. 3.14. For the axial bar, which is in a state of uniaxial stress and strain, Hooke's law can be stated as

$$\sigma = E\varepsilon \tag{3.4}$$

In other words, the stress at any point is proportional to the strain at that point and the constant of proportionality is Young's modulus E. This material property represents the *stiffness* of the material (not the strength). The elastic modulus tells us how much

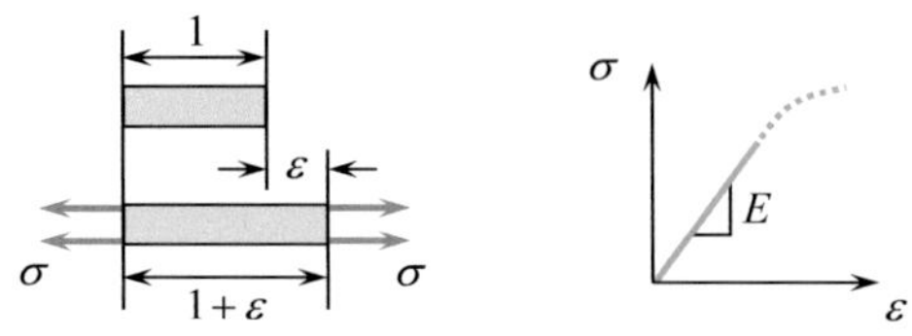

Fig. 3.14 *Hooke's law.* Hooke's law for uniaxial tension and compression. Stress is proportional to strain. The constant of proportionality is E called *Young's modulus*

Table 3.1 *Young's modulus and density for various materials.*

Type of material	Young's modulus (GPa)	Density (kg/m^3)
Silicon carbide	450	3200
Steel	200	7800
Titanium	115	4500
Aluminum	68	2700
Concrete (high strength)	30–50	3500
Concrete (normal)	15–40	2400
Wood (Douglas fir)	13	530
Wood (yellow pine)	9	420

the material will stretch if subjected to tension forces. It does not tell us how much force is required to break the material or to cause it to yield (if it is a ductile metal). The *strength* of the material σ_o is a material property independent of the elastic stiffness. In simplest terms, if $\sigma = \sigma_o$, then the material will fail. It is not possible to have a stress greater than the strength of the material, i.e., $\sigma > \sigma_o$, because the bar will break when it reaches the strength limit.

As Fig. 3.14 suggests, eventually the response ceases to be linear and stress is no longer proportional to strain. The reasons for this departure from linearity depend upon the material. Steel, for example, yields, and inelastic deformations occur due to changes in the crystalline structure of the material. Concrete cracks if it is in tension and crushes in compression. The mechanics behind nonlinear material response, including material failure, is complex and mostly beyond the scope of this book (see Chap. 12 for a brief introduction to these ideas). Our focus is on elastic behavior. Typical values for the elastic modulus and density of various materials are given in Table 3.1.

3.1.6 Axial Force-Displacement Relationship

We can connect the net axial force to the displacement through the following derivation:

$$\begin{aligned} N(x) &= \int_A \sigma \, dA \\ &= \int_A E\varepsilon \, dA \\ &= \int_A Eu'(x) \, dA \\ &= \int_A E \, dA \, u'(x) \end{aligned}$$

The derivation starts with the definition of the stress resultant $N(x)$. In the next step, we substitute the constitutive equation $\sigma = E\varepsilon$. We then substitute the strain-displacement relationship $\varepsilon = u'(x)$, where the prime indicates derivative with respect to x. Finally, we pull $u'(x)$ out of the integral over the area because it does not depend on either y or z. The term in parentheses is called the *effective stiffness* of the cross section. We will designate it as

$$\overline{EA} = \int_A E \, dA \tag{3.5}$$

Often, the elastic modulus is constant over the cross section and can be pulled out of the integral. In such a case, $\overline{EA} = EA$, i.e., the elastic modulus E multiplied by the cross-sectional area A. The relationship between the internal axial force and the displacement is

$$u'(x) = \frac{N(x)}{\overline{EA}} \tag{3.6}$$

This differential equation gives a way to find the displacement function $u(x)$.

3.1.7 Stress Formula

The kinematic hypothesis is the key to the kinematics of motion and the strain-displacement relationship. It also provides a route to finding a formula for the internal stress distribution from the internal axial force. Recall that the internal axial force is the *resultant* of the stresses acting on a cross section (see Fig. 3.6). In other words, $N(x)$ is the aggregate effect of the field $\sigma(x, y, z)$ at x. In general, one cannot determine a distribution from its net effect. For example, if one knows all the individual scores on an exam, one can compute the class average. However, if one knows only the class average, it is not possible to determine individual scores from it. More information is required. In this case, the kinematic hypothesis provides the extra information needed to determine stress from its net effect.

Start the derivation with the constitutive relationship

$$\sigma(x, y, z) = E(y, z)\varepsilon(x)$$

The strain only depends upon x because the kinematic hypothesis forces all fibers in the bar to strain by the same amount. This equation keeps open the possibility that Young's modulus might vary throughout the cross section. We can substitute $\varepsilon = u'$ and then use Eq. 3.6 in place of u' to get

$$\sigma(x, y, z) = \frac{E(y, z)N(x)}{\overline{EA}} \tag{3.7}$$

If E is constant over the cross section, then $\overline{EA} = EA$ and the E in the numerator cancels the one in the denominator, leaving

$$\sigma = \frac{N}{A} \tag{3.8}$$

In other words, the stress is the net axial force divided by the area of the cross section and is constant over the cross section. This relationship is called a *stress formula*. It is important to realize that we can compute the stress from the net axial force only because of the kinematic hypothesis. The derivation above is the proof.

If the elastic modulus is not constant, then the stresses at a cross section are also not constant, as the following example shows.

Example 3.5
Consider the square cross section with variable modulus shown in Fig. 3.15. This is called a *sandwich construction*. The outside layers are made of a stiffer material, with modulus $3E_o$, than the core, which has modulus E_o, where E_o is a reference value. Find the effective modulus $\overline{EA}$ and the stress if the axial force is $N = 10\,\text{kN}$.

Solution. The effective elastic modulus can be determined from Eq. 3.5 as

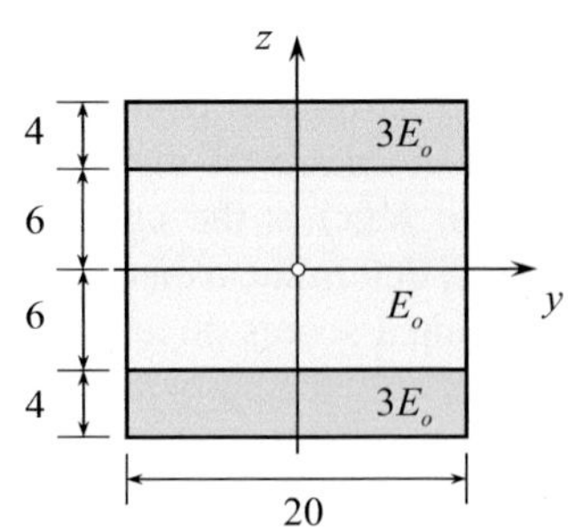

Fig. 3.15 *Example.* Square cross section with variable elastic modulus. The center part has modulus E_o and the outer parts have modulus $3E_o$. Dimensions are in millimeters

$$
\begin{aligned}
\overline{EA} &= \int_A E(z)\,dA \\
&= \int_{-10}^{-6} (3E_o)\,20\,dz + \int_{-6}^{6} (E_o)\,20\,dz + \int_{6}^{10} (3E_o)\,20\,dz \\
&= 240E_o + 240E_o + 240E_o = 720E_o
\end{aligned}
$$

where we have used $dA = 20\,dz$. The stress distribution is

$$
\sigma(z) = \frac{E(z)N}{\overline{EA}} = \begin{cases} (3E_o)(10\,\text{kN})/(720\,\text{mm}^2\,E_o) & 6 \le z \le 10 \\ (E_o)(10\,\text{kN})/(720\,\text{mm}^2\,E_o) & -6 \le z \le 6 \\ (3E_o)(10\,\text{kN})/(720\,\text{mm}^2\,E_o) & -10 \le z \le -6 \end{cases}
$$

Note that $720\,\text{mm}^2 = 0.720 \times 10^{-3}\,\text{m}^2$. Thus, the stress reduces to

$$
\sigma(z) = \begin{cases} 41{,}667\,\text{kPa} & 6 \le z \le 10 \\ 13{,}889\,\text{kPa} & -6 \le z \le 6 \\ 41{,}667\,\text{kPa} & -10 \le z \le -6 \end{cases}
$$

as shown in Fig. 3.16. The unit of stress is $1\,\text{kPa} = 1\,\text{kN/m}^2$. The reference value E_o does not show up in the final expression for the stress. Only the relative stiffnesses of the parts affect the stress.

Observe that the stresses in the regions with larger elastic modulus are greater than the stresses in the region with lower elastic modulus. This result is a direct consequence of the kinematic hypothesis. Because all fibers in the cross section must strain by the same amount, and because the stress is Young's modulus times the strain, the resulting stresses are proportional to their respective moduli.

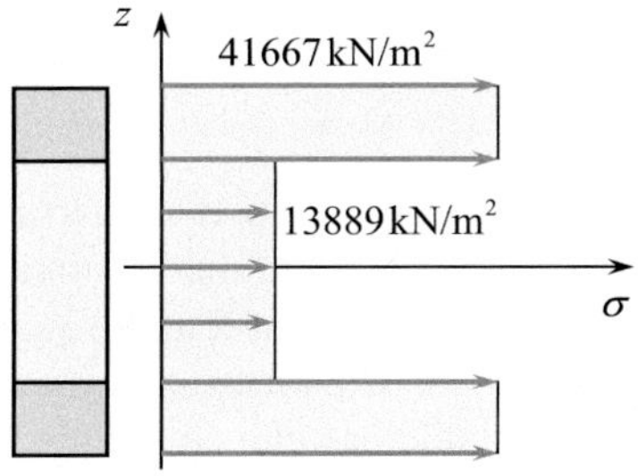

Fig. 3.16 *Example.* Stress distribution on a cross section with variable modulus. The stresses in the regions with higher modulus are greater than the stresses in the region with lower modulus

3.2 Solving Axial Bar Problems

In this section we will show how to solve axial bar problems by walking through some specific examples. These problems illustrate a range of important issues, including problems with static indeterminacy and nonprismatic bars.

We can find $N(x)$ either by solving Eq. 3.3 or by establishing equilibrium on a finite-sized free body diagram with a cut at x, which exposes $N(x)$. We can then substitute that expression into Eq. 3.6 and directly integrate the differential equation, as the following example illustrates.

Example 3.6
Consider the fixed-free bar shown in Fig. 3.11. Find the displacement function $u(x)$. The internal axial force was determined in previous examples to be $N(x) = p_o\big(L - x\big)$. How much longer does the bar get?

Solution. We can determine the axial displacement function by solving the differential equation Eq. 3.6. This is a separable differential equation. Hence, we can integrate it as follows:

$$u'(x) = \frac{p_o\big(L - x\big)}{EA}$$
$$u(x) = \frac{p_o}{EA}\big(Lx - \tfrac{1}{2}x^2\big) + C$$

The constant of integration C can be determined from the boundary condition $u(0) = 0$. To wit,

$$u(0) = \frac{p_o}{EA}\big(L(0) - \tfrac{1}{2}(0)^2\big) + C = 0$$

We can solve this equation to get $C = 0$. Substituting this value into the general expression gives the final result for the axial displacement:

$$u(x) = \frac{p_o}{EA}\big(Lx - \tfrac{1}{2}x^2\big)$$

It is always a good idea to check the units. First, p_o has units of force per length. The modulus has units of force per area or force per length squared. Area has units of length squared. Hence, the effective modulus EA has units of force. The terms in parentheses all have units of length squared. Thus, the right side has units of length, which is appropriate for the displacement $u(x)$.

The displacement at the right end of the bar is $u(L) = p_o L^2/2EA$. Since the bar was fixed at the left end, this amount is how much longer the bar is due to the applied loads.

3.2.1 Static Indeterminacy

A structure is called *statically determinate* if the reactions and internal forces can be determined with the equations of static equilibrium alone. A structure for which this is not possible is called *statically indeterminate*. In this section we will show how the complete set of equations including equilibrium, strain-displacement, Hooke's law, and boundary conditions provide precisely the needed information to solve statically indeterminate problems.

In the previous example, the internal axial force $N(x)$ was completely determined from the equations of statics. That function was then used in Eq. 3.6 to solve for the displacement $u(x)$. For the axial bar problem, the fixed-fixed boundary conditions lead to a statically indeterminate system (see Fig. 3.8a). In that case, it is not possible to uniquely determine the reaction forces from equations of static equilibrium alone. There are two unknown reactions and only one equation of equilibrium for the free body diagram of the entire bar. Each time we take a cut, we get one new equation, but we also get one new unknown, i.e., $N(x)$. Hence, there is no statics strategy that will allow us to find both reaction forces.

To solve a statically indeterminate problem, we need to use the equations of kinematics along with the boundary conditions on displacement to find the additional reaction force. So far, the boundary condition on displacement was used to compute the constant of integration that comes from solving the differential equation Eq. 3.6. In statically indeterminate problems, the number of boundary conditions on displacement is greater than the statically determinate case. For the fixed-fixed axial bar, there are *two* displacement boundary conditions. The additional condition allows us to find the unknown reaction force, as the following example demonstrates.

Example 3.7
Consider the fixed-fixed bar of length L subjected to a linearly increasing load with magnitude p_o at the right end, shown in Fig. 3.17 (on the left). The effective modulus of the bar is EA. Determine the reaction forces, the internal axial force $N(x)$, and the axial displacement $u(x)$.

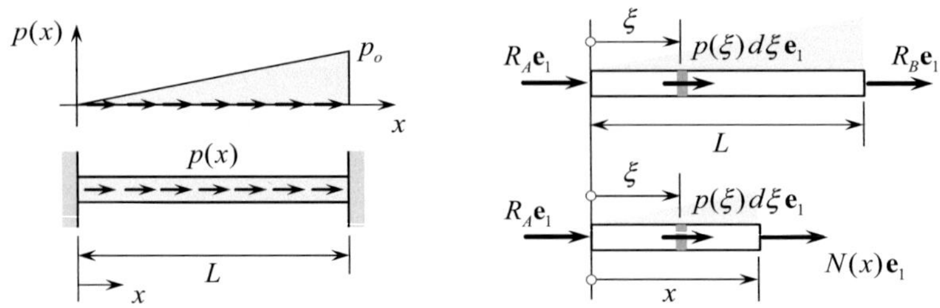

Fig. 3.17 *Example.* Fixed-fixed bar of length L with linearly distributed applied load of magnitude p_o at the right end. The effective modulus of the bar is EA and the magnitude of the load at the right end is p_o. The sketch on the left is the problem definition (showing the load distribution). The sketches on the right are free body diagrams

Solution. We need a mathematical expression for the loading function that represents what the picture of the loading indicates. It is a linear function, i.e., $p(x) = ax + b$. We need to determine the values of a and b for this particular case. We know that $p(0) = 0$ and $p(L) = p_o$. Substituting these two points into the equation gives two equations in the unknowns a and b. Solving these equations and substituting back into the original expression gives

$$p(x) = p_o \frac{x}{L}$$

If we need $p(\xi)$, we will just substitute ξ for x in this equation.

We can establish equilibrium of the whole bar using the free body diagram of the entire system (the top right diagram in Fig. 3.17). To wit,

$$\begin{aligned} R_A \mathbf{e}_1 + \int_0^L p_o \frac{\xi}{L} \mathbf{e}_1 \, d\xi + R_B \mathbf{e}_1 &= \mathbf{0} \\ R_A + \tfrac{1}{2} p_o L + R_B &= 0 \end{aligned}$$

In the second step, we have dotted the equation with $\mathbf{e}_1$ and evaluated the integral. This equation relates the two unknown reaction forces R_A and R_B, but this equation is not adequate to uniquely determine both. We can establish equilibrium using the second free body diagram as follows:

$$\begin{aligned} R_A \mathbf{e}_1 + \int_0^x p_o \frac{\xi}{L} \mathbf{e}_1 \, d\xi + N(x) \mathbf{e}_1 &= \mathbf{0} \\ R_A + \tfrac{1}{2} p_o \frac{x^2}{L} + N(x) &= 0 \end{aligned}$$

With this equation, we can solve for the internal axial force $N(x)$ as

$$N(x) = -R_A - \tfrac{1}{2}p_o\frac{x^2}{L}$$

We have determined $N(x)$ in terms of the reaction force R_A. We use $N(x)$ in the strain-displacement equation, which we will integrate to get the displacement. Having R_A unknown will not be a problem for integrating the differential equation because while it is unknown, it is constant. Thus, we can compute

$$u'(x) = \frac{1}{EA}\Big(-R_A - \tfrac{1}{2}p_o\frac{x^2}{L}\Big)$$
$$u(x) = \frac{1}{EA}\Big(-R_A x - \tfrac{1}{6}p_o\frac{x^3}{L}\Big) + C$$

The final equation has two unknowns: the constant of integration C and the reaction force R_A. We have two boundary conditions $u(0)=0$ and $u(L)=0$. The first gives

$$u(0) = \frac{1}{EA}\Big(-R_A(0) - \tfrac{1}{6}p_o\frac{(0)^3}{L}\Big) + C = 0$$

which we can solve to get $C=0$. The second condition gives

$$u(L) = \frac{1}{EA}\Big(-R_A(L) - \tfrac{1}{6}p_o\frac{(L)^3}{L}\Big) = 0$$

where we have already noted that $C=0$. Solve this equation for the reaction force

$$R_A = -\tfrac{1}{6}p_o L$$

Now, with the reaction force known, we can complete the solution by substituting it back to find R_B, $N(x)$, and $u(x)$. To wit,

$$R_B = -\tfrac{1}{3}p_o L$$
$$N(x) = p_o\Big(\tfrac{1}{6}L - \tfrac{1}{2}\frac{x^2}{L}\Big)$$
$$u(x) = \frac{p_o}{EA}\Big(\tfrac{1}{6}Lx - \tfrac{1}{6}\frac{x^3}{L}\Big)$$

Once you have the solution, it is a good idea to differentiate to verify that $u'(x)=N(x)/EA$ and $N'(x)=-p(x)$ as a check. Also, notice that both reactions are negative, which means that they both point in the $-\mathbf{e}_1$ direction. That makes sense because the applied loads are in the positive $\mathbf{e}_1$ direction and the

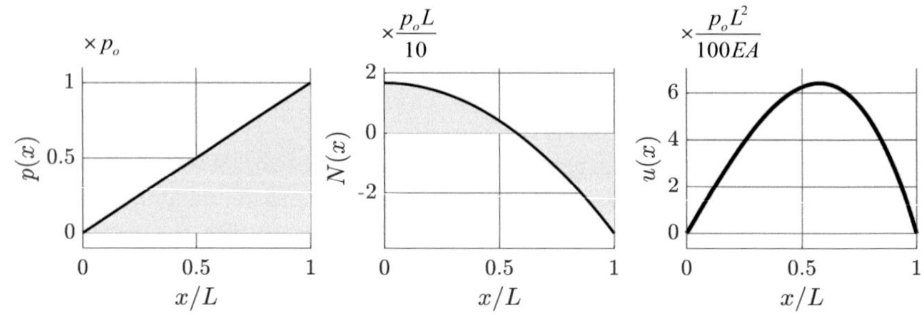

Fig. 3.18 *Example.* Results of the analysis of the fixed-fixed bar shown in Fig. 3.17. Note the scaling of the axes

reactions must resist those loads. The total applied load (the area under the $p(x)$ curve) is $p_o L/2$, equal to the sum of the two reaction forces.

The final stage of any solution is to plot the results. The results are shown in Fig. 3.18. The slope of the $N(x)$ diagram is equal to the negative of the value of the $p(x)$ diagram for every value of x. Since $p(x)$ is positive over the entire length, the slope of the axial force diagram must be negative at every point. The slope is zero at $x = 0$ because $p(0) = 0$, and it gets increasingly negative with the largest slope at the right end. The slope of $u(x)$ is proportional to the value of $N(x)$. Hence, the point where $N = 0$ corresponds to the maximum value of $u(x)$. The location of this maximum can be found by setting the expression for N equal to zero and solving for x. In this case, we get $x = 0.577L$. Finally, observe that the values of the internal axial force at the ends, i.e., $N(0)$ and $N(L)$, are equal to the reaction forces.

This example demonstrates that there is a balance between the unknowns (reactions and constants of integration) and the boundary conditions available to determine them. This fact suggests that there is really no difference between a statically determinate and statically indeterminate problem beyond the nature of the boundary conditions and the need to carry the unknown reaction force all the way through the solution for $u(x)$ to determine it.

3.2.2 Nonprismatic Bars

If the cross section varies along the length of the bar, then it is called *nonprismatic*. There is no change to the preceding derivation of the governing equations except to note that the effective modulus EA is now a function of x. There are two basic kinds of problems that arise for nonprismatic bars: (1) continuous variation of the effective modulus, and (2) prismatic segments with abrupt changes in the effective modulus between them. We will show what is at stake through two examples. The first example shows how the response of the axial bar is affected by a *continuous* variable effective stiffness.

Example 3.8
Let us reconsider the loading and boundary conditions of the previous example, shown in Fig. 3.17, but now with a nonprismatic bar with cross-sectional area that varies according to the equation $A(x) = A_o\big(1 + x/L\big)$. In other words, the area increases linearly from a value A_o at the left end to $2A_o$ at the right end. Find the reaction forces, internal axial force, and displacement.

Solution. First, notice that the free body diagrams and the equations of equilibrium do not change. Hence, we have

$$N(x) = -R_A - \tfrac{1}{2} p_o \frac{x^2}{L}$$

as before. Now, we need to solve for the displacement from Eq. 3.6, which, in this case, takes the form

$$u(x) = \int \frac{-R_A - \frac{1}{2} p_o x^2/L}{EA_o\big(1 + x/L\big)}\, dx + C$$

where C is the constant of integration. It will be convenient to do a change of variable to make progress with the integral. Let $\zeta = 1 + x/L$, which means that $x = L(\zeta - 1)$ and $dx = L\, d\zeta$. Substituting the change of variable gives

$$\begin{aligned} \hat{u}(\zeta) &= -\frac{R_A L}{EA_o} \int \frac{d\zeta}{\zeta} - \frac{p_o L^2}{EA_o} \int \tfrac{1}{2}\Big(\zeta - 2 + \frac{1}{\zeta}\Big)\, d\zeta + C \\ &= -\frac{R_A L}{EA_o} \ln\zeta - \frac{p_o L^2}{EA_o}\Big(\tfrac{1}{4}\zeta^2 - \zeta + \tfrac{1}{2}\ln\zeta\Big) + C \end{aligned}$$

We have renamed the function $\hat{u}(\zeta)$ so as not to confuse which function has which arguments. The two functions satisfy $\hat{u}(\zeta) = u(x)$ when the value of ζ corresponds with the value of x. Before we transform back to the original coordinate x, we can implement the boundary conditions to solve for C and R_A. Note that at $x = 0$, $\zeta = 1$ and at $x = L$, $\zeta = 2$. Thus, we can compute

$$\hat{u}(1) = -\frac{p_o L^2}{EA_o}\Big(\tfrac{1}{4} - 1\Big) + C = 0 \quad \rightarrow \quad C = -\tfrac{3}{4}\frac{p_o L^2}{EA_o}$$

Using the other boundary condition gives

$$\hat{u}(2) = -\frac{R_A L}{EA_o}\ln 2 - \frac{p_o L^2}{EA_o}\Big(1 - 2 + \tfrac{1}{2}\ln 2 + \tfrac{3}{4}\Big) = 0$$

Solving this equation for the reaction force gives

$$R_A = -p_o L\Big(\frac{2\ln 2 - 1}{4\ln 2}\Big) = -0.1393\, p_o L$$

Thus, $R_B = -0.3607\, p_o L$. Contrast these values with $R_A = -0.1667\, p_o L$ and $R_B = -0.3333\, p_o L$ for the uniform fixed-fixed bar. The nonprismatic bar is stiffer at the right end, and that end attracts a larger fraction of the applied load as a result.

We can substitute the reaction force and constant of integration and transform back to the original coordinates to get the final expressions for the internal axial force and axial displacement. To wit,

$$N(\xi) = p_o L\Big(0.1393 - \tfrac{1}{2}\xi^2\Big)$$
$$u(\xi) = \frac{p_o L^2}{EA_o}\Big(\tfrac{1}{2}\xi - \tfrac{1}{4}\xi^2 - \frac{\ln(1+\xi)}{4\ln 2}\Big)$$

where $\xi = x/L$. We can differentiate these functions to prove that they are compatible with the differential equations they come from.

Figure 3.19 compares the axial force and displacement functions for the prismatic and nonprismatic cases. The dotted line is the prismatic case with $A = 1.5A_o$ and the solid line is the nonprismatic case. The reason for using $1.5A_o$ for the area for the prismatic case is to make the comparison between two bars with the same total amount of material in them. The plot of $N(x)$ shows that the nonprismatic case is a constant shift from the prismatic case. The variation of the cross section does not affect the differential equation for $N(x)$ and, therefore, does not change the shape of the curve. However, as we have seen, the distribution of material *does* affect the reaction forces. That is why the plots of $N(x)$ are shifted by a constant amount.

The displacement $u(x)$ is affected by the distribution of the cross section. The thinner part of the bar is more prone to displacement than the thicker part. Thus, the curve is shifted left a bit. Even with the adjustment of the areas, the peak of the nonprismatic case is slightly lower than the prismatic case.

This example serves to illustrate the additional complexity that arises in the nonprismatic bar due to the variable area. All aspects of the solution except for finding the antiderivative of the functions were similar to the example with the prismatic bar.

The next example shows how to solve a problem with an abrupt *discontinuity* in the cross-sectional area. This approach requires the definition of *piecewise functions*. Because discontinuity in the internal axial force caused by point loads also requires the same treatment, we will consider an example that has both forms of discontinuity.

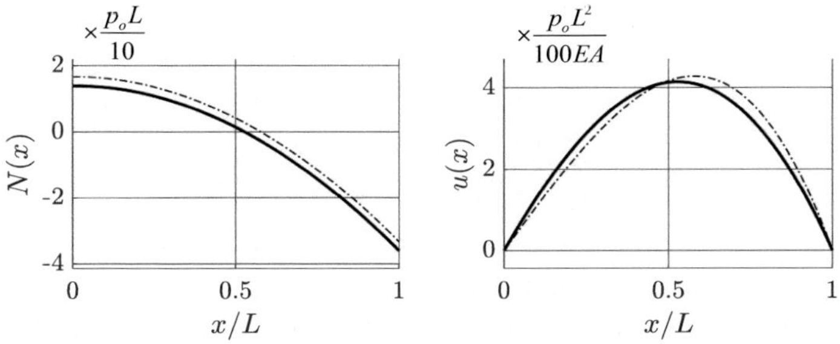

Fig. 3.19 *Example.* Results of the analysis of the fixed-fixed bar shown in Fig. 3.17. The dotted line is the prismatic bar with $A = 1.5A_o$ and the solid line is the nonprismatic bar

Example 3.9
Consider the fixed-fixed bar of length $3L$ subjected to a concentrated force P at the one-third point with an abrupt change in the cross-sectional area from A_o to $2A_o$ at the two-thirds point, shown in Fig. 3.20. Find the internal axial force, displacement, and stress.

Solution. Because there is a point load, the internal axial force $N(x)$ is discontinuous, and because the cross section has an abrupt change, the strain $\varepsilon = u'(x)$ is discontinuous. To deal with these discontinuities, we will divide the bar into three segments and define the internal axial force and displacement with piecewise functions as follows:

$$N(x) = \begin{cases} N_1(x) & 0 \le x \le L \\ N_2(x) & L \le x \le 2L \\ N_3(x) & 2L \le x \le 3L \end{cases}, \qquad u(x) = \begin{cases} u_1(x) & 0 \le x \le L \\ u_2(x) & L \le x \le 2L \\ u_3(x) & 2L \le x \le 3L \end{cases}$$

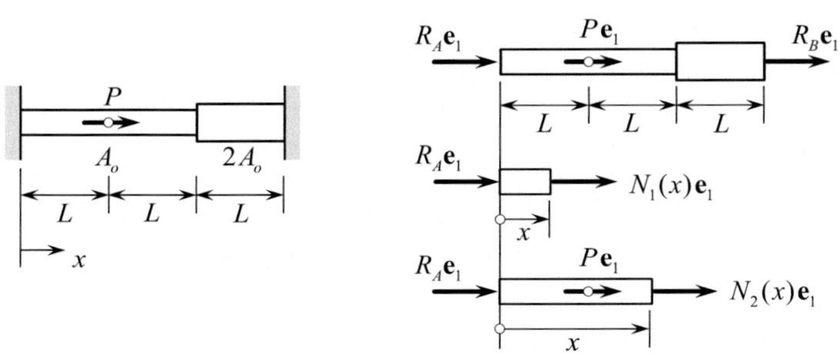

Fig. 3.20 *Example.* Fixed-fixed bar of length $3L$ with point load P and abrupt change in cross-sectional area. The problem definition is on the left. Free body diagrams are on the right

Figure 3.20 shows the free body diagrams needed to solve the problem. The one on top is of the entire bar and is useful for relating the two reaction forces to each other. The second free body diagram has a cut to the left of the point load, thereby defining the component function $N_1(x)$. The bottom free body diagram has a cut to the right of the point load, thereby defining the component functions $N_2(x) = N_3(x)$. The functions are equal because there are no additional forces in the second and third regions that would change equilibrium. Thus, the presence of the discontinuity in the cross section does not affect the equations of equilibrium. You can verify this fact by taking a cut in region three and showing that you get exactly the same internal force as when you take the cut in region two.

Establishing equilibrium of the entire bar results in the equation

$$R_A + P + R_B = 0$$

which we can view as an equation to find R_B once we have determined R_A. This problem is statically indeterminate, so we will carry R_A as an unknown until we can find it using boundary conditions on the displacements. The equilibrium equations that result from the other two free body diagrams determines the internal axial force to be

$$N(x) = \begin{cases} -R_A & 0 \le x \le L \\ -(R_A + P) & L \le x \le 2L \\ -(R_A + P) & 2L \le x \le 3L \end{cases}$$

Again, the internal axial force is known in terms of the unknown constant R_A. The force-displacement relationship must be integrated for each segment. To wit,

$$u_1' = \frac{-R_A}{EA_o} \qquad u_2' = \frac{-(R_A + P)}{EA_o} \qquad u_3' = \frac{-(R_A + P)}{2EA_o}$$

$$u_1 = \frac{-R_A x}{EA_o} + C_1 \quad u_2 = \frac{-(R_A + P)x}{EA_o} + C_2 \quad u_3 = \frac{-(R_A + P)x}{2EA_o} + C_3$$

Each segment has a different constant of integration. Thus, there are four unknown quantities: the three constants of integration plus the reaction force R_A. We have two boundary conditions and two *continuity conditions* associated with the piecewise function $u(x)$. While the derivative of $u(x)$ is discontinuous at the location of the abrupt change in cross section, the displacement itself is continuous. Therefore, we have the following conditions

$$u_1(0) = 0, \qquad u_1(L) = u_2(L), \qquad u_2(2L) = u_3(2L), \qquad u_3(3L) = 0$$

The first and last are the boundary conditions, and the two in the middle are continuity conditions. We do not know the value of the displacement at $x = L$ or $x = 2L$, but we do know that whatever it is, the component functions must give the same value at that point.

Implementing the first boundary condition $u_1(0) = 0$ gives

$$\frac{-R_A(0)}{EA_o} + C_1 = 0, \qquad \rightarrow \qquad C_1 = 0$$

The second condition, $u_1(L) = u_2(L)$, holds if (noting that $C_1 = 0$)

$$\frac{-R_A(L)}{EA_o} = \frac{-(R_A + P)(L)}{EA_o} + C_2, \qquad \rightarrow \qquad C_2 = \frac{PL}{EA_o}$$

The third condition, $u_2(2L) = u_3(2L)$, holds if

$$\frac{-(R_A + P)(2L)}{EA_o} + \frac{PL}{EA_o} = \frac{-(R_A + P)(2L)}{2EA_o} + C_3$$

where we have used the value of C_2 found previously. We can solve this equation to give

$$C_3 = -\frac{R_A L}{EA_o}$$

Finally, the fourth condition $u_3(3L) = 0$, substituting the value of C_3, gives

$$\frac{-(R_A + P)(3L)}{2EA_o} - \frac{R_A L}{EA_o} = 0, \qquad \rightarrow \qquad R_A = -\tfrac{3}{5}P$$

Now that we have all the constants and the reaction force, we can go back and complete the solution for the reaction $R_B = -\frac{2}{5}P$, the internal axial force, and the displacement. To wit,

$$N(x) = \begin{cases} \frac{3}{5}P & 0 \le x \le L \\ -\frac{2}{5}P & L \le x \le 2L \\ -\frac{2}{5}P & 2L \le x \le 3L \end{cases}, \quad u(x) = \frac{P}{EA_o}\begin{cases} \frac{3}{5}x & 0 \le x \le L \\ L - \frac{2}{5}x & L \le x \le 2L \\ \frac{3}{5}L - \frac{1}{5}x & 2L \le x \le 3L \end{cases}$$

The results are shown in Fig. 3.21. Note that the stress is the axial force divided by the area. Since the third segment has an area twice as large as the second one, the stress is lower even though the internal axial force is the same. The displacement is linear within each segment. The displacement is continuous, but the derivative of displacement (i.e., the strain) is discontinuous. In reality, the stress and strain fields are complicated in the region near an abrupt change

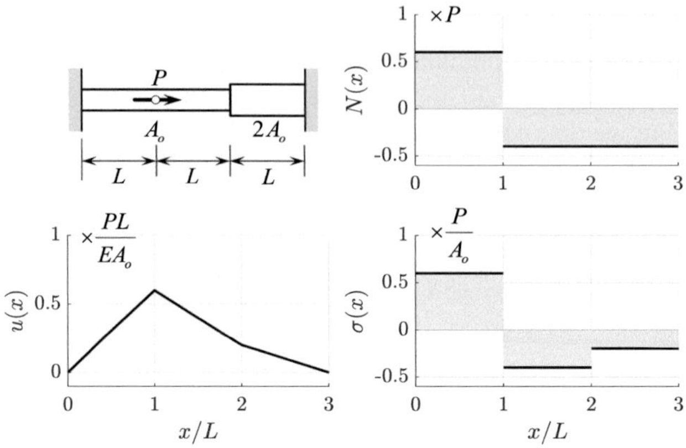

Fig. 3.21 *Example.* Fixed-fixed bar of length $3L$ with point load P and abrupt change in cross-sectional area. Plots of internal axial force $N(x)$, axial stress $\sigma(x)$, and displacement $u(x)$

in cross section and depend upon the precise nature of the geometry. In practice, a design that has a smooth transition will better avoid stress concentrations.

3.3 Computational Strategy

In this section we outline a computational strategy that will solve a broad range of prismatic axial bar problems with any boundary conditions and any load distribution. The resulting code `AxialBar` will provide a means of exploring the effects of loading and boundary conditions on the distribution of internal axial force, reactions, and displacement. We will omit concentrated loads and cross-section discontinuities, as those will be discussed in detail in Chap. 10 on computational strategies for beams. This section will set the stage for that more complex problem.

3.3.1 Formulation and Code

Recall that the differential equations we need to solve to determine the internal axial force and displacement are

$$N'(x) + p(x) = 0, \qquad u'(x) = \frac{N(x)}{EA}$$

Integrating the first equation gives

$$N(x) = N_o - \int_0^x p(\xi)\, d\xi \tag{3.9}$$

where the constant of integration is N_o and ξ is a dummy variable of integration. To understand why $N_o = N(0)$, just plug $x = 0$ into Eq. 3.9 and observe that the integral has the same upper and lower limits and hence evaluates to zero.

The displacement equation can be integrated to give

$$u(x) = u_o + \int_0^x \frac{N(\xi)}{EA}\, d\xi \tag{3.10}$$

where $u_o = u(0)$ is the constant of integration. Since we have already solved for $N(x)$ in Eq. 3.9, we can substitute it into Eq. 3.10. To wit,

$$u(x) = u_o + \frac{1}{EA}\int_0^x \left(N_o - \int_0^\xi p(\eta)\, d\eta\right) d\xi \tag{3.11}$$

Notice that we use a different dummy variable η in the integral to avoid confusion with ξ, which is in the upper limit of that integral. The term with N_o can be integrated directly because N_o is constant. The double integral can be simplified using integration by parts to give

$$\int_0^x \int_0^\xi p(\eta)\, d\eta\, d\xi = \int_0^x (x - \xi)\, p(\xi)\, d\xi \tag{3.12}$$

To see why this equation is true, define the functions $v(\xi)$ and $w(\xi)$ as follows:

$$v(\xi) = \xi, \qquad w(\xi) = \int_0^\xi p(\eta)\, d\eta$$

Observe that by the product rule for differentiation, $(vw)' = v'w + vw'$. Also, we can compute $w' = p(\xi)$ and $v' = 1$ from the definition of the functions. Now, we can integrate from 0 to x as follows:

$$\begin{aligned} \int_0^x (vw)'\, d\xi &= \int_0^x v'w\, d\xi + \int_0^x vw'\, d\xi \\ \Big[v(\xi)w(\xi)\Big]_0^x &= \int_0^x v'(\xi)w(\xi)\, d\xi + \int_0^x v(\xi)w'(\xi)\, d\xi \end{aligned}$$

The left side is the integral of an exact differential. We can substitute the original functions, noting that

$$v(0) = 0, \qquad v(x) = x, \qquad w(0) = 0, \qquad w(x) = \int_0^x p(\xi)\, d\xi$$

to give the final expression

$$x \int_0^x p(\xi)\, d\xi = \int_0^x \int_0^\xi p(\eta)\, d\eta\, d\xi + \int_0^x \xi\, p(\xi)\, d\xi$$

Rearranging terms, noting that x can be moved inside the integral on the left side since the dummy variable of integration is ξ, gives the final result in Eq. 3.12. This proof shows very clearly why we need to be careful in the naming of the dummy variable of integration.

Finally, we can write Eq. 3.11 as

$$u(x) = u_o + \frac{N_o\, x}{EA} - \frac{1}{EA} \int_0^x (x - \xi)\, p(\xi)\, d\xi \tag{3.13}$$

The strategy that we will adopt to solve for $u(x)$ and $N(x)$ numerically will be to evaluate the two state variables at the ends of the bar and then numerically integrate with the generalized trapezoidal rule from the left end to the right end, starting with the known values at the left end. The numerical values can then be plotted. In essence, we are converting the boundary value problem to an initial value problem and using what is often called a *shooting method* to create the plots.

Computing the end states In order to carry out this strategy, we will need to be able to compute u_o and N_o for any boundary conditions. Equations 3.9 and 3.13 can be evaluated at $x = L$ to give

$$\begin{aligned} N_L &= N_o - \int_0^L p(\xi)\, d\xi \\ u_L &= u_o + \frac{N_o\, L}{EA} - \frac{1}{EA} \int_0^L (L - \xi)\, p(\xi)\, d\xi \end{aligned}$$

where we have introduced the notation $N_L = N(L)$ and $u_L = u(L)$ to represent the values of the axial force and displacement at the right end of the bar. Let us also give names to the two definite integrals that appear in these equations:

$$\mathcal{I}_0 = \int_0^L p(\xi)\, d\xi \qquad \mathcal{I}_1 = \frac{1}{EA} \int_0^L (L - \xi)\, p(\xi)\, d\xi$$

In the code, we will compute these integrals numerically to make it easy to include a library of loading functions. We can now write our two equations in matrix form as

$$\begin{bmatrix} 1 & 0 & -1 & 0 \\ d & 1 & 0 & -1 \end{bmatrix} \begin{Bmatrix} N_o \\ u_o \\ N_L \\ u_L \end{Bmatrix} = \begin{Bmatrix} \mathcal{I}_0 \\ \mathcal{I}_1 \end{Bmatrix} \tag{3.14}$$

where $d = L/EA$. This system of two equations relates the four end-state variables $\{N_o, u_o, N_L, u_L\}$. For any set of boundary conditions, two of the end-state variables are known, one at each end of the bar. For example, the fixed-fixed bar has the boundary conditions $u_o = 0$ and $u_L = 0$. If we substitute those values into Eq. 3.14, it reduces to

$$\begin{bmatrix} 1 & -1 \\ d & 0 \end{bmatrix} \begin{Bmatrix} N_o \\ N_L \end{Bmatrix} = \begin{Bmatrix} \mathcal{I}_0 \\ \mathcal{I}_1 \end{Bmatrix} \tag{3.15}$$

because, by the rules of matrix multiplication, the zeros multiply quantities in their associated column in the coefficient matrix and give zero. This step amounts to simply deleting the columns of the coefficient matrix associated with the variables that are known to be equal to zero. The result is a 2×2 system of equations that can be solved for the unknown end-state variables.

There are four boundary conditions possible: fixed-fixed, fixed-free, free-fixed, and free-free. Each one of those cases has different prescribed end-state variables. There is a simple way to enforce the boundary conditions in MATLAB code, as we will soon see. Let us define the arrays

$$\mathbf{B} = \begin{bmatrix} 1 & 0 & -1 & 0 \\ d & 1 & 0 & -1 \end{bmatrix} \qquad \mathbf{z} = \begin{Bmatrix} \mathcal{I}_0 \\ \mathcal{I}_1 \end{Bmatrix}$$

Let $\mathbf{C}$ be the 2×2 matrix that results from deleting the appropriate columns of $\mathbf{B}$. We can write a simple few lines of code that will allow us to get $\mathbf{C}$ directly from $\mathbf{B}$ because all we need to do is pick out the correct columns of $\mathbf{B}$. We can accomplish the task by defining an index array `Var` with four entries (one for each of the four end-state variables). An entry of 1 means keep that column, and an entry of 0 means do not keep it (i.e., that column is associated with one of the known zero values). For the example with fixed ends, the array would look like:

```
Var = [ 1, 0, 1, 0 ];
```

Now, with one line of MATLAB code, we can compute `C` from `B` as

```
C = B(:,Var==1);
```

In MATLAB, that line means take all rows (the colon in the first slot of the `B` array) and take any column that is associated with a value of 1 in the array `Var`. In this case, it picks out the first and third columns of `B`. Now, all we need to do is to articulate the four cases of the `Var` matrix in accord with the four possible boundary conditions. Code 3.1 will show a convenient way to do that.

Aside. There are generally multiple ways to formulate any given problem. As an alternative to the approach mentioned above, we could augment the two equations with the boundary condition equations to give, for the fixed-fixed example, the system of equations

$$\begin{bmatrix} 1 & 0 & -1 & 0 \\ d & 1 & 0 & -1 \\ 0 & 1 & 0 & 0 \\ 0 & 0 & 0 & 1 \end{bmatrix} \begin{Bmatrix} N_o \\ u_o \\ N_L \\ u_L \end{Bmatrix} = \begin{Bmatrix} \mathcal{I}_0 \\ \mathcal{I}_1 \\ 0 \\ 0 \end{Bmatrix} \tag{3.16}$$

Observe that the third equation is $u_o = 0$ and the fourth is $u_L = 0$, which are the boundary conditions for the fixed-fixed bar. Solving these equations will give the same result as the preceding strategy.

Computing the functions To plot the functions $N(x)$ and $u(x)$, we will generate an adequate number of discrete points along the curve by numerical integration and then connect the dots. Now that we have determined all the state variables at the ends of the bar, we can integrate the differential equations numerically, as if this were an initial value problem. We will use the generalized trapezoidal rule to carry out the integration. The basic framework for numerical integration by the generalized trapezoidal rule is outlined in Appendix A.3.

The recursive numerical integration equations for this case are

$$\begin{aligned} N_{n+1} &= N_n - \Delta x\big[\beta\, p_n + \big(1-\beta\big)\, p_{n+1}\big] \\ u_{n+1} &= u_n + \frac{\Delta x}{EA}\big[\beta\, N_n + \big(1-\beta\big)\, N_{n+1}\big] \end{aligned}$$

where Δx is the distance between adjacent points along the bar, β is the generalized trapezoidal rule parameter (the best value is 0.5), and the subscripts on the state variables and load indicate the location, e.g., $N_n = N(x_n)$. We can get this algorithm started because we know N_o and u_o from the previous analysis and we can compute $p_0 = p(0)$.

If we carry the numerical integration from 0 to L, then we will get state variables at $x = L$. The version of the values of $N(L)$ and $u(L)$ computed from the generalized trapezoidal rule can be compared to the values N_L and u_L, obtained from the end-states problem. The difference between these values is a measure of accuracy of the generalized trapezoidal rule integration. If the values are not the same, then the step length Δx should be reduced. Note also that the accuracy of the evaluation of $\mathcal{I}_0$ and $\mathcal{I}_1$ will affect this comparison. If those values are computed numerically, then that integration can also be a source of error and, again, the solution is to use more integration points.

Implementation The equations developed in the preceding section can be organized into a MATLAB program. Code 3.1 solves the prismatic axial bar problem for any boundary conditions and any distributed loading.

Code 3.1 MATLAB code `AxialBar` to solve the axial bar problem

```
   clear; clc;

%. Physical parameteters
   L = 4;                              % Bar length
   EA = 20;                            % Bar axial stiffness
   po = 0.3;                           % Load magnitude

%. Boundary conditions =1 fixed, =2 free
   BCL = 1;                            % Left BC
   BCR = 1;                            % Right BC

%. Establish loading function
   LoadType = 4;                       % =0 Menu input, >0 specify type
   [~,~,LoadType] = LoadFunction(0,L,po,LoadType);

%. Compute z=[I0,I1] by numerical integration
   nPts = 21;                          % Number of Simpson points
   [wt,st,nPts] = NumInt(3,nPts);      % Simpson weights and stations
   z = [0; 0];
   for i=1:nPts
     x = st(i)*L;
     [p,~,~] = LoadFunction(x,L,po,LoadType);
     z = z + p*L*[ 1; (L-x)/EA ]*wt(i);
   end

%. BC:      N  u        =0 for prescribed value, =1 unknown value
   BCs = [ 1, 0   ;     % fixed
           0, 1 ];      % free

   Var = [BCs(BCL,:), BCs(BCR,:)];

%. Construct Matrix B
   d = L/EA;
   B = [ 1,    0, -1,   0 ;...
         d,    1,  0,  -1 ];

%. Compute C matrix by selecting columns of B
   C = B(:,Var==1);

%. Compute missing end state values s, place results in f
   s = C\z;
   f = zeros(4,1);
   f(Var==1) = s;

%. Compute N and u by numerical integration
   nGTR = 51;                          % Number of GTR points
   beta = 0.5;                         % GTR parameter
   dx = L/(nGTR-1);                    % Step size
   Store = zeros(nGTR,4);              % Storage for results

%. Initialize
   x = 0;
   Nold = f(1);
   Uold = f(2);
   [Pold,~,~] = LoadFunction(x,L,po,LoadType);

   for i=1:nGTR
     Store(i,:) = [x, Uold, Nold, Pold];
```

```
%... Compute new state
     x = x + dx;
     [Pnew,~,~] = LoadFunction(x,L,po,LoadType);
     Nnew = Nold - dx*(beta*Pold + (1-beta)*Pnew);
     Unew = Uold + dx*(beta*Nold + (1-beta)*Nnew)/EA;

%... Update
     Nold = Nnew;
     Uold = Unew;
     Pold = Pnew;

   end

%. Output
   AxialBarEcho(L,EA,po,BCL,BCR,LoadType,nPts,nGTR,beta);
   AxialBarOutput(Store,L,BCL,BCR,f,z)
```

First, the physical inputs of length `L`, axial modulus `EA`, and load magnitude `po` are established. We specify the boundary conditions at the left and right ends with `BCL` and `BCR`, respectively. If those values are set to 1, it indicates a fixed end; if set to 2, it indicates a free end. These values are used later in the code to select the appropriate row of `BCs` when forming the `Var` array. Code 3.2, `LoadFunction`, takes care of establishing a library of loading functions. If `LoadType` is initially set to zero, then the selection of the loading function is from an interactive menu. If `LoadType` is greater than zero, then that value is the load type. It should be evident that new load distributions can be added to this function very easily. Note that any additional parameters (e.g., the end points for a patch load) associated with a particular loading are set in `LoadFunction`.

The code computes $\mathbf{z} = (\mathcal{I}_0, \mathcal{I}_1)$ by composite Simpson integration. The call to Code A.2, `NumInt`, discussed in Appendix A.4, establishes the weights and stations for the numerical integration scheme. The call `NumInt(3,nPts)` returns the `nPts` weights, `wt`, and stations, `st`, for the composite Simpson rule. The 3 is the key for the composite Simpson rule (1 gives Gauss-Legendre, 2 gives Gauss-Lobatto, and 4 gives composite trapezoidal rule). Simpson integration is used in the code to establish the equations for the end-state variables. The weights and stations for numerical integration from `NumInt` are for $\xi \in [0, 1]$. Thus, the integrals are done using the change of variable $\xi = x/L$, which is where the extra factor of L comes from in the computation of `z`.

Next, the code sets up the `Var` array using the boundary condition codes `BCL` and `BCR` for the boundary conditions at the left and right ends, respectively. A value of 1 means a *fixed* end and a value of 2 means a *free* end. The boundary conditions can be changed by changing these values. The strategy is to define an array `BCs` whose first row is associated with a fixed end and second row with a free end. Then, `Var` is created by taking the appropriate row from `BCs` in accord with the boundary condition codes.

The end state values are computed by first establishing the **B** matrix, reducing it to **C** by eliminating the columns associated with prescribed values, and solving the resulting equations. Those values are then stored in the appropriate place in **f**, which contains all the end state values in the prescribed order. The reason this is important is becausc we will always take the first one to be the axial force at the left end and the second to be the displacement at the left end as a starting point for the integration by generalized trapezoidal rule.

To generate points along the curve for plotting the response, we use the generalized trapezoidal rule with `nGTR` integration points. The most accurate integrator uses $\beta = 0.5$, but any value between zero and one is valid. The variables are initialized to their values at the left end of the bar, and the integration proceeds forward from left to right. To compute the new state variables, we only need the values from the previous step. Hence, we operate with `old` and `new` values to represent steps n and $n+1$. The values are stored in the array `Store` for later plotting.

The user establishes the number of integration points `nPts` and `nGTR`. There must be enough points to give accurate results. Both numerical integrators must be accurate. For smooth loading functions, the number of Simpson integration points can be relatively few. The number of generalized trapezoidal points will then be determined by how smooth you want the plot to be, which may be more than needed to accurately integrate the equations. For functions that change rapidly or are confined to small patches, these numerical integration schemes do not perform as well. One approach is to take lots of integration points. The easiest way to tell if the numerical integration is accurate is to examine the values of $N(L)$ and $u(L)$ that come from the generalized trapezoidal rule and compare them with the actual values dictated by the boundary conditions. If they do not match, then the integration is not accurate.

The last two lines of the code control the output. The codes for these functions are not included here. The function `AxialBarEcho` simply writes the input values to the command window for a concise summary. The function `AxialBarOutput` writes some computed results to the command window (e.g., $\mathcal{I}_0$, $\mathcal{I}_1$, the values of N and u at the end states, and the values of $N(L)$ and $u(L)$ that come from the generalized trapezoidal rule as a check on accuracy of integration) and generates the graphics shown in the example that follows.

Code 3.2 `LoadFunction` defines a library of loading functions that can be called by the main program. The input is the specific value of x where the function value is needed, along with the length of the bar, the load magnitude, and the type of load function. The function passes back the value of the load at x along with the name of the load type and the load type index (which is set if the input is interactive).

Code 3.2 MATLAB code to define loading functions

```
function [p,LoadType,Type] = LoadFunction(x,L,po,Type)

%. Library of loading functions
   LoadCase = {'Constant';...
               'Linear ramp up';...
               'Linear ramp down';...
               'Half sinusoid';...
               'Sinusoidal patch load';...
               'Trapezoidal load';...
               'Cosine ramp up';...
               'Quarter sinusoidal load';...
               'Quarter cosinusoidal load';...
               'Self-equilibrating load'};
```

```
%. Select load type by menu
   if Type==0
     msg = 'Select Load Function';
     Type = listdlg('PromptString',msg,'ListString',LoadCase);
   end
   LoadType = char(LoadCase(Type));

%. Convert x to relative value on range [0,1]
   s = x/L;

%. Evaluate load
   switch Type

     case 1  % Constant load
       p = 1;

     case 2  % Linear ramp up
       p = s;

     case 3  % Linear ramp down
       p = 1 - s;

     case 4  % Sinusoidal load
       p = sin(pi*s);

     case 5  % Sinusoidal patch load
       a = 0.4; b = 0.6;
       p = 0;
       if (s > a) && (s < b)
          p = 1 - cos(2*pi*(s-a)/(b-a));
       end

     case 6  % Trapezoidal load
       p1 = 0.5; p2 = 1.5;
       p =  p1*(1-s) + p2*s;

    case 7  % Cosine ramp up
       a = 0.3; b = 0.7;
       if s <= a
          p = 0;
       elseif (s > a) && (s < b)
          p = (1 - cos(pi*(s-a)/(b-a)))/2;
       else
          p = 1;
       end

    case 8  % Quarter sinusoidal load
       p = sin(pi*s/2);

    case 9  % Quarter cosinusoidal load
       p = cos(pi*s/2);

    case 10 % Self-equilibrating load
       p = cos(2*pi*s);

   end % switch

%. Scale load
   p = p*po;
```

The current version of Code 3.2 has some common distributed load forms. Additional distributed loads can be added as new cases in the switch statement. The parameters associated with the specific load distributions are set in this function. Those values can also be changed by the user. For example, the patch loading case 5

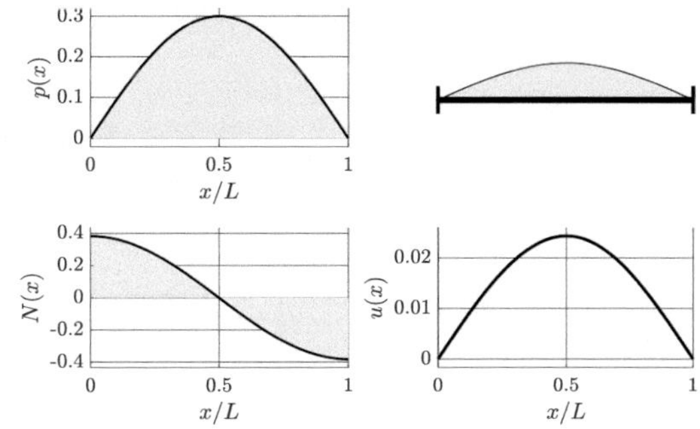

Fig. 3.22 *Example.* Fixed-fixed bar of length $L = 4$ m and axial modulus $EA = 20$ kN, subjected to a half-sine load of magnitude $p_o = 0.3$ kN/m. Plots of applied load $p(x)$, axial force $N(x)$, and displacement $u(x)$

has a load distributed between $x = a$ and $x = b$ with zero value outside of that range. Moving a and b will spread out or sharpen the distribution.

Example 3.10
Compute the internal axial force and displacement for a fixed-fixed bar of length $L = 4$ m and axial modulus $EA = 20$ kN, subjected to a load of magnitude $p_o = 0.3$ kN/m.

Solution. The results from the code are shown in Fig. 3.22. The computation uses `nPts=21` Simpson points and `nGTR=51` generalized trapezoidal rule points (with $\beta = 0.5$). Observe that the maximum displacement occurs at the location where the internal axial force is zero. The boundary conditions on $u(x)$ appear to be met visually, but $u(L) = 2.54 \times 10^{-5}$ from the generalized trapezoidal rule calculation, the difference being due to the error in numerical integration.

3.3.2 Study: What Is a Point Load?

In mechanics we often make use of a concentrated load acting at a point. We call these *point loads*, but they do not exist in nature. All forces in nature are distributed over an area or volume. The concept of point load, then, is a mathematical approximation to a distributed load. One might imagine that the quality of the approximation would be better if the region over which the distributed load acted was small. In this section we use the `AxialBar` code to study this issue.

Consider the fixed-fixed axial bar of length L and modulus EA, subjected to a point load of magnitude P acting at midspan, shown in Fig. 3.23 (bottom sketch). The distributed load $p(x)$ shown in the top figure can be represented by the function

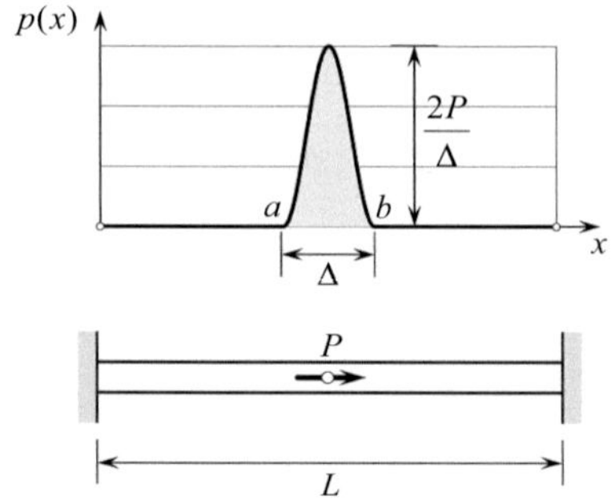

Fig. 3.23 *Point load study.* Fixed-fixed bar of length L and axial modulus EA, subjected to a concentrated load P at midspan or a distributed load $p(x)$ distributed over a patch in the neighborhood of midspan. The shaded area under the distributed load is equal to P

$$p(x) = \frac{P}{\Delta}\left[1 - \cos\left(2\pi\frac{x-a}{\Delta}\right)\right]$$

where $\Delta = b - a$ is the distance between the start of the non-zero part of the function at a and the end at b. The peak value of $p(x)$ occurs at $x = L/2$ and is equal to $2P/\Delta$. The area under the curve is P. Hence, the total force represented by the distributed load is P, which you can demonstrate by integrating the function $p(x)$ from a to b. In some sense, then, the distributed load is similar to the point load.

The exact solution of the problem for the point load is

$$N(x) = \frac{P}{2}\begin{cases} 1 & 0 \le x \le \frac{1}{2}L \\ -1 & \frac{1}{2}L \le x \le L \end{cases}, \quad u(x) = \frac{P}{2EA}\begin{cases} x & 0 \le x \le \frac{1}{2}L \\ L - x & \frac{1}{2}L \le x \le L \end{cases}$$

Thus, the internal axial force $N(x)$ is piecewise constant with a jump equal to P at $x = L/2$. The displacement ramps up linearly to $PL/4EA$ and then ramps back down linearly to zero. The results for the distributed load (generated by the `AxialBar` code) are shown in Fig. 3.24 for values of $\Delta = 0.1, 0.2, 0.4, 0.6, 0.8$, and 1.0. From the plot of $p(x)$, it is clear that the smaller the region on which the distributed force acts, the larger the magnitude needs to be to maintain the area equal to P (which is taken to be one in this example).

The internal axial force $N(x)$ starts and ends at 0.5 and -0.5, respectively, in all cases. That makes sense because the internal axial force must be equal to the reaction forces at the ends of the bar. Since the total load is the same in all cases, the reaction forces are the same. For the broadly distributed loads, $N(x)$ makes a smooth and gradual transition from the top value to the bottom value. As the loading becomes more concentrated, the response curves come closer and closer to the solution for the point load. The displacement $u(x)$ shows the same trends. For the value $\Delta = 0.1$, shown with a thicker line in the plots, the distributed load response is almost indistinguishable from the point load response.

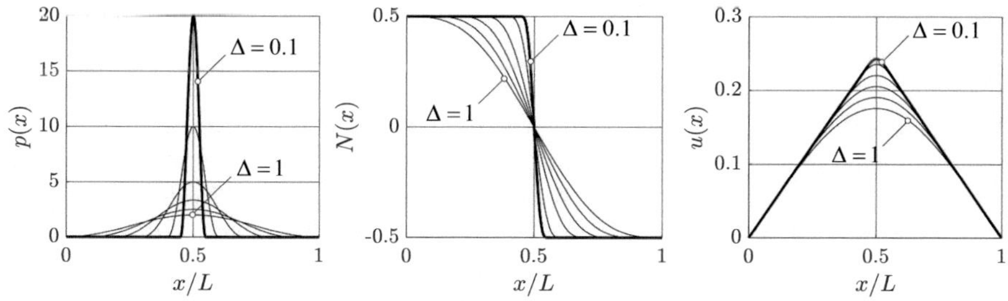

Fig. 3.24 *Point load study.* Results for the patch loading defined in Fig. 3.23 for values of $\Delta = 0.1, 0.2, 0.4, 0.6, 0.8$, and 1.0. The load is $P = 1$, the length is $L = 1$, and the axial stiffness is $EA = 1$

This example shows the difference between a point load and a distributed load. Through this study one can begin to appreciate why using the artifice of point loads is justified. To create the plots for this study, the code was modified by making `AxialBar` into a function that is called from a main program. The call can then be put into a loop so the plots can accumulate in the same figure. For each case, the number of integration points is made inversely proportional to Δ to ensure that the results are accurate for even the most peaked case.

3.4 Summary

This chapter represents the official launch of our study of the mechanics of deformable bodies. While the axial bar itself has practical applications (e.g., foundation piles, trusses, reinforcement bars in concrete, etc.), the main purpose was to lay out the elements of the theory—the three legs of the mechanics stool (equilibrium, constitution, and kinematics)—in the simplest possible mathematical context. Equilibrium provides a way to relate externally applied loads to the internal axial force. Kinematics hinges on the *kinematic hypothesis* and results in a differential equation relating strain and displacement. The force and deformation sides of the picture connect to form a complete theory through the constitutive model, which relates stress to strain. The chapter outlined strategies for solving axial bar problems, including a computational context that completely solves an entire class of problems.

The remainder of the book builds on the concepts introduced here, in some cases explaining why certain assumptions (e.g., the particular form of Hooke's law used for the axial bar) were appropriate, and in other cases (e.g., beam bending and torsion) showing how to extend the ideas to more general and practically useful engineering problems.

3.5 Problems

3.1 Bar $\mathcal{AB}$ has length L and is fixed at the left end and free at the right end. A triangular axial load of magnitude p_o (force per unit length) in the middle is applied along the bar. The bar has a square cross section with area A and Young's Modulus of E. Determine the internal axial force and displacement as a function of x. Express results in terms of E, L, A and p_o. Plot the results. Neglect the weight of the bar.

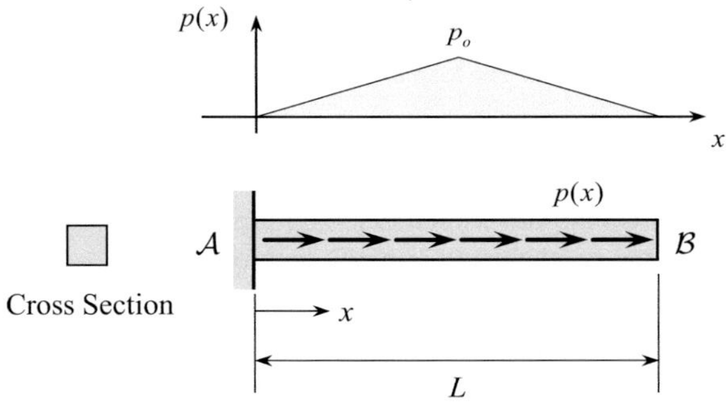

3.2 A hollow circular post $\mathcal{ABC}$ supports a load P_o acting at $\mathcal{C}$, and a load $3P_o$ is acting at $\mathcal{B}$. The length, centerline diameter, and thickness of bar $\mathcal{AB}$ are L_o, d_o, and $0.2d_o$, respectively. The length, centerline diameter, and thickness of bar $\mathcal{BC}$ are $0.75L_o$, $0.7d_o$, and $0.1d_o$, respectively. Calculate the internal axial force $N(x)$, the stress $\sigma(x)$, and displacement $u(x)$ throughout the post. Young's modulus E is the same for both parts. Plot the response fields. Neglect the weight of the bar.

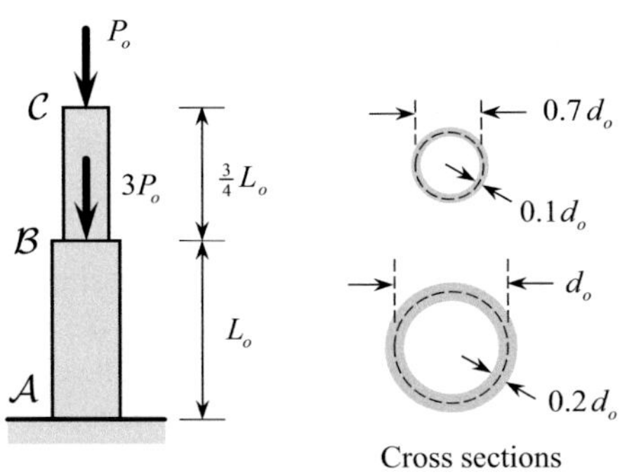

3.3 The post $\mathcal{ABCD}$ consists of three prismatic segments with circular cross sections, all with Young's modulus E. The left segment has a cross-sectional area A_o and length $2L$. The middle segment has a cross-sectional area of $3A_o$ and length L. The right segment has a cross-sectional area $2A_o$ and length $1.5L$. A load of $2P$ is applied at point $\mathcal{B}$ and a load of P is applied at point $\mathcal{C}$, in the directions shown. Determine the axial stress field and displacement field for the bar in terms of E, A_o, L and P. Plot the net axial force distribution and displacement as a function of distance from the left end. Neglect the weight of the bar.

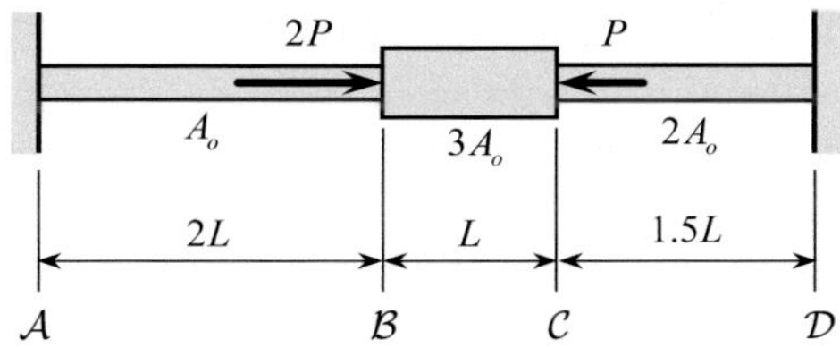

3.4 Bars $\mathcal{AB}$ and $\mathcal{BC}$ are welded together at point $\mathcal{B}$. The system is fixed at both ends ($\mathcal{A}$ and $\mathcal{C}$) and subjected to a concentrated axial force P applied at point $\mathcal{B}$. Both bars have length L and Young's modulus E. Bar $\mathcal{AB}$ has cross-sectional area A_o, and bar $\mathcal{BC}$ has cross-sectional area $2A_o$. Calculate the reaction forces. Calculate and plot the axial force $N(x)$, the stress $\sigma(x)$, and displacement $u(x)$ along the bars. Express the results in terms of P, L, E, and A_o. Neglect the weight of the bar.

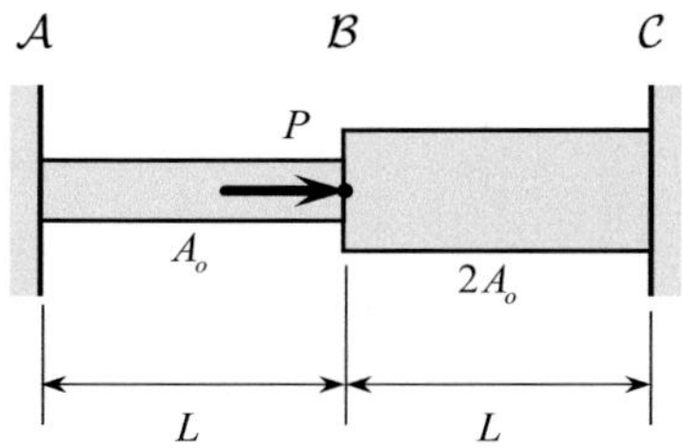

3.5 Bar $\mathcal{AB}$ has a circular cross section, and the radius varies linearly from R at $\mathcal{A}$ to $2R$ at $\mathcal{B}$. The bar is fixed at both ends and subjected to a uniformly distributed axial force of magnitude p_o along its length. The bar has length L and Young's modulus E. Calculate the reaction forces. Calculate the axial force $N(x)$, the stress $\sigma(x)$, and displacement $u(x)$ along the bar. Find the maximum displacement. Express the results in terms of p_o, L, E, and R. Plot the axial force, stress, and displacement. Neglect the weight of the bar.

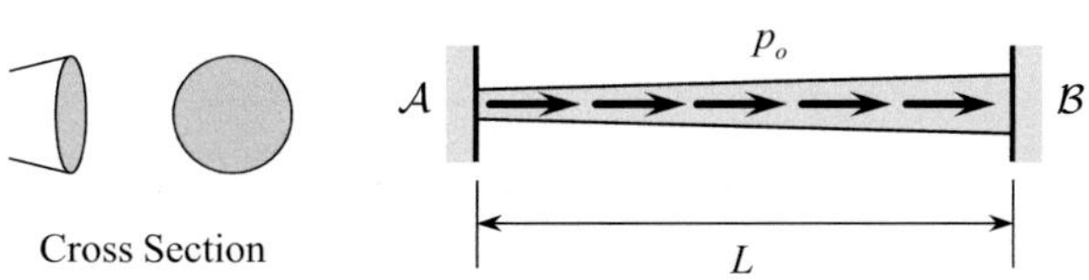

3.6 The prismatic bar $\mathcal{AB}$ is loaded with a uniformly distributed axial load of magnitude p_o along its entire length L. The bar is fixed at both ends, has Young's modulus E, and cross-sectional area A. Determine the internal axial force and the axial displacement as a function of x. Express results in terms of E, A, L, and p_o. Plot the internal axial force distribution and displacement. Neglect the weight of the bar.

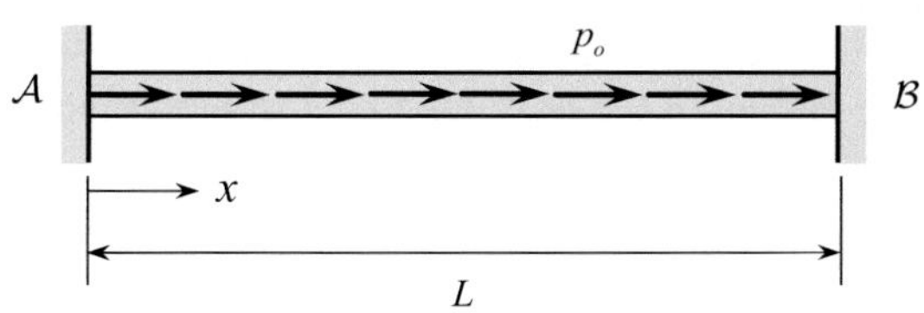

3.7 Bar $\mathcal{AB}$ has length L and is fixed at the left end and free at the right end. A linearly decreasing axial load with magnitude p_o (force per unit length) at the left end is applied along the length of the bar. The bar has a uniform circular cross section with area A and Young's Modulus E. Determine the internal axial force and displacement as a function of x. Express results in terms of E, L, A, and p_o. Plot the results. Neglect the weight of the bar.

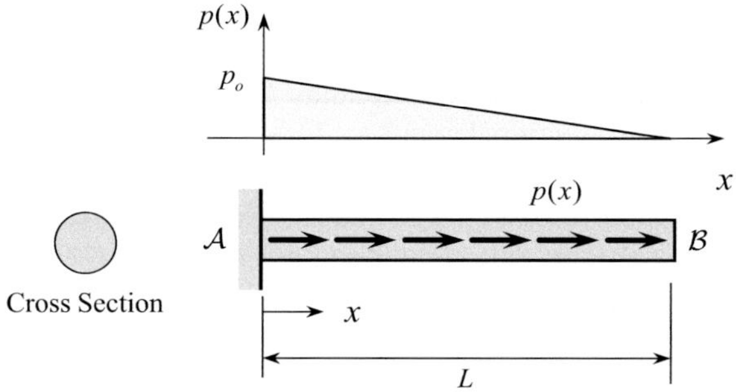

3.8 Bar $\mathcal{ABC}$ consists of segment $\mathcal{AB}$ of cross-sectional area A_o and segment $\mathcal{BC}$ of cross-sectional area $2A_o$. Segment $\mathcal{BC}$ is subjected to a linearly increasing axial load with magnitude p_o at point $\mathcal{B}$ and $2p_o$ at end $\mathcal{C}$. The system is fixed at end $\mathcal{A}$ and subjected to a concentrated axial force of magnitude $3p_oL$ at point $\mathcal{C}$, acting as shown. Both bars have length L and Young's modulus E. Calculate the reaction forces. Calculate the axial force $N(x)$, the stress $\sigma(x)$, strain $\varepsilon(x)$, and displacement $u(x)$ along the bars. Express the results in terms of p_o, L, E, and A_o. Plot the axial force, stress, and displacement. Find the maximum displacement. Neglect the weight of the bar.

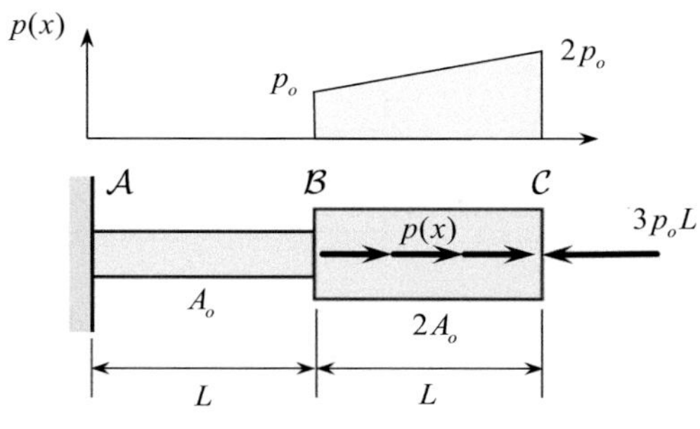

3.9 Post $\mathcal{AB}$ with square cross section is fixed at the base and supports its own weight. The side dimensions of the bar vary linearly along its height L, with dimension b_o at the top and $1.5b_o$ at the base. The weight per unit volume of the material in the post is w. Derive the expression for the net internal axial force N as a function of distance from the bottom of the post due to the weight. Find the displacement at the top.

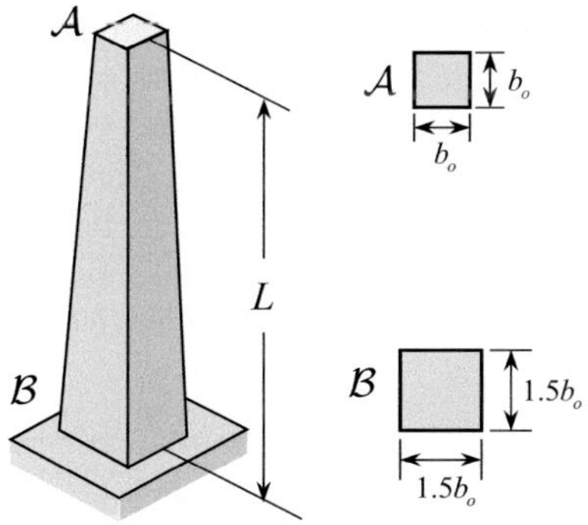

3.10 Bar $\mathcal{ABC}$ consists of segment $\mathcal{AB}$ of cross-sectional area $2A_o$ and segment $\mathcal{BC}$ of cross-sectional area A_o. The bar is subjected to a linearly decreasing axial load with magnitude p_o at $\mathcal{A}$ going to zero at $\mathcal{C}$. The bar is fixed at $\mathcal{A}$. Both bar segments have length L and Young's modulus E. Calculate the reaction force, the axial force $N(x)$, the stress $\sigma(x)$, and displacement $u(x)$. Express the results in terms of p_o, L, E, and A_o. Plot the axial force, stress, and displacement. Find the maximum displacement. Neglect the weight of the bar.

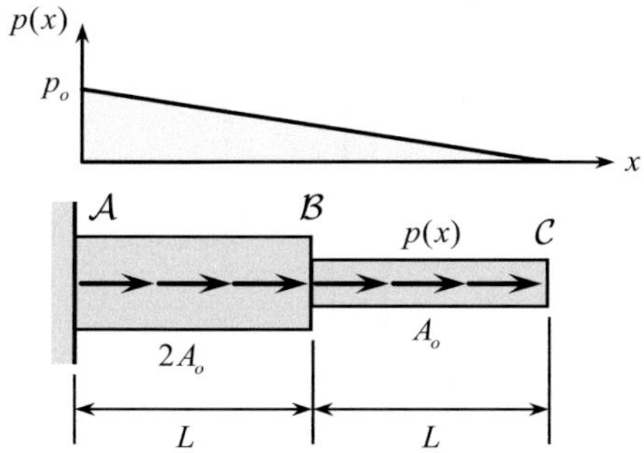

3.11 The bar $\mathcal{ABCD}$ consists of segments $\mathcal{AB}$ and $\mathcal{CD}$ with circular cross section of radius $0.5R$ and segment $\mathcal{BC}$ with circular cross section of radius R. The system is loaded with concentrated loads with magnitudes and directions shown on the sketch at points $\mathcal{A}$, $\mathcal{B}$, $\mathcal{C}$, and $\mathcal{D}$. If all bars have length L and Young's modulus E, determine the displacement of end $\mathcal{D}$ with respect to end $\mathcal{A}$. Express your results in terms of P, L, E, and R. Plot the axial force, stress, and displacement. Neglect the weight of the bar.

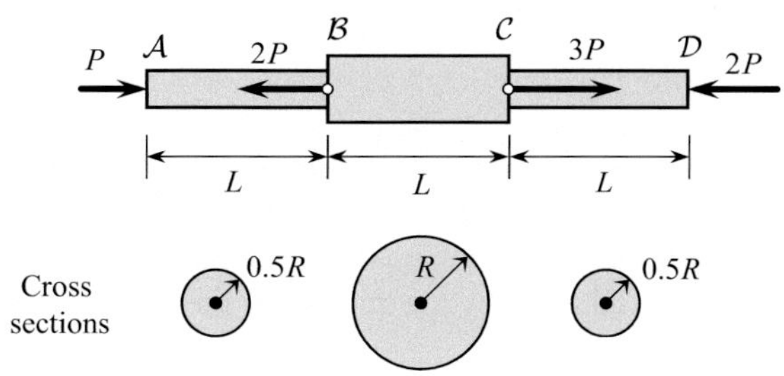

3.12 Post $\mathcal{ABC}$ consists of two segments, each with length L and Young's Modulus of E. Segment $\mathcal{AB}$ is a truncated cone with cross section radius varying linearly, beginning with a radius of R and transitioning to $0.5R$. Segment $\mathcal{BC}$ is loaded with a constant distributed force around its exterior of τ_o (force per unit area oriented in the axial direction) and has a constant circular cross section with a radius of $0.5R$. The post is fixed at end $\mathcal{A}$. Determine the internal axial force and displacement along the bar. Express results in terms of E, L, R, and τ_o. Neglect the weight of the bar.

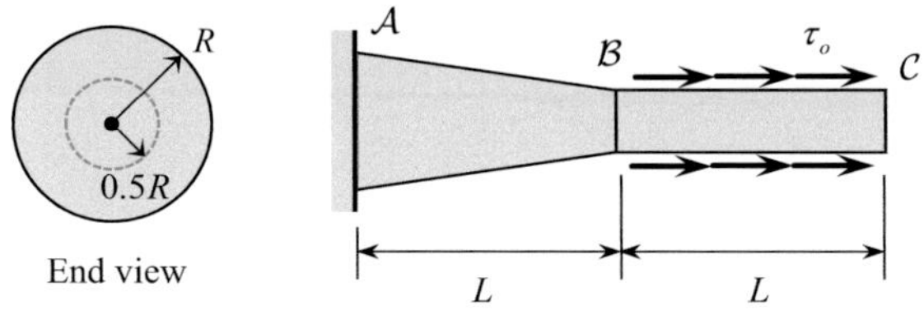

3.13 Post $\mathcal{ABC}$ consists of two segments of different length. Segment $\mathcal{AB}$ has a mass density of ρ, a cross-sectional area of $2A_o$, and Young's Modulus of E. Segment $\mathcal{BC}$ has a mass density of ρ, a cross-sectional area of A_o, and Young's Modulus of E. The post is fixed at both ends. Determine the internal axial force, normal stress, and displacement along the length of the bar. Express results in terms of E, A_o, L, and ρ.

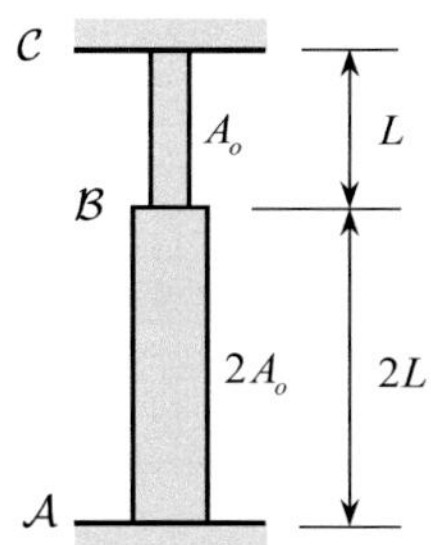

3.14 *Computation.* Modify Code 3.2, `LoadFunction`, to have start and end values of the distributed loads (like a and b for the patch loads) passed in as arguments in the function call rather than specifying them directly in as lines of code in the function. Convert all cases to be patch loads (note that $a = 0$ and $b = L$ are possible inputs). For example, `case 2` is the linear ramp up. Make it ramp from zero to its maximum between $x = a$ and $x = b$. Add some additional load cases to the function to increase the library of available loads.

3.15 *Computation.* Modify Code 3.1, `AxialBar`, to include point loads along the length of the bar. Create a function similar to Code 3.2, `LoadFunction`, to specify cases of point load (e.g., one load in the middle of the beam or three loads equally spaced). Modify the calculations of $\mathcal{I}_0$ and $\mathcal{I}_1$ to include the point loads. Modify the generalized trapezoidal rule integration by looping over each segment between point loads and adjusting the axial force by the amount of the point load before moving to the next segment.

3.16 *Computation.* Use Code 3.1, `AxialBar`, to explore the relationship between the applied load, the internal axial force, and the axial displacement. These functions are related to each other through differential equations. The slope of the $N(x)$ plot is equal to the negative of the value of the $p(x)$ plot. The slope of the $u(x)$ plot is proportional to the value of the $N(x)$ plot. Examine the plots produced by the code for several different load cases and boundary conditions. How can these observations be used to qualitatively sketch the curves by hand without doing calculations?

Chapter 4
Motion, Deformation, and Strain

Kinematics is the study of motion apart from the cause of that motion. In the previous chapter, we encountered motion and deformation in the one-dimensional context of the axial bar problem. In this chapter we extend those basic ideas to motion and deformation in three-dimensional solid bodies. Throughout this book motion is considered to be quasi-static (i.e., the motion happens slowly enough that inertial forces can be neglected). Our aim is to characterize multiaxial deformation in a way that we can eventually use in a model of material behavior. As it was for the axial bar, the concept of *strain*, a relative measure of deformation, will be ideal for that purpose. Hence, this chapter is all about extending the concept of strain to the multidimensional setting.

For most of the chapter, we will consider a very special case of *homogeneous* deformation where all parts of the body experience the same deformation. This simple case, which one can argue is true in a small enough neighborhood of a point in a body, will allow us to think through the problem without needing vector calculus. Further, we will start out looking mostly at *planar* problems, because they are much easier to visualize and the math in two dimensions is more manageable. Once we develop a basic understanding of the issues in two dimensions, we will extend the ideas to three dimensions and to nonhomogeneous deformations. In the developments of the simpler cases, we will establish a mathematical notation that extends directly to the more complex cases.

4.1 Motion and Deformation

In the study of deformable body mechanics, we seek to describe how the material in a body deforms and how that deformation induces internal stresses. To help us think about this problem, we will consider two states of the body, as illustrated in

K. D. Hjelmstad, *Mechanics of Deformable Solids: An Introduction*,
https://doi.org/10.1007/978-3-031-89873-0_4

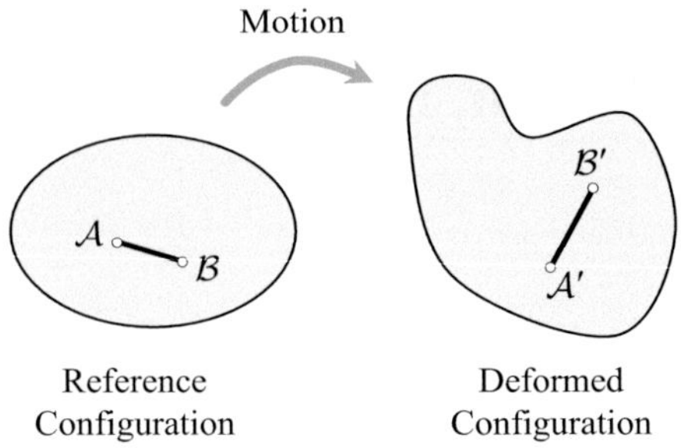

Fig. 4.1 *Motion.* To characterize the motion of a body, we will refer to the *reference configuration* and the *deformed configuration*. The line $\mathcal{AB}$ changes length and orientation due to the motion of the body

Fig. 4.1. The *reference configuration*[1] is the position of the body before loads are applied. The *deformed configuration* is the position of the body after the loads have been applied. The motion of the body is the path it takes to get from the undeformed state to the deformed state.

Deformation Map The *deformation map* is a function that describes the motion, as illustrated in Fig. 4.2. In essence, the deformation map will take as input the location of a point like $\mathcal{P}$ in the reference configuration and give as output its location $\mathcal{P}'$ in the deformed configuration. We establish the location of $\mathcal{P}$ with the vector $\mathbf{z}$ relative to the origin $\mathcal{O}$. The position of that same point, marked as $\mathcal{P}'$ in the figure, is given by the vector

$$\mathbf{x} = \boldsymbol{\varphi}(\mathbf{z}) \tag{4.1}$$

relative to the origin $\mathcal{O}'$. The function $\boldsymbol{\varphi}(\mathbf{z})$ is the *deformation map*. If we know the mapping function, then we know everything about the motion. Each point will have unique positions $\mathbf{z}$ and $\mathbf{x}$ in the reference and deformed configurations, respectively. Thus, we will be able to examine relative motions of those points to answer questions like whether or not line $\mathcal{A}'\mathcal{B}'$ in Fig. 4.1 is longer or shorter than line $\mathcal{AB}$.

The location of the origin $\mathcal{O}'$ in the deformed configuration does not need to be coincident with the location of point $\mathcal{O}$ in the reference configuration, but often it is. If $\mathcal{O}$ and $\mathcal{O}'$ are coincident, then the vector that goes from $\mathcal{P}$ to $\mathcal{P}'$ is called the *displacement vector* and has the expression

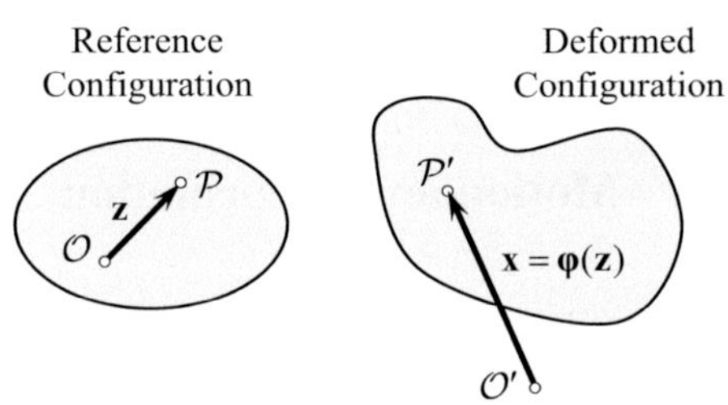

Fig. 4.2 *Deformation map.* In the reference configuration, the location of point $\mathcal{P}$ is $\mathbf{z}$, relative to origin $\mathcal{O}$. The location of the same point in the deformed configuration, i.e., $\mathcal{P}'$, is $\mathbf{x} = \boldsymbol{\varphi}(\mathbf{z})$, relative to origin $\mathcal{O}'$

[1] The reference configuration is often called the *undeformed configuration*. In this book we will use the terms interchangeably.

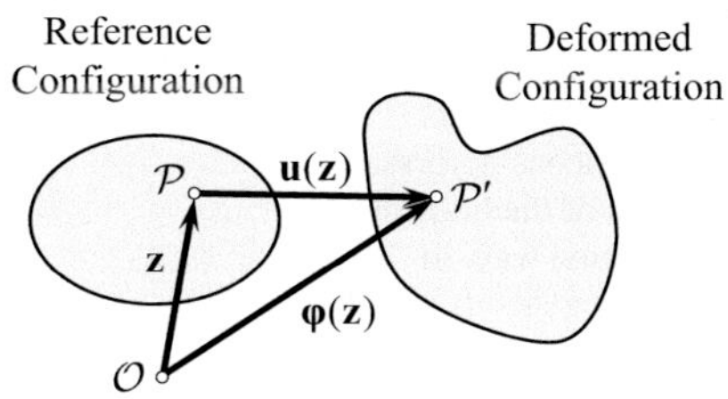

Fig. 4.3 *Displacement map.* If the reference and deformed configurations use the same origin, then the displacement vector can be defined as the difference between the position in the deformed configuration and the reference configuration

$$\mathbf{u}(\mathbf{z}) = \boldsymbol{\varphi}(\mathbf{z}) - \mathbf{z}$$

by vector addition, as shown in Fig. 4.3. The origins must be at the same location in space for this equation to make sense. The vector $\boldsymbol{\varphi}(\mathbf{z})$ gives the *position* of a point relative to an arbitrarily chosen origin. The *displacement* of a point is the location of that point relative to where it was in the reference configuration. From that perspective, the mapping function and the displacement function measure essentially the same thing, just from different reference perspectives.

Implicit in this description of motion is the idea that the material is a *continuous medium*. This assumption is at odds with the fact that at the microscopic level, all materials have an atomic structure that is decidedly not continuous. At macroscopic scales, though, the assumption provides a reasonable model for real solid bodies. The main advantage of the continuum hypothesis is that it opens up the possibility of using the tools of calculus to do the mathematical analysis. The mechanics of a continuous medium is often called *continuum mechanics*.

In this chapter we will assume that we *know* the mapping function, or that we have enough data from the reference and deformed configurations to construct it. That will allow us to focus on understanding what the deformation map implies about the relative motions of points in the body, and thus to define deformation and strain. Later in the book we will embrace the more common engineering task of *finding* the mapping function given the reference configuration, applied loads, and boundary conditions.

4.2 Homogeneous Deformation

The difference between *homogeneous deformation* and *nonhomogeneous deformation* is that in the former, all regions in the body experience the same deformation, while in the latter, deformation varies from one region of the body to the next. Figure 4.4 illustrates this difference. For the homogeneous deformation case on the left, the two squares in opposite corners of the body in the reference configuration deform into parallelograms that are exactly the same size and shape in the deformed configuration. In contrast, those same two squares deform into very different shapes and

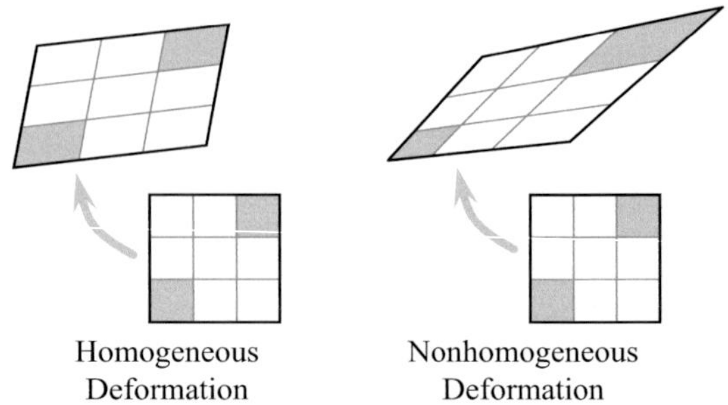

Fig. 4.4 *Homogeneous and nonhomogeneous deformation.* With homogeneous deformation, all parts of the body deform in the same way. In nonhomogeneous deformation, different regions deform by different amounts

sizes in the deformed configuration for the nonhomogeneous deformation case. We will begin our exploration of deformation by considering homogeneous deformation.

The mapping function that represents a homogeneous deformation can be written in the general form

$$\boldsymbol{\varphi}(\mathbf{z}) = \mathbf{x}_o + \mathbf{F}\mathbf{z} \tag{4.2}$$

where $\mathbf{x}_o$ is the translational motion of the reference point $\mathcal{A}$, as shown in Fig. 4.5. We will soon see that the *deformation* is not affected by the translation of the reference point. The constant tensor $\mathbf{F}$ is called the *deformation gradient*. The reason for using this name will be evident when we look at nonhomogeneous deformation. The tensor $\mathbf{F}$ captures both rigid rotation *and* deformation. The position vector $\mathbf{z}$ is measured relative to point $\mathcal{A}$ in the reference configuration, which thus serves as the origin.

To get an idea of what the deformation map implies about the straining of the material, consider the following example. Imagine that we have a flexible body on which we etch a light 7×5 grid in the reference configuration and then subject it to a homogeneous deformation, as shown in Fig. 4.6. Each square in the reference configuration grid represents a unit value for (z_1, z_2). On the undeformed grid, we etch two bigger squares oriented at different angles. Those squares are shown in heavier lines and are shaded. As the body deforms, the squares etched on it do, too.

The mapping function applied to the body in Fig. 4.6 is given by

$$\boldsymbol{\varphi}(\mathbf{z}) = \begin{bmatrix} 1.2726 & 0.2830 \\ 0.2830 & 0.8487 \end{bmatrix} \begin{Bmatrix} z_1 \\ z_2 \end{Bmatrix}$$

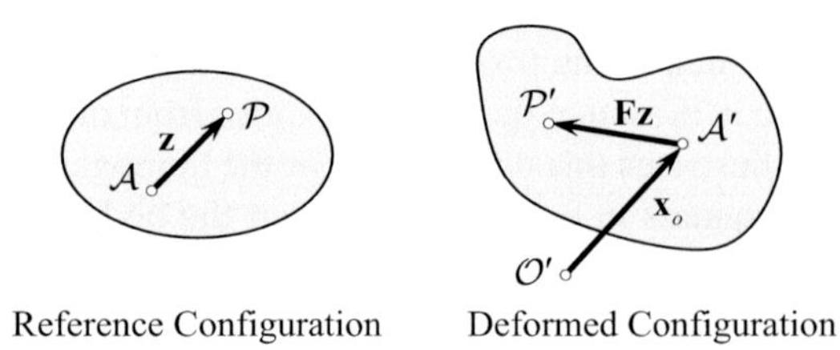

Fig. 4.5 *Homogeneous deformation.* The deformation map for homogeneous deformation can be written as Eq. 4.2. Point $\mathcal{A}$ is the reference point for tracking translation

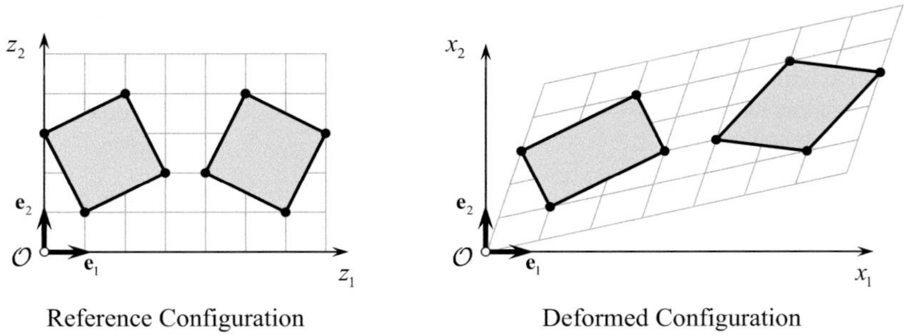

Fig. 4.6 *Example.* Two square regions are etched on a body in the reference configuration. The body is subjected to a homogeneous deformation map. The deformed regions are shown in the deformed configuration

Every point is mapped with the same function. Observe the background grids on the reference and deformed configurations. The deformation map is homogeneous, so every little square in the undeformed grid becomes a parallelogram in the deformed grid. Every point on the substrate grid gets mapped the same way. Under the homogeneous deformation map, originally straight lines are still straight in the deformed configuration.

Now, look at what happens to the bigger squares that were etched on the reference grid. The corners of the deformed squares are located at the same grid intersections. The first square deforms into a rectangle. The other one deforms into a parallelogram. Observe that the sides of the rectangle associated with the square on the left are not equal. Thus, the amount of stretching must depend upon the orientation of the line. The lines that were on the top and bottom got longer, and the lines that were on the left and right got shorter. The lines that define the sides of the other square are initially perpendicular, but the angle between those lines is changed by the deformation. Change in angle between two lines is a different kind of deformation (we call it *shearing*) than deformation that is associated with change in length of lines.

If you think about the motion in terms of what happens to different lines and the angles between lines, then you can see that the outcome depends upon the initial orientation of the etched bigger square. This observation forms the basis for our study of motion, deformation, and strain. The stress induced in a material depends upon how much it stretches. Therefore, knowing how lines stretch is fundamental to our understanding and modeling of material response to deformation. We will characterize the deformation in terms of normal and shear strains. *Normal strain* is related to relative change in length of a line due to the motion. *Shear strain* is related to change in the angle between two lines due to the motion.

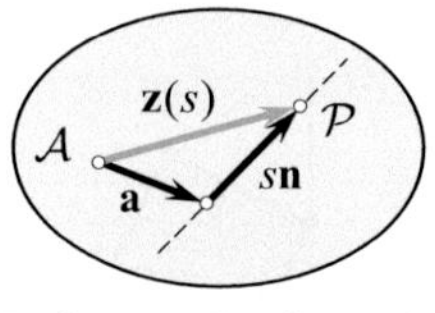

Fig. 4.7 *Example.* A straight line in the reference configuration will deform into a straight line in the deformed configuration

Example 4.1
Prove that a straight line in the reference configuration is still a straight line in the deformed configuration for a homogeneous deformation map.

Solution. Consider the straight line (shown as a dotted line) in Fig. 4.7. This line can be described with the function

$$\mathbf{z}(s) = \mathbf{a} + s\mathbf{n}$$

where **a** is constant and the vector **n**, which is also constant, points along the line. Any value of s will generate a point that lies along the line (e.g., point $\mathcal{P}$ in the figure). We can now apply the map to points $\mathbf{z}(s)$ along the line. To wit,

$$\begin{aligned}\varphi(s) &= \mathbf{x}_o + \mathbf{F}\mathbf{z}(s)\\ &= \mathbf{x}_o + \mathbf{F}\big(\mathbf{a} + s\mathbf{n}\big)\\ &= \mathbf{x}_o + \mathbf{F}\mathbf{a} + s\mathbf{F}\mathbf{n}\end{aligned}$$

Define $\mathbf{b} = \mathbf{x}_o + \mathbf{F}\mathbf{a}$ and $\mathbf{m} = \mathbf{F}\mathbf{n}$. Substituting these definitions into the previous equation gives

$$\varphi(s) = \mathbf{b} + s\mathbf{m}$$

which is the equation of a line passing through point **b** with constant direction **m**. Again, any value of s produces the location of a point that lies on the line. This derivation applies to any vectors **a** and **n**. Therefore, all straight lines in the reference configuration become straight lines in the deformed configuration for a homogeneous deformation.

4.2.1 Rigid Body Motion

The motion of a body is generally composed of a rigid motion and a motion causing deformation. A rigid body motion includes a translation and a rotation, but no line in

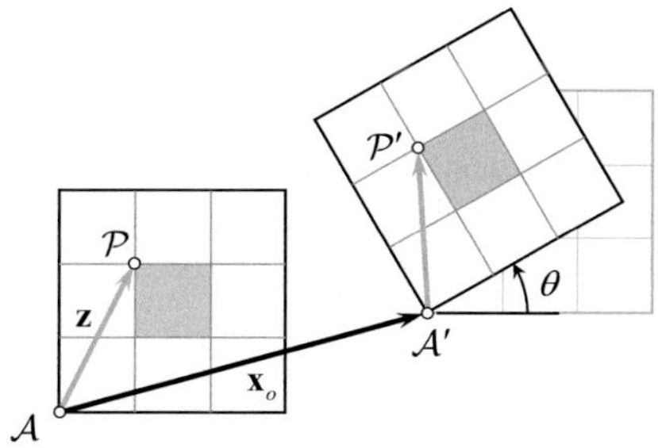

Fig. 4.8 *Rigid body motion in a plane* can be characterized with a translation $\mathbf{x}_o$ and a rotation θ. The reference point for translation is $\mathcal{A}$. Point $\mathcal{P}$ is located at position $\mathbf{z}$ relative to $\mathcal{A}$ in the reference configuration

the body changes length, and no two lines in the body change their relative angle with respect to each other. For rigid body motion, there is no deformation of the material.

A case of rigid body motion in a plane is illustrated in Fig. 4.8. We can characterize the final position of each particle $\mathbf{z}$ in the reference configuration by keeping track of the translation of the lower left corner $\mathbf{x}_o$ and the angle of rotation of the entire body θ. In this case, the upper left corner of the gray square, point $\mathcal{P}$, is one unit right and two units up from the lower left corner of the body. Thus, $z_1 = 1$ and $z_2 = 2$.

The mapping function for rigid body motion can be written as

$$\boldsymbol{\varphi}(\mathbf{z}) = \mathbf{x}_o + \mathbf{Q}\mathbf{z}$$

where $\mathbf{Q}$ is an *orthogonal tensor*, i.e., it has the property $\mathbf{Q}^T\mathbf{Q} = \mathbf{I}$. For two-dimensional motion, the rotation tensor $\mathbf{Q}$ can be written in components as

$$\mathbf{Q} = \begin{bmatrix} \cos\theta & -\sin\theta \\ \sin\theta & \cos\theta \end{bmatrix}$$

where θ is the angle of rotation (see Fig. 4.8). A positive value of θ gives a counterclockwise rotation, and a negative value gives a clockwise rotation. It is clear in Fig. 4.8 that the gray vector in the deformed configuration is merely a translated and rotated version of vector $\mathbf{z}$.

4.2.2 *Determining* F *from the Motion*

If the motion is known to be homogeneous, and therefore can be represented by the mapping function in Eq. 4.2, then we can determine the components of $\mathbf{x}_o$ and $\mathbf{F}$ if we know the locations of points in the reference *and* deformed configurations.

The vector $\mathbf{x}_o$ is simply the location of the reference point $\mathcal{A}$ in the deformed configuration, as that is the point with $\mathbf{z} = \mathbf{0}$. Consider a planar motion and let us assume that we know the locations of two points, $\mathbf{z}_i$ and $\mathbf{x}_i$ $(i=1,2)$ in the reference and deformed configurations, respectively. This knowledge constitutes two vector equations $\mathbf{x}_i - \mathbf{x}_o = \mathbf{F}\mathbf{z}_i$ in four unknowns (i.e., the four components of $\mathbf{F}$). Hence, the knowledge of the locations of two points in the reference and deformed configurations gives enough information to find the four components of the tensor $\mathbf{F}$. We can write these equations in matrix form as

$$\mathbf{X} = \mathbf{F}\mathbf{Z} \tag{4.3}$$

where the matrices $\mathbf{X}$ and $\mathbf{Z}$ are formed by taking the components of the two vectors as the columns of the matrices. To wit,

$$\mathbf{X} = \big[\,(\mathbf{x}_1 - \mathbf{x}_o),\ (\mathbf{x}_2 - \mathbf{x}_o)\big], \qquad \mathbf{Z} = \big[\,\mathbf{z}_1,\ \mathbf{z}_2\big]$$

where we are representing $\mathbf{x}_o$, $\mathbf{x}_i$, and $\mathbf{z}_i$ $(i=1,2)$ as column vectors. We can find $\mathbf{F}$ by solving the linear system of equations in Eq. 4.3. To wit,

$$\mathbf{F} = \mathbf{X}\mathbf{Z}^{-1} \tag{4.4}$$

where $\mathbf{Z}^{-1}$ is the inverse of the matrix $\mathbf{Z}$. For this computation to succeed, the matrix $\mathbf{Z}$ must have an inverse. For problems in two dimensions, only two points are needed to make $\mathbf{Z}$ and $\mathbf{X}$ into 2×2 matrices. The vectors selected must be independent, which means that they cannot have the same direction, even if they have different magnitudes. There is a good physical reason for the mathematical requirement that $\mathbf{Z}$ be invertible. The tensor $\mathbf{F}$ contains all the information about the deformation in all directions. If we sample the same direction twice, we do not learn anything new about the deformation map in other directions, and therefore we cannot find $\mathbf{F}$. Since two independent vectors in two-dimensional space form a basis, like $\{\mathbf{e}_1, \mathbf{e}_2\}$ do, linear combinations of those two vectors can represent any other vector in the space. Hence, two independent vectors are exactly what we need to find the components of $\mathbf{F}$.

In three dimensions, $\mathbf{Z}$ and $\mathbf{X}$ must be 3×3 matrices, which will be true only if three points are included. If the three points selected are not coplanar (i.e., they do not lie in a single plane), then the three vectors are independent, and the deformation gradient can be found.

Example 4.2
Consider the mapping problem shown in Fig. 4.9, where right triangle $\mathcal{ABC}$ is mapped to right triangle $\mathcal{A}'\mathcal{B}'\mathcal{C}'$ with a homogeneous deformation. Find the deformation tensor $\mathbf{F}$ associated with the mapping.

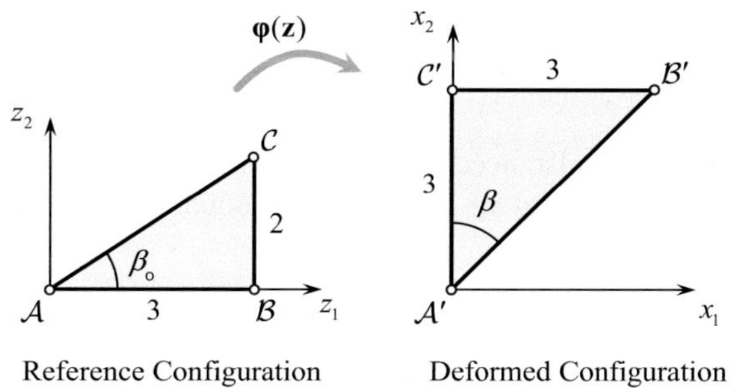

Fig. 4.9 *Example.* Right triangle $\mathcal{ABC}$ in the reference configuration is mapped to right triangle $\mathcal{A}'\mathcal{B}'\mathcal{C}'$ in the deformed configuration with a homogeneous deformation

Solution. Because $\mathcal{A}$ and $\mathcal{A}'$ are both at the origin of their respective coordinate systems, $\mathbf{x}_o = \mathbf{0}$. The position vectors for points $\mathcal{B}$ and $\mathcal{C}$ in the reference configuration are

$$\mathbf{z}_{\mathcal{B}} = \begin{Bmatrix} 3 \\ 0 \end{Bmatrix}, \quad \mathbf{z}_{\mathcal{C}} = \begin{Bmatrix} 3 \\ 2 \end{Bmatrix}$$

The position vectors for points $\mathcal{B}'$ and $\mathcal{C}'$ in the deformed configuration are

$$\mathbf{x}_{\mathcal{B}'} = \begin{Bmatrix} 3 \\ 3 \end{Bmatrix}, \quad \mathbf{x}_{\mathcal{C}'} = \begin{Bmatrix} 0 \\ 3 \end{Bmatrix}$$

These vectors comprise the columns of the matrices $\mathbf{Z}$ and $\mathbf{X}$, respectively. Thus,

$$\mathbf{Z} = \begin{bmatrix} 3 & 3 \\ 0 & 2 \end{bmatrix}, \quad \mathbf{X} = \begin{bmatrix} 3 & 0 \\ 3 & 3 \end{bmatrix}$$

Compute the deformation gradient $\mathbf{F}$ from Eq. 4.4:

$$\mathbf{F} = \mathbf{X}\mathbf{Z}^{-1} = \begin{bmatrix} 3 & 0 \\ 3 & 3 \end{bmatrix} \left(\frac{1}{6} \begin{bmatrix} 2 & -3 \\ 0 & 3 \end{bmatrix} \right) = \begin{bmatrix} 1 & -1.5 \\ 1 & 0 \end{bmatrix}$$

where we have used the formula for the inverse of a 2×2 matrix:

$$\mathbf{A} = \begin{bmatrix} a & b \\ c & d \end{bmatrix} \quad \rightarrow \quad \mathbf{A}^{-1} = \frac{1}{ad - bc} \begin{bmatrix} d & -b \\ -c & a \end{bmatrix}$$

We have treated this example as a two-dimensional problem, even though every mapping lives in three-dimensional space. In a state of *plane strain*, the z-coordinate of every point in the reference configuration maps to the same value in the deformed configuration. We could use a third point $\mathcal{D}$ (which maps to $\mathcal{D}'$). If that point is located a distance d in the $\mathbf{e}_3$ direction in the reference and deformed coordinates, then

$$\mathbf{z}_{\mathcal{D}} = \begin{Bmatrix} 0 \\ 0 \\ d \end{Bmatrix}, \qquad \mathbf{x}_{\mathcal{D}'} = \begin{Bmatrix} 0 \\ 0 \\ d \end{Bmatrix}$$

We would also need to include the zero element for the $\mathbf{e}_3$ component of the vectors $\mathbf{z}_{\mathcal{B}}$, $\mathbf{z}_{\mathcal{C}}$, $\mathbf{x}_{\mathcal{B}'}$, and $\mathbf{x}_{\mathcal{C}'}$. The mapping equations $\mathbf{F} = \mathbf{X}\mathbf{Z}^{-1}$ would then be

$$\mathbf{F} = \begin{bmatrix} 3 & 0 & 0 \\ 3 & 3 & 0 \\ 0 & 0 & d \end{bmatrix} \begin{bmatrix} 3 & 3 & 0 \\ 0 & 2 & 0 \\ 0 & 0 & d \end{bmatrix}^{-1} = \begin{bmatrix} 1 & -1.5 & 0 \\ 1 & 0 & 0 \\ 0 & 0 & 1 \end{bmatrix}$$

Hence, it does not matter what d is because the motion in the z direction is uncoupled from the motions in the plane. That is why we can treat planar problems with 2×2 matrices. Observe that the (3, 3) component of $\mathbf{F}$ is one.

4.3 Definition of Strain

Strain is a relative measure of deformation that is useful for developing mathematical models for the response of materials. Consider the line $\mathcal{AB}$ in the undeformed configuration in Fig. 4.1. After deformation that same line has translated, rotated, and stretched to become the line $\mathcal{A}'\mathcal{B}'$ in the deformed configuration. In this section we will try to put that idea into a context that will allow us to faithfully describe the concept of strain as a means of completely describing what happens to a body when it deforms. Strain is, roughly speaking, the change in the length of a line relative to the original length of that line, as we asserted in Chap. 3 for the axial bar. The concept of strain in multiple dimensions must also characterize changes in angles between lines, which was not a feature present in the axial bar problem. Our task is to figure out what strain means for two- and three-dimensional states of deformation.

4.3.1 Common Strain Measures

Let L_o be the length of line $\mathcal{AB}$ in Fig. 4.1, and let L be the length of line $\mathcal{A}'\mathcal{B}'$. The *stretch* of the line is the ratio of the deformed length to the original (reference) length, which we define as

$$\lambda = \frac{L}{L_o} \tag{4.5}$$

The change in length of the line is $\Delta = L - L_o$. Recall that the measure of strain that we used for the axial bar problem was $\varepsilon = \Delta/L_o$, a measure usually called *engineering strain*. We can write the engineering strain in terms of the stretch as $\varepsilon = \lambda - 1$. It turns out that engineering strain is not the only way to characterize

Table 4.1 *Strain measures.* Various definitions of strain, in terms of the stretch ratio $\lambda = L/L_o$, and the name most commonly associated with the strain

ε	Common name
$\frac{1}{2}(\lambda^2 - 1)$	Lagrangian strain
$\lambda - 1$	Engineering strain
$\ln \lambda$	Logarithmic strain
$1 - 1/\lambda$	True strain
$\frac{1}{2}(1 - 1/\lambda^2)$	Eulerian strain

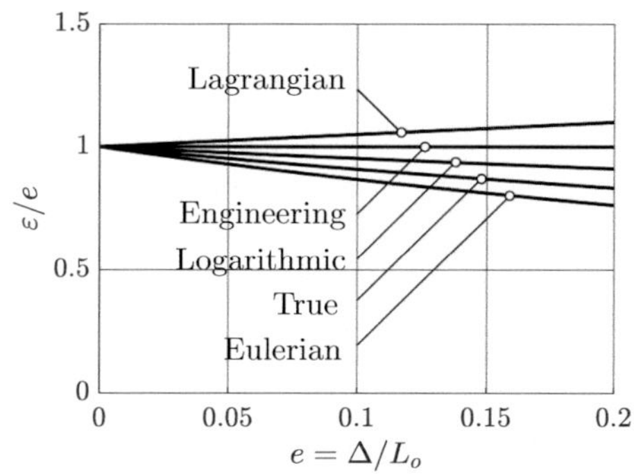

Fig. 4.10 *Strain measures.* Comparison of different strain measures, defined in Table 4.1, normalized by engineering strain $e = \Delta/L_o$, over a range of engineering strains from zero to $e = 0.2$

the concept of strain. In fact, there are several useful definitions, summarized in Table 4.1.

Figure 4.10 shows all five of the strain measures plotted versus $e = \Delta/L_o$. The strains are normalized by the value of the engineering strain. There are a few things to notice about these different strain measures. First, they are all equal to each other in the neighborhood of $e = 0$, which corresponds to $\lambda = 1$ (i.e., $L = L_o$). Second, all the strain measures produce roughly the same value for small values of strain, and they are all equal to zero when there is no deformation.

Is any one of the strain measures better than the others? No. But they are obviously not the same. The difference manifests when we establish constitutive models of materials. Let's say we want to determine the parameters of a constitutive model for a certain material. In a laboratory test (e.g., a tension test) we measure the applied force and the change in length between two points along the axis of the test piece. To plot the results as stress versus strain, we first divide the applied axial force by the cross-sectional area of the test piece to get stress. Then we take the current length data, divide it by the original gauge length, and compute any one of the strain measures. The measured data, presented in this form, can then be fit to a curve. If the data plot along a straight line, then we would select a linear function as the curve, and the parameter we fit to the data would be the slope of the line. Obviously, with five different strain measures, we would get five different curves. Each one would give a different value of the model parameter. Thus, the choice of strain measure matters when we use it in a constitutive model.

Example 4.3
Find the final length of a line given the Lagrangian strain ε.

Solution. To solve this problem, we go back to the definition of the Lagrangian strain in terms of the original and deformed length of the line

$$\varepsilon = \tfrac{1}{2}(\lambda^2 - 1) = \tfrac{1}{2}\left(\frac{L^2}{L_o^2} - 1\right) = \frac{L^2 - L_o^2}{2L_o^2}$$

Solve this equation for L by multiplying through by $2L_o^2$, adding L_o^2 to both sides, and taking the square root to get

$$L = L_o\sqrt{1+2\varepsilon} \tag{4.6}$$

Of course, the same computation can be done for any of the strain measures, with the results given in Table 4.2. Engineering strain is the easiest to interpret. The final length is the original length plus εL_o. That means the original length amplifies the effect of the strain. Strain on a long bar will result in a change in length that is larger than the same strain on a shorter bar.

To see the similarities between all the formulas given in the table, one can express the functions as Taylor series to show that for small strains, they all predict roughly the same deformed length. The Taylor series are all of the form $L_o(1+\varepsilon+c\,\varepsilon^2+\cdots)$. The coefficient of the ε^2 term is different for each one. For Lagrangian, engineering, logarithmic, true, and Eulerian strain, the values are $c = -0.5$, 0, 0.5, 1, and 1.5, respectively.

Table 4.2 *Deformed length of line.* Computation of the deformed length L of a line given the reference length L_o and the strain ε for various strain measures

L	Strain measure
$L_o\sqrt{1+2\varepsilon}$	Lagrangian strain
$L_o(1+\varepsilon)$	Engineering strain
$L_o\exp(\varepsilon)$	Logarithmic strain
$L_o/(1-\varepsilon)$	True strain
$L_o/\sqrt{1-2\varepsilon}$	Eulerian strain

4.3.2 The Strain Tensor

Strain in three-dimensional solid bodies can be viewed through the lens of change in length of lines. However, in three dimensions, there are lines in all directions. All those lines are affected by the deformation map, possibly in different ways. In this section we develop a general strategy for determining the strain of a line and the change in angle between two lines based upon the original orientations of the lines. Along the way, we encounter the strain *tensor*.

Consider two points, $\mathcal{B}$ and $\mathcal{C}$, in a solid body in the reference configuration in Fig. 4.11. The length of the line joining these two points is s_o and the unit vector that points from $\mathcal{B}$ to $\mathcal{C}$ is $\mathbf{n}$. Under a *homogeneous* deformation map $\boldsymbol{\varphi}(\mathbf{z})$, the two points move to locations $\mathcal{B}'$ and $\mathcal{C}'$. The length of the line joining the two deformed points is s, and the unit vector pointing from $\mathcal{B}'$ to $\mathcal{C}'$ is $\mathbf{m}$. The motion can include rotation, so the two unit vectors $\mathbf{n}$ and $\mathbf{m}$ do not necessarily point in the same direction.

For homogeneous deformation, the mapping is given by Eq. 4.2. Let $\mathbf{z}_B$ and $\mathbf{z}_C$ be the locations of points $\mathcal{B}$ and $\mathcal{C}$, respectively, relative to the reference point $\mathcal{A}$ for translation. Thus, the length of the line joining the points and the unit vector pointing from one to the other is

$$s_o = \|\mathbf{z}_C - \mathbf{z}_B\|, \qquad \mathbf{n} = \frac{\mathbf{z}_C - \mathbf{z}_B}{s_o} \tag{4.7}$$

Hence, if we know the location of both points in the reference configuration, then we can compute the unit vector and the length of the line between them. In the deformed configuration, the locations of points $\mathcal{B}'$ and $\mathcal{C}'$ are, according to Eq. 4.2,

$$\boldsymbol{\varphi}(\mathbf{z}_B) = \mathbf{x}_o + \mathbf{F}\mathbf{z}_B, \qquad \boldsymbol{\varphi}(\mathbf{z}_C) = \mathbf{x}_o + \mathbf{F}\mathbf{z}_C$$

where $\mathbf{x}_o$ is the translation of point $\mathcal{A}$. The vector from $\mathcal{B}'$ to $\mathcal{C}'$ is the difference between the mapped locations of the two points. Hence,

$$s\mathbf{m} = \boldsymbol{\varphi}(\mathbf{z}_C) - \boldsymbol{\varphi}(\mathbf{z}_B) = \mathbf{F}\big(\mathbf{z}_C - \mathbf{z}_B\big) = s_o\mathbf{F}\mathbf{n}$$

Divide through by s_o to get the relationship

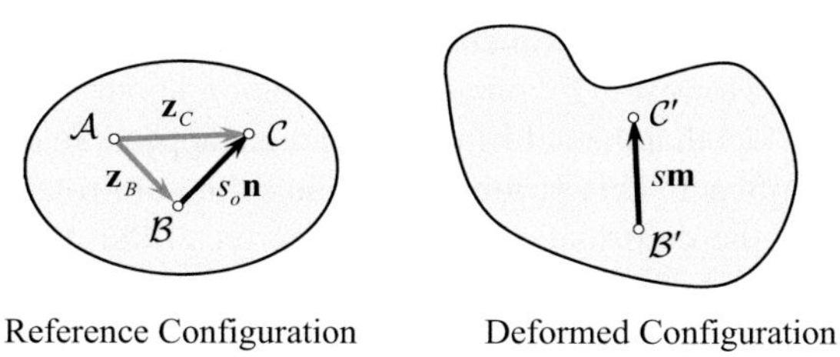

Fig. 4.11 *Stretch.* Solid body subjected to mapping $\boldsymbol{\varphi}(\mathbf{z})$. Find the length of the line between two points in the reference and deformed configurations

$$\lambda \mathbf{m} = \mathbf{F}\mathbf{n} \tag{4.8}$$

where $\lambda = s/s_o$ is the ratio of the length of the deformed line to the length of the line in the reference configuration. Thus, λ is the *stretch* of the line originally connecting points $\mathcal{B}$ and $\mathcal{C}$. To isolate λ, dot each side of the equation with itself. Noting that $\mathbf{m} \cdot \mathbf{m} = 1$, we get

$$\lambda^2 = \mathbf{F}\mathbf{n} \cdot \mathbf{F}\mathbf{n} = \mathbf{n} \cdot \mathbf{F}^T\mathbf{F}\mathbf{n} \tag{4.9}$$

where we have noted that moving the $\mathbf{F}$ across the dot product requires that we transpose it.[2]

Now, we can compute the scalar *Lagrangian strain* as

$$\varepsilon = \tfrac{1}{2}(\lambda^2 - 1) = \tfrac{1}{2}(\mathbf{n} \cdot \mathbf{F}^T\mathbf{F}\mathbf{n} - \mathbf{n} \cdot \mathbf{I}\mathbf{n})$$

where $\mathbf{I}$ is the identity tensor and $\mathbf{n} \cdot \mathbf{I}\mathbf{n} = \mathbf{n} \cdot \mathbf{n} = 1$. Factoring $\mathbf{n}$ out of the above expression gives

$$\varepsilon = \mathbf{n} \cdot \tfrac{1}{2}\big[\mathbf{F}^T\mathbf{F} - \mathbf{I}\big]\mathbf{n}$$

The quantity sandwiched between the unit vectors does not depend upon the direction of the line (i.e., the direction $\mathbf{n}$). Therefore, it applies to *any* line. Let us define this special tensor as

$$\mathbf{E} = \tfrac{1}{2}\big[\mathbf{F}^T\mathbf{F} - \mathbf{I}\big] \tag{4.10}$$

We call this tensor the *Lagrangian strain tensor*. We can now think of the (scalar) Lagrangian strain as a function of the direction $\mathbf{n}$. To wit,

$$\varepsilon(\mathbf{n}) = \mathbf{n} \cdot \mathbf{E}\mathbf{n} \tag{4.11}$$

It should be evident why we selected the Lagrangian strain from the five choices we had. Because the dot product of a vector with itself is the square of the length of the vector, the simplest equations are those that involve the squares of lengths.

Equations 4.10 and 4.11 have several important implications. First, if we consider the state of strain to be a property of the deformation independent of direction, then strain *must* be defined as a tensor (not a scalar and not a vector). With that tensor, we can investigate the strain of any line in any direction, using Eq. 4.11. Second, we can compute the strain tensor directly from the deformation gradient $\mathbf{F}$, which comes from the mapping function $\boldsymbol{\varphi}(\mathbf{z})$. Hence, all the information regarding strain is present in $\mathbf{F}$. However, $\mathbf{F}$ contains information about both rigid rotation *and* deformation. The fact that the deformation gradient appears in the strain tensor in the form $\mathbf{F}^T\mathbf{F}$ is significant. This product eliminates the pure rotation part of the motion and leaves only the deformation part. In fact, recall that for a rigid body motion, $\mathbf{F} = \mathbf{Q}$ and $\mathbf{Q}^T\mathbf{Q} = \mathbf{I}$. Therefore, the strain associated with a rigid body motion is exactly zero.

[2] The definition of the transpose of a tensor $\mathbf{A}$ comes from the dot product of vectors. If we have two vectors $\mathbf{u}$ and $\mathbf{v}$, then we can prove that $\mathbf{A}\mathbf{u} \cdot \mathbf{v} = \mathbf{u} \cdot \mathbf{A}^T\mathbf{v}$.

We can also define the *Green deformation tensor* as

$$\mathbf{C} = \mathbf{F}^T\mathbf{F} \tag{4.12}$$

and observe that the square of the stretch in direction $\mathbf{n}$ can be computed as

$$\lambda^2(\mathbf{n}) = \mathbf{n} \cdot \mathbf{Cn} \tag{4.13}$$

Thus, an alternative expression for the Lagrangian strain tensor is

$$\mathbf{E} = \tfrac{1}{2}[\,\mathbf{C} - \mathbf{I}\,] \tag{4.14}$$

Note that the tensors $\mathbf{E}$ and $\mathbf{C}$ are always *symmetric* while $\mathbf{F}$ is not. Any tensor created by taking a tensor transposed and multiplying by that same tensor is guaranteed to be symmetric.

4.3.3 Shear Strain

So far, we have only considered the changes in lengths of lines in the solid body. We saw in Fig. 4.6 that the angle between two lines can also change. That change is independent of any changes in the lengths of the lines. The mode of deformation that causes angles between lines to change is called *shear strain*.

Consider the situation shown in Fig. 4.12. There are two lines: $\mathcal{BC}$ and $\mathcal{BD}$ in the reference configuration. Vector $s_1\mathbf{n}_1$ is the vector from point $\mathcal{B}$ to point $\mathcal{C}$. Similarly, vector $s_2\mathbf{n}_2$ is the vector from point $\mathcal{B}$ to point $\mathcal{D}$. The lengths of these two vectors are s_1 and s_2, respectively, and the unit vectors along the lines connecting the points are $\mathbf{n}_1$ and $\mathbf{n}_2$. As before, if we know the locations of all three points in the reference configuration, then we can compute the lengths of the lines and the unit vectors. The angle between the vectors in the reference configuration is β_o. We can use the definition of the dot product to compute the cosine of the angle between two vectors. Thus,

$$\cos\beta_o = \mathbf{n}_1 \cdot \mathbf{n}_2$$

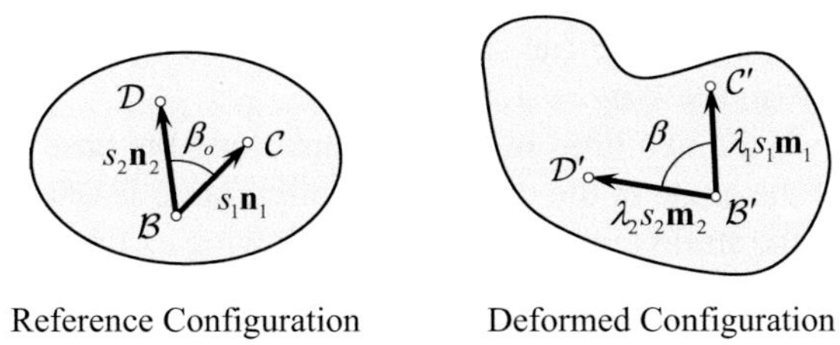

Fig. 4.12 *Shear strain.* Shear strain is related to the change in angles between two lines $\mathcal{BC}$ and $\mathcal{BD}$

If $\mathbf{n}_1$ and $\mathbf{n}_2$ are perpendicular, then $\beta_o = \pi/2$ and $\cos\beta_o = 0$.

The deformation causes the points to move and the vectors to rotate and change length to $\lambda_1 s_1 \mathbf{m}_1$ and $\lambda_2 s_2 \mathbf{m}_2$, respectively, where λ_1 and λ_2 are the stretch ratios of the two lines. Therefore, the deformed lengths of the two lines are $\lambda_1 s_1$ and $\lambda_2 s_2$, respectively. We know from Eq. 4.8 that

$$\lambda_1 \mathbf{m}_1 = \mathbf{F}\mathbf{n}_1, \qquad \lambda_2 \mathbf{m}_2 = \mathbf{F}\mathbf{n}_2 \tag{4.15}$$

Dotting the left and right sides of these two equations with each other gives

$$\lambda_1 \mathbf{m}_1 \cdot \lambda_2 \mathbf{m}_2 = \mathbf{F}\mathbf{n}_1 \cdot \mathbf{F}\mathbf{n}_2 = \mathbf{n}_1 \cdot \mathbf{C}\mathbf{n}_2 \tag{4.16}$$

The cosine of the angle between the deformed lines is $\cos\beta = \mathbf{m}_1 \cdot \mathbf{m}_2$, therefore

$$\lambda_1 \lambda_2 \cos\beta = \mathbf{n}_1 \cdot \mathbf{C}\mathbf{n}_2 \tag{4.17}$$

We can rewrite Eq. 4.17 as

$$\cos\beta(\mathbf{n}_1, \mathbf{n}_2) = \frac{\mathbf{n}_1 \cdot \mathbf{C}\mathbf{n}_2}{\lambda(\mathbf{n}_1)\lambda(\mathbf{n}_2)} = \frac{\mathbf{n}_1 \cdot [\,\mathbf{I} + 2\mathbf{E}\,]\mathbf{n}_2}{\lambda(\mathbf{n}_1)\lambda(\mathbf{n}_2)}$$

where we have substituted $\mathbf{C} = \mathbf{I} + 2\mathbf{E}$. We have also added the notation $\beta(\mathbf{n}_1, \mathbf{n}_2)$ to remind us that the angle we are computing is the one between deformed lines that were *originally* oriented in the $\mathbf{n}_1$ and $\mathbf{n}_2$ directions in the reference configuration. We have also adopted the notation introduced in Eq. 4.13 to indicate, for example, that $\lambda_1 = \lambda(\mathbf{n}_1)$, again reminding us that the stretch is computed from the unit vector in the reference configuration. Noting that $\cos\beta_o = \mathbf{n}_1 \cdot \mathbf{I}\mathbf{n}_2 = \mathbf{n}_1 \cdot \mathbf{n}_2$, we get the final form

$$\cos\beta(\mathbf{n}_1, \mathbf{n}_2) = \frac{\cos\beta_o + 2\,\mathbf{n}_1 \cdot \mathbf{E}\mathbf{n}_2}{\lambda(\mathbf{n}_1)\lambda(\mathbf{n}_2)} \tag{4.18}$$

What this equation shows is that the change in angle can be computed from the Lagrangian strain tensor $\mathbf{E}$. If the two lines are perpendicular in the reference configuration, then $\cos\beta_o = 0$. Equations 4.11 and 4.18 are the keys to computing normal and shear strains in a deformed body. Equation 4.18 shows that, while the angle between the deformed lines is affected by the change in length of the lines (i.e., the presence of $\lambda(\mathbf{n}_1)$ and $\lambda(\mathbf{n}_2)$ in the denominator), it is not determined by them. In particular, there can still be a change in angle if the lines do not change length, i.e., if $\lambda(\mathbf{n}_1) = \lambda(\mathbf{n}_2) = 1$.

If the two lines in Eq. 4.18 are along the base vectors $\mathbf{e}_1$ and $\mathbf{e}_2$, then the cosine of the angle in the deformed configuration is related to the off-diagonal component of the strain tensor, ε_{xy}, as the following example shows.

Example 4.4
Compute the cosine of the angle between the base vectors $\mathbf{e}_1$ and $\mathbf{e}_2$ for a given mapping with deformation gradient $\mathbf{F}$ and corresponding Lagrangian strain tensor $\mathbf{E}$.

Solution. We can represent the components of the Lagrangian strain tensor and the base vectors in three dimensions as follows:

$$\mathbf{E} = \begin{bmatrix} \varepsilon_{xx} & \varepsilon_{xy} & \varepsilon_{xz} \\ \varepsilon_{xy} & \varepsilon_{yy} & \varepsilon_{yz} \\ \varepsilon_{xz} & \varepsilon_{yz} & \varepsilon_{zz} \end{bmatrix}, \qquad \mathbf{e}_1 = \begin{Bmatrix} 1 \\ 0 \\ 0 \end{Bmatrix}, \qquad \mathbf{e}_2 = \begin{Bmatrix} 0 \\ 1 \\ 0 \end{Bmatrix}$$

Now, we can compute

$$\mathbf{e}_1 \cdot \mathbf{E}\mathbf{e}_2 = \begin{bmatrix} 1 & 0 & 0 \end{bmatrix} \begin{bmatrix} \varepsilon_{xx} & \varepsilon_{xy} & \varepsilon_{xz} \\ \varepsilon_{xy} & \varepsilon_{yy} & \varepsilon_{yz} \\ \varepsilon_{xz} & \varepsilon_{yz} & \varepsilon_{zz} \end{bmatrix} \begin{Bmatrix} 0 \\ 1 \\ 0 \end{Bmatrix} = \varepsilon_{xy}$$

Thus, from Eq. 4.18, we have

$$2\,\varepsilon_{xy} = \lambda(\mathbf{e}_1)\lambda(\mathbf{e}_2)\cos\beta(\mathbf{e}_1, \mathbf{e}_2)$$

Hence, the component ε_{xy} is a measure of the angle between lines initially oriented along the base vectors $\mathbf{e}_1$ and $\mathbf{e}_2$. If the stretch ratios are approximately equal to one (i.e., small strains), then

$$2\,\varepsilon_{xy} \approx \cos\beta(\mathbf{e}_1, \mathbf{e}_2)$$

Finally, we can note that the angle β is the angle between the two lines in the deformed configuration. If we want to compute the *departure* of the two lines from perpendicularity, then we can define the angle $\gamma = \pi/2 - \beta$. A trigonometric identity proves that $\cos\beta = \sin\gamma$. If the angle is small, then $\sin\gamma \approx \gamma$, and we end up with

$$\gamma \approx 2\,\varepsilon_{xy}$$

There is a lot going on here, but this computation shows that the change in angle is essentially equal to twice the off-diagonal element of the strain tensor $\mathbf{E}$. For a three-dimensional case, there are three off-diagonal elements. Those would each measure an angular departure from perpendicularity of the three pairings of the three lines along the coordinate axes.

The following example shows the computation of the Green deformation tensor and the Lagrangian strain tensor, and shows how the developments of this section comport with ordinary trigonometry and the Pythagorean theorem.

Example 4.5
Consider the deformation shown in Fig. 4.9 from one of the previous examples. The deformation gradient was found to be

$$\mathbf{F} = \begin{bmatrix} 1 & -1.5 \\ 1 & 0 \end{bmatrix}$$

Compute the Green deformation tensor **C** and the Lagrangian strain tensor **E**. Demonstrate that the changes in lengths of lines $\mathcal{AB}$ and $\mathcal{AC}$ and the angles between them before and after deformation comport with elementary geometric notions.

Solution. The Green deformation tensor **C** can be computed as

$$\mathbf{C} = \mathbf{F}^T\mathbf{F} = \begin{bmatrix} 1 & 1 \\ -1.5 & 0 \end{bmatrix}\begin{bmatrix} 1 & -1.5 \\ 1 & 0 \end{bmatrix} = \begin{bmatrix} 2 & -1.5 \\ -1.5 & 2.25 \end{bmatrix}$$

The Lagrangian strain tensor is

$$\mathbf{E} = \tfrac{1}{2}\big[\mathbf{C} - \mathbf{I}\big] = \begin{bmatrix} 0.5 & -0.75 \\ -0.75 & 0.625 \end{bmatrix}$$

With the Green deformation tensor, we can compute the deformed lengths of lines $\mathcal{AB}$ and $\mathcal{AC}$. The square of the stretch in the $\mathbf{e}_1$ direction (along line $\mathcal{AB}$) is, from Eq. 4.13,

$$\lambda^2(\mathbf{e}_1) = \begin{bmatrix} 1 & 0 \end{bmatrix}\begin{bmatrix} 2 & -1.5 \\ -1.5 & 2.25 \end{bmatrix}\begin{Bmatrix} 1 \\ 0 \end{Bmatrix} = 2$$

The deformed length of line $\mathcal{AB}$ is $L_f = L_o\lambda(\mathbf{e}_1) = 3\sqrt{2}$, which can be verified by the Pythagorean theorem for this simple geometry. Similarly, the square of the stretch in the **n** direction (along line $\mathcal{AC}$) is

$$\lambda^2(\mathbf{n}) = \begin{bmatrix} \frac{3}{\sqrt{13}} & \frac{2}{\sqrt{13}} \end{bmatrix}\begin{bmatrix} 2 & -1.5 \\ -1.5 & 2.25 \end{bmatrix}\begin{Bmatrix} \frac{3}{\sqrt{13}} \\ \frac{2}{\sqrt{13}} \end{Bmatrix} = \frac{9}{13}$$

The deformed length of line $\mathcal{AC}$ is $L_f = L_o\lambda(\mathbf{n}) = \sqrt{13}\,(\sqrt{9/13}) = 3$, which is obvious from the sketch of the deformed configuration and confirmed by our calculation.

The triangle rotates and deforms. The angle between line $\mathcal{AC}$ from vertical is $\cos^{-1}(2/\sqrt{13}) = 0.9828$ rad or 56.31°. In the deformed configuration, that line is vertical. Hence, under the deformation it rotated by 56.31°. Line $\mathcal{A}'\mathcal{B}'$ makes an angle of 45° with horizontal. Hence, its rotation is 45°. The fact that these two lines did not rotate by the same amount is evidence that the motion is not a rigid body rotation about point $\mathcal{A}$ because every line would have rotated by the same amount if that were the case. The difference is explained by *shearing*.

The cosine of the angle between lines $\mathcal{AB}$ and $\mathcal{AC}$ is $\cos\beta_o = 3/\sqrt{13}$, giving an angle of $\beta_o = 33.69°$. The cosine of the angle between lines $\mathcal{A'B'}$ and $\mathcal{A'C'}$ is

$$
\begin{aligned}
\cos\beta &= \frac{\mathbf{e}_1 \cdot \mathbf{Cn}}{\lambda(\mathbf{e}_1)\lambda(\mathbf{n})} \\
&= \frac{1}{(\sqrt{2})\,(\sqrt{9/13})}\,[1\;\;0]\begin{bmatrix} 2 & -1.5 \\ -1.5 & 2.25 \end{bmatrix}\begin{Bmatrix} \frac{3}{\sqrt{13}} \\ \frac{2}{\sqrt{13}} \end{Bmatrix} \\
&= 0.7071
\end{aligned}
$$

Thus, $\beta = \cos^{-1}(0.7071) = 0.7845$. In other words, $\beta = 45°$. This fact is obvious from the sketch of the deformed configuration. The change in angle between lines $\mathcal{AB}$ and $\mathcal{AC}$ due to the deformation is $45 - 33.69 = 11.31°$, which is equal to the difference in the rotation of the two lines, $56.31 - 45 = 11.31°$. The difference is due to shearing.

The previous example shows that Eqs. 4.13 and 4.18 are a simple means of doing ordinary Euclidean geometry on the deformed configuration using knowledge only of the geometry of the reference configuration and the deformation gradient or the Lagrangian strain tensor.

4.4 Plane Strain

In Fig. 4.6, we saw that the orientation of the two squares in the reference configuration made a difference in how the deformed squares turned out. In this section we will examine how that orientation affects the components of the strain tensor $\mathbf{E}$. We will start with a deformation where the action is in the (x, y) plane with no deformation in the z-direction, i.e., a *plane strain* condition.

The state of strain characterized by the strain tensor $\mathbf{E}$ does not depend upon a coordinate system (just like a vector $\mathbf{v}$ exists independent of a coordinate system). When we lay out a coordinate system, we can describe the tensor in terms of its components relative to that coordinate system. For example, when we write

$$
\{\mathbf{E}\} = \begin{bmatrix} \varepsilon_{xx} & \varepsilon_{xy} \\ \varepsilon_{xy} & \varepsilon_{yy} \end{bmatrix}
$$

the components ε_{xx}, ε_{xy}, and ε_{yy} refer to the (x, y) coordinates shown in Fig. 4.13 on the left. This coordinate system has base vectors $\{\mathbf{e}_1, \mathbf{e}_2\}$, which point along the

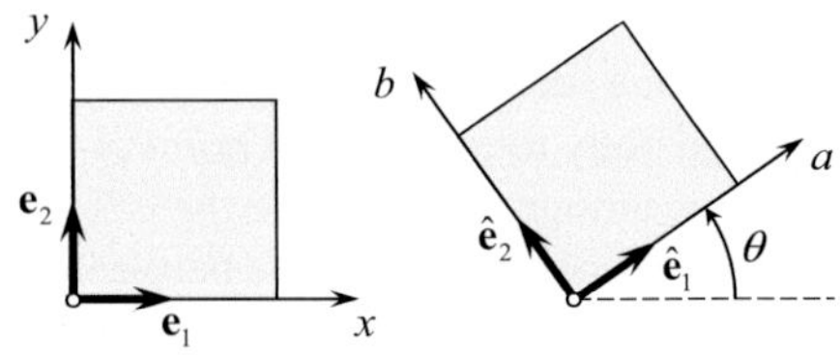

Fig. 4.13 *Transformation of coordinates*. Two coordinate systems (x, y), with base vectors $\{\mathbf{e}_1, \mathbf{e}_2\}$, and (a, b), with base vectors $\{\hat{\mathbf{e}}_1, \hat{\mathbf{e}}_2\}$. The second coordinate system is rotated by an angle θ from the first

coordinate axes. We can define a different coordinate system (a, b) with base vectors $\{\hat{\mathbf{e}}_1, \hat{\mathbf{e}}_2\}$, as illustrated on the right side of Fig. 4.13. The (a, b) coordinate system is rotated counterclockwise by an angle θ from the (x, y) system.

The base vectors in the (a, b) system can be expressed in terms of the base vectors in the (x, y) system as

$$\begin{aligned}\hat{\mathbf{e}}_1 &= \cos\theta\,\mathbf{e}_1 + \sin\theta\,\mathbf{e}_2\\ \hat{\mathbf{e}}_2 &= -\sin\theta\,\mathbf{e}_1 + \cos\theta\,\mathbf{e}_2\end{aligned}$$

We can also express the rotated base vectors in terms of the standard base vectors as

$$\hat{\mathbf{e}}_1 = \mathbf{Q}^T\mathbf{e}_1, \qquad \hat{\mathbf{e}}_2 = \mathbf{Q}^T\mathbf{e}_2$$

where $\mathbf{Q}$ is the orthogonal rotation tensor with components

$$\{\mathbf{Q}\} = \begin{bmatrix} \cos\theta & \sin\theta \\ -\sin\theta & \cos\theta \end{bmatrix}$$

relative to the base vectors $\{\mathbf{e}_1, \mathbf{e}_2\}$. The components of $\mathbf{Q}$ are also the cosines of the angles between the base vectors, i.e., $Q_{ij} = \hat{\mathbf{e}}_i \cdot \mathbf{e}_j$, as defined in Eq. 1.36.

Since we can relate the base vectors to each other, we can relate the components of the strain tensor in the two coordinate systems, too. First, recall that we can extract the components of the tensor by using the base vectors. To wit,

$$\varepsilon_{xx} = \mathbf{e}_1 \cdot \mathbf{E}\mathbf{e}_1, \qquad \varepsilon_{xy} = \mathbf{e}_1 \cdot \mathbf{E}\mathbf{e}_2, \qquad \varepsilon_{yy} = \mathbf{e}_2 \cdot \mathbf{E}\mathbf{e}_2$$

Second, note that each of these has a physical meaning through Eqs. 4.11 and 4.18. In particular, ε_{xx} and ε_{yy} are the strains of lines parallel to the x and y axes, respectively, and ε_{xy} is related to the change in angle between lines originally along the x and y axes. The components of the tensor $\mathbf{E}$ relative to the rotated basis can be computed as

$$\varepsilon_{aa} = \hat{\mathbf{e}}_1 \cdot \mathbf{E}\,\hat{\mathbf{e}}_1, \qquad \varepsilon_{ab} = \hat{\mathbf{e}}_1 \cdot \mathbf{E}\,\hat{\mathbf{e}}_2, \qquad \varepsilon_{bb} = \hat{\mathbf{e}}_2 \cdot \mathbf{E}\,\hat{\mathbf{e}}_2$$

with similar interpretations. Rendering the first equation with matrix components relative to $\{\mathbf{e}_1, \mathbf{e}_2\}$ shows that

$$
\begin{aligned}
\varepsilon_{aa} &= \begin{bmatrix} \cos\theta & \sin\theta \end{bmatrix} \begin{bmatrix} \varepsilon_{xx} & \varepsilon_{xy} \\ \varepsilon_{xy} & \varepsilon_{yy} \end{bmatrix} \begin{Bmatrix} \cos\theta \\ \sin\theta \end{Bmatrix} \\
&= \begin{bmatrix} \cos\theta & \sin\theta \end{bmatrix} \begin{Bmatrix} \varepsilon_{xx}\cos\theta + \varepsilon_{xy}\sin\theta \\ \varepsilon_{xy}\cos\theta + \varepsilon_{yy}\sin\theta \end{Bmatrix} \\
&= \varepsilon_{xx}\cos^2\theta + 2\varepsilon_{xy}\sin\theta\cos\theta + \varepsilon_{yy}\sin^2\theta
\end{aligned}
$$

A similar computation holds for the other two components, a task left to the reader. The result is the *strain transformation equations*, which can be summarized as

$$
\begin{aligned}
\varepsilon_{aa} &= \varepsilon_{xx}\cos^2\theta + 2\varepsilon_{xy}\sin\theta\cos\theta + \varepsilon_{yy}\sin^2\theta \\
\varepsilon_{ab} &= \big(\varepsilon_{yy} - \varepsilon_{xx}\big)\sin\theta\cos\theta + \varepsilon_{xy}\big(\cos^2\theta - \sin^2\theta\big) \\
\varepsilon_{bb} &= \varepsilon_{xx}\sin^2\theta - 2\varepsilon_{xy}\sin\theta\cos\theta + \varepsilon_{yy}\cos^2\theta
\end{aligned}
\tag{4.19}
$$

These equations provide a way to compute the components of $\mathbf{E}$ in the (a, b) coordinate system from the components of $\mathbf{E}$ in the (x, y) coordinate system. Of course, the components represent the same strain tensor $\mathbf{E}$. It is quite evident that the values of the components ε_{aa}, ε_{ab}, and ε_{bb} depend upon the angle θ. It is also clear that for $\theta = 0$, $\varepsilon_{aa} = \varepsilon_{xx}$, $\varepsilon_{ab} = \varepsilon_{xy}$, and $\varepsilon_{bb} = \varepsilon_{yy}$.

We can substitute $\theta + \pi/2$ into the equation for ε_{aa}. The result is the equation for ε_{bb} because

$$
\sin(\theta + \pi/2) = \cos\theta, \qquad \cos(\theta + \pi/2) = -\sin\theta
$$

4.4.1 Strain Invariants

As mentioned in Sect. 1.2.3, all tensors have *invariants*, which are functions of the tensor components that do *not* depend upon the coordinate system. For the planar problem represented by the strain transformation equations in Eq. 4.19, that would mean functions that do not depend upon θ. The strain transformation equations provide a simple framework for recognizing invariance in a two-dimensional setting.

If we add the first and third of Eq. 4.19 together, we find that

$$
\varepsilon_{aa} + \varepsilon_{bb} = \varepsilon_{xx} + \varepsilon_{yy} = \mathrm{tr}(\mathbf{E}) \tag{4.20}
$$

The sum of diagonal element of a tensor is called the *trace* of the tensor. The reason this sum does not depend upon θ is because the middle term cancels out and the outer terms are both multiplied by $\sin^2\theta + \cos^2\theta$, which is equal to one by the Pythagorean identity of trigonometry. The calculation is a bit more tedious, but we can show that the determinant, i.e., $\det(\mathbf{E})$, is also invariant. To wit,

$$
\varepsilon_{aa}\varepsilon_{bb} - \varepsilon_{ab}^2 = \varepsilon_{xx}\varepsilon_{yy} - \varepsilon_{xy}^2 = \det(\mathbf{E}) \tag{4.21}
$$

To prove this result, simply substitute Eq. 4.19 in the left side of Eq. 4.21 and simplify the result to show that it gives the expression on the right side of Eq. 4.21. The details of this calculation are left to the reader.

As mentioned in Sect. 1.2.3, any function of invariants is also invariant. For example, the function

$$\begin{aligned}\mathcal{R}(\mathbf{E}) &= \left(\tfrac{1}{2}\text{tr}(\mathbf{E})\right)^2 - \det(\mathbf{E}) \\ &= \tfrac{1}{4}\left(\varepsilon_{xx} + \varepsilon_{yy}\right)^2 - \left(\varepsilon_{xx}\varepsilon_{yy} - \varepsilon_{xy}^2\right) \\ &= \tfrac{1}{4}\left(\varepsilon_{xx} - \varepsilon_{yy}\right)^2 + \varepsilon_{xy}^2\end{aligned} \tag{4.22}$$

is invariant. This invariant will play a role in constructing the Mohr circle later in the chapter. In fact, we will discover that $\mathcal{R}(\mathbf{E})$ is the square of the radius of the Mohr circle.

4.4.2 The Principal Angle and Maximum Shear Angle

Equation 4.19 shows us that the strain components depend upon the angle θ, which characterizes the orientation of the (a, b) coordinate system. It is reasonable to ask if there are values of θ that lead to the largest or smallest values of the components. We could, for example, determine the value of θ that maximizes or minimizes $\varepsilon_{aa}(\theta)$ by taking the derivative of the first of Eq. 4.19, setting the result equal to zero, and solving for the angle. First, take the derivative

$$\frac{d\varepsilon_{aa}}{d\theta} = 2\left(\varepsilon_{yy} - \varepsilon_{xx}\right)\cos\theta\sin\theta + 2\varepsilon_{xy}\left(\cos^2\theta - \sin^2\theta\right)$$

Next, substitute the trigonometric identities

$$2\cos\theta\sin\theta = \sin 2\theta, \qquad \cos^2\theta - \sin^2\theta = \cos 2\theta \tag{4.23}$$

and set the equation to zero to find the maximum. Denoting θ_p as the solution, we can put the equation in the form

$$\left(\varepsilon_{yy} - \varepsilon_{xx}\right)\sin 2\theta_p + 2\varepsilon_{xy}\cos 2\theta_p = 0$$

and simplify to get

$$\tan 2\theta_p = \frac{2\varepsilon_{xy}}{\varepsilon_{xx} - \varepsilon_{yy}} \tag{4.24}$$

The angle θ_p is called the *principal angle*. It is the angle at which the strain component ε_{aa} achieves its maximum or minimum value. If we follow the same process for ε_{bb}, we get exactly the same equation. Hence, the same angle θ_p gives the maximum or minimum value of ε_{bb}, too. In fact, one of the two values achieves a maximum while the other gives a minimum. Observe, also, that

$$\frac{d\varepsilon_{aa}}{d\theta} = 2\varepsilon_{ab}$$

Therefore, the condition for a maximum of ε_{aa} also implies that $\varepsilon_{ab}(\theta_p) = 0$. In other words, if we etch a square on the body at the principal angle in the reference configuration, then there will be no shearing in the deformed configuration. The square will deform into a rectangle. That is the only angle at which that happens. However, angles of $\theta_p + n\pi/2$ also satisfy Eq. 4.24. When $n = 1$, the roles of ε_{bb} and ε_{aa} switch. If one of them was maximum at θ_p, then the other is maximum at $\theta_p + \pi/2$, and vice versa. For $n = 2$ the square has turned all the way around, and that configuration cannot be distinguished from the one at θ_p.

Principal Strains The maximum and minimum normal strains are called the *principal strains*, which can be computed by substituting the value θ_p into the strain transformation equations for ε_{aa} and ε_{bb}. The first one gives

$$\begin{aligned}
\varepsilon_{aa}(\theta_p) &= \varepsilon_{xx}\cos^2\theta_p + 2\varepsilon_{xy}\sin\theta_p\cos\theta_p + \varepsilon_{yy}\sin^2\theta_p \\
&= \tfrac{1}{2}\varepsilon_{xx}\big(1+\cos 2\theta_p\big) + \varepsilon_{xy}\sin 2\theta_p + \tfrac{1}{2}\varepsilon_{yy}\big(1-\cos 2\theta_p\big) \\
&= \tfrac{1}{2}\big(\varepsilon_{xx}+\varepsilon_{yy}\big) + \tfrac{1}{2}\big(\varepsilon_{xx}-\varepsilon_{yy}\big)\cos 2\theta_p + \varepsilon_{xy}\sin 2\theta_p
\end{aligned}$$

In the second line of this derivation, we used the trigonometric identities

$$\cos^2\theta = \tfrac{1}{2}\big(1+\cos 2\theta\big), \qquad \sin^2\theta = \tfrac{1}{2}\big(1-\cos 2\theta\big) \tag{4.25}$$

and the first of Eq. 4.23. We can eliminate θ_p from this equation using the triangle shown in Fig. 4.14 (with $\theta = \theta_p$). Let

$$\alpha = \tan 2\theta_p = \frac{2\varepsilon_{xy}}{\varepsilon_{xx} - \varepsilon_{yy}}$$

From the triangle, $\sin 2\theta_p = \alpha/\sqrt{1+\alpha^2}$ and $\cos 2\theta_p = 1/\sqrt{1+\alpha^2}$. Substituting these expressions gives

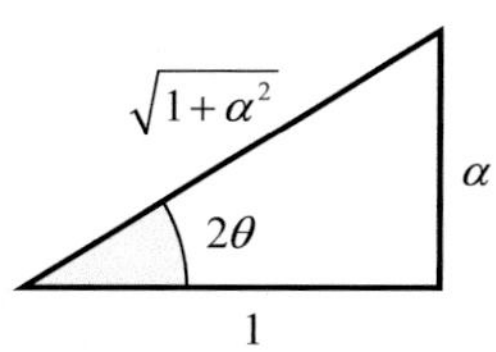

Fig. 4.14 *Triangle construction.* Triangle used to compute the principal angle and the maximum shear angle

$$\begin{aligned}\varepsilon_{aa}(\theta_p) &= \tfrac{1}{2}\big(\varepsilon_{xx}+\varepsilon_{yy}\big)+\tfrac{1}{2}\big(\varepsilon_{xx}-\varepsilon_{yy}\big)\frac{1}{\sqrt{1+\alpha^2}}+\varepsilon_{xy}\frac{\alpha}{\sqrt{1+\alpha^2}}\\ &= \tfrac{1}{2}\big(\varepsilon_{xx}+\varepsilon_{yy}\big)+\tfrac{1}{2}\big(\varepsilon_{xx}-\varepsilon_{yy}\big)\frac{1}{\sqrt{1+\alpha^2}}+\tfrac{1}{2}\big(\varepsilon_{xx}-\varepsilon_{yy}\big)\frac{\alpha^2}{\sqrt{1+\alpha^2}}\\ &= \tfrac{1}{2}\big(\varepsilon_{xx}+\varepsilon_{yy}\big)+\tfrac{1}{2}\big(\varepsilon_{xx}-\varepsilon_{yy}\big)\sqrt{1+\alpha^2}\end{aligned}$$

In the second line of this derivation, we used $2\varepsilon_{xy}=\alpha(\varepsilon_{yy}-\varepsilon_{xx})$ from the definition of α. Substituting the expression for α gives the final form of the equation

$$\varepsilon_{aa}(\theta_p)=\tfrac{1}{2}\big(\varepsilon_{xx}+\varepsilon_{yy}\big)+\sqrt{\tfrac{1}{4}\big(\varepsilon_{xx}-\varepsilon_{yy}\big)^2+\varepsilon_{xy}^2}=\varepsilon_1$$

Following the same process for ε_{bb} gives

$$\varepsilon_{bb}(\theta_p)=\tfrac{1}{2}\big(\varepsilon_{xx}+\varepsilon_{yy}\big)-\sqrt{\tfrac{1}{4}\big(\varepsilon_{xx}-\varepsilon_{yy}\big)^2+\varepsilon_{xy}^2}=\varepsilon_2$$

The values ε_1 and ε_2 are called the *principal values* of strain. They represent the largest and smallest possible values of the normal strain for a given strain tensor **E** for plain strain.

Maximum shear angle To find the angle at which the component ε_{ab} is maximum, we can follow the same process using the second of Eq. 4.19. To wit,

$$\begin{aligned}\frac{d\varepsilon_{ab}}{d\theta} &= \big(\varepsilon_{yy}-\varepsilon_{xx}\big)\big(\cos^2\theta-\sin^2\theta\big)-4\varepsilon_{xy}\cos\theta\sin\theta\\ &= \big(\varepsilon_{yy}-\varepsilon_{xx}\big)\cos 2\theta-2\varepsilon_{xy}\sin 2\theta=0\end{aligned}$$

This equation is satisfied at the *maximum shear angle* θ_m. Thus,

$$\tan 2\theta_m=-\frac{\varepsilon_{xx}-\varepsilon_{yy}}{2\varepsilon_{xy}} \tag{4.26}$$

This equation shows that the tangent of twice the maximum shear angle is the negative inverse of the tangent of twice the principal angle. To discover what this relationship implies, multiply Eqs. 4.24 and 4.26 to get

$$\tan 2\theta_p\tan 2\theta_m=-1$$

Converting the tangents to sine divided by cosine, and multiplying through by the cosine terms, we can rewrite this equation as

$$\cos 2\theta_p\cos 2\theta_m+\sin 2\theta_p\sin 2\theta_m=0$$

Finally, using the trigonometric identity $\cos(x-y) = \cos x \cos y + \sin x \sin y$, the equation reduces to $\cos(2\theta_p - 2\theta_m) = 0$. Because the cosine function passes through zero at the angle $\pm\pi/2$, the solution is $\theta_p - \theta_m = \pm\,\pi/4$. In other words, the absolute value of the difference between the principal angle and the maximum shear angle is always 45°.

The maximum value of the shear strain can be obtained by substituting the angle θ_m into the second of Eq. 4.19:

$$\begin{aligned}\varepsilon_{ab}(\theta_m) &= \big(\varepsilon_{yy} - \varepsilon_{xx}\big)\sin\theta_m\cos\theta_m + \varepsilon_{xy}\big(\cos^2\theta_m - \sin^2\theta_m\big)\\ &= \tfrac{1}{2}\big(\varepsilon_{yy} - \varepsilon_{xx}\big)\sin 2\theta_m + \varepsilon_{xy}\cos 2\theta_m\end{aligned}$$

Again, to eliminate θ_m from this equation, consider the triangle defined in Fig. 4.14 (with $\theta = \theta_m$), and let

$$\alpha = \tan 2\theta_m = \frac{\varepsilon_{yy} - \varepsilon_{xx}}{2\varepsilon_{xy}}$$

From the triangle, $\sin 2\theta_m = \alpha/\sqrt{1+\alpha^2}$ and $\cos 2\theta_m = 1/\sqrt{1+\alpha^2}$. Substituting these expressions gives

$$\begin{aligned}\varepsilon_{ab}(\theta_m) &= \tfrac{1}{2}\big(\varepsilon_{yy} - \varepsilon_{xx}\big)\frac{\alpha}{\sqrt{1+\alpha^2}} + \varepsilon_{xy}\frac{1}{\sqrt{1+\alpha^2}}\\ &= \alpha\varepsilon_{xy}\frac{\alpha}{\sqrt{1+\alpha^2}} + \varepsilon_{xy}\frac{1}{\sqrt{1+\alpha^2}}\\ &= \varepsilon_{xy}\sqrt{1+\alpha^2}\end{aligned}$$

Upon substituting for α in terms of the strain components, the maximum shear is

$$\gamma_{\max} = \varepsilon_{ab}(\theta_m) = \sqrt{\tfrac{1}{4}(\varepsilon_{xx} - \varepsilon_{yy})^2 + \varepsilon_{xy}^2} \tag{4.27}$$

Observe that the quantity under the square root is the invariant defined in Eq. 4.22. It makes sense that the maximum shear is invariant because it was the result of finding the maximum over all angles θ. Hence, it cannot depend upon θ.

The following example illustrates many of the features of the strain transformation equations.

Example 4.6
Show the variation of ε_{aa}, ε_{ab}, and ε_{bb} with θ for $\varepsilon_{xx} = 10$, $\varepsilon_{xy} = 15$, and $\varepsilon_{yy} = -5$ in accord with the strain transformation equations. Plot the values of the strains at the principal angle and maximum shear angle.

Solution. A plot of the strains given by Eq. 4.19 is shown in Fig. 4.15. The gray dots at $\theta = 0$ are ε_{xy}, ε_{xx}, and ε_{yy}, from top to bottom. The solid line is

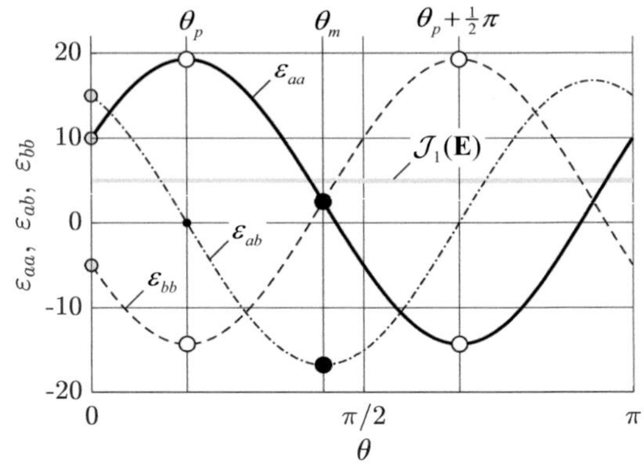

Fig. 4.15 *Strain transformation equations.* Plots of the strain transformation equations, Eq. 4.19, with values $\varepsilon_{xx} = 10$, $\varepsilon_{xy} = 15$, and $\varepsilon_{yy} = -5$

ε_{aa}, the dotted line is ε_{bb}, and the dot-dash line is ε_{ab}. The open circles are associated with the principal angle θ_p, and the black dots are associated with the maximum shear angle θ_m.

At the principal angle θ_p, one can see that $\varepsilon_{ab} = 0$. At that angle, ε_{aa} is maximum (hence, it is ε_1) and ε_{bb} is minimum (hence, it is ε_2). Clearly, at $\theta_p + \pi/2$, those roles reverse. At the maximum shear angle θ_m, we have $\varepsilon_{aa} = \varepsilon_{bb}$, but they are not equal to zero. Observe that the principal value is larger than any of the components in the (x, y) coordinates, which makes sense because we specifically sought out a maximum.

It is evident that all the curves are sinusoids with period equal to π. The strain invariant $\mathcal{J}_1(\mathbf{E}) = \varepsilon_{aa} + \varepsilon_{bb}$ is shown as a gray line, verifying that it is, indeed, independent of θ. The second invariant, $\det(\mathbf{E})$, which is not shown on this plot, is also constant.

4.4.3 The Mohr Circle

There are several different ways of visualizing the strain transformation relationships. One of those is the plot of the strain components versus angle, as shown in Fig. 4.15. Another is the *Mohr circle*.[3] To derive the Mohr circle, let us start with the first two strain transformation equations

$$\begin{aligned} \varepsilon &= \varepsilon_{xx}\cos^2\theta + 2\varepsilon_{xy}\sin\theta\cos\theta + \varepsilon_{yy}\sin^2\theta \\ \gamma &= \big(\varepsilon_{yy} - \varepsilon_{xx}\big)\sin\theta\cos\theta + \varepsilon_{xy}\big(\cos^2\theta - \sin^2\theta\big) \end{aligned} \tag{4.28}$$

[3] The Mohr circle is named after Christian Otto Mohr (1835–1918), the German civil engineer who conceived it.

where we are calling $\varepsilon = \varepsilon_{aa}$ and $\gamma = \varepsilon_{ab}$ for short. Noting the trigonometric identities in Eq. 4.25, we can rewrite the strain transformation equations in the form

$$\begin{aligned} \varepsilon - c &= \varepsilon_{xy} \sin 2\theta + \tfrac{1}{2}\big(\varepsilon_{xx} - \varepsilon_{yy}\big)\cos 2\theta \\ \gamma &= \varepsilon_{xy} \cos 2\theta - \tfrac{1}{2}\big(\varepsilon_{xx} - \varepsilon_{yy}\big)\sin 2\theta \end{aligned}$$

where $c = \frac{1}{2}\big(\varepsilon_{xx} + \varepsilon_{yy}\big)$. If we square both sides of both equations and then add them together, we arrive at the result

$$\big(\varepsilon - c\big)^2 + \gamma^2 = \tfrac{1}{4}\big(\varepsilon_{xx} - \varepsilon_{yy}\big)^2 + \varepsilon_{xy}^2$$

Finally, let us define the right side to be R^2. Now, the strain transformation equation is reduced to

$$\big(\varepsilon - c\big)^2 + \gamma^2 = R^2 \tag{4.29}$$

which is the equation of a circle in (ε, γ) space, centered at the point $(c, 0)$ with radius R. The center and radius are defined in terms of the strain components in the (x, y) coordinate system as

$$c = \tfrac{1}{2}\big(\varepsilon_{xx} + \varepsilon_{yy}\big), \qquad R = \sqrt{\tfrac{1}{4}(\varepsilon_{xx} - \varepsilon_{yy})^2 + \varepsilon_{xy}^2} \tag{4.30}$$

The quantities c and R are both invariants of the strain tensor. The Mohr circle is shown in Fig. 4.16. Observe that the components of the strain tensor in the (x, y) coordinate system are plotted in pairs $(\varepsilon_{yy}, \varepsilon_{xy})$ and $(\varepsilon_{xx}, -\varepsilon_{xy})$. The dotted line joining these two points passes through the center of the circle $(c, 0)$. Because we derived the Mohr circle from the strain transformation equations, the pairs $(\varepsilon_{bb}, \varepsilon_{ab})$ and $(\varepsilon_{aa}, -\varepsilon_{ab})$ will also plot on the circle, for any value of θ. The principal values of strain can be computed simply as

$$\varepsilon_1 = c + R, \qquad \varepsilon_2 = c - R \tag{4.31}$$

The maximum shear $\gamma_{\max}$ is equal to the radius R. Twice the principal angle $2\theta_p$ is the angle between the dotted line and the horizontal axis, while twice the maximum

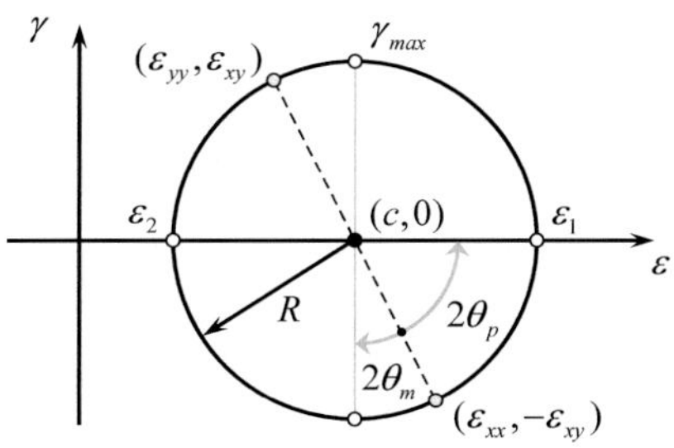

Fig. 4.16 *The Mohr circle.* The Mohr circle is a graphical representation of the state of strain

shear angle $2\theta_m$ is the angle between the dotted line and the vertical axis. It is clear from the Mohr circle that $|2\theta_p - 2\theta_m| = \pi/2$, a result we proved earlier.

Example 4.7
Construct the Mohr circle for a state of strain with $\varepsilon_{xx} = 20$, $\varepsilon_{xy} = 15$, and $\varepsilon_{yy} = 5$ microstrain.[4]

Solution. The plot of the Mohr circle is shown in Fig. 4.17. The values of the center and radius of the circle are

$$c = \tfrac{1}{2}(20 + 5) = 12.5, \qquad R = \tfrac{1}{2}\sqrt{(20-5)^2 + 4(15)^2} = 16.77$$

The center of the circle is at (12.5, 0). The maximum shear strain is

$$\gamma_{max} = 16.77$$

The principal values are

$$\varepsilon_1 = 12.5 + 16.77 = 29.27, \qquad \varepsilon_2 = 12.5 - 16.77 = -4.27$$

The principal angle can be computed from Eq. 4.24 to be $\theta_p = 31.7°$. All these values can be identified on the plot.

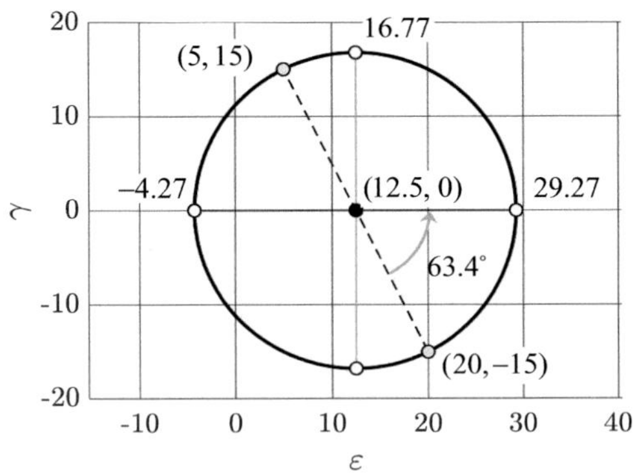

Fig. 4.17 *Example.* The Mohr circle for the state of strain with $\varepsilon_{xx} = 20$, $\varepsilon_{xy} = 15$, and $\varepsilon_{yy} = 5$

[4] The unit microstrain, often abbreviated as $\mu\epsilon$, means multiply by 10^{-6}.

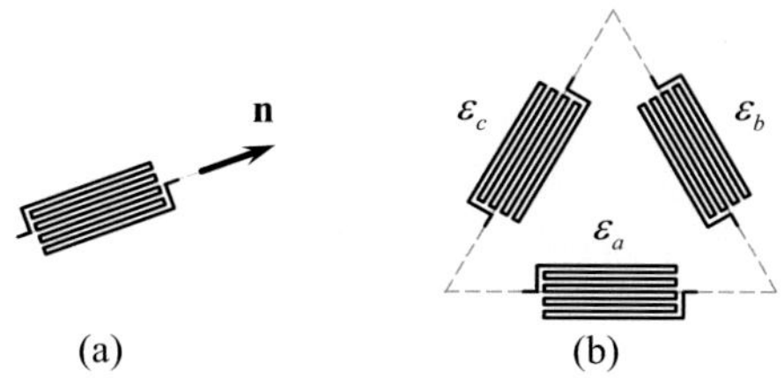

Fig. 4.18 *Strain gauges.* Electrical resistance strain gauges. **a** Single gauge oriented in the direction **n**. **b** Strain gauge rosette with gauges oriented along the sides of an equilateral triangle

4.4.4 Strain Gauges

Strain along a line oriented in a certain direction can be measured in different ways. One popular device is the electrical resistance *strain gauge*, illustrated in Fig. 4.18a. The strain gauge operation is based on the fact that the electrical resistance in a wire changes if the wire changes length. The amount of change is small, so the strain gauge amplifies the change by wrapping wires back and forth over a small distance. The voltage drop across the gauge is proportional to the change in the length of the wires, so the gauge can be calibrated to directly measure the strain in the direction of the axis, **n**, of the gauge.

Strain gauges can be arranged into a *strain gauge rosette*, as illustrated in Fig. 4.18b, which shows three individual gauges oriented along the sides of an equilateral triangle. Since all the gauges are oriented in different directions, they can be used to construct the components of the strain tensor **E** in the plane of the gauges and, therefore, the gauge readings ε_a, ε_b, and ε_c can be used to find the principal strains and principal directions in that plane, as outlined in the following example.

Example 4.8
Consider a strain gauge rosette with the gauges oriented along the sides of an equilateral triangle, as shown in Fig. 4.18b. The gauges read $\varepsilon_a = 120$, $\varepsilon_b = -80$, and $\varepsilon_c = 150$ microstrain. Find the components of the strain tensor in the plane and analyze the state of strain.

Solution. Let us assume that the rosette lies in the (x, y) plane, with gauge a oriented along the x-axis. The unit vectors in the direction of the gauges relative to this coordinate system have components

$$\mathbf{n}_a = \begin{Bmatrix} 1 \\ 0 \end{Bmatrix}, \qquad \mathbf{n}_b = \begin{Bmatrix} -\frac{1}{2} \\ \frac{\sqrt{3}}{2} \end{Bmatrix}, \qquad \mathbf{n}_c = \begin{Bmatrix} \frac{1}{2} \\ \frac{\sqrt{3}}{2} \end{Bmatrix}$$

To find the components of **E**, we will apply Eq. 4.11 in the directions of the gauges. For gauge a, we get

$$\varepsilon_a = \mathbf{n}_a \cdot \mathbf{E}\mathbf{n}_a = \begin{bmatrix} 1 & 0 \end{bmatrix} \begin{bmatrix} \varepsilon_{xx} & \varepsilon_{xy} \\ \varepsilon_{xy} & \varepsilon_{yy} \end{bmatrix} \begin{Bmatrix} 1 \\ 0 \end{Bmatrix} = \varepsilon_{xx} = 120$$

For gauge b, we get

$$\begin{aligned} \varepsilon_b = \mathbf{n}_b \cdot \mathbf{E}\mathbf{n}_b &= \begin{bmatrix} -\frac{1}{2} & \frac{\sqrt{3}}{2} \end{bmatrix} \begin{bmatrix} \varepsilon_{xx} & \varepsilon_{xy} \\ \varepsilon_{xy} & \varepsilon_{yy} \end{bmatrix} \begin{Bmatrix} -\frac{1}{2} \\ \frac{\sqrt{3}}{2} \end{Bmatrix} \\ &= \tfrac{1}{4}\varepsilon_{xx} - \tfrac{\sqrt{3}}{2}\varepsilon_{xy} + \tfrac{3}{4}\varepsilon_{yy} = -80 \end{aligned}$$

For gauge c, we get

$$\begin{aligned} \varepsilon_c = \mathbf{n}_c \cdot \mathbf{E}\mathbf{n}_c &= \begin{bmatrix} \frac{1}{2} & \frac{\sqrt{3}}{2} \end{bmatrix} \begin{bmatrix} \varepsilon_{xx} & \varepsilon_{xy} \\ \varepsilon_{xy} & \varepsilon_{yy} \end{bmatrix} \begin{Bmatrix} \frac{1}{2} \\ \frac{\sqrt{3}}{2} \end{Bmatrix} \\ &= \tfrac{1}{4}\varepsilon_{xx} + \tfrac{\sqrt{3}}{2}\varepsilon_{xy} + \tfrac{3}{4}\varepsilon_{yy} = 150 \end{aligned}$$

These equations constitute a system of three equations in three unknowns, which can be solved by substitution. First, notice that we have the value of ε_{xx} directly. Substituting that value into the other two equations gives

$$\begin{aligned} -\tfrac{\sqrt{3}}{2}\varepsilon_{xy} + \tfrac{3}{4}\varepsilon_{yy} &= -110 \\ \tfrac{\sqrt{3}}{2}\varepsilon_{xy} + \tfrac{3}{4}\varepsilon_{yy} &= 120 \end{aligned}$$

Adding the two equations eliminates ε_{xy} and subtracting them eliminates ε_{yy}. The result is

$$\varepsilon_{xx} = 120, \qquad \varepsilon_{xy} = 133, \qquad \varepsilon_{yy} = 6.7$$

These values can now be used to analyze the state of strain. The principal values are $\varepsilon_1 = 208$ and $\varepsilon_2 = -81$. The principal angle is $\theta_p = 0.584$ rad. The maximum shear strain is $\gamma_{\max} = 144$. All strains are in units of microstrain.

4.5 Analysis of Strain in Three Dimensions

Most of the concepts developed for planar problems extend to states of strain in three dimensions. Some do not. In this section we reconsider the concepts of principal values, principal directions, and the Mohr circle for states of strain in three dimensions. We will see that, while it is still possible to define strain transformation equations in three dimensions, that is not the simplest route to get to principal values and directions. Instead, we will find that the eigenvalue problem is the key to solving this problem.

4.5.1 Principal Values and Principal Directions

For the two-dimensional problem, we found that the principal angle θ_p was sufficient to describe the orientation of the small square that gave the maximum and minimum values of the strain $\varepsilon(\mathbf{n})$. There were two directions that appeared to be relevant, θ_p and $\theta_p + \pi/2$. One of these directions gave the maximum and one gave the minimum. At that time, we did not form unit vectors to describe those orientations, but we certainly can. They are

$$\begin{aligned}\mathbf{n}_1 &= \cos\theta_p \mathbf{e}_1 + \sin\theta_p \mathbf{e}_2 \\ \mathbf{n}_2 &= -\sin\theta_p \mathbf{e}_1 + \cos\theta_p \mathbf{e}_2\end{aligned}$$

These vectors are the same as $\{\hat{\mathbf{e}}_1, \hat{\mathbf{e}}_2\}$ from Fig. 4.13 at the specific angle θ_p. In three dimensions, a single angle is not adequate to describe orientation. Rather than think of the principal directions in terms of principal angles, we will directly employ unit vectors to describe the directions.

What we observed in two dimensions was that the principal orientation not only maximized or minimized the normal strain ε, it was also an orientation on which there was no shearing. In other words, right angles in the reference configuration remained right angles in the deformed configuration. Consider three orthogonal unit vectors **n**, **p**, and **q** in the reference configuration, as shown in Fig. 4.19. The change in angles among them caused by the deformation can be computed from Eq. 4.18. Rearranging the equation and noting that $\cos\beta_o$ is zero between any two of the three vectors, we can write

$$\begin{aligned}\lambda(\mathbf{p})\lambda(\mathbf{n})\cos\beta(\mathbf{p},\mathbf{n}) &= 2\,\mathbf{p}\cdot\mathbf{En} \\ \lambda(\mathbf{q})\lambda(\mathbf{n})\cos\beta(\mathbf{q},\mathbf{n}) &= 2\,\mathbf{q}\cdot\mathbf{En}\end{aligned} \tag{4.32}$$

If the angles in the deformed configuration are right angles, which is the case if the shear deformation is zero, then $\cos\beta = 0$ in both cases. Thus,

$$\mathbf{p}\cdot\mathbf{En} = \mathbf{0}, \qquad \mathbf{q}\cdot\mathbf{En} = \mathbf{0} \tag{4.33}$$

The vector **En** is perpendicular to both **p** and **q**. From the reference configuration, we know that the vector **n** is orthogonal to both **p** and **q**. Therefore, while we do not know the magnitude of **En**, we know that it must be a scalar multiple of **n**, which we can express mathematically as

$$\mathbf{En} = \varepsilon\mathbf{n} \tag{4.34}$$

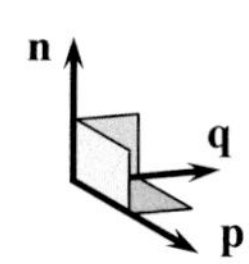

Fig. 4.19 *Orthogonal unit vectors.* The unit vectors **n**, **p**, and **q** are normal to each other in the reference configuration

This is an *eigenvalue problem*, whose solution $\mathbf{n}$ is the principal direction. We will also find the value of ε and determine that it is the principal value. The reason for selecting the symbol ε is because if you dot Eq. 4.34 with $\mathbf{n}$, you get $\varepsilon = \mathbf{n} \cdot \mathbf{E}\mathbf{n}$, which is consistent with Eq. 4.11.

To solve the eigenvalue problem, let us put it into the form

$$\left[\mathbf{E} - \varepsilon\mathbf{I}\right]\mathbf{n} = \mathbf{0}$$

This equation is a homogeneous system of linear equations for the unknown components of the vector $\mathbf{n}$. A theorem of linear algebra tells us that a solution other than $\mathbf{n} = \mathbf{0}$ is possible only if the determinant of the coefficient matrix is equal to zero, i.e., $\det[\mathbf{E} - \varepsilon\mathbf{I}] = 0$. Written out in components, this equation takes the form

$$\det\begin{bmatrix} \varepsilon_{xx} - \varepsilon & \varepsilon_{xy} & \varepsilon_{xz} \\ \varepsilon_{xy} & \varepsilon_{yy} - \varepsilon & \varepsilon_{yz} \\ \varepsilon_{xz} & \varepsilon_{yz} & \varepsilon_{zz} - \varepsilon \end{bmatrix} = 0$$

Carrying out the computation of the determinant and collecting terms results in a cubic equation for ε in the form

$$f(\varepsilon) = \varepsilon^3 - \mathcal{J}_1(\mathbf{E})\,\varepsilon^2 + \mathcal{J}_2(\mathbf{E})\,\varepsilon - \mathcal{J}_3(\mathbf{E}) = 0 \tag{4.35}$$

where the coefficients are the *principal invariants* of the strain tensor $\mathbf{E}$ given by (see Eq. 1.45)

$$\begin{aligned} \mathcal{J}_1(\mathbf{E}) &= \mathrm{tr}(\mathbf{E}) \\ \mathcal{J}_2(\mathbf{E}) &= \tfrac{1}{2}\left[\left(\mathrm{tr}(\mathbf{E})\right)^2 - \mathrm{tr}(\mathbf{E}^2)\right] \\ \mathcal{J}_3(\mathbf{E}) &= \det(\mathbf{E}) \end{aligned} \tag{4.36}$$

where $\mathrm{tr}(\bullet)$ is the trace of the tensor $(\bullet)$, the sum of the diagonal elements. Thus, $\mathrm{tr}(\mathbf{E}) = \varepsilon_{xx} + \varepsilon_{yy} + \varepsilon_{zz}$. The square of a tensor is simply the tensor multiplied by itself, e.g., $\mathbf{E}^2 = \mathbf{E}\mathbf{E}$. We can write out the invariants as

$$\begin{aligned} \mathcal{J}_1(\mathbf{E}) &= \varepsilon_{xx} + \varepsilon_{yy} + \varepsilon_{zz} \\ \mathcal{J}_2(\mathbf{E}) &= \varepsilon_{xx}\varepsilon_{yy} + \varepsilon_{yy}\varepsilon_{zz} + \varepsilon_{zz}\varepsilon_{xx} - \varepsilon_{xy}^2 - \varepsilon_{yz}^2 - \varepsilon_{xz}^2 \\ \mathcal{J}_3(\mathbf{E}) &= \varepsilon_{xx}\varepsilon_{yy}\varepsilon_{zz} - \varepsilon_{xx}\varepsilon_{yz}^2 - \varepsilon_{yy}\varepsilon_{xz}^2 - \varepsilon_{zz}\varepsilon_{xy}^2 + 2\varepsilon_{xy}\varepsilon_{yz}\varepsilon_{xz} \end{aligned}$$

These invariants are analogous to the ones we encountered in the two-dimensional problem. We will show the relationship between them in Sect. 4.5.3, when we revisit the plane strain problem from the three-dimensional perspective.

In Eq. 4.35, $f(\varepsilon)$ is called the *characteristic function*, and the equation is called the *characteristic equation*. The solution of the characteristic equation yields the principal values ε_1, ε_2, and ε_3 of the strain tensor $\mathbf{E}$. We will order the principal values so that

$$\varepsilon_1 \geq \varepsilon_2 \geq \varepsilon_3$$

This ordering is arbitrary but will help to clarify any comments related to the principal values. Each one of the principal values ε_i is associated with a principal direction $\mathbf{n}_i$, which we will take as a unit vector. Eigenvectors do not need to be unit vectors to satisfy Eq. 4.34, but we will take them as unit vectors in anticipation of how we will use them later. The approach to finding the eigenvectors is outlined in Appendix A.5. The eigenvalue problem is discussed in more detail in Sect. 1.2.4.

Example 4.9
Find the principal values and principal directions of the strain tensor with components relative to the standard base vectors

$$\{\mathbf{E}\} = \begin{bmatrix} 0.10 & 0.00 & -0.10 \\ 0.00 & 0.20 & 0.05 \\ -0.10 & 0.05 & 0.15 \end{bmatrix}$$

Solution. The invariants of **E** can be computed from Eq. 4.36 to give the values

$$\mathscr{I}_1(\mathbf{E}) = 0.45 \qquad \mathscr{I}_2(\mathbf{E}) = 0.0025 \qquad \mathscr{I}_3(\mathbf{E}) = 0.00075$$

The characteristic equation is

$$\varepsilon^3 - 0.45\varepsilon^2 + 0.0025\varepsilon - 0.00075 = 0$$

This equation can be solved by Newton's method (see Code 4.1 below) to give the principal strains

$$\varepsilon_1 = 0.2573, \qquad \varepsilon_2 = 0.1762, \qquad \varepsilon_3 = 0.01655$$

The first eigenvector can be found by solving $[\,\mathbf{E} - \varepsilon_1 \mathbf{I}\,]\mathbf{n}_1 = \mathbf{0}$. Substituting ε_1, we get

$$\begin{bmatrix} -0.1573 & 0.0000 & -0.1000 \\ 0.0000 & -0.0573 & 0.0500 \\ -0.1000 & 0.0500 & -0.1073 \end{bmatrix} \begin{Bmatrix} 1 \\ a \\ b \end{Bmatrix} = \begin{Bmatrix} 0 \\ 0 \\ 0 \end{Bmatrix}$$

where we have assumed the first component of the vector $\mathbf{n}_1$ is equal to one because the equations are singular. This choice will work only if that first component, in reality, is not zero. Taking the second and third equations, we get

$$\begin{bmatrix} -0.0573 & 0.0500 \\ 0.0500 & -0.1073 \end{bmatrix} \begin{Bmatrix} a \\ b \end{Bmatrix} = \begin{Bmatrix} 0.0 \\ 0.1 \end{Bmatrix} \quad \rightarrow \quad \begin{Bmatrix} a \\ b \end{Bmatrix} = \begin{Bmatrix} -1.3733 \\ -1.5726 \end{Bmatrix}$$

Thus, the first eigenvector is $\mathbf{n}_1 = [\,1, -1.3733, -1.5726\,]$, which can be scaled to a unit vector by dividing by its own length. If the 2×2 matrix is singular, then the assumption that the first component is one was wrong. If that is the case, then make a different assumption on which component is one. At least one of the components must be nonzero. The second and third eigenvectors can be found by the same process. In summary, the three (unit) eigenvectors are

$$\mathbf{n}_1 = \begin{Bmatrix} 0.4320 \\ -0.5932 \\ -0.6793 \end{Bmatrix}, \quad \mathbf{n}_2 = \begin{Bmatrix} 0.4913 \\ 0.7864 \\ -0.3744 \end{Bmatrix}, \quad \mathbf{n}_3 = \begin{Bmatrix} 0.7563 \\ -0.1720 \\ 0.6312 \end{Bmatrix}$$

The solution process for this example is implemented in the following MATLAB code.

Code 4.1 Code `Eigen` to find the roots of the characteristic equation by Newton's method and to compute the eigenvectors once eigenvalues have been found

```
   clear; clc;

%. Input strain tensor
   E = [0.1, 0.0, -0.1; 0.0, 0.2, 0.05; -0.1, 0.05, 0.15];

%. Compute primary invariants
   I = [trace(E); trace(E*E); trace(E*E*E)];

%. Compute principal invariants
   J1 = I(1);
   J2 = (I(1)^2-I(2))/2;
   J3 = (I(1)^3 - 3*I(1)*I(2) + 2*I(3))/6;

%. Set tolerance and starting value for Newton's method
   tol = 1.e-8;
   StartValues = [0.4,0.2,0.0];

%. Newton iteration to solve for principal strains
   for i=1:3
     err = 1; e = StartValues(i);
     while err>tol
       g = e^3 - J1*e^2 + J2*e - J3;
       A = 3*e^2 - 2*J1*e + J2;
       e = e - g/A;
       err = norm(g);
     end
     fprintf('\n Solution %2i: %12.7f, n = ',i,e)

%... Compute eigenvector associated with e
     C = E - e*eye(3);
     B = C(2:3,2:3);  b = -C(2:3,1);
     n = [1; B\b];    n = n/norm(n);
     fprintf('%9.4f',n);

   end
```

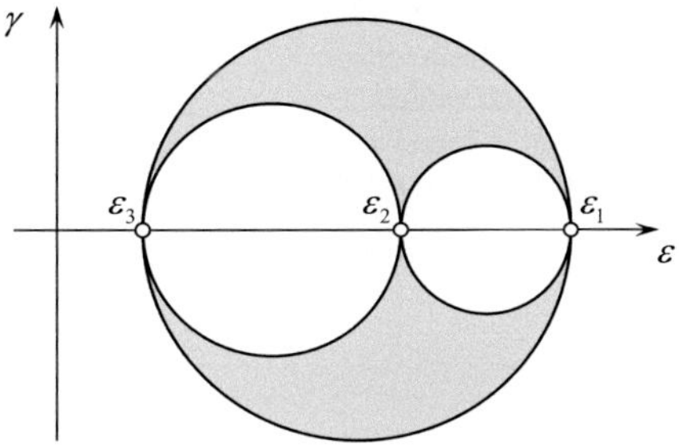

Fig. 4.20 *The Mohr circle.* For three-dimensional states of strain, there are three Mohr circles, one between each pair of principal strains. A state of strain (ε, γ) at some arbitrary orientation $\mathbf{n}$ must lie in the gray region

Of course, the eigenvalues and eigenvectors can also be obtained with the MATLAB `eig` function.

4.5.2 The Mohr Circle in Three Dimensions

In three dimensions there are three individual Mohr circles (not a Mohr sphere), as illustrated in Fig. 4.20. Each circle is the planar Mohr circle in the plane perpendicular to one of the principal directions. The simplest way to construct the circles is to recognize that each circle crosses the ε axis at two of the three principal values. There are three pairings of the three values, so the result is three circles. The interesting feature of the three-dimensional Mohr circle is that for any random direction $\mathbf{n}$ you can generate a value of ε and γ, and those can be represented on the plot. These values can be computed from

$$\varepsilon = \mathbf{n} \cdot \mathbf{En}, \qquad \gamma = \mathbf{m} \cdot \mathbf{En} \tag{4.37}$$

where the vector $\mathbf{m}$ points in the direction perpendicular to $\mathbf{n}$ and lies in the plane containing $\mathbf{En}$, which can be computed from $\mathbf{n}$ as $\hat{\mathbf{m}} = \mathbf{PEn}$, where $\mathbf{P} = \mathbf{I} - \mathbf{nn}^T$ is the projection tensor. Then, $\mathbf{m} = \hat{\mathbf{m}}/\|\hat{\mathbf{m}}\|$. Projecting $\mathbf{En}$ essentially finds the shadow of that vector on the plane with normal $\mathbf{n}$. The selection of $\mathbf{m}$ is different from the two-dimensional case, where there was only one vector perpendicular to $\mathbf{n}$. It is possible to prove that *any* (ε, γ) pair will fall in the shaded region between the outer circle and the two inner circles, as the following example demonstrates.

Example 4.10
Show that the values of ε and γ plot in the gray region of the Mohr circle by selecting random directions $\mathbf{n}$ for the strain tensor $\mathbf{E}$ with principal values $\varepsilon_1 = 11$, $\varepsilon_2 = 7$, and $\varepsilon_3 = 1$.

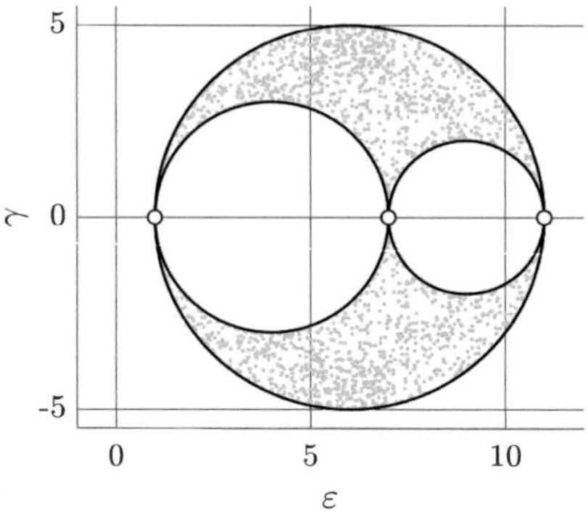

Fig. 4.21 *Example.* Mohr circles for a strain tensor with principal values $\varepsilon_1 = 11$, $\varepsilon_2 = 7$, and $\varepsilon_3 = 1$ with 1000 random **n** vectors

Solution. The results for 1000 random **n** vectors is shown in Fig. 4.21. Note that in order to get all the points shown, we have plotted both (ε, γ) and $(\varepsilon, -\gamma)$. The plot was generated from Eq. 4.37 with a MATLAB code, which is essentially the same as Code 5.1 for the three-dimensional Mohr circle for stress.

From the plot it is clear that $\gamma_{\max} = (\varepsilon_1 - \varepsilon_3)/2$, the radius of the largest circle. If two of the three principal values are equal, then the shaded region in Fig. 4.20 vanishes, and the plot reduces to a single circle. More accurately, one of the circles shrinks to a point and the other two lie on top of each other. If all three principal values are equal, then the Mohr circle reduces to a single point.

4.5.3 Plane Strain Revisited

The state of strain called *plane strain* is characterized by

$$\varepsilon_{zz} = \varepsilon_{xz} = \varepsilon_{yz} = 0$$

Substituting these values into the expressions for the strain invariants in Eq. 4.36 gives

$$\hat{\mathcal{J}}_1(\mathbf{E}) = \varepsilon_{xx} + \varepsilon_{yy}, \qquad \hat{\mathcal{J}}_2(\mathbf{E}) = \varepsilon_{xx}\varepsilon_{yy} - \varepsilon_{xy}^2, \qquad \hat{\mathcal{J}}_3(\mathbf{E}) = 0$$

where the hat on the symbols is there to remind us that this is a special case of a three-dimensional problem. It is interesting to note that the determinant of a 2×2 matrix is the same as $\hat{\mathcal{J}}_2$ for this special case (see Eq. 4.21), which is why we called that same quantity $\det(\mathbf{E})$ for the two-dimensional case. The characteristic equation reduces to

$$f(\varepsilon) = \varepsilon\left(\varepsilon^2 - \hat{\mathcal{J}}_1\varepsilon + \hat{\mathcal{J}}_2\right) = 0$$

One of the solutions to this cubic is $\varepsilon_3 = 0$.[5] The other two can be found by the quadratic formula as

$$\begin{aligned}
\varepsilon_{1,2} &= \tfrac{1}{2}\hat{\mathcal{J}}_1 \pm \tfrac{1}{2}\sqrt{\hat{\mathcal{J}}_1^2 - 4\hat{\mathcal{J}}_2} \\
&= \tfrac{1}{2}\left(\varepsilon_{xx} + \varepsilon_{yy}\right) \pm \tfrac{1}{2}\sqrt{\left(\varepsilon_{xx} + \varepsilon_{yy}\right)^2 - 4\left(\varepsilon_{xx}\varepsilon_{yy} - \varepsilon_{xy}^2\right)} \\
&= \tfrac{1}{2}\left(\varepsilon_{xx} + \varepsilon_{yy}\right) \pm \sqrt{\tfrac{1}{4}\left(\varepsilon_{xx} - \varepsilon_{yy}\right)^2 + \varepsilon_{xy}^2}
\end{aligned}$$

From the final result of this calculation, we can identify $\varepsilon_1 = c + R$ and $\varepsilon_2 = c - R$, where c and R are the center and radius of the Mohr circle from Eq. 4.30, which we found considering only a two-dimensional formulation.

Example 4.11

Find the principal directions for the plane strain condition from the eigenvalue problem. Show how the principal directions $\mathbf{n}_1$ and $\mathbf{n}_2$ relate to the principal angle θ_p from Eq. 4.24.

Solution. Once the principal values have been found, we can substitute them back into the eigenvalue problem to solve for the directions $\mathbf{n}$. First, recognize that the eigenvector associated with $\varepsilon_3 = 0$ is $\mathbf{n}_3 = \mathbf{e}_3$. The remaining two eigenvectors must be orthogonal to $\mathbf{e}_3$, which means that the $\mathbf{e}_3$ component of those vectors will be zero. Thus, all we need to do is solve for the other two components. The vector $\mathbf{n} = \cos\theta\mathbf{e}_1 + \sin\theta\mathbf{e}_2$ is a unit vector orthogonal to $\mathbf{e}_3$ for any value of θ.

First consider ε_1 with corresponding eigenvector $\mathbf{n}_1 = \cos\theta_p\mathbf{e}_1 + \sin\theta_p\mathbf{e}_2$, where θ_p is the principal angle. This vector must satisfy the equation (the eigenvalue problem)

$$\begin{bmatrix} \varepsilon_{xx} - \varepsilon_1 & \varepsilon_{xy} \\ \varepsilon_{xy} & \varepsilon_{yy} - \varepsilon_1 \end{bmatrix} \begin{Bmatrix} \cos\theta_p \\ \sin\theta_p \end{Bmatrix} = \begin{Bmatrix} 0 \\ 0 \end{Bmatrix}$$

Because ε_1 was specifically chosen to make the 2×2 matrix singular, both of the equations will give the same result for θ_p. Hence, we can use either equation to find its value. Taking the first equation, we get

$$\left(\varepsilon_{xx} - \varepsilon_1\right)\cos\theta_p + \varepsilon_{xy}\sin\theta_p = 0$$

[5] We are labeling this principal value as ε_3 even though we do not yet know the values of the other roots. It may turn out that this choice violates our arbitrary ordering system because one or both ε_1 and ε_2 could be less than zero. However, this labeling will be consistent with the two-dimensional formulation of plane strain.

Thus,

$$\tan\theta_p = \frac{\varepsilon_1 - \varepsilon_{xx}}{\varepsilon_{xy}} = \frac{\frac{1}{2}(\varepsilon_{xx} + \varepsilon_{yy}) + \sqrt{\frac{1}{4}(\varepsilon_{xx} - \varepsilon_{yy})^2 + \varepsilon_{xy}^2} - \varepsilon_{xx}}{\varepsilon_{xy}}$$

which we can write as

$$\tan\theta_p = -\alpha + \sqrt{1+\alpha^2}$$

if we define

$$\alpha = \frac{\varepsilon_{xx} - \varepsilon_{yy}}{2\,\varepsilon_{xy}}$$

Now, we can use the trigonometric identity

$$\begin{aligned}\tan 2\theta_p &= \frac{2\tan\theta_p}{1-\tan^2\theta_p}\\ &= \frac{2(-\alpha+\sqrt{1+\alpha^2})}{1-(\alpha^2 - 2\alpha\sqrt{1+\alpha^2} + 1 + \alpha^2)}\\ &= \frac{2(-\alpha+\sqrt{1+\alpha^2})}{2\alpha(-\alpha+\sqrt{1+\alpha^2})} = \frac{1}{\alpha} = \frac{2\varepsilon_{xy}}{\varepsilon_{xx} - \varepsilon_{yy}}\end{aligned}$$

which is the same result as Eq. 4.24, showing that the two approaches agree. The principal direction for the second principal value must be orthogonal to the first one. Thus, $\mathbf{n}_2 = -\sin\theta_p\mathbf{e}_1 + \cos\theta_p\mathbf{e}_2$. The details are left to the reader.

4.5.4 Volume Change

Deformation of a solid body that involves pure expansion in all directions but no shearing is a condition we call *dilatation*. Consider the deformation shown in Fig. 4.22. The initial cuboid of dimensions $a \times b \times c$ with edges oriented along the principal directions experiences a deformation. We can see that a change in volume involves a change in length of each side with no shearing. The deformed length of each side is given by Eq. 4.6, with the strains along the respective sides being the principal strains ε_1, ε_2, and ε_3. Using the equation for the length of a deformed line in terms of strain, Eq. 4.6, we can write the volume of the deformed cuboid as

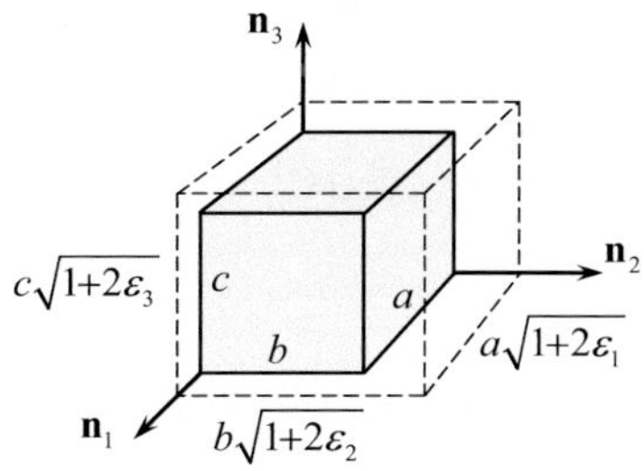

Fig. 4.22 *Volume change.* Rectangular parallelepiped of dimensions $a \times b \times c$ with edges oriented along the principal directions, expanding under the deformation

$$
\begin{aligned}
V &= \left(a\sqrt{1+2\varepsilon_1}\,\right)\left(b\sqrt{1+2\varepsilon_2}\,\right)\left(c\sqrt{1+2\varepsilon_3}\,\right) \\
&= abc\sqrt{(1+2\varepsilon_1)(1+2\varepsilon_2)(1+2\varepsilon_3)} \\
&= V_o\sqrt{1+2\mathcal{J}_1(\mathbf{E})+4\mathcal{J}_2(\mathbf{E})+8\mathcal{J}_3(\mathbf{E})}
\end{aligned}
$$

where $V_o = abc$ is the undeformed volume of the cuboid and the strain invariants in principal coordinates are

$$
\begin{aligned}
\mathcal{J}_1(\mathbf{E}) &= \varepsilon_1 + \varepsilon_2 + \varepsilon_3 \\
\mathcal{J}_2(\mathbf{E}) &= \varepsilon_1\varepsilon_2 + \varepsilon_2\varepsilon_3 + \varepsilon_3\varepsilon_1 \\
\mathcal{J}_3(\mathbf{E}) &= \varepsilon_1\varepsilon_2\varepsilon_3
\end{aligned}
\tag{4.38}
$$

because in the principal orientation, the strain tensor $\mathbf{E}$ is diagonal with the principal values on the diagonal, as shown in Sect. 1.2.5. Observe that the first invariant is a sum of strains, the second invariant is the sum of the products of two strains, and the third invariant is the product of three strains.

If the strains are small, then the third invariant is much smaller than the second, which is much smaller than the first. If we neglect the two smallest terms, then we can find an approximation to the deformed volume. Noting that if $x \ll 1$, then $\sqrt{1+2x} \approx 1+x$. Therefore, we can write

$$
V \approx V_o\left[\,1+\mathcal{J}_1(\mathbf{E})\,\right] = V_o\left[\,1+\mathrm{tr}(\mathbf{E})\,\right]
$$

Thus, the ratio of the change in volume to the original volume can be expressed as

$$
\frac{\Delta V}{V_o} = \frac{V - V_o}{V_o} \approx \mathrm{tr}(\mathbf{E}) \tag{4.39}
$$

Equation 4.39 is an approximation that works only for small deformations. The exact relationship between the undeformed and deformed volumes can be expressed in terms of the Green deformation tensor

$$
V = V_o\sqrt{\det(\mathbf{C})} \tag{4.40}
$$

where $\mathbf{C} = \mathbf{F}^T\mathbf{F}$ and $\mathbf{F}$ is the deformation gradient from Eq. 4.2. The reason this is true is because $\mathbf{C} = \mathbf{I} + 2\mathbf{E}$, and we can show that

$$\det(\mathbf{C}) = \det\left(\mathbf{I} + 2\mathbf{E}\right) = 1 + 2\mathcal{J}_1(\mathbf{E}) + 4\mathcal{J}_2(\mathbf{E}) + 8\mathcal{J}_3(\mathbf{E})$$

4.6 Nonhomogeneous Deformation

Solid bodies subjected to loads generally experience *nonhomogeneous deformation*. While the consideration of homogeneous deformations allowed us to explore the concept of strain without using calculus, it does not allow the exploration of strain gradients. In many applications, spatial gradients in strain (and the associated stresses) are central to understanding the system. For example, high strain gradients are present at reentrant corners of welded assemblies and in the vicinity of crack tips of fractured materials. In more common scenarios like axial extension, beam bending, and torsion, strain gradients develop as the main mode of response to loading. In this section we revisit the developments done for homogeneous deformations and generalize them for nonhomogeneous deformations.

Let us return to the *deformation map* illustrated in Fig. 4.2. Recall that the deformation map takes as input the location of a point like $\mathcal{P}$ in the reference configuration and gives as output the location of that same point in the deformed configuration. The mapping is described by Eq. 4.1, which we repeat here for convenience:

$$\mathbf{x} = \boldsymbol{\varphi}(\mathbf{z}) \tag{4.41}$$

The vector $\mathbf{z}$ is the location of $\mathcal{P}$ relative to the origin of the reference configuration, which is often taken as the reference point for translational motion. The vector $\mathbf{x}$ is the position of that same point, marked as $\mathcal{P}'$, relative to the origin in the deformed configuration. There is no need for these two origins to be in the same place. Similarly, each configuration can have its own base vectors to establish a coordinate system. However, for many problems it will be convenient to have a common origin and common base vectors for both configurations.

4.6.1 *The Deformation Gradient*

Consider the two points $\mathcal{A}$ and $\mathcal{B}$ in the reference configuration, as shown in Fig. 4.23. Point $\mathcal{A}$ is located at $\mathbf{z}$ relative to the origin $\mathcal{O}$. Point $\mathcal{B}$ is located at $\mathbf{z} + s\mathbf{n}$ (by vector addition). The distance between the two points is s, and the unit vector $\mathbf{n}$ points from $\mathcal{A}$ to $\mathcal{B}$. One can think of these two points as lying along a straight line connecting the two points, which is shown in light gray. In the deformed configuration, these two points have moved to locations $\mathcal{A}'$ and $\mathcal{B}'$ and are located at positions $\boldsymbol{\varphi}(\mathbf{z})$ and $\boldsymbol{\varphi}(\mathbf{z} + s\mathbf{n})$, relative to origin $\mathcal{O}'$. The difference between these two vectors is

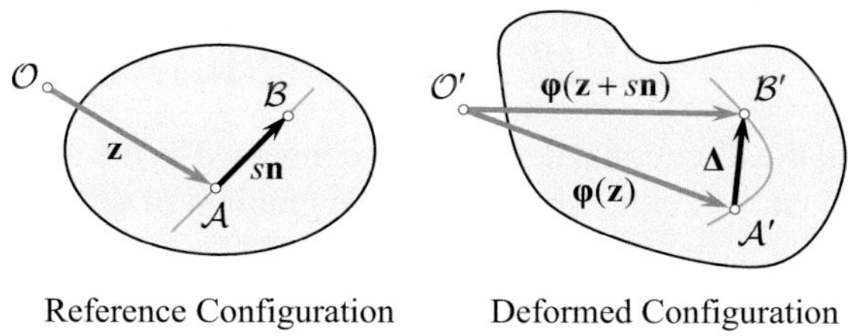

Fig. 4.23 *Motion of two points.* The derivation of the deformation gradient considers two points, $\mathcal{A}$ and $\mathcal{B}$, that lie along a line in direction $\mathbf{n}$ a distance s apart. The two points are mapped with the mapping function $\mathbf{x} = \boldsymbol{\varphi}(\mathbf{z})$

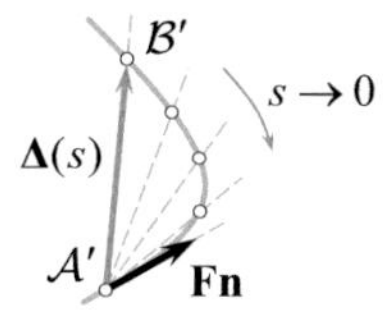

Fig. 4.24 *Limit of secants.* In the limit, the vector $\boldsymbol{\Delta}(s)/s \to \mathbf{Fn}$ as $s \to 0$. The secant aligns with the tangent to the curve at $\mathcal{A}'$

$$\boldsymbol{\Delta}(s) = \boldsymbol{\varphi}(\mathbf{z} + s\mathbf{n}) - \boldsymbol{\varphi}(\mathbf{z}) \tag{4.42}$$

which is a vector that lies along the secant to the deformed line connecting the two points (again, shown in light gray). If s is small, then the length of the secant is very nearly the same as the length of the arc of the curve between the two points. In the limit as $s \to 0$, the length of the secant and arc become identical, as shown in Fig. 4.24. This setup gives us a route to finding out how much the line stretches in the neighborhood of $\mathcal{A}$.

The distance between the two points in the reference configuration is s. The distance between the two points in the deformed configuration is $\|\boldsymbol{\Delta}(s)\|$, which converges to the length of the curve in the limit as $s \to 0$. The ratio of the deformed length of the line to the original length is the *stretch*. Thus,

$$\lambda = \lim_{s \to 0} \frac{\|\boldsymbol{\Delta}(s)\|}{s} \tag{4.43}$$

By the definition of the derivative of a function, we can write

$$\lim_{s \to 0} \frac{\boldsymbol{\varphi}(\mathbf{z} + s\mathbf{n}) - \boldsymbol{\varphi}(\mathbf{z})}{s} = \mathcal{D}\boldsymbol{\varphi}(\mathbf{z}) \cdot \mathbf{n} \tag{4.44}$$

This quantity is known as the *directional derivative* of the vector-valued function $\boldsymbol{\varphi}(\mathbf{z})$. It is the rate of change of $\boldsymbol{\varphi}$ in the direction $\mathbf{n}$ at location $\mathbf{z}$, i.e., at point $\mathcal{A}$. If $\mathbf{z}$ and $\mathbf{n}$ are held fixed, then this is just the ordinary derivative with respect to s. As shown in Appendix A.2, we can write the directional derivative as

$$\mathcal{D}\boldsymbol{\varphi}(\mathbf{z}) \cdot \mathbf{n} = \frac{d}{ds}\Big[\boldsymbol{\varphi}(\mathbf{z}+s\mathbf{n})\Big]_{s=0} \tag{4.45}$$

We also know that the directional derivative is connected to the *gradient* through the relationship $\mathcal{D}(\boldsymbol{\varphi}) \cdot \mathbf{n} = [\nabla \boldsymbol{\varphi}]\mathbf{n}$, where $\nabla \boldsymbol{\varphi}$ is the gradient of $\boldsymbol{\varphi}$

$$\nabla \boldsymbol{\varphi} = \frac{d\boldsymbol{\varphi}}{d\mathbf{z}} \tag{4.46}$$

This gradient is the *deformation gradient* for nonhomogeneous deformations. As we did for homogeneous deformations, we will call the deformation gradient $\mathbf{F} = \nabla \boldsymbol{\varphi}$. The components of the deformation gradient are

$$\mathbf{F} = \frac{d\boldsymbol{\varphi}}{d\mathbf{z}}, \qquad F_{ij} = \frac{\partial \varphi_i}{\partial z_j} \tag{4.47}$$

where φ_i is the ith component of $\boldsymbol{\varphi}$ and z_j is the jth component of $\mathbf{z}$. Now, we can write

$$\mathcal{D}\boldsymbol{\varphi}(\mathbf{z}) \cdot \mathbf{n} = \mathbf{F}\mathbf{n} \tag{4.48}$$

The square of the stretch in the direction $\mathbf{n}$ is

$$\lambda^2(\mathbf{n}) = \mathbf{F}\mathbf{n} \cdot \mathbf{F}\mathbf{n} = \mathbf{n} \cdot \mathbf{F}^T\mathbf{F}\mathbf{n} = \mathbf{n} \cdot \mathbf{C}\mathbf{n} \tag{4.49}$$

where, as before, $\mathbf{C} = \mathbf{F}^T\mathbf{F}$ is the *Green deformation tensor*.

In the limit, as point $\mathcal{B}'$ approaches $\mathcal{A}'$, the secant line gets closer and closer to the tangent line at $\mathcal{A}'$. Thus, the unit vector $\mathbf{n}$ in the reference configuration maps to $\mathbf{Fn}$ in the deformed configuration, as shown in Fig. 4.25. The length of the vector $\|\mathbf{Fn}\|$ is equal to the stretch $\lambda(\mathbf{n})$, since the length of $\mathbf{n}$ is one. The vector $\mathbf{Fn}$ is tangent to the deformed curve. In the infinitesimal limit, the curved lines look more and more like straight lines, which justifies the observation that in a local region the deformation is essentially homogeneous.

The connection between the deformation gradient in the current formulation and the one in the homogeneous deformation case is illustrated in the following example.

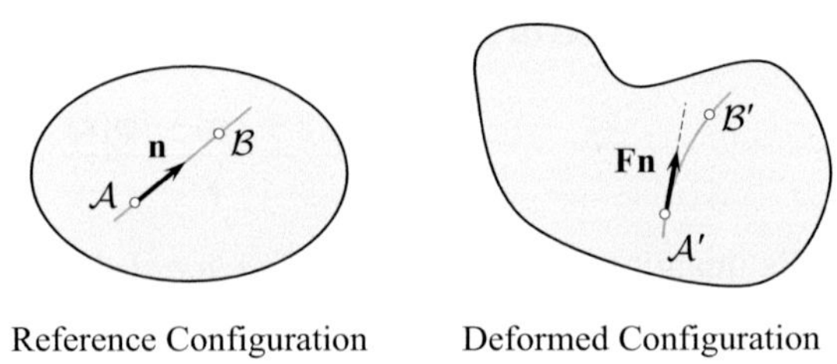

Fig. 4.25 *Tangent vector.* In the limit, the vector **Fn** is tangent to the curve in the deformed configuration

Example 4.12
Compute the deformation gradient for the homogeneous map

$$\boldsymbol{\varphi}(\mathbf{z}) = \mathbf{a} + \mathbf{A}\mathbf{z}$$

where **a** is a constant vector and **A** is a constant tensor.

Solution. The ith component of the map can be written in component form as

$$\varphi_i = a_i + \sum_{k=1}^{3} A_{ik} z_k$$

where a_i is the ith component of **a**, A_{ik} is the ikth component of **A**, and z_k is the kth component of **z**. The components of the deformation gradient can be computed as

$$F_{ij} = \frac{\partial \varphi_i}{\partial z_j} = \frac{\partial}{\partial z_j}\Big(a_i + \sum_{k=1}^{3} A_{ik} z_k\Big) = \sum_{k=1}^{3} A_{ik} \frac{\partial z_k}{\partial z_j} = A_{ij}$$

The last equality is due to the fact that $\partial z_k / \partial z_j = \delta_{kj}$, i.e., it is one when $j = k$ and zero otherwise. Thus, we have shown that $\mathbf{F} = \mathbf{A}$. It should also be evident that we can identify $\mathbf{a} = \mathbf{x}_o$ so that the map can be written as

$$\boldsymbol{\varphi}(\mathbf{z}) = \mathbf{x}_o + \mathbf{F}\mathbf{z}$$

as it was for the homogeneous map in Eq. 4.2.

4.6.2 Strain

Once we have the deformation gradient **F**, the derivations related to the definition of the strain tensor follow exactly along the lines of the homogeneous deformation case. In this section we present those concepts and highlight the differences in interpretation that result from the deformation not being homogeneous.

Normal strain The calculation of the strain follows exactly along the lines of the homogeneous deformation case. Specifically, the Lagrangian strain is

$$\varepsilon(\mathbf{n}) = \tfrac{1}{2}\big(\lambda^2 - 1\big) = \mathbf{n} \cdot \tfrac{1}{2}\big[\mathbf{F}^T\mathbf{F} - \mathbf{I}\big]\mathbf{n}$$

Therefore, we can define the Lagrangian strain tensor as

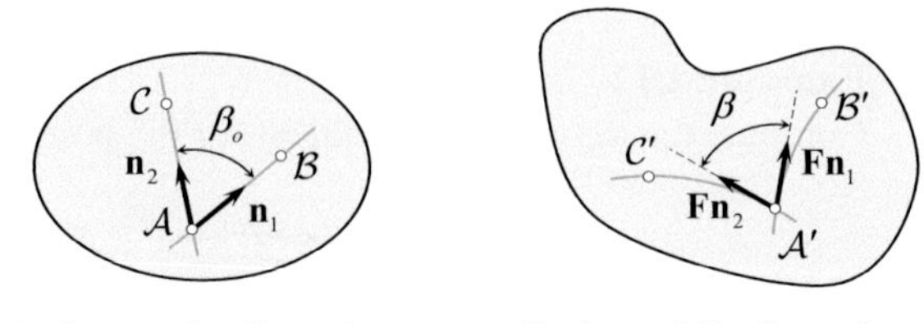

Fig. 4.26 *Shearing.* The angle between lines $\mathcal{AB}$ and $\mathcal{AC}$ in the reference configuration is β_o. After deformation the angle between lines tangent to the curves at $\mathcal{A}'$ is β

$$\mathbf{E} = \tfrac{1}{2}\big[\mathbf{F}^T\mathbf{F} - \mathbf{I}\,\big] \qquad \rightarrow \qquad \varepsilon(\mathbf{n}) = \mathbf{n}\cdot\mathbf{En} \tag{4.50}$$

The strain tensor $\mathbf{E}$ depends only on the deformation gradient $\mathbf{F}$, and Eq. 4.50 allows us to compute the strain $\varepsilon(\mathbf{n})$ in *any* direction $\mathbf{n}$. The difference between the homogeneous case and the nonhomogeneous case is that in the former, the strain tensor is constant throughout the body, and in the latter, it characterizes the strain only in the neighborhood of a point located at $\mathbf{z}$ in the reference configuration.

Shearing Deformation Shear is related to the change in angle between two lines. Consider the lines passing through $\mathcal{AB}$ and $\mathcal{AC}$ in the reference configuration, shown in gray in Fig. 4.26. Unit vectors $\mathbf{n}_1$ and $\mathbf{n}_2$ lie along these lines. Under the mapping $\boldsymbol{\varphi}(\mathbf{z})$, the straight lines get deformed into curved lines $\mathcal{A}'\mathcal{B}'$ and $\mathcal{A}'\mathcal{C}'$. We found in the previous development that the vectors $\mathbf{Fn}_1$ and $\mathbf{Fn}_2$ are tangent to the two curves at the point $\mathcal{A}'$. The secant line from $\mathcal{A}'$ to $\mathcal{B}'$ approaches the tangent direction in the limit as the point $\mathcal{B}'$ gets closer to $\mathcal{A}'$. Contrast this situation with homogeneous deformation, where the deformed lines are straight.

The cosine of the angle between the two lines in the reference configuration is simply $\cos\beta_o = \mathbf{n}_1\cdot\mathbf{n}_2$. The angle between the tangents to the lines in the deformed configuration is β, which is the appropriate measure of shearing deformation. Note that $\|\mathbf{Fn}_i\| = \lambda(\mathbf{n}_i)$ for $i = 1, 2$. Therefore, the formula for the dot product gives

$$\lambda(\mathbf{n}_1)\lambda(\mathbf{n}_2)\cos\beta = \mathbf{Fn}_1\cdot\mathbf{Fn}_2 = \mathbf{n}_1\cdot\mathbf{Cn}_2$$

Noting that $\mathbf{C} = 2\mathbf{E} + \mathbf{I}$, we can write this equation in the form

$$\cos\beta(\mathbf{n}_1, \mathbf{n}_2) = \frac{\beta_o + 2\,\mathbf{n}_1\cdot\mathbf{En}_2}{\lambda(\mathbf{n}_1)\lambda(\mathbf{n}_2)} \tag{4.51}$$

This equation is identical to the one for the homogeneous case, Eq. 4.18. Therefore, all observations about the shearing angle apply equally to the nonhomogeneous case. Also, all aspects of the strain analysis (i.e., principal values, principal directions, maximum shear strain, etc.) carry over from the homogeneous case. The key difference between the homogeneous and nonhomogeneous case is the computation of the deformation gradient $\mathbf{F}$ and the interpretation of the results as being confined to a small neighborhood of the point of computation. The following example illustrates the computation of the deformation gradient and the Lagrangian strain in a fixed direction.

Example 4.13
Compute the deformation gradient for the nonhomogeneous map

$$\boldsymbol{\varphi}(\mathbf{z}) = \big(z_1 + 0.1 z_1 z_2\big)\mathbf{e}_1 + \big(z_2 + 0.3 z_1 z_2\big)\mathbf{e}_2 + z_3\mathbf{e}_3$$

Compute the Lagrangian strain in the direction of the vector $\mathbf{n} = \frac{1}{\sqrt{2}}\big(\mathbf{e}_1 + \mathbf{e}_2\big)$ at the point with $z_1 = 1$, $z_2 = 0.5$, $z_3 = 0$.

Solution. The components of the mapping function can be written as

$$\varphi_1(\mathbf{z}) = z_1 + 0.1 z_1 z_2, \qquad \varphi_2(\mathbf{z}) = z_2 + 0.3 z_1 z_2, \qquad \varphi_3(\mathbf{z}) = z_3$$

The components of the deformation gradient are $F_{ij} = \partial\varphi_i/\partial z_j$. Thus,

$$\{\mathbf{F}\} = \begin{bmatrix} 1+0.1z_2 & 0.1z_1 & 0 \\ 0.3z_2 & 1+0.3z_1 & 0 \\ 0 & 0 & 1 \end{bmatrix}$$

The vector $\mathbf{Fn}$ at the point $\mathbf{z} = (1, 0.5, 0)$ can be computed in components as

$$\{\mathbf{Fn}\} = \begin{bmatrix} 1+0.1(0.5) & 0.1(1) & 0 \\ 0.3(0.5) & 1+0.3(1) & 0 \\ 0 & 0 & 1 \end{bmatrix} \frac{1}{\sqrt{2}} \begin{Bmatrix} 1 \\ 1 \\ 0 \end{Bmatrix} = \begin{Bmatrix} 0.8132 \\ 1.0253 \\ 0 \end{Bmatrix}$$

The stretch in direction $\mathbf{n}$ can be computed as

$$\lambda(\mathbf{n}) = \sqrt{\mathbf{Fn}\cdot\mathbf{Fn}} = \sqrt{(0.8132)^2 + (1.0253)^2 + (0)^2} = 1.3086$$

Finally, the Lagrangian strain in direction $\mathbf{n}$ can be computed as

$$\varepsilon(\mathbf{n}) = \tfrac{1}{2}\big(\lambda^2(\mathbf{n}) - 1\big) = \tfrac{1}{2}\big((1.3086)^2 - 1\big) = 0.3562$$

4.6.3 Strain-Displacement Relationship

We can write the strain tensor in terms of the *displacement* if we take the origin in the reference and deformed configurations at the same place, as shown in Fig. 4.3. Now, the mapping function can be written as

$$\boldsymbol{\varphi}(\mathbf{z}) = \mathbf{z} + \mathbf{u}(\mathbf{z}) \tag{4.52}$$

where $\mathbf{u}(\mathbf{z})$ is the displacement. The deformation gradient can be written as

$$\mathbf{F} = \frac{d\boldsymbol{\varphi}}{d\mathbf{z}} = \mathbf{I} + \nabla\mathbf{u} \tag{4.53}$$

where the gradient of the displacement function can be computed in components as

$$\nabla\mathbf{u} = \sum_{i=1}^{3}\sum_{j=1}^{3} \frac{\partial u_i}{\partial z_j}[\,\mathbf{e}_i \otimes \mathbf{e}_j\,] \quad \rightarrow \quad [\,\nabla\mathbf{u}\,]_{ij} = \frac{\partial u_i}{\partial z_j} \tag{4.54}$$

Now, we can substitute Eq. 4.53 into the definition of the strain tensor, Eq. 4.10, to get

$$\mathbf{E} = \tfrac{1}{2}\big[\mathbf{F}^T\mathbf{F} - \mathbf{I}\big] = \tfrac{1}{2}\big[(\mathbf{I} + \nabla\mathbf{u})^T(\mathbf{I} + \nabla\mathbf{u}) - \mathbf{I}\big]$$

Distributing the products and simplifying gives the strain tensor in terms of displacements as

$$\mathbf{E} = \tfrac{1}{2}\big[\,\nabla\mathbf{u} + \nabla\mathbf{u}^T + \nabla\mathbf{u}^T\nabla\mathbf{u}\,\big] \tag{4.55}$$

If the displacement gradients are small, then the last term is much smaller than the first two. If we neglect that term, then we arrive at the *linearized strain* tensor

$$\bar{\mathbf{E}} = \tfrac{1}{2}\big[\nabla\mathbf{u} + \nabla\mathbf{u}^T\big] \tag{4.56}$$

The components of the linearized strain tensor are

$$\bar{E}_{ij} = \tfrac{1}{2}\left[\frac{\partial u_i}{\partial z_j} + \frac{\partial u_j}{\partial z_i}\right] \tag{4.57}$$

This strain-displacement relationship is used to solve problems in linear elasticity.

Example 4.14

Write out the strain components ε_{xx}, ε_{xy}, and ε_{yy} in terms of the displacement $\mathbf{u} = u(x, y)\mathbf{e}_1 + v(x, y)\mathbf{e}_2$ with coordinates (x, y, z).

Solution. Using Eq. 4.57, with the notation $z_1 = x$, $z_2 = y$, $u_1 = u$, and $u_2 = v$, we can write

$$\begin{aligned}
\varepsilon_{xx} &= \tfrac{1}{2}\left(\frac{\partial u}{\partial x} + \frac{\partial u}{\partial x}\right) = \frac{\partial u}{\partial x} \\
\varepsilon_{xy} &= \tfrac{1}{2}\left(\frac{\partial u}{\partial y} + \frac{\partial v}{\partial x}\right) \\
\varepsilon_{yy} &= \tfrac{1}{2}\left(\frac{\partial v}{\partial y} + \frac{\partial v}{\partial y}\right) = \frac{\partial v}{\partial y}
\end{aligned} \tag{4.58}$$

Because the displacement does not have a component in the $\mathbf{e}_3$ direction and the displacement components u and v are not functions of z, the remaining terms in the strain tensor are zero, i.e.,

$$\varepsilon_{zz} = \varepsilon_{xz} = \varepsilon_{yz} = 0$$

which is a state of *plane strain*.

4.7 Summary

Strain is a measure of the distortion or deformation of the material. It is a normalized measure of the change in the length of a line (normal strain) or the change in the relative angle between two lines (shear strain). The analysis of strain revolves around the description of the deformation map. In this chapter we first considered homogeneous deformation, in which all points in the body feel the same state of strain. While this is generally not the case for solid bodies, it is approximately true in a small neighborhood of a point in the body.

This chapter also presented the analysis needed to explore the state of strain at a point in a body. We encountered the concept of strain transformation, which showed how the strain tensor components change with the orientation of a square etched in the material. From the strain transformation equations, we found the conditions of maximum normal and shear strain, and we derived formulas to compute them from the strain components. We found that the principal strains (maximum/minimum normal strains) occur at orientations where the shear strain is zero. The orientation of maximum shear strain has equal normal strains. The Mohr circle provided a graphical way of characterizing the state of strain.

Some of the techniques used for the plane strain problem do not generalize to three-dimensional states of strain. The eigenvalue problem is the most effective tool for finding the principal values and principal directions for these problems. We showed how this approach is applied and how it connects with the strategies used for plane strain problems.

Most of the developments of the chapter were based upon the assumption of homogeneous deformation. The chapter ends with the generalization of the mapping concept to nonhomogeneous deformations. While the computation of the deformation gradient is different from the homogeneous case, the analysis of strain is the same.

4.8 Problems

4.1 A sheet $\mathcal{ABCD}$ of width 1.5 cm and height 1.0 cm is subjected to a homogeneous deformation as shown. The coordinates of points $\mathcal{B}$ and $\mathcal{C}$ in the deformed configuration are given. Find the deformation map, the location of the deformed point $\mathcal{D}$, and the components of the strain tensor **E** associated with the deformation (i.e., ε_{xx}, ε_{xy}, ε_{yy}). Find the principal strains and the maximum shear strain. Draw a Mohr circle for the state of strain.

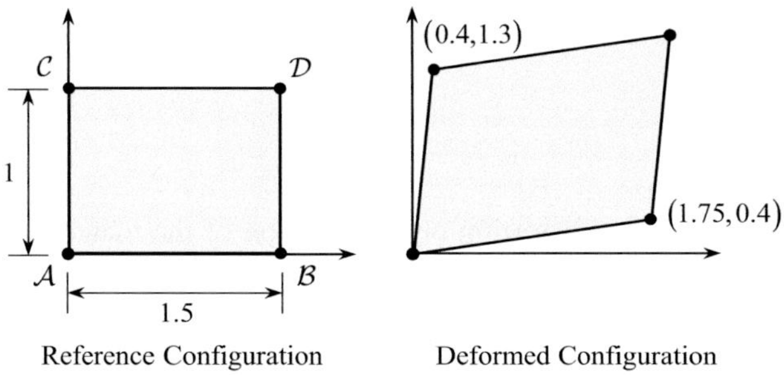

4.2 A solid body is in a state of plane strain. The normal strains at point $\mathcal{A}$ are measured using a strain gauge rosette with the strain gauges oriented at 0°, 45°, and 90° from horizontal, relative to the $\{\mathbf{e}_1, \mathbf{e}_2\}$ coordinate system. The measured values of the strains are $\varepsilon_a = 120 \times 10^{-4}$, $\varepsilon_b = 200 \times 10^{-4}$, and $\varepsilon_c = -100 \times 10^{-4}$. Find the principal strains and the principal angle. Find the maximum shear strain and the angle of maximum shear strain. Draw a Mohr circle for the state of strain.

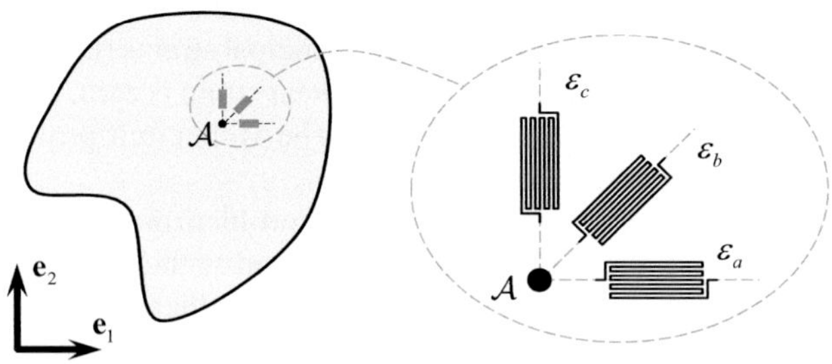

4.3 A square sheet is subjected to homogeneous deformation as shown. The values of a and b are unknown, but the two strain gauges have readings of $\varepsilon_n = 0.255$ and $\varepsilon_m = -0.178$. The strain gauges are oriented at 30° from horizontal. Find the deformation map, the location of the deformed point $\mathcal{C}$, and the components of the strain tensor **E** associated with the deformation. Find the principal strains and the maximum shear strain. Draw a Mohr circle for the state of strain.

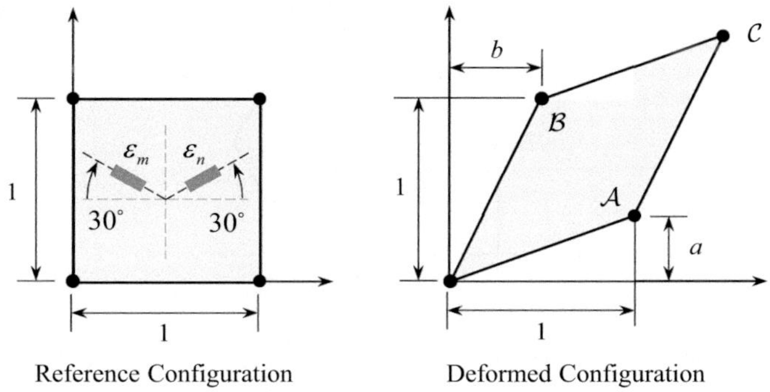

4.4 Consider an equilateral triangular sheet of side length a, oriented as shown in the first sketch. The sheet is subjected to a homogeneous deformation in the (x, y) plane to create the triangle shown in the second sketch. Find the deformation map, the location of the centroid of the triangle in the reference configuration and its location in the deformed configuration, and the components of the strain tensor associated with the deformation. Find the principal angle, principal strains, and maximum shear strain. Draw a Mohr circle for the state of strain.

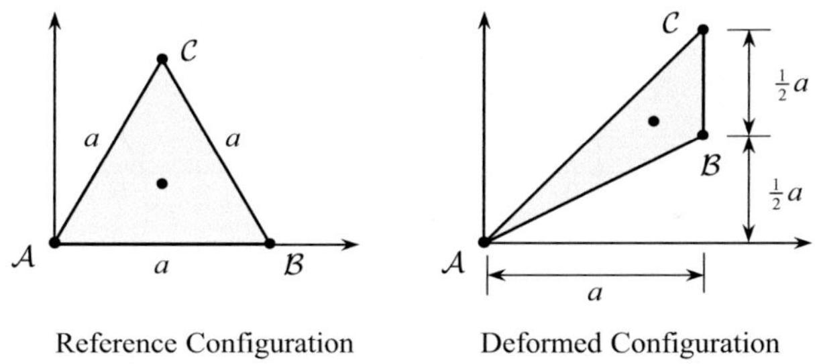

4.5 A regular hexagon with unit length sides undergoes a uniform deformation with the deformation tensor

$$\mathbf{F} = \begin{bmatrix} 1.00 & 0.50 \\ 0.00 & 0.75 \end{bmatrix}$$

Point $\mathcal{B}$ lies on the y-axis and point $\mathcal{A}$ lies on the x-axis in the reference configuration (pictured). There are two strain gauges on the hexagon that produce strain readings, ε_a and ε_b. Strain gauge b is at an angle of $\pi/6$ relative to the horizontal, and strain gauge a is initially in the $\mathbf{e}_2$ direction. Find the location of the deformed point $\mathcal{C}$ and the length of the line joining $\mathcal{B}$ and $\mathcal{C}$ in the deformed configuration. Find the components of the strain tensor associated with the deformation. Find the values of ε_a and ε_b and the principal strains and the maximum shear strain. Draw a Mohr circle for the state of strain.

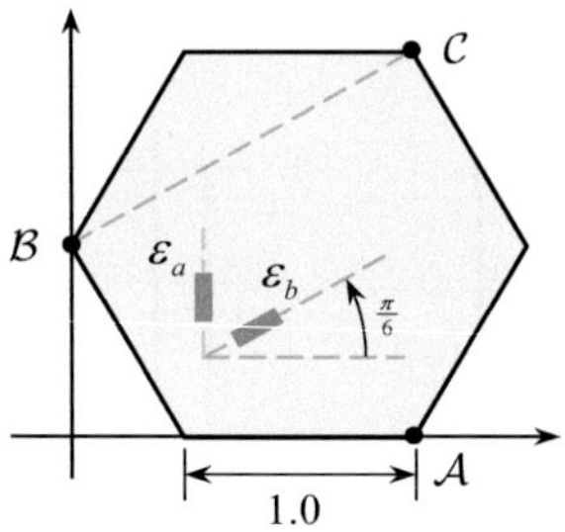

4.6 A thin sheet of width 0.5 in., height 1.0 in., and thickness 0.1 in. is subjected to a homogeneous deformation as shown. Point $\mathcal{B}$ moves horizontally by an amount 0.5, and Point $\mathcal{C}$ moves horizontally by an amount 0.5. From the strain gauge initially oriented in the vertical direction, we know that $\varepsilon_n = 0$. Find the deformation map, the location of the deformed point $\mathcal{D}$, and the components of the strain tensor associated with the deformation. Find the principal strains and the maximum shear strain. Draw a Mohr circle for the state of strain.

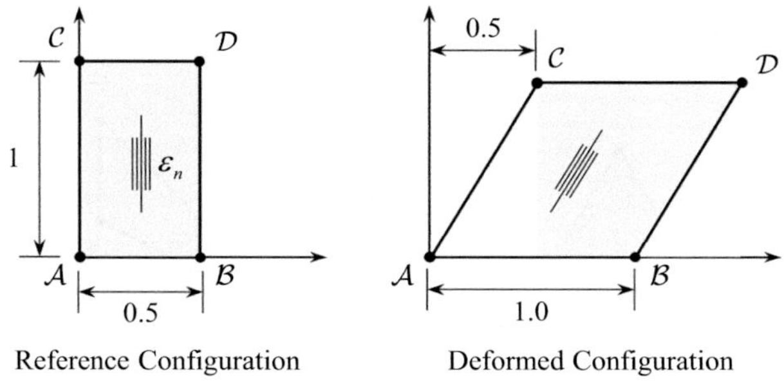

4.7 A solid body is in a state of plane strain. The normal strains at point $\mathcal{A}$ are measured using a strain gage rosette formed in an equilateral triangle. The measured values of the strains are $\varepsilon_a = 0.0095$, $\varepsilon_b = 0.0125$, and $\varepsilon_c = -0.0020$. Find the components of the strain tensor associated with the deformation relative to the $\{\mathbf{e}_1, \mathbf{e}_2\}$ coordinate system. Find the principal strains and the principal angle. Find the maximum shear strain and the angle of maximum shear strain. Draw a Mohr circle for the state of strain. Draw a picture of an undeformed and deformed piece in the principal orientation.

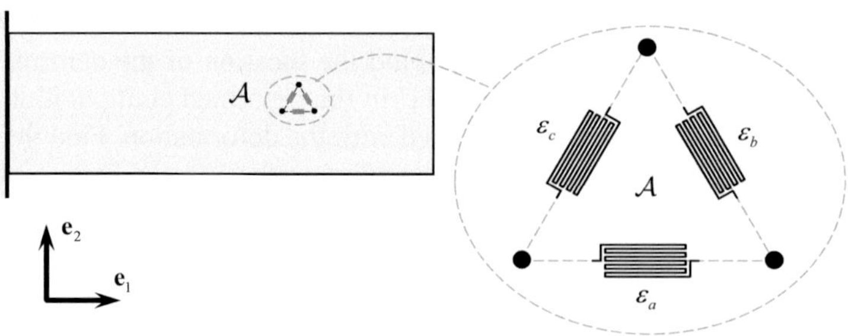

4.8 A circle of unit radius is subjected to a homogeneous deformation. The principal angle for the deformation is $\theta_p = -0.41442$ radians (clockwise). The locations of points $\mathcal{B}$ and $\mathcal{D}$ in the deformed configuration are known, in terms of the value of a. Find the components of the strain tensor. Find the deformed lengths of lines $\mathcal{AB}$ and $\mathcal{AC}$. Find the angle β. Find the reading in the strain gauge ε_g, which is oriented at an angle of 30° from horizontal in the reference configuration. Find the principal strains and the maximum shear strain. Draw a Mohr circle for the strains.

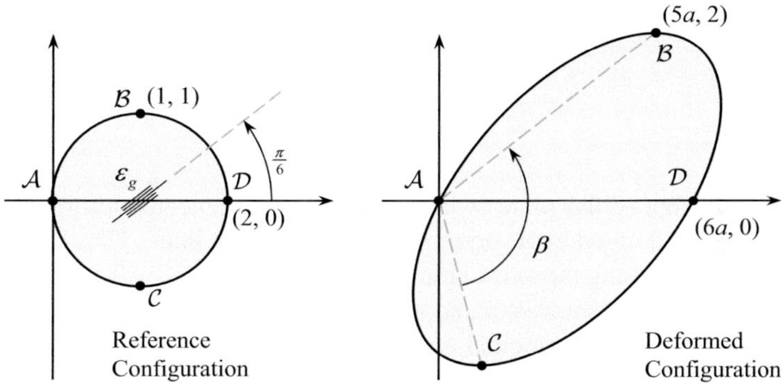

4.9 A square sheet is in a state of homogeneous deformation. The deformed length of side $\mathcal{AD}$ is 1.1, the deformed length of side $\mathcal{AB}$ is 1.2, and the angle $\mathcal{DAB}$ in the deformed configuration is 80°. Determine the length of the deformed line $\mathcal{AC}$ and the strain tensor associated with the deformation. Find the principal strains and the maximum shear strain. Draw a Mohr circle for the strains.

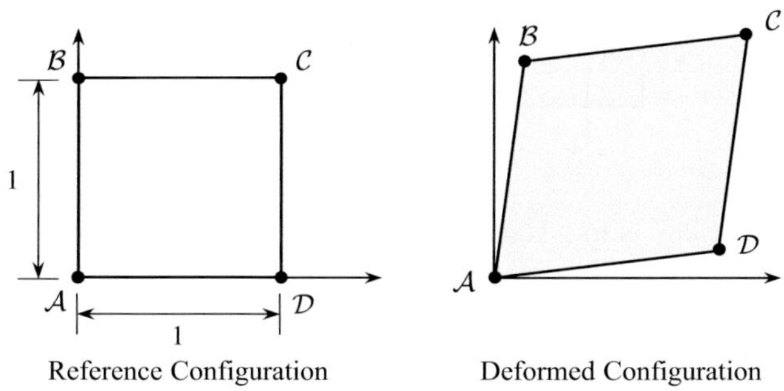

4.10 A thin regular pentagonal plate of side length 5 units in the reference configuration is in a state of homogeneous deformation. We have observed that the little square oriented at $-15°$ from the horizontal becomes a rectangle in the deformed configuration. We have also noted that point $\mathcal{E}$ is located at (5.1, 0.0) in the deformed configuration. The strain $\varepsilon_a = 0.03$ is measured by the strain gauge shown on the sketch, which is oriented at 45° from horizontal. Find the strain tensor $\mathbf{E}$ associated with the deformation. Find the deformed length of line $\mathcal{AC}$. Compute the Lagrangian strain of the line $\mathcal{AC}$. Find the location of points $\mathcal{C}$ and $\mathcal{D}$ in the deformed configuration. Find the principal strains and the maximum shearing strain. Draw a Mohr circle for the strains. Note: A regular pentagon has interior angles of 108°.

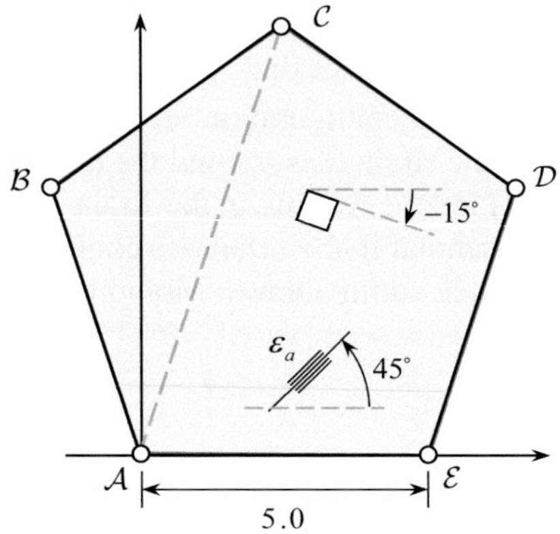

4.11 A thin trapezoidal plate is in a state of homogeneous deformation. Three strain gauges are arranged in the orientations shown along lines $\mathcal{AD}$, $\mathcal{AC}$, and $\mathcal{CB}$ and measure the strains along the axis of the gauge of $\varepsilon_a = 0.02$, $\varepsilon_b = 0.01$, and $\varepsilon_c = 0.07$. Determine the strain tensor associated with the deformation and the deformed length of the line $\mathcal{AB}$. Find the principal strains and the maximum shear strain. Draw a Mohr circle for the strains.

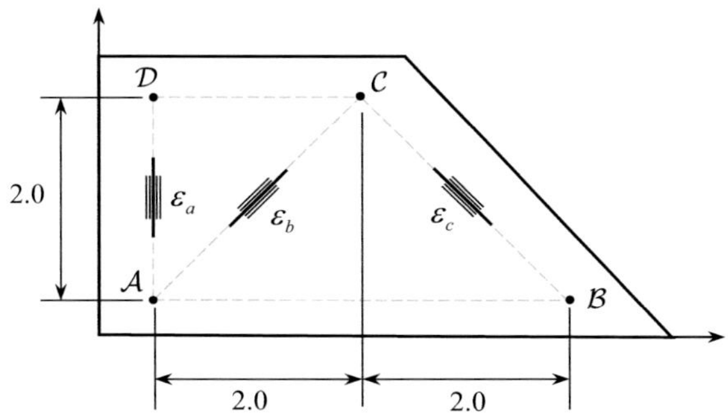

4.12 A square plate is in a state of homogeneous deformation, and it deforms in a way that line $\mathcal{OB}$ stretches by 1% and the angle $\mathcal{AOB}$ reduces by 0.01 radians. The angle of zero shear strain is 30° counterclockwise from the orientation shown. Find the components of the strain tensor. Find the deformed length of line $\mathcal{AB}$. Find the principal strains and the maximum shearing strain. Draw a Mohr circle for the strains. The reference configuration is shown in the sketch.

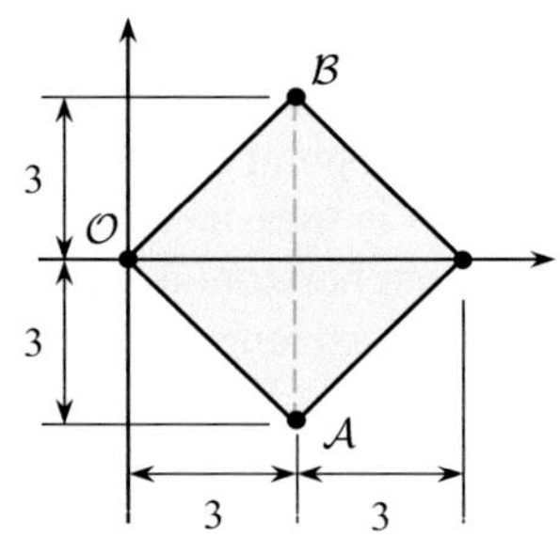

4.13 A thin rhombic plate is in a state of homogeneous deformation. Load is applied to the perimeter of the rhombus that causes diagonal $\mathcal{BD}$ to stretch by 0.5%, line $\mathcal{CA}$ to shrink by 0.2%, and the angle $\mathcal{AOB}$ (a right angle in the reference configuration) to reduce by 0.0092 radians. Find the components of the strain tensor. Find the deformed length of line $\mathcal{AB}$. Find the principal strains and the maximum shear strain. Draw a Mohr circle for the strains. The reference configuration is shown in the sketch.

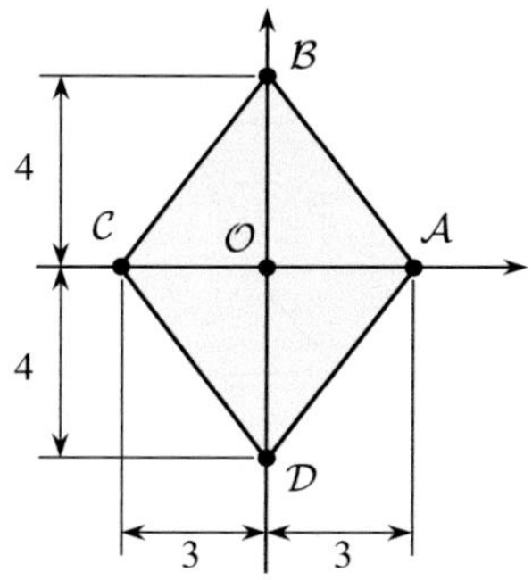

4.14 A body is subjected to a nonhomogeneous deformation described by the mapping function

$$\boldsymbol{\varphi}(\mathbf{z}) = \big(z_1 + \gamma z_2 z_3\big)\mathbf{e}_1 + \big(z_2 + \gamma z_1 z_3\big)\mathbf{e}_2 + \big(z_3 + \gamma z_1 z_2\big)\mathbf{e}_3$$

where γ is a constant. Find the deformation gradient $\mathbf{F}$, the Green deformation tensor $\mathbf{C}$, and the Lagrangian strain tensor $\mathbf{E}$ as a function of $\mathbf{z}$. Find the stretch of a line pointing in the $\mathbf{e}_1$ direction at the point with coordinates $z_1 = 1$, $z_2 = 1$, and $z_3 = 2$ when $\gamma = 0.1$. Find the deformed angle between two lines at that same point that were oriented in the $\mathbf{e}_1$ and $\mathbf{e}_2$ directions in the reference configuration.

4.15 A body is subjected to a deformation described by the mapping function

$$\boldsymbol{\varphi}(\mathbf{z}) = \big(1 + \gamma\big)\mathbf{z} + \gamma\big(\mathbf{n} \cdot \mathbf{z}\big)\mathbf{n}$$

where γ is a constant and $\mathbf{n}$ is a constant direction. Find the deformation gradient $\mathbf{F}$, the Green deformation tensor $\mathbf{C}$, and the Lagrangian strain tensor $\mathbf{E}$ as a function of $\mathbf{z}$. Is the deformation homogeneous or nonhomogeneous?

4.16 A bar of length L and depth $2h$ is bent into a circular arc of radius ρ centered at $\mathcal{C}$. The deformation does not decrease the depth, and the length of the axis of the bar does not change. The mapping function that describes this deformation is

$$\boldsymbol{\varphi}(\mathbf{z}) = \rho\,\mathbf{e}_2 + \big(\rho - z_2\big)\mathbf{n}(z_1)$$

where $\mathbf{n}(z_1) = \sin\theta(z_1)\,\mathbf{e}_1 - \cos\theta(z_1)\,\mathbf{e}_2$, with $\theta(z_1) = z_1/\rho$. Find the deformation gradient $\mathbf{F}$, the Green deformation tensor $\mathbf{C}$, and the Lagrangian strain tensor $\mathbf{E}$ as a function of $\mathbf{z}$.

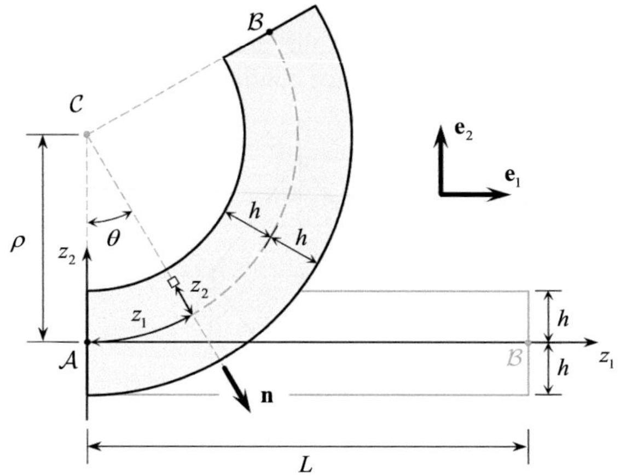

Chapter 5
Force and Stress

In Newtonian mechanics, forces are the cause of motion. Because we start with particle mechanics, we often first visualize force as a single vector acting at a point. Summing forces, then, is an act of adding discrete vectors. From there, it may seem logical to conceptualize a distributed force as a discrete force acting over some area. However, concentrated forces are a useful mathematical construct that works only in the idealized world of rigid bodies. In nature, forces are always distributed, and real materials fail if subjected to forces that are highly concentrated. A knife blade cuts because the force is concentrated at the sharp edge. Therefore, it makes more sense to think of distributed forces as being the primary building block, and discrete (i.e., concentrated) force vectors as simply the aggregate effect of distributed forces over some area or volume. The wind on a sail exerts a distributed pressure. The force on the sail is the net effect of the wind *pressure* multiplied by the area of the sail on which it acts. The force associated with the weight of an object is its *density* (mass per unit volume) multiplied by the acceleration of gravity multiplied by its volume.

In the study of the mechanics of deformable bodies, we seek to describe how forces get transmitted through solid bodies. For our purposes, forces come in only two varieties: *body forces* and *surface tractions*. A body force is a force per unit volume and acts at a distance (e.g., gravity). A surface traction is a force per unit area and acts through contact of one body with another. When we take a cut through a solid body to form a free body diagram, the forces acting at the cut are tractions. From that perspective, we can think of the two pieces as if they were bodies in contact. In this chapter we demonstrate that an internal state called *stress* is an ideal way to characterize how forces flow through a solid body, and that stress must be represented mathematically as a tensor. Ultimately, the objective of the chapter is to show how the concept of stress relates to equilibrium of a solid body.

K. D. Hjelmstad, *Mechanics of Deformable Solids: An Introduction*,
https://doi.org/10.1007/978-3-031-89873-0_5

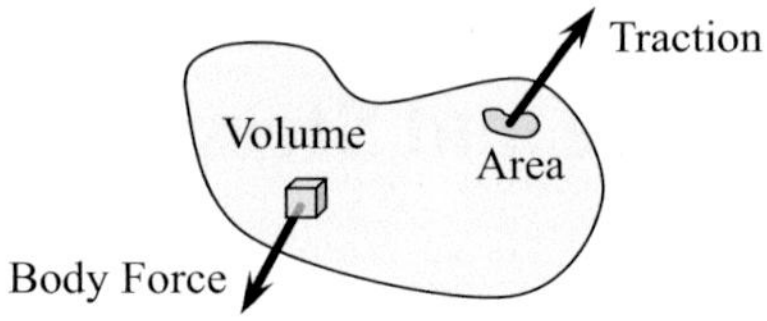

Fig. 5.1 *Traction and body force.* Traction is a measure of force per unit area acting on the surface of a body. Body force is a force per unit volume acting at a distance

5.1 Cauchy's Formula and the Stress Tensor

At the most basic level, *traction* is a quantity that represents force per unit area, and *body force* is a quantity that represents force per unit volume, as illustrated in Fig. 5.1. The net force acting on a body (or some region of a body) is the traction multiplied by the area or the body force multiplied by the volume. If the force distributions are variable, then the net effects will be found as integrals over the region of forces like $\mathbf{t}\,dA$ or $\mathbf{b}\,dV$ acting on typical particles, where $\mathbf{t}$ and $\mathbf{b}$ are tractions and body forces, respectively, and dA and dV are their corresponding infinitesimal areas and volumes. These forces are represented by vectors and hence, they obey all the rules of vector algebra laid out in Chap. 1.

In a multiaxial state of stress, there are tractions on all exposed surfaces. The infinitesimal cube shown in Fig. 5.2, for example, is a free body diagram cut from a larger body. There are tractions acting on each face. The combined effect of those tractions creates the internal state of stress in the cube. In this chapter we seek a way to characterize the internal state in such a way that assures equilibrium of any free body we might take. Through equilibrium considerations, the stress tensor will emerge as the most natural way to characterize the internal state. Note that in our sketches, we show a single traction vector acting at a typical particle to represent the distributed effects to simplify the graphic.

The stress tensor, as a means of characterizing the internal state of force flow in a body, emanates from an argument first brought forward by Cauchy,[1] based on the hypothesis that the traction acting on a surface in the neighborhood of a point depends upon the direction of the normal to the surface. Then, from purely geometric arguments, in conjunction with equilibrium, the notion of stress as a tensor emerges naturally. In essence, Cauchy's principle will tell us how to deal with free body diagrams in three-dimensional solid bodies.

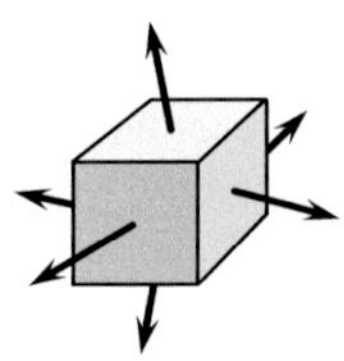

Fig. 5.2 *Multiaxial state of stress.* In a multiaxial state of stress, traction forces act on each face of the infinitesimal cube, each with its own direction and magnitude

[1] Augustin-Louis Cauchy (1789–1857) was a French mathematician, engineer, and physicist who laid the foundations of modern mechanics of deformable solids.

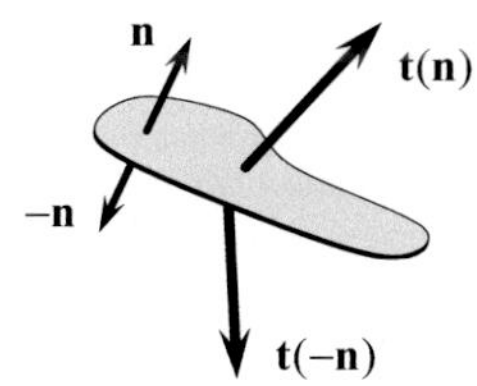

Fig. 5.3 *Cauchy wafer.* Infinitesimally thin wafer of area A and normal vectors $\mathbf{n}$ and $-\mathbf{n}$ subjected to tractions $\mathbf{t}(\mathbf{n})$ and $\mathbf{t}(-\mathbf{n})$

Consider the free body diagram, which we call the Cauchy *wafer*, shown in Fig. 5.3. We will designate the traction on a surface with unit normal vector $\mathbf{n}$ as $\mathbf{t}(\mathbf{n})$. On the top face the normal is $\mathbf{n}$, and on the bottom face it is $-\mathbf{n}$. Hence, the traction on top is $\mathbf{t}(\mathbf{n})$, and the traction on the bottom is $\mathbf{t}(-\mathbf{n})$. The sides of the wafer are so thin that even if there were tractions along the edges, the area is so small as to make their contribution to equilibrium negligible. The forces on the free body diagram must sum to zero. Thus,

$$\mathbf{t}(\mathbf{n})A + \mathbf{t}(-\mathbf{n})A = \mathbf{0}$$

Since the area A is the same for the top and bottom face, we get Cauchy's first principle:

$$\mathbf{t}(\mathbf{n}) = -\mathbf{t}(-\mathbf{n}) \tag{5.1}$$

We have already encountered this basic idea in the axial bar problem. When we took a cut with unit normal $\mathbf{e}_1$, the net force was represented on the free body diagram as $N(x)\mathbf{e}_1$. For a cut with unit normal vector $-\mathbf{e}_1$, the net force was represented as $-N(x)\mathbf{e}_1$. That allowed us to characterize the internal axial force with a single scalar variable $N(x)$, which was either positive (tension) or negative (compression). When we form a free body diagram, the actions at a cut must be force vectors. Having the force at a cut depend upon the direction of the normal vector at the cut gives a simple way to assert equilibrium of an infinitesimal piece for any given value of the internal force.

Now, consider the Cauchy *tetrahedron* shown in Fig. 5.4a, which shows features of the geometry of the tetrahedron. The vertices of the tetrahedron are at the origin and at distances a, b, and c along the x, y, and z axes, respectively. Each of the three sides of the tetrahedron associated with the coordinate planes is a right triangle. Those triangles have areas

$$A_1 = \tfrac{1}{2}bc, \qquad A_2 = \tfrac{1}{2}ac, \qquad A_3 = \tfrac{1}{2}ab$$

where the subscript on A_i indicates the normal direction to that face (e.g., A_1 is the area of the face with normal $-\mathbf{e}_1$). We can compute the area and normal vector to the oblique face using the cross product between a vector from a to b (i.e., $-a\,\mathbf{e}_1 + b\,\mathbf{e}_2$) and a vector from a to c (i.e., $-a\,\mathbf{e}_1 + c\,\mathbf{e}_3$). The magnitude of this cross product will be twice the area A_n and the direction will be normal to the plane of the oblique face. Thus,

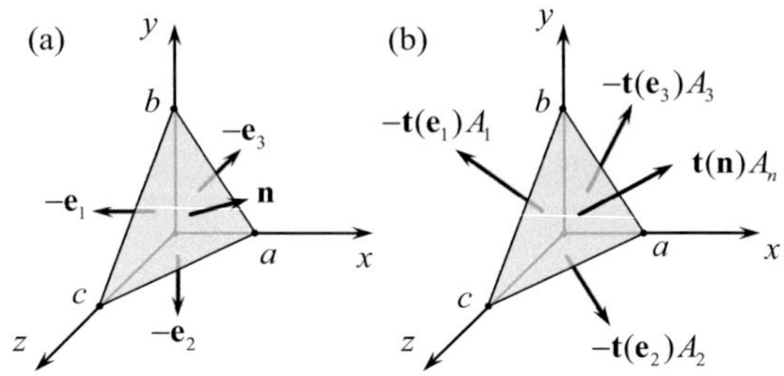

Fig. 5.4 *Cauchy tetrahedron.* Tetrahedron with three sides along the coordinate planes, subjected to tractions on each face. The oblique face has unit normal vector **n**. **a** Geometry. **b** Free body diagram

$$\begin{aligned} 2A_n\mathbf{n} &= \big(-a\,\mathbf{e}_1 + b\,\mathbf{e}_2\big) \times \big(-a\,\mathbf{e}_1 + c\,\mathbf{e}_3\big) \\ &= bc\,\mathbf{e}_1 + ac\,\mathbf{e}_2 + ab\,\mathbf{e}_3 \end{aligned}$$

Dividing through by 2 and recognizing the expressions for the areas of the coordinate faces, we get

$$A_n\mathbf{n} = A_1\mathbf{e}_1 + A_2\mathbf{e}_2 + A_3\mathbf{e}_3 \tag{5.2}$$

This remarkable geometric fact is the key to the Cauchy tetrahedron construction. Observe that we can dot Eq. 5.2 with $\mathbf{e}_i$ to find the area of the ith face. To wit,

$$A_i = A_n\big(\mathbf{e}_i \cdot \mathbf{n}\big)$$

for $i = 1, 2, 3$. Thus, the area of each coordinate face A_i is related to the area of the oblique face A_n and the normal vector $\mathbf{n}$.

Now, we can establish equilibrium of the tetrahedron using the free body diagram shown in Fig. 5.4b. Notice that we have already used Eq. 5.1 on the tractions on the coordinate faces, e.g., $\mathbf{t}(-\mathbf{e}_1) = -\mathbf{t}(\mathbf{e}_1)$. We neglect body forces here because the volume vanishes an order of magnitude faster than the surface area in the limit as the tetrahedron shrinks to infinitesimal size. Setting the sum of the forces equal to zero gives

$$\mathbf{t}(\mathbf{n})A_n - \mathbf{t}(\mathbf{e}_1)A_1 - \mathbf{t}(\mathbf{e}_2)A_2 - \mathbf{t}(\mathbf{e}_3)A_3 = \mathbf{0}$$

Substituting the expressions for the areas of the coordinate faces and moving the last three terms to the right side gives

$$\mathbf{t}(\mathbf{n})A_n = \mathbf{t}(\mathbf{e}_1)\big(\mathbf{e}_1 \cdot \mathbf{n}\big)A_n + \mathbf{t}(\mathbf{e}_2)\big(\mathbf{e}_2 \cdot \mathbf{n}\big)A_n + \mathbf{t}(\mathbf{e}_3)\big(\mathbf{e}_3 \cdot \mathbf{n}\big)A_n$$

Finally, dividing through by A_n, we get

$$\mathbf{t}(\mathbf{n}) = \sum_{i=1}^{3} \mathbf{t}(\mathbf{e}_i)\big(\mathbf{e}_i \cdot \mathbf{n}\big) \tag{5.3}$$

This result has two important implications. First, the measure of the areas does not matter, only the ratios. To alter the direction of the normal vector to the oblique face, we can adjust the relative values of a, b, and c, thereby changing the proportions of the tetrahedron. In the end, the only thing we need to know about the geometry of the tetrahedron is the direction of the unit vector $\mathbf{n}$, which embodies the values a, b, and c. The equation holds at a point as we shrink the tetrahedron to the infinitesimal limit.

The second implication is that keeping track of the tractions on the three coordinate faces, $\mathbf{t}(\mathbf{e}_i)$, is sufficient to compute any other traction on any other face. We can factor out the $\mathbf{n}$ from this equation and use the rule for the tensor product from Eq. 1.11 to get

$$\mathbf{t}(\mathbf{n}) = \sum_{i=1}^{3} \big[\, \mathbf{t}(\mathbf{e}_i) \otimes \mathbf{e}_i \,\big]\mathbf{n} \tag{5.4}$$

where $\otimes$ is the tensor product defined as $[\,\mathbf{u} \otimes \mathbf{v}\,]\mathbf{w} = \mathbf{u}\big(\mathbf{v} \cdot \mathbf{w}\big)$ for any vectors $\mathbf{u}$, $\mathbf{v}$, and $\mathbf{w}$. Thus, the result of the sum in Eq. 5.4 is a tensor. We will call this tensor the *stress tensor* and define it as

$$\mathbf{S} = \sum_{i=1}^{3} \big[\, \mathbf{t}(\mathbf{e}_i) \otimes \mathbf{e}_i \,\big] \tag{5.5}$$

What the operation really does is put the components of the vector $\mathbf{t}(\mathbf{e}_i)$ in the ith column of the matrix representing the components of the tensor $\mathbf{S}$. Each of the tractions on the coordinate faces is a vector, and each of these vectors has three components relative to the base vectors $\{\mathbf{e}_1, \mathbf{e}_2, \mathbf{e}_3\}$, as illustrated in Fig. 5.5. Write those components in matrix form as

$$\mathbf{t}(\mathbf{e}_1) = \begin{Bmatrix} \sigma_{xx} \\ \sigma_{yx} \\ \sigma_{zx} \end{Bmatrix}, \qquad \mathbf{t}(\mathbf{e}_2) = \begin{Bmatrix} \sigma_{xy} \\ \sigma_{yy} \\ \sigma_{zy} \end{Bmatrix}, \qquad \mathbf{t}(\mathbf{e}_3) = \begin{Bmatrix} \sigma_{xz} \\ \sigma_{yz} \\ \sigma_{zz} \end{Bmatrix}$$

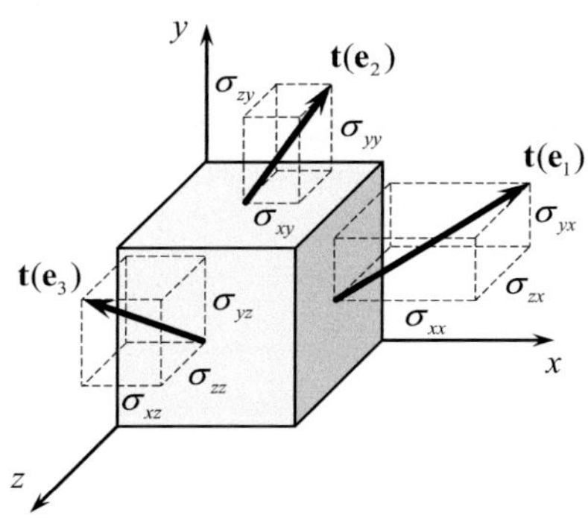

Fig. 5.5 *Stress components.* The traction vectors acting on the coordinate faces each have components relative to the basis $\{\mathbf{e}_1, \mathbf{e}_2, \mathbf{e}_3\}$

where the first subscript indicates the coordinate direction associated with the component of the vector, and the second subscript indicates the coordinate direction associated with the normal to the face. Thus, we can define the stress tensor as having components

$$\{\mathbf{S}\} = \begin{bmatrix} \sigma_{xx} & \sigma_{xy} & \sigma_{xz} \\ \sigma_{yx} & \sigma_{yy} & \sigma_{yz} \\ \sigma_{zx} & \sigma_{zy} & \sigma_{zz} \end{bmatrix} \tag{5.6}$$

relative to the (x, y, z) coordinate system with base vectors $\{\mathbf{e}_1, \mathbf{e}_2, \mathbf{e}_3\}$. Now, we can express the *Cauchy formula* as

$$\mathbf{t}(\mathbf{n}) = \mathbf{S}\mathbf{n} \tag{5.7}$$

What this simple and elegant formula says is that we can compute the traction on a face with unit normal vector **n** from the stress tensor. At a given point in a solid body, we can imagine taking free body diagrams with cuts through the point in question, each with a different normal vector. Cauchy's formula tells us how to deal with replacing the force on the free body diagram using the stress tensor. Implicit in this derivation is that we must keep track of three independent tractions in order to characterize the internal state. The simplest way to keep the books on the internal state, then, is with a tensor. The tensor **S** is the extension of the concept of internal force (e.g., $N(x)$ for the axial bar) to multiple dimensions. The components of the stress tensor will be either positive or negative. Cauchy's formula takes care of converting those values to tractions at a cut on a free body diagram.

Once we establish that **S** is a tensor with components given by Eq. 5.6, then keeping track of those components is completely equivalent to keeping track of three traction vectors. In fact, Eq. 5.7 applied to the ith unit base vector gives $\mathbf{t}(\mathbf{e}_i) = \mathbf{S}\mathbf{e}_i$. As an example, the unit base vector $\mathbf{e}_1$ has the component representation $\mathbf{e}_1 = (1, 0, 0)$. Therefore, by the rules of matrix multiplication, $\mathbf{S}\mathbf{e}_1$ is simply the first column of **S**. Also, according to the rules of matrix algebra, $\mathbf{S}(-\mathbf{n}) = -\mathbf{S}\mathbf{n}$. Therefore, the Cauchy formula faithfully represents the requirement of the Cauchy wafer construction that $\mathbf{t}(-\mathbf{n}) = -\mathbf{t}(\mathbf{n})$.

Example 5.1

Consider the axial bar with square cross section shown in Fig. 5.6a. The bar is in a uniaxial state of stress with $\sigma_{xx} = \sigma$ and all other components equal to zero. Find the traction acting on the oblique face with unit normal **n** exposed by the cut shown in Fig. 5.6b.

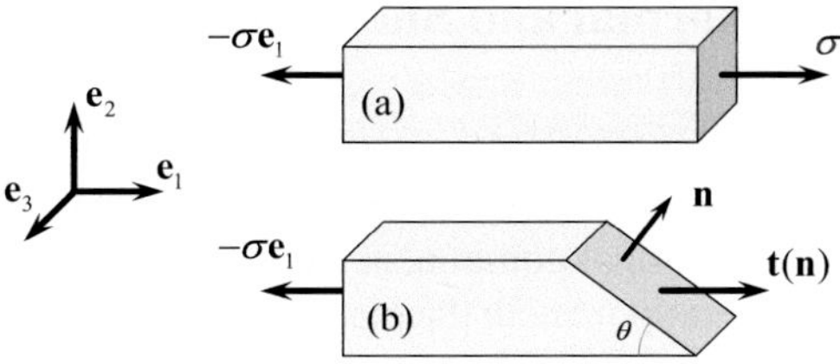

Fig. 5.6 *Example.* Bar in a state of uniaxial stress. **a** The magnitude of the stress is σ and the cross-sectional area is A. **b** Bar with cut exposing face with normal $\mathbf{n}$

Solution. The stress tensor associated with a uniaxial state of stress is

$$\mathbf{S} = \begin{bmatrix} \sigma & 0 & 0 \\ 0 & 0 & 0 \\ 0 & 0 & 0 \end{bmatrix}$$

The cut makes an angle θ with horizontal. Thus, the unit normal vector has components $\mathbf{n} = \sin\theta \mathbf{e}_1 + \cos\theta \mathbf{e}_2$. Apply the Cauchy formula to find the traction on the oblique face as

$$\mathbf{t}(\mathbf{n}) = \mathbf{S}\mathbf{n} = \begin{bmatrix} \sigma & 0 & 0 \\ 0 & 0 & 0 \\ 0 & 0 & 0 \end{bmatrix} \begin{Bmatrix} \sin\theta \\ \cos\theta \\ 0 \end{Bmatrix} = \begin{Bmatrix} \sigma \sin\theta \\ 0 \\ 0 \end{Bmatrix} = \sigma \sin\theta \mathbf{e}_1$$

Notice that the magnitude of the traction on the oblique face is $\sin\theta$ times the original value, but it still points in the $\mathbf{e}_1$ direction. That value goes from zero for a horizontal cut to one for a vertical cut, with values between those limits for other cuts. You can show that equilibrium still holds, though, as the area on which the traction acts also changes. The area of the oblique face is

$$A_n = \frac{A}{\mathbf{n} \cdot \mathbf{e}_1} = \frac{A}{\sin\theta}$$

which is an area larger than A. For the special case where $\theta = 30°$, the stress of the face would be 0.5σ and the area of the face would be $2A$. Summing forces on the free body diagram in Fig. 5.6 gives

$$-\sigma A \mathbf{e}_1 + \big(\sigma \sin\theta\, \mathbf{e}_1\big)\Big(\frac{A}{\sin\theta}\Big) = \mathbf{0}$$

Thus, equilibrium of forces holds.

5.2 Normal and Shear Stresses

The traction vector $\mathbf{t}(\mathbf{n})$ can be resolved into two parts, with one component along the normal $\mathbf{n}$ and the other in the plane perpendicular to the normal, as shown in Fig. 5.7. The normal component $\sigma\mathbf{n}$ has magnitude σ, which is called the *normal stress*, and the component in the plane $\tau\mathbf{m}$ has magnitude τ, which is called the *shear stress*. By vector addition, we can write

$$\mathbf{t}(\mathbf{n}) = \sigma\mathbf{n} + \tau\mathbf{m} \tag{5.8}$$

If we take the dot product of this equation with the unit vector $\mathbf{n}$, then we find that

$$\sigma = \mathbf{n} \cdot \mathbf{t}(\mathbf{n}) = \mathbf{n} \cdot \mathbf{S}\mathbf{n} \tag{5.9}$$

The vector $\tau\mathbf{m}$ is simply the shadow of $\mathbf{t}(\mathbf{n})$ cast on the plane. We can write

$$\tau\mathbf{m} = \mathbf{t}(\mathbf{n}) - \sigma\mathbf{n}$$

The lengths of the two sides of the right triangle in Fig. 5.7 are σ and τ. The length of the hypotenuse is $\|\mathbf{t}(\mathbf{n})\|$. By the Pythagorean theorem,

$$\tau = \sqrt{\|\mathbf{t}(\mathbf{n})\|^2 - \sigma^2}$$

Note that we can also dot Eq. 5.8 with $\mathbf{m}$ to get

$$\tau = \mathbf{m} \cdot \mathbf{t}(\mathbf{n}) = \mathbf{m} \cdot \mathbf{S}\mathbf{n} \tag{5.10}$$

This version of the shear stress requires the unit vector $\mathbf{m}$, which can be found by first projecting out the normal component of $\mathbf{t}(\mathbf{n})$ and then dividing the result by its own length. To wit,

$$\hat{\mathbf{m}} = \left[\mathbf{I} - \mathbf{n}\mathbf{n}^T\right]\mathbf{t}(\mathbf{n}), \qquad \mathbf{m} = \frac{\hat{\mathbf{m}}}{\|\hat{\mathbf{m}}\|}$$

where $\mathbf{I} - \mathbf{n}\mathbf{n}^T$ is the projection tensor described in Eq. 1.28 on Sect. 1.1.

We will make use of the normal and shear stresses in the Mohr circle construction and for the definition and computation of the principal values and principal directions

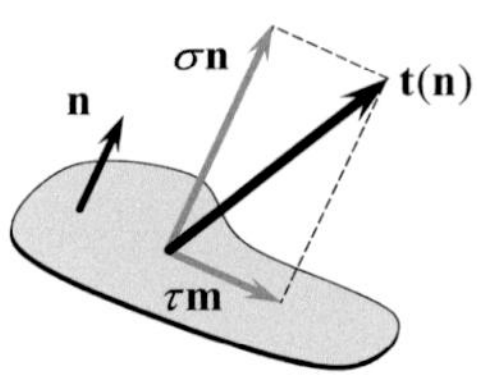

Fig. 5.7 *Normal and shear stresses*. The normal stress σ is the component of the traction $\mathbf{t}(\mathbf{n})$ in the $\mathbf{n}$ direction and the shear stress τ is the component that lies in the plane

of the stress tensor. These concepts are simplest to visualize in the context of the two-dimensional setting of plane stress, as outlined in the following section.

5.3 Plane Stress

Plane stress is a two-dimensional state with no stress in the direction normal to the plane. When we take the plane to be the (x, y) plane, then the plane stress condition is

$$\sigma_{xz} = \sigma_{yz} = \sigma_{zz} = 0$$

For this class of problems, we can represent the stress tensor as

$$\mathbf{S} = \begin{bmatrix} \sigma_{xx} & \sigma_{xy} \\ \sigma_{xy} & \sigma_{yy} \end{bmatrix} \tag{5.11}$$

with all of the remaining components being equal to zero. The Cauchy tetrahedron reduces to the Cauchy wedge, shown in Fig. 5.8. For the plane problem, vectors can be represented in terms of their components relative to $\mathbf{e}_1$ and $\mathbf{e}_2$. The unit vector normal to the oblique plane, for example, can be expressed in terms of the angle θ as $\mathbf{n} = \cos\theta\mathbf{e}_1 + \sin\theta\mathbf{e}_2$. Represented as a matrix, the components are

$$\mathbf{n} = \begin{Bmatrix} \cos\theta \\ \sin\theta \end{Bmatrix}$$

For the Cauchy wedge, we will take the uniform thickness to be c over the entire triangle. Hence, the area of the face with normal $-\mathbf{e}_1$ is $A_1 = bc$ and the area of the face with normal $-\mathbf{e}_2$ is $A_2 = ac$. The area of the oblique face is $A_n = c\sqrt{a^2 + b^2}$. From the triangle, we have $A_1 = A_n \cos\theta$ and $A_2 = A_n \sin\theta$. Equilibrium of the wedge, using the free body diagram in Fig. 5.8, requires

$$\mathbf{t}(\mathbf{n})A_n - \mathbf{t}(\mathbf{e}_1)A_1 - \mathbf{t}(\mathbf{e}_2)A_2 = \mathbf{0}$$

Dividing out A_n gives

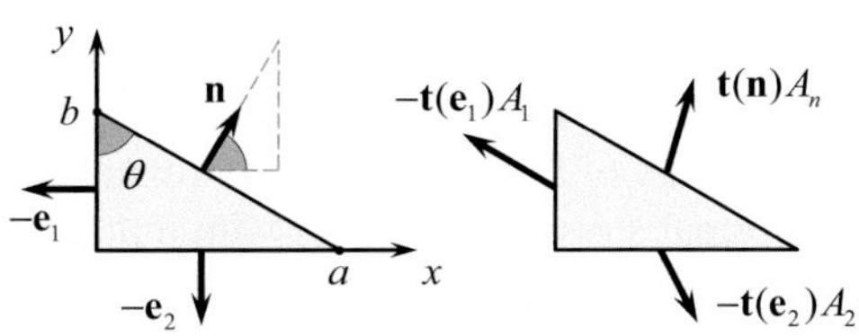

Fig. 5.8 *Cauchy wedge.* The Cauchy tetrahedron in two dimensions is a Cauchy wedge

$$\begin{aligned}\mathbf{t}(\mathbf{n}) &= \mathbf{t}(\mathbf{e}_1)\cos\theta + \mathbf{t}(\mathbf{e}_2)\sin\theta \\ &= \begin{bmatrix}\mathbf{t}(\mathbf{e}_1), & \mathbf{t}(\mathbf{e}_2)\end{bmatrix}\begin{Bmatrix}\cos\theta \\ \sin\theta\end{Bmatrix} = \mathbf{Sn}\end{aligned}$$

confirming that the Cauchy formula reduces gracefully to two dimensions. The columns of the stress tensor are the components of the traction vectors on the coordinate faces. Since the problem is planar, those vectors have only two components.

Now, we can compute the normal stress from Eq. 5.9 as follows:

$$\begin{aligned}\sigma &= \mathbf{n}^T\mathbf{Sn} \\ &= \begin{bmatrix}\cos\theta & \sin\theta\end{bmatrix}\begin{bmatrix}\sigma_{xx} & \sigma_{xy} \\ \sigma_{xy} & \sigma_{yy}\end{bmatrix}\begin{Bmatrix}\cos\theta \\ \sin\theta\end{Bmatrix} \\ &= \sigma_{xx}\cos^2\theta + 2\sigma_{xy}\cos\theta\sin\theta + \sigma_{yy}\sin^2\theta\end{aligned}$$

We can define a unit vector $\mathbf{m}$ perpendicular to $\mathbf{n}$ that lies in the plane of the oblique face. The components of this vector are

$$\mathbf{m} = \begin{Bmatrix}-\sin\theta \\ \cos\theta\end{Bmatrix}$$

Notice that for the two-dimensional problem, there is only one choice for the vector $\mathbf{m}$, unlike the three-dimensional case where there are an infinite number of vectors perpendicular to $\mathbf{n}$, and the one we seek is the projection of the traction onto the plane. With this vector, we can compute the shear stress as

$$\begin{aligned}\tau &= \mathbf{m}^T\mathbf{Sn} \\ &= \begin{bmatrix}-\sin\theta & \cos\theta\end{bmatrix}\begin{bmatrix}\sigma_{xx} & \sigma_{xy} \\ \sigma_{xy} & \sigma_{yy}\end{bmatrix}\begin{Bmatrix}\cos\theta \\ \sin\theta\end{Bmatrix} \\ &= \left(\sigma_{yy} - \sigma_{xx}\right)\cos\theta\sin\theta + \sigma_{xy}\left(\cos^2\theta - \sin^2\theta\right)\end{aligned}$$

5.3.1 *Stress Transformation Equations*

We can use the normal and shear stress equations to find the stress transformation equations, which give the answer to the question of what the values of the components of $\mathbf{S}$ would be in the (a, b) coordinate system, shown in Fig. 5.9, in terms of the components of $\mathbf{S}$ in the (x, y) coordinate system. The components σ_{aa} and σ_{bb} are the normal stress on the faces with unit normal vectors $\mathbf{e}_a$ and $\mathbf{e}_b$, respectively. The component σ_{ab} is the shear stress on the face with unit normal vector $\mathbf{e}_a$. Thus,

$$\sigma_{aa} = \mathbf{e}_a \cdot \mathbf{S}\mathbf{e}_a, \qquad \sigma_{ab} = \mathbf{e}_b \cdot \mathbf{S}\mathbf{e}_a, \qquad \sigma_{bb} = \mathbf{e}_b \cdot \mathbf{S}\mathbf{e}_b$$

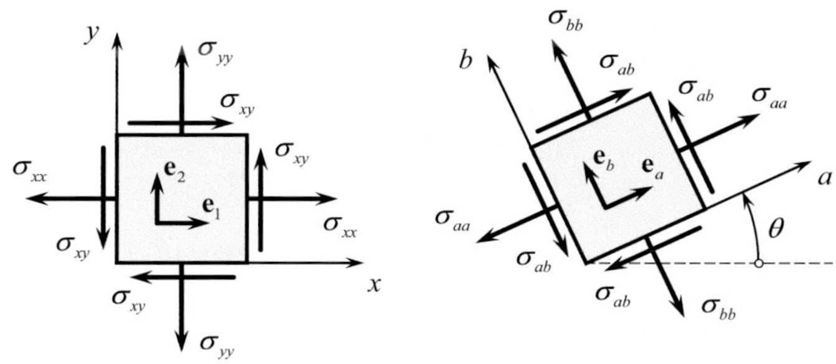

Fig. 5.9 *Change of coordinates.* The components of the stress tensor in the (x, y) and (a, b) coordinate systems. The coordinate system (a, b) is rotated counterclockwise by the angle θ from (x, y)

The unit vectors $\mathbf{e}_a$ and $\mathbf{e}_b$ have the same component representations as $\mathbf{n}$ and $\mathbf{m}$ from the previous section. Written out in components, the equations for the components in the (a, b) coordinate system take the form

$$\begin{aligned} \sigma_{aa} &= \sigma_{xx}\cos^2\theta + 2\sigma_{xy}\cos\theta\sin\theta + \sigma_{yy}\sin^2\theta \\ \sigma_{ab} &= \big(\sigma_{yy} - \sigma_{xx}\big)\cos\theta\sin\theta + \sigma_{xy}\big(\cos^2\theta - \sin^2\theta\big) \\ \sigma_{bb} &= \sigma_{xx}\sin^2\theta - 2\sigma_{xy}\cos\theta\sin\theta + \sigma_{yy}\cos^2\theta \end{aligned} \tag{5.12}$$

These equations are commonly called the *stress transformation equations* because they transform the components from one coordinate system to another. In the two-dimensional case, this transformation only involves the single angle θ. Notice the similarity between these equations and the *strain* transformation equations given in Eq. 4.19.

5.3.2 Principal Angle and Maximum Shear Angle

As was the case for the strain tensor, we can find a particular orientation of the (a, b) coordinate system that gives the maximum and minimum values of the normal stress. Using Eq. 5.12a as our starting point, we can find the extreme value σ_{aa} with respect to θ by taking the derivative, setting it equal to zero, and solving for the angle that gives the maximum or minimum. The derivative of σ_{aa} is

$$\frac{d\sigma_{aa}}{d\theta} = 2\sigma_{xy}\cos 2\theta - \big(\sigma_{xx} - \sigma_{yy}\big)\sin 2\theta \tag{5.13}$$

where we have invoked the trigonometric identities from Eq. 4.23. Observe that this derivative is equal to twice the shear stress σ_{ab} (compare it with Eq. 5.12b). Hence, the angle that maximizes or minimizes the normal stress σ_{aa} is also the angle at which the shear stress σ_{ab} is equal to zero. Setting the derivative of σ_{aa} in Eq. 5.13 equal to zero gives the equation for the principal angle θ_p

$$\tan 2\theta_p = \frac{2\sigma_{xy}}{\sigma_{xx} - \sigma_{yy}} \tag{5.14}$$

The *principal stresses* are $\sigma_{aa}(\theta_p)$ and $\sigma_{aa}(\theta_p + \pi/2) = \sigma_{bb}(\theta_p)$. One of these is the maximum stress and one is the minimum.

To find the angle at which the shear stress is maximum, we take the derivative of σ_{ab} with respect to θ, set it equal to zero, and solve for the angle. The derivative of σ_{ab} is

$$\frac{d\sigma_{ab}}{d\theta} = -2\sigma_{xy} \sin 2\theta - \left(\sigma_{xx} - \sigma_{yy}\right) \cos 2\theta \tag{5.15}$$

Setting this expression equal to zero and solving for the angle gives the maximum shear angle

$$\tan 2\theta_m = -\frac{\sigma_{xx} - \sigma_{yy}}{2\sigma_{xy}} \tag{5.16}$$

The maximum shear stress is $\sigma_{ab}(\theta_m)$:

$$\tau_{\max} = \sqrt{\tfrac{1}{4}\left(\sigma_{xx} - \sigma_{yy}\right)^2 + \sigma_{xy}^2} \tag{5.17}$$

One can show that the derivative of σ_{ab} with respect to θ in Eq. 5.15 is equal to $\sigma_{bb} - \sigma_{aa}$ from Eq. 5.12. Therefore, at θ_m we have $\sigma_{aa} = \sigma_{bb}$. The equations that govern the principal angle and maximum shear angle for stress are exactly analogous to the equations for strain, i.e., Eqs. 4.24 and 4.26. The details of the derivations are given there.

5.3.3 *The Mohr Circle*

Like strain, stress is described by a tensor. Thus, the analysis of stress parallels the analysis of strain. In fact, the developments are so similar that we can go back and replace each ε in the previous chapter with a σ in this one to get the analogous results. The Mohr circle is a case in point. To derive the Mohr circle for stress, let us start with the first two stress transformation equations (Eq. 5.12)

$$\begin{aligned} \sigma &= \sigma_{xx} \cos^2\theta + 2\sigma_{xy} \sin\theta\cos\theta + \sigma_{yy} \sin^2\theta \\ \tau &= \left(\sigma_{yy} - \sigma_{xx}\right) \sin\theta\cos\theta + \sigma_{xy}\left(\cos^2\theta - \sin^2\theta\right) \end{aligned} \tag{5.18}$$

where we are calling $\sigma = \sigma_{aa}$ and $\tau = \sigma_{ab}$ for short. Noting the trigonometric identities in Eq. 4.25, we can rewrite the stress transformation equations in the form

$$\begin{aligned}\sigma - c &= \sigma_{xy}\sin 2\theta + \tfrac{1}{2}\big(\sigma_{xx} - \sigma_{yy}\big)\cos 2\theta \\ \tau &= \sigma_{xy}\cos 2\theta - \tfrac{1}{2}\big(\sigma_{xx} - \sigma_{yy}\big)\sin 2\theta\end{aligned} \tag{5.19}$$

where $c = \frac{1}{2}\big(\sigma_{xx} + \sigma_{yy}\big)$. If we square both sides of both equations and then add them together, we arrive at the result

$$\big(\sigma - c\big)^2 + \tau^2 = \tfrac{1}{4}\big(\sigma_{xx} - \sigma_{yy}\big)^2 + \sigma_{xy}^2$$

Finally, let us define the right side to be R^2. Now, the stress transformation equation is reduced to

$$\big(\sigma - c\big)^2 + \tau^2 = R^2 \tag{5.20}$$

which is the equation of a circle in (σ, τ) space, centered at the point $(c, 0)$ with radius R. The center and radius are defined in terms of the stress components as

$$c = \tfrac{1}{2}\big(\sigma_{xx} + \sigma_{yy}\big), \qquad R = \sqrt{\tfrac{1}{4}\big(\sigma_{xx} - \sigma_{yy}\big)^2 + \sigma_{xy}^2} \tag{5.21}$$

The quantities c and R are both invariants of the stress tensor. The Mohr circle is shown in Fig. 5.10. Observe that the components of the stress tensor in the (σ, τ) coordinates are plotted in pairs $(\sigma_{yy}, \sigma_{xy})$ and $(\sigma_{xx}, -\sigma_{xy})$. The dotted line joining these two points passes through the center of the circle $(c, 0)$. The stress components in any other orientation, i.e., in the coordinate system (a, b) would also plot as diametrically opposed points on the circle. The principal value of stress can be computed simply as

$$\sigma_1 = c + R, \qquad \sigma_2 = c - R \tag{5.22}$$

The maximum shear stress $\tau_{\max}$ is equal to the radius R. Twice the principal angle $2\theta_p$ is the angle between the dotted line and the horizontal axis, while twice the maximum shear angle $2\theta_m$ is the angle between the dotted line and the vertical axis. It is clear from the Mohr circle that $|2\theta_p - 2\theta_m| = \pi/2$, a result we proved for strain.

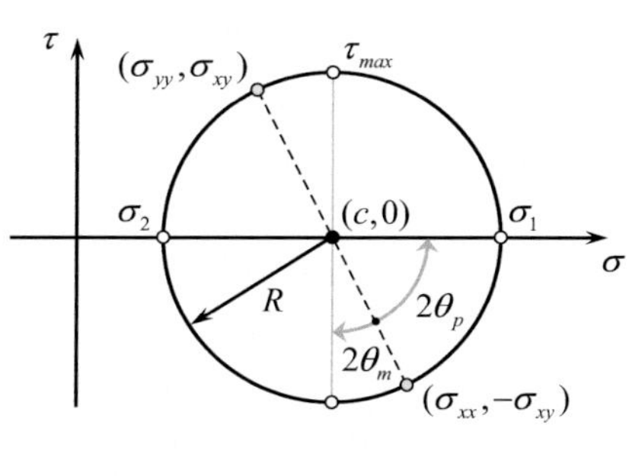

Fig. 5.10 *The Mohr circle.* The Mohr circle is a graphical representation of the state of stress

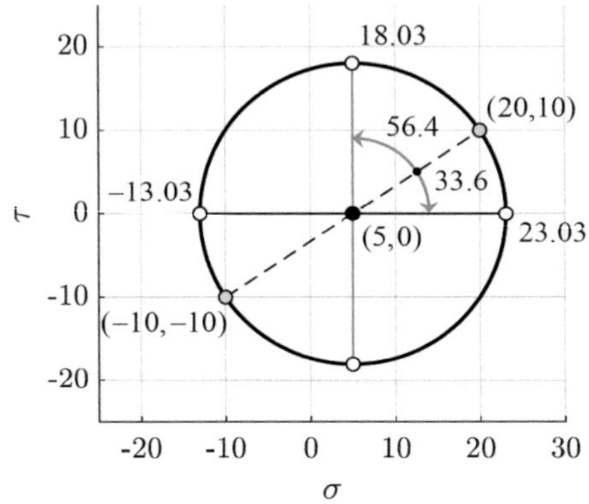

Fig. 5.11 *Example.* The Mohr circle for the state of stress with $\sigma_{xx} = -10$, $\sigma_{xy} = 10$, and $\sigma_{yy} = 20$

Example 5.2
Construct the Mohr circle for a two-dimensional state of stress with $\sigma_{xx} = -10$, $\sigma_{xy} = 10$, and $\sigma_{yy} = 20$.

Solution. The values of the center and radius of the circle can be computed from Eq. 5.21 as

$$c = \tfrac{1}{2}\big(-10 + 20\big) = 5, \qquad R = \tfrac{1}{2}\sqrt{(-10 - 20)^2 + 4(10)^2} = 18.03$$

The center of the circle is at (5, 0). The maximum shear stress is equal to the radius of the circle, i.e., $\tau_{\max} = 18.03$. The principal values can be computed from Eq. 5.10 as

$$\sigma_1 = 5 + 18.03 = 23.03, \qquad \sigma_2 = 5 - 18.03 = -13.03$$

The principal angle can be computed from Eq. 5.14 to be $\theta_p = -16.8°$. The maximum shear angle is $\theta_m = 28.2°$. Note that positive angles on the Mohr circle plot are counterclockwise. All these values can be identified on the plot of the Mohr circle shown in Fig. 5.11.

5.4 Analysis of Stress in Three Dimensions

The extension of the concepts of principal values and principal directions to three-dimensional states of stress parallels the same developments for strain. The physical interpretations that go along with the derivations are different, though. In this section we reconsider the concepts of principal values, principal directions, and the Mohr circle for states of stress in three dimensions. We will again see that the eigenvalue problem is the key to solving this problem.

5.4.1 Principal Values and Principal Directions

For the planar case, the orientation that gave the maximum and minimum values of the normal stress was associated with zero shear stress. The principal value problem for a given stress tensor $\mathbf{S}$ can be stated either as finding a direction $\mathbf{n}$ that yields the extreme values of the function $\sigma = \mathbf{n} \cdot \mathbf{S}\mathbf{n}$ (with the constraint that $\mathbf{n}$ is a unit vector) or finding a direction $\mathbf{n}$ that results in a zero value of the shear stress $\tau = \mathbf{m} \cdot \mathbf{S}\mathbf{n}$.

The vector $\tau\mathbf{m}$ is the projection of the traction $\mathbf{t}(\mathbf{n})$ onto the plane with normal $\mathbf{n}$. The projection tensor is defined in Eq. 1.28 as $\mathbf{P} = \mathbf{I} - \mathbf{n} \otimes \mathbf{n}$. Therefore, if the shearing component of the traction is zero, then

$$\tau\mathbf{m} = \left[\mathbf{I} - \mathbf{n} \otimes \mathbf{n}\right]\mathbf{t}(\mathbf{n}) = \mathbf{0}$$

Substituting $\mathbf{t}(\mathbf{n}) = \mathbf{S}\mathbf{n}$ and noting that $\sigma = \mathbf{n} \cdot \mathbf{t}(\mathbf{n})$ gives the equation

$$\mathbf{S}\mathbf{n} = \sigma\mathbf{n} \tag{5.23}$$

Equation 5.23 is an eigenvalue problem, and it governs the direction $\mathbf{n}$ that gives the maximum or minimum value of σ and also the direction associated with $\tau = 0$. Geometrically, what the eigenvalue problem asks is: What is the direction $\mathbf{n}$ for which multiplying the stress tensor by $\mathbf{n}$ gives the same result as multiplying $\mathbf{n}$ by a scalar?

As we did for the principal strain problem, we can write the eigenvalue problem in the form

$$\left[\mathbf{S} - \sigma\mathbf{I}\right]\mathbf{n} = \mathbf{0} \tag{5.24}$$

which is a linear homogeneous system of equations. A theorem of linear algebra tells us that there is a nontrivial solution only if the determinant of the coefficient matrix is zero. Taking the determinant of $[\,\mathbf{S} - \sigma\mathbf{I}\,]$ and setting it equal to zero gives the *characteristic equation*

$$\sigma^3 - \mathcal{J}_1(\mathbf{S})\,\sigma^2 + \mathcal{J}_2(\mathbf{S})\,\sigma - \mathcal{J}_3(\mathbf{S}) = 0 \tag{5.25}$$

where the stress invariants are defined as (see Eqs. 1.40 and 1.45)

$$\begin{aligned} \mathcal{J}_1(\mathbf{S}) &= \mathrm{tr}(\mathbf{S}) \\ \mathcal{J}_2(\mathbf{S}) &= \tfrac{1}{2}\left[\left(\mathrm{tr}(\mathbf{S})\right)^2 - \mathrm{tr}(\mathbf{S}^2)\right] \\ \mathcal{J}_3(\mathbf{S}) &= \det(\mathbf{S}) \end{aligned} \tag{5.26}$$

We can write out the expressions for the invariants in terms of the components as

$$\begin{aligned}
\mathcal{J}_1(\mathbf{S}) &= \sigma_{xx} + \sigma_{yy} + \sigma_{zz} \\
\mathcal{J}_2(\mathbf{S}) &= \sigma_{xx}\sigma_{yy} + \sigma_{yy}\sigma_{zz} + \sigma_{zz}\sigma_{xx} - \sigma_{xy}^2 - \sigma_{yz}^2 - \sigma_{xz}^2 \\
\mathcal{J}_3(\mathbf{S}) &= \sigma_{xx}\sigma_{yy}\sigma_{zz} - \sigma_{xx}\sigma_{yz}^2 - \sigma_{yy}\sigma_{xz}^2 - \sigma_{zz}\sigma_{xy}^2 + 2\sigma_{xy}\sigma_{yz}\sigma_{xz}
\end{aligned}$$

parallel to the results that we got for strain in the last chapter.

The roots of the characteristic equation are the principal values σ_1, σ_2, and σ_3. If all three roots to the cubic equation are distinct principal values, then the three principal directions $\mathbf{n}_1$, $\mathbf{n}_2$, and $\mathbf{n}_3$ associated with the principal values are unique and they are orthogonal. If one of the roots is distinct and the other two are repeated, then the principal direction associated with the distinct root is unique, and any vector in the plane perpendicular to that direction is a principal direction associated with the repeated principal value. Uniaxial tension is an example of such a stress state, with two of the principal values being zero. Finally, if all three principal values are the same, then any vector in space is an eigenvector, meaning that all directions are principal directions. Hydrostatic pressure is an example of this stress state.

Example 5.3
Find the principal values and principal directions of the stress tensor with components

$$\{\mathbf{S}\} = \begin{bmatrix} 12 & 6 & 6 \\ 6 & 12 & 6 \\ 6 & 6 & 12 \end{bmatrix}$$

Solution. The invariants of **S** can be computed from Eq. 5.26 to give the values

$$\mathcal{J}_1(\mathbf{S}) = 36 \qquad \mathcal{J}_2(\mathbf{S}) = 324 \qquad \mathcal{J}_3(\mathbf{S}) = 864$$

The characteristic equation is

$$\sigma^3 - 36\,\sigma^2 + 324\,\sigma - 864 = 0$$

This equation can be solved by Newton's method (see Code 4.1 for a similar example) to give the principal stresses:

$$\sigma_1 = 24 \qquad \sigma_2 = 6 \qquad \sigma_3 = 6$$

The first eigenvector can be found by solving the linear system of equations $[\,\mathbf{S} - \sigma_1 \mathbf{I}\,]\mathbf{n}_1 = \mathbf{0}$ for the vector $\mathbf{n}_1$. Substituting σ_1, we get

$$\begin{bmatrix} -12 & 6 & 6 \\ 6 & -12 & 6 \\ 6 & 6 & -12 \end{bmatrix} \begin{Bmatrix} 1 \\ a \\ b \end{Bmatrix} = \begin{Bmatrix} 0 \\ 0 \\ 0 \end{Bmatrix}$$

Because the coefficient matrix is singular (that was the criterion for a solution in the first place), we must specify one of the components of $\mathbf{n}_1$. Here we are assuming that the first component of the vector $\mathbf{n}_1$ is equal to one and using the second and third equations to solve for the other two. Now, the second and third equations reduce to

$$\begin{bmatrix} -12 & 6 \\ 6 & -12 \end{bmatrix} \begin{Bmatrix} a \\ b \end{Bmatrix} = \begin{Bmatrix} -6 \\ -6 \end{Bmatrix} \quad \rightarrow \quad \begin{Bmatrix} a \\ b \end{Bmatrix} = \begin{Bmatrix} 1 \\ 1 \end{Bmatrix}$$

Thus, the first eigenvector is $\mathbf{n}_1 = (1, 1, 1)$, which can be scaled to a unit vector by dividing by $\sqrt{3}$. Thus, the unit vector is $\mathbf{n}_1 = (1, 1, 1)/\sqrt{3}$.

The second and third eigenvalues are equal. Therefore, the associated eigenvectors are not unique. *Any* vector perpendicular to $\mathbf{n}_1$ is an eigenvector. We can find a vector perpendicular to $\mathbf{n}_1$ by taking any random vector and projecting out $\mathbf{n}_1$ using the projection tensor $\mathbf{I} - \mathbf{n}_1 \otimes \mathbf{n}_1$, which can be computed in matrix form as

$$\mathbf{I} - \mathbf{n}_1 \otimes \mathbf{n}_1 = \begin{bmatrix} 1 & 0 & 0 \\ 0 & 1 & 0 \\ 0 & 0 & 1 \end{bmatrix} - \frac{1}{3} \begin{bmatrix} 1 & 1 & 1 \\ 1 & 1 & 1 \\ 1 & 1 & 1 \end{bmatrix} = \frac{1}{3} \begin{bmatrix} 2 & -1 & -1 \\ -1 & 2 & -1 \\ -1 & -1 & 2 \end{bmatrix}$$

Now, take a random vector, e.g., $\mathbf{v} = (1, 2, 3)$, and compute a vector perpendicular to $\mathbf{n}_1$ as

$$\frac{1}{3} \begin{bmatrix} 2 & -1 & -1 \\ -1 & 2 & -1 \\ -1 & -1 & 2 \end{bmatrix} \begin{Bmatrix} 1 \\ 2 \\ 3 \end{Bmatrix} = \begin{Bmatrix} -1 \\ 0 \\ 1 \end{Bmatrix}$$

This vector, call it $\mathbf{n}_2$, can be divided by $\sqrt{2}$ to make it a unit vector. Finally, we know that the cross product between $\mathbf{n}_1$ and $\mathbf{n}_2$ is perpendicular to both of them. Using the unscaled versions of the two vectors for simplicity, we get

$$\mathbf{n}_1 \times \mathbf{n}_2 = \det \begin{bmatrix} \mathbf{e}_1 & \mathbf{e}_2 & \mathbf{e}_3 \\ 1 & 1 & 1 \\ -1 & 0 & 1 \end{bmatrix} = \begin{Bmatrix} 1 \\ -2 \\ 1 \end{Bmatrix}$$

Again, this vector can be divided by $\sqrt{6}$ to make it a unit vector. In summary, the three (unit) eigenvectors are

$$\mathbf{n}_1 = \frac{1}{\sqrt{3}} \begin{Bmatrix} 1 \\ 1 \\ 1 \end{Bmatrix} \qquad \mathbf{n}_2 = \frac{1}{\sqrt{2}} \begin{Bmatrix} -1 \\ 0 \\ 1 \end{Bmatrix} \qquad \mathbf{n}_3 = \frac{1}{\sqrt{6}} \begin{Bmatrix} 1 \\ -2 \\ 1 \end{Bmatrix}$$

The vectors $\mathbf{n}_2$ and $\mathbf{n}_3$ can be plugged into the eigenvalue problem to verify that they, indeed, satisfy it. Of course, the eigenvalues and eigenvectors can also be obtained with the MATLAB `eig` function. The eigenvectors will not necessarily match the ones we have just computed because any vector in the plane spanned by the second and third eigenvectors is an eigenvector. The vector $\mathbf{n}_1$ should match, though, because it is unique.

5.4.2 The Mohr Circle in Three Dimensions

As we discovered for strain, in three dimensions there are three individual Mohr circles, as illustrated in Fig. 5.12. The simplest way to construct the circles is to recognize that each circle crosses the σ axis at two of the three principal values. There are three pairings of the three values, so the result is three circles. With any unit vector $\mathbf{n}$, one can generate values of σ and τ as

$$\mathbf{t}(\mathbf{n}) = \mathbf{S}\mathbf{n}, \qquad \sigma = \mathbf{n} \cdot \mathbf{t}(\mathbf{n}), \qquad \tau = \pm\sqrt{\|\mathbf{t}(\mathbf{n})\| - \sigma^2} \tag{5.27}$$

and that individual point can be represented on the plot. In three dimensions, the point will fall into the gray region in Fig. 5.12.

Example 5.4
Show that the values of σ and τ plot in the gray region of the Mohr circle by selecting random directions $\mathbf{n}$ in Eq. 5.27. Use the stress tensor $\mathbf{S}$

$$\mathbf{S} = \begin{bmatrix} 9 & -3 & 2 \\ -3 & -6 & -5 \\ 2 & -5 & -5 \end{bmatrix}$$

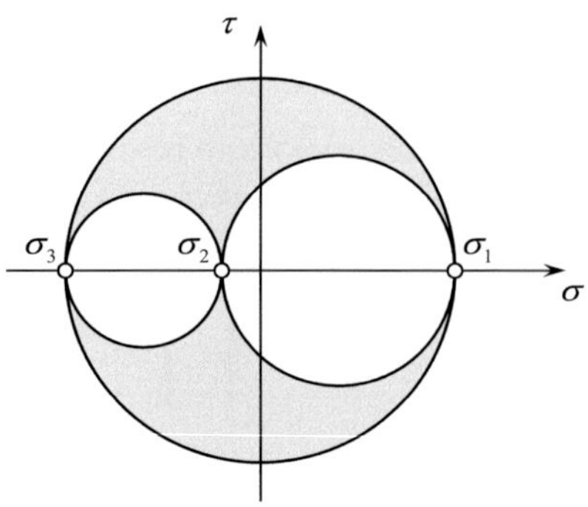

Fig. 5.12 *The Mohr circle.* For three-dimensional states of stress, there are three Mohr circles, one between each pair of principal stresses. The normal and shear stresses must lie in the gray region

to make the demonstration.

Solution. We can solve the eigenvalue problem to show that the stress tensor has principal values

$$\sigma_1 = 10.1842, \qquad \sigma_2 = -1.6168, \qquad \sigma_3 = -10.5674$$

The results for (σ, τ) pairs with 1000 random $\mathbf{n}$ vectors is shown in Fig. 5.13. The plot was generated with the following MATLAB code:

Code 5.1 `Mohr3D` code to plot 1000 (σ, τ) pairs for random $\mathbf{n}$ vectors for tensor $\mathbf{S}$

```
% Illustrate Mohr's circle in 3D
   clear; clc;

%. Establish stress tensor, compute principal values
   S = [9, -3, 2; -3, -6, -5; 2, -5, -5];
   PrincipalValues = eig(S);

%. Set up figure for Mohr's circle
   figure(1); clf; grid on; axis equal; hold on;
   figsize = [0.0, 0.2, 0.3, 0.4];
   set(gcf,'Units','Normalized','OuterPosition',figsize);

%. Compute and plot random points
   gray = 0.75*[1,1,1];
   nPts = 1000;
   sig = zeros(nPts,1);
   tau = zeros(nPts,1);
   for i=1:nPts
     n = 1-2*rand(3,1);   n = n/norm(n);
     tn = S*n;
     sig(i) = dot(n,tn);
     tau(i) = sqrt(dot(tn,tn)-sig(i)^2);
   end
   plot([sig,sig],[tau,-tau],'o','Color','k','LineWidth',1,...
        'MarkerEdgeColor','n','MarkerFaceColor',gray,'MarkerSize',2);

%. Center and radius of Mohr's Circle
   pairs = [2,1; 3,2; 3,1];
   th = linspace(0,2*pi,150);
   for i=1:3
     s1 = PrincipalValues(pairs(i,1));
     s2 = PrincipalValues(pairs(i,2));
     c = (s1 + s2)/2;
     R = (s1 - s2)/2;
     x = c + R*cos(th);
     y = R*sin(th);
     plot(x,y,'-','Color','k','LineWidth',1.5);
     plot([s2,s1],[0,0],'o','Color','k','LineWidth',1,...
          'MarkerEdgeColor','k','MarkerFaceColor','w','MarkerSize',7);
   end

%. Write axis labels
   axis([-12,12,-12,12]);
   xlabel('$\sigma$','interpreter','latex','FontSize',16);
   ylabel('$\tau$','interpreter','latex','FontSize',16);
   set(gca,'TickLabelInterpreter','latex','FontSize',16)
```

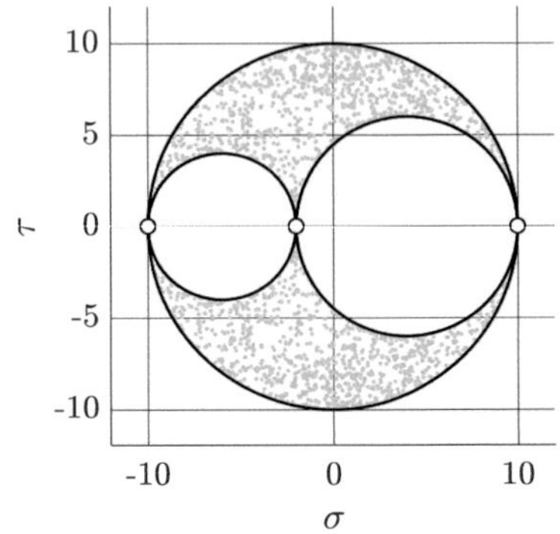

Fig. 5.13 *Example.* Mohr circles for a stress tensor **S** with 1000 random **n** vectors generated from Code 5.1

As expected, all the points are inside the large circle and outside the two smaller circles. Recall from Fig. 1.16 that the tip of every vector **Sn** lies on an ellipsoid with major axis dimensions σ_1, σ_2, and σ_3. This alternative visualization is often called the *Lamé stress ellipsoid*.[2]

As was the case for strain, if two of the three principal values are equal, then the shaded region in Fig. 5.12 vanishes and the plot reduces to a single circle (i.e., one circle shrinks to a point and the other two are the same). If all three principal values are equal, as is the case for hydrostatic pressure, then the three Mohr circles all reduce to a single point.

5.4.3 Plane Stress Revisited

The state of stress called *plane stress* is characterized by

$$\sigma_{xz} = \sigma_{yz} = \sigma_{zz} = 0$$

Substituting these values into the expressions for the stress invariants gives

$$\hat{\mathcal{J}}_1(\mathbf{S}) = \sigma_{xx} + \sigma_{yy}, \qquad \hat{\mathcal{J}}_2(\mathbf{S}) = \sigma_{xx}\sigma_{yy} - \sigma_{xy}^2, \qquad \hat{\mathcal{J}}_3(\mathbf{S}) = 0$$

where the hat on the invariants simply reminds us that this is a special case. The characteristic equation reduces to

$$\sigma\left(\sigma^2 - \hat{\mathcal{J}}_1\sigma + \hat{\mathcal{J}}_2\right) = 0$$

[2] Gabriel Lamé (1795–1870) was a French mathematician who made important contributions to elasticity theory.

One of the solutions to this cubic is $\sigma_3 = 0$. The other two can be found by the quadratic formula as

$$\begin{aligned}\sigma_{1,2} &= \tfrac{1}{2}\hat{\mathcal{J}}_1 \pm \tfrac{1}{2}\sqrt{\hat{\mathcal{J}}_1^2 - 4\,\hat{\mathcal{J}}_2} \\ &= \tfrac{1}{2}\left(\sigma_{xx} + \sigma_{yy}\right) \pm \tfrac{1}{2}\sqrt{\left(\sigma_{xx} + \sigma_{yy}\right)^2 - 4\left(\sigma_{xx}\sigma_{yy} - \sigma_{xy}^2\right)} \\ &= \tfrac{1}{2}\left(\sigma_{xx} + \sigma_{yy}\right) \pm \sqrt{\tfrac{1}{4}\left(\sigma_{xx} - \sigma_{yy}\right)^2 + \sigma_{xy}^2}\end{aligned}$$

From the final result of this calculation, we can identify $\sigma_1 = c + R$ and $\sigma_2 = c - R$, where c and R are the center and radius of the Mohr circle from Eq. 5.21, which we found considering only a two-dimensional formulation.

5.5 Differential Equations of Equilibrium

To establish the differential equations governing the stress tensor, let us consider the small cuboid of dimensions Δx_1 by Δx_2 by Δx_3 shown in Fig. 5.14a. The lower back corner of the cuboid is located at position $\mathbf{x}$. The area of the faces with normal vectors $\pm\mathbf{e}_i$ is called A_i. Thus, the areas are

$$A_1 = \Delta x_2 \Delta x_3, \qquad A_2 = \Delta x_1 \Delta x_3, \qquad A_3 = \Delta x_1 \Delta x_2$$

The traction acting on each face can be related to the stress tensor through the Cauchy formula in Eq. 5.7. Since the region is small and we will eventually pass to the infinitesimal limit, we can assume that the tractions on each face are constant and that the net force on a face acts at its center, as shown in Fig. 5.14b. For example, the net force on the face with unit normal vector $\mathbf{e}_1$ is $\mathbf{S}(\mathbf{x} + \Delta x_1\mathbf{e}_1)\mathbf{e}_1 A_1$ because the face is located at position $\mathbf{x} + \Delta x_1\mathbf{e}_1$, has unit normal $\mathbf{e}_1$, and area A_1. The net force on the face with unit normal vector $-\mathbf{e}_1$ is $-\mathbf{S}(\mathbf{x})\mathbf{e}_1 A_1$ because the face is located at position $\mathbf{x}$, has unit normal $-\mathbf{e}_1$, and area A_1. The body force $\mathbf{b}$ gives rise to a total force of $\mathbf{b}V$, where the volume of the cuboid is $V = \Delta x_1 \Delta x_2 \Delta x_3$. The free body diagram of the cuboid is shown in Fig. 5.14b.

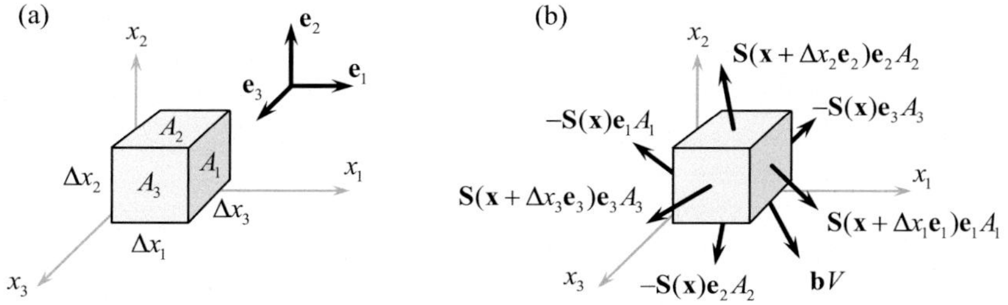

Fig. 5.14 *Equilibrium.* Equilibrium of tractions and body forces on a cuboid. **a** Geometry. **b** Free body diagram

Balance of forces Equilibrium of the forces acting on the cuboid gives

$$\sum_{j=1}^{3}\Big(\mathbf{S}(\mathbf{x}+\Delta x_j\mathbf{e}_j)\mathbf{e}_j-\mathbf{S}(\mathbf{x})\mathbf{e}_j\Big)A_j+\mathbf{b}V=\mathbf{0}$$

Dividing through by the volume V, noting that $V=\Delta x_j A_j$, and taking the limit as the volume shrinks to zero gives

$$\lim_{V\to 0}\sum_{j=1}^{3}\left(\frac{\mathbf{S}(\mathbf{x}+\Delta x_j\mathbf{e}_j)-\mathbf{S}(\mathbf{x})}{\Delta x_j}\right)\mathbf{e}_j+\mathbf{b}=\mathbf{0}$$

The term in parentheses is the *partial derivative* of $\mathbf{S}$ with respect to x_j. Thus, equilibrium requires the stress tensor to satisfy the equation

$$\sum_{j=1}^{3}\frac{\partial\mathbf{S}}{\partial x_j}\mathbf{e}_j+\mathbf{b}=\mathbf{0} \tag{5.28}$$

The first term is called the *divergence* of $\mathbf{S}$ and is often written as div$(\mathbf{S})$. We can write the stress tensor in terms of its components. To find the ith component equation, we will dot Eq. 5.28 with $\mathbf{e}_i$. To prepare for that, compute

$$\mathbf{e}_i\cdot\frac{\partial\mathbf{S}}{\partial x_j}\mathbf{e}_j=\frac{\partial}{\partial x_j}\Big(\mathbf{e}_i\cdot\mathbf{S}\mathbf{e}_j\Big)=\frac{\partial S_{ij}}{\partial x_j}$$

since $\mathbf{e}_i$ is constant and $\mathbf{e}_i\cdot\mathbf{S}\mathbf{e}_j=S_{ij}$. Using this result, the ith equation from Eq. 5.28 is

$$\sum_{j=1}^{3}\frac{\partial S_{ij}}{\partial x_j}+b_i=0 \tag{5.29}$$

where S_{ij} is the ijth component of stress tensor $\mathbf{S}$, and b_i is the ith component of the body force vector $\mathbf{b}$. This equation holds for $i=1,2,3$. Therefore, Eq. 5.29 is actually three scalar equations. The equations specialized for a two-dimensional state of stress are written out in the following example.

Example 5.5
Write out the equilibrium equations for a two-dimensional state of stress.

Solution. The components of the stress tensor for a two-dimensional state of stress are

$$\mathbf{S}=\begin{bmatrix}\sigma_{xx} & \sigma_{xy}\\ \sigma_{yx} & \sigma_{yy}\end{bmatrix}$$

The equations of equilibrium, then, take the form

$$\frac{\partial \sigma_{xx}}{\partial x} + \frac{\partial \sigma_{xy}}{\partial y} + b_x = 0$$
$$\frac{\partial \sigma_{yx}}{\partial x} + \frac{\partial \sigma_{yy}}{\partial y} + b_y = 0$$

where b_x and b_y are the x and y components of the body force $\mathbf{b}$.

Solving the partial differential equations of equilibrium is a challenging task that is beyond the scope of this book. However, checking to see if a given stress field satisfies the equations of equilibrium is simply a matter of computing the divergence of the stress tensor to verify that the equilibrium equations are satisfied at every point in the body and checking the tractions on the boundaries of the region to verify that the stresses match the applied tractions on the boundary, as the following example illustrates.

Example 5.6
Show that the stress in a hollow sphere of inside radius $r_i = 1$ and outside radius $r_o = 2$, subjected to a pressure p inside, is

$$\mathbf{S}(\mathbf{x}) = \frac{p}{7}\Big[\Big(1 + \frac{4}{r^3}\Big)\mathbf{I} - \frac{12}{r^3}(\mathbf{n} \otimes \mathbf{n})\Big]$$

where position $\mathbf{x}$ is measured from the center of the sphere, r is the radial distance to any point $\mathbf{x}$ in the sphere, and $\mathbf{n} = \mathbf{x}/r$ is a unit vector pointing radially outward. Note that $r^2 = \mathbf{x} \cdot \mathbf{x}$ is the square of the radial distance at any point in the sphere (and, thus, r is a function of $\mathbf{x}$). Neglect body forces (i.e., $\mathbf{b} = \mathbf{0}$). Find the principal values and directions.

Solution. First, we can verify that the conditions at the inner radius and outer radius are satisfied. The unit normal on the outside surface is $\mathbf{n}$, and the unit normal on the inside surface is $-\mathbf{n}$. Compute the tractions $\mathbf{S}(\pm\mathbf{n})$

$$\mathbf{t} = \mathbf{S}(\pm\mathbf{n}) = \frac{p}{7}\Big(1 + \frac{4}{r^3} - \frac{12}{r^3}\Big)(\pm\mathbf{n})$$

since $\mathbf{I}\mathbf{n} = \mathbf{n}$ and $[\,\mathbf{n} \otimes \mathbf{n}\,]\mathbf{n} = \mathbf{n}$. The inside radius is $r = 1$ and the outside radius is $r = 2$. So, the tractions on those surfaces are

$$\mathbf{t}(r=1) = \frac{p}{7}\Big(1 + 4 - 12\Big)(-\mathbf{n}) = p\,\mathbf{n}$$
$$\mathbf{t}(r=2) = \frac{p}{7}\Big(1 + \frac{4}{8} - \frac{12}{8}\Big)\mathbf{n} = \mathbf{0}$$

The traction on the inside radius has the same magnitude as the internal pressure, i.e., p, and points in direction $\mathbf{n}$, which means it is in compression, pushing on the inside of the hollow sphere.

To prepare to compute the divergence of the stress tensor from Eq. 5.28, compute

$$\mathbf{S}\mathbf{e}_j = \frac{p}{7}\Big[\Big(1+\frac{4}{r^3}\Big)\mathbf{e}_j - \frac{12}{r^5}(x_j\mathbf{x})\Big]$$

where we have substituted $\mathbf{x}=r\mathbf{n}$ and noted that $\mathbf{x}\cdot\mathbf{e}_j=x_j$. To take the derivative of this expression with respect to x_j, we will need the following derivatives

$$\frac{\partial x_j}{\partial x_j}=\delta_{jj}, \qquad \frac{\partial \mathbf{x}}{\partial x_j}=\mathbf{e}_j, \qquad \frac{\partial r}{\partial x_j}=\frac{x_j}{r}$$

To prove the last one, start with $r^2=\mathbf{x}\cdot\mathbf{x}$ and take the derivative of both sides with respect to x_j. Next, compute the derivative of $\mathbf{S}$ with respect to x_j using product rule and chain rule. To wit,

$$\begin{aligned}\frac{\partial \mathbf{S}\mathbf{e}_j}{\partial x_j} &= \frac{p}{7}\Big[-\frac{12}{r^4}\frac{\partial r}{\partial x_j}\mathbf{e}_j + \frac{60}{r^6}\frac{\partial r}{\partial x_j}(x_j\mathbf{x}) - \frac{12}{r^5}\Big(\frac{\partial x_j}{\partial x_j}\mathbf{x} + x_j\frac{\partial \mathbf{x}}{\partial x_j}\Big)\Big]\\ &= \frac{p}{7}\Big[-\frac{12}{r^5}x_j\mathbf{e}_j + \frac{60}{r^7}x_j(x_j\mathbf{x}) - \frac{12}{r^5}\Big(\delta_{jj}\mathbf{x}+x_j\mathbf{e}_j\Big)\Big]\end{aligned}$$

Finally, sum on j to get the divergence

$$\begin{aligned}\sum_{j=1}^{3}\frac{\partial \mathbf{S}\mathbf{e}_j}{\partial x_j} &= \sum_{j=1}^{3}\frac{p}{7}\Big[-\frac{12}{r^5}x_j\mathbf{e}_j + \frac{60}{r^7}x_j(x_j\mathbf{x}) - \frac{12}{r^5}\Big(\delta_{jj}\mathbf{x}+x_j\mathbf{e}_j\Big)\Big]\\ &= \frac{p}{7}\Big[-\frac{12}{r^5}\mathbf{x} + \frac{60}{r^5}\mathbf{x} - \frac{12}{r^5}(4\mathbf{x})\Big] = \mathbf{0}\end{aligned}$$

where we have noted

$$\sum_{j=1}^{3}x_j\mathbf{e}_j=\mathbf{x}, \qquad \sum_{j=1}^{3}x_jx_j=r^2, \qquad \sum_{j=1}^{3}\delta_{jj}=3$$

Thus, the divergence of $\mathbf{S}$ is zero and, hence, equilibrium is assured at every point in the solid body.

Due to the form of the stress tensor, we can guess that one of the eigenvectors is the radial direction $\mathbf{n}$. Let us compute $\mathbf{S}\mathbf{n}$ to see if it turns out to be in the form $\sigma\mathbf{n}$. If it is, then $\mathbf{n}$ is an eigenvector and σ is the eigenvalue. To wit,

$$\begin{aligned}\mathbf{Sn} &= \frac{p}{7}\Big[\Big(1+\frac{4}{r^3}\Big)\mathbf{I} - \frac{12}{r^3}(\mathbf{n}\otimes\mathbf{n})\Big]\mathbf{n} \\ &= \frac{p}{7}\Big[\Big(1+\frac{4}{r^3}\Big)\mathbf{n} - \frac{12}{r^3}(\mathbf{n}\cdot\mathbf{n})\mathbf{n}\Big] = \frac{p}{7}\Big(1-\frac{8}{r^3}\Big)\mathbf{n}\end{aligned}$$

We can also try any vector $\mathbf{m}$ that is perpendicular to $\mathbf{n}$. To wit,

$$\begin{aligned}\mathbf{Sm} &= \frac{p}{7}\Big[\Big(1+\frac{4}{r^3}\Big)\mathbf{I} - \frac{12}{r^3}(\mathbf{n}\otimes\mathbf{n})\Big]\mathbf{m} \\ &= \frac{p}{7}\Big[\Big(1+\frac{4}{r^3}\Big)\mathbf{m} - \frac{12}{r^3}(\mathbf{n}\cdot\mathbf{m})\mathbf{n}\Big] = \frac{p}{7}\Big(1+\frac{4}{r^3}\Big)\mathbf{m}\end{aligned}$$

since $\mathbf{n}\cdot\mathbf{m}=0$. Therefore, both $\mathbf{n}$ and $\mathbf{m}$ are eigenvectors with associated eigenvalues, respectively, of

$$\sigma_1 = \frac{p}{7}\Big(1-\frac{8}{r^3}\Big), \qquad \sigma_{2,3} = \frac{p}{7}\Big(1+\frac{4}{r^3}\Big)$$

We know that the second principal value is repeated because any vector $\mathbf{m}$ in the plane perpendicular to $\mathbf{n}$ is an eigenvector. Both principal values depend only on the radial distance r due to the spherical symmetry. The principal value σ_1 is the stress in the radial direction. Its maximum value is $\sigma_1=-p$ (compression) at the inner surface ($r=1$), and it decreases to zero at the outer surface. The principal value $\sigma_{2,3}$ is the stress in the tangential direction. Its maximum value is $\sigma_{2,3}=5p/7$ (tension), also at the inner surface. The tangential stress is $\sigma_{2,3}=3p/14$ at the outer surface ($r=2$).

Balance of moments Every free body diagram must also satisfy balance of moments about any point in space. For the cuboid in Fig. 5.14, we can take moments about the center. Again, assume that the body force acts at the center and the tractions act at the centers of their respective faces (the limiting process will take care of any discrepancy). The body force creates no moment about the center of the cuboid. The moment arms for the forces on the faces are $\pm\frac{1}{2}\Delta x_j\mathbf{e}_j$. Balance of moments gives

$$\begin{aligned}\sum\mathbf{M} = \sum_{j=1}^{3}\Big[&\tfrac{1}{2}\Delta x_j\mathbf{e}_j\times\mathbf{S}(\mathbf{x}+\Delta x_j\mathbf{e}_j)\mathbf{e}_jA_j \\ &+\big(-\tfrac{1}{2}\Delta x_j\mathbf{e}_j\big)\times\mathbf{S}(\mathbf{x})\big(-\mathbf{e}_j\big)A_j\Big]=\mathbf{0}\end{aligned}$$

Rearranging terms gives

$$\sum_{j=1}^{3}\mathbf{e}_j\times\tfrac{1}{2}\Big[\mathbf{S}(\mathbf{x}+\Delta x_j\mathbf{e}_j)+\mathbf{S}(\mathbf{x})\Big]\mathbf{e}_j\,\Delta x_jA_j=\mathbf{0}$$

Each term has $\Delta x_j A_j = V$, which does not depend upon the index j. Also, note that

$$\lim_{V\to 0} \tfrac{1}{2}\Big[\mathbf{S}(\mathbf{x}+\Delta x_j\mathbf{e}_j) + \mathbf{S}(\mathbf{x})\Big] = \mathbf{S}(\mathbf{x})$$

Dividing through by V and taking the limit as $V \to 0$ shows that in order for moments to balance, the stress tensor must satisfy the equation

$$\sum_{i=1}^{3} \mathbf{e}_i \times \mathbf{S}\mathbf{e}_i = \mathbf{0} \tag{5.30}$$

To see what this equation requires, write it out in terms of the components of the stress tensor in Eq. 5.6. To wit,

$$\big(\sigma_{yz} - \sigma_{zy}\big)\mathbf{e}_1 + \big(\sigma_{xz} - \sigma_{zx}\big)\mathbf{e}_2 + \big(\sigma_{xy} - \sigma_{yx}\big)\mathbf{e}_3 = \mathbf{0}$$

To be equal to zero, each coefficient of this vector must be zero. Thus,

$$\sigma_{yz} = \sigma_{zy}, \qquad \sigma_{xz} = \sigma_{zx}, \qquad \sigma_{xy} = \sigma_{yx}$$

In other words, the stress tensor must be *symmetric*, i.e.,

$$\mathbf{S} = \mathbf{S}^T \tag{5.31}$$

To enforce the symmetry of the stress tensor, we will write the stress tensor components as

$$\mathbf{S} = \begin{bmatrix} \sigma_{xx} & \sigma_{xy} & \sigma_{xz} \\ \sigma_{xy} & \sigma_{yy} & \sigma_{yz} \\ \sigma_{xz} & \sigma_{yz} & \sigma_{zz} \end{bmatrix} \tag{5.32}$$

where we have simply replaced the terms below the diagonal with their mirror counterparts above the diagonal. This form of the stress tensor automatically guarantees symmetry and, therefore, balance of moments. It implies that there are only six independent components of the stress tensor.

The symmetry requirement for the stress tensor can be illustrated in two dimensions. Consider Fig. 5.15, which shows the stress components acting on a square free

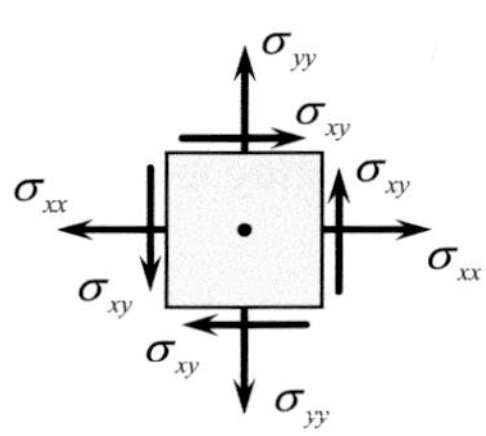

Fig. 5.15 *Symmetry of the stress tensor.* The stress tensor must be symmetric for moment equilibrium to hold

body diagram. The components are all shown in their positive directions according to the Cauchy formula (i.e., the arrows point in the negative coordinate directions on the faces with normal vectors $-\mathbf{e}_1$ and $-\mathbf{e}_2$). Sum moments about the center of the square. The components σ_{xx} and σ_{yy} do not create any moment about any point because they both have equal and opposite counterparts on opposite faces (and for moments about the center of the square, they have no moment arm, either). The shear stress components σ_{xy} do create moment. If the stress components on the top face were σ_{yx} and it was not equal to σ_{xy}, then the moments they cause would not balance. Hence, they must be equal. You can repeat this argument with the other two coordinate faces to deduce that the stress tensor must be symmetric. Thus, the four arrows associated with shear stresses must be either in the directions shown or all of them must be in the opposite direction. The direction shown in the figure is our convention for positive shear.

Solving the full three-dimensional differential equations of equilibrium of stress is beyond the scope of this book. In the remainder of the book, these equilibrium equations will show up in special forms for beams and torsion, simplified through the introduction of stress resultants. We have already seen the case for the axial bar, as the following example reminds us.

Example 5.7
Show that the equations of equilibrium for the axial bar problem in Eq. 3.3, i.e.,

$$\frac{dN}{dx} + p(x) = 0$$

can be obtained from the general equations of equilibrium for a uniaxial state of stress.

Solution. A state of uniaxial stress with the x-axis being the direction of the stress has $\sigma_{xx} \neq 0$, and all other components of the stress tensor equal to zero. With no shear stresses, the first of Eq. 5.29 reduces to

$$\frac{\partial \sigma_{xx}}{\partial x} + b_x = 0$$

The terms with σ_{xy} and σ_{xz} are zero. If we integrate this equation over the area of the cross section at x, we get

$$\int_A \left(\frac{\partial \sigma_{xx}}{\partial x} + b_x \right) dA = \frac{\partial}{\partial x} \int_A \sigma_{xx}\, dA + \int_A b_x\, dA = 0$$

The first term is the derivative of the net axial force N and the second term is the applied axial load, defined as the net effect of the body force b_x over the cross section. Thus, the equation can be written as

$$\frac{dN}{dx} + p(x) = 0$$

where we have noted that the integral of σ_{xx} over the area depends only on x, so the derivative is an ordinary derivative rather than a partial derivative. The second and third of Eq. 5.29 are satisfied only if $b_y = b_z = 0$, which is consistent with the restrictions placed on the axial bar problem.

5.6 Summary

In the axial bar problem, there was only one way that forces could flow through the bar, and we captured that with the internal axial force. This chapter has considered the issue of how forces flow through a general three-dimensional body. We found that the concept of *stress*, which can be represented as a tensor, was the key to formulating a mathematical description of this more general flow of forces. In a sense, we learned how to form free body diagrams for three-dimensional solid bodies.

Because stress is represented as a tensor, we found that all the machinery developed for analyzing the state of strain could be directly applied to stress. Concepts like principal values of stress, maximum shear stress, and tensor invariants issued directly from developments of the previous chapter. While stress and strain could not be more different physically, they are very similar mathematically.

The number of chapter end problems for this chapter is few, but the concepts of this chapter will show up in most of the problems of the following chapter on constitutive models, where it will finally be possible to consider problems with all three of the legs of the mechanics stool present. Look at the problems at the end of the next chapter to find more opportunities to exercise the ideas of this one.

5.7 Problems

5.1 An element in plane stress is subjected to stresses of 3.5 MPa tension in the horizontal direction and 12.2 MPa tension in the vertical direction. Also, shear stresses of magnitude 3.3 MPa act in the directions shown. Use the stress transformation equations to determine the principal stresses for the element. Determine the stresses acting on an element oriented at a counterclockwise angle of $\theta = 51^\circ$ from the horizontal. Show both on a sketch of an element oriented at the given angle.

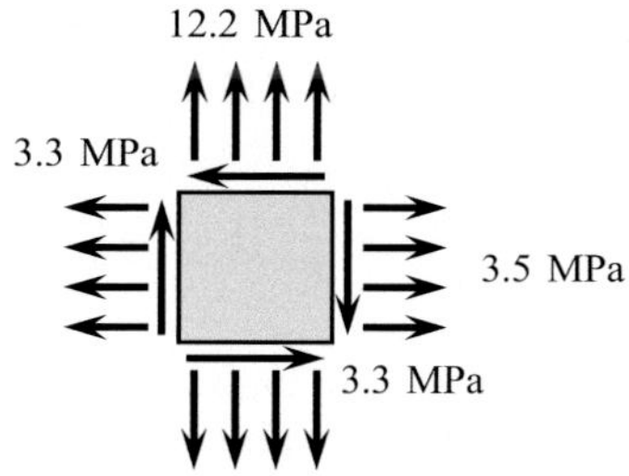

5.2 Consider the long pressure vessel of radius r (to the middle of the wall), thickness t, and length L shown in the sketch. The dimensions of the vessel are such that L is much greater than r, which is much greater than t. The vessel has an internal pressure of p_o. Find the state of stress for the small block of material at point $\mathcal{A}$. What are the principal stresses? The shell of the pressure vessel is constructed by a spiral weld that makes an angle of 42° with the horizontal. If the weld material fails at a maximum normal stress of $\sigma_{\text{allow}} = 10$ MPa, find the maximum allowable pressure p_o. For this calculation, take $L = 5$ m, $r = 1.2$ m, and $t = 1$ cm.

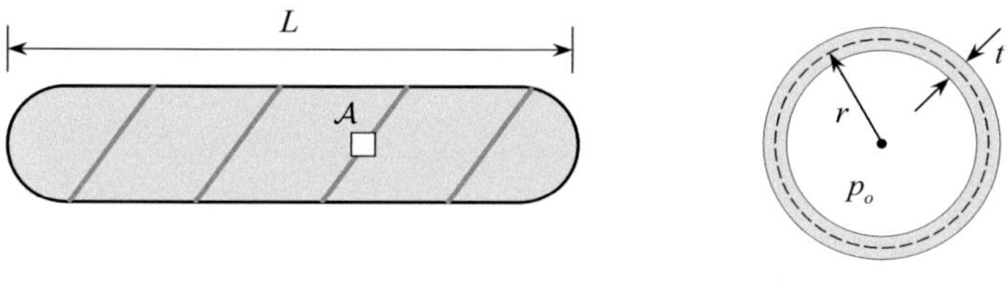

5.3 An element in plane stress is subjected to stresses of 8.5 MPa compression in the horizontal direction and 4.5 MPa tension in the vertical direction. Also, shear stresses of magnitude 2.2 MPa act in the directions shown. Find principal angle, principal stresses, maximum shear angle, and maximum shear stress. Find the stresses for an element rotated 24° counterclockwise. Draw a Mohr circle including the given stresses, the rotated stresses, the principal stresses, and the maximum shear stress.

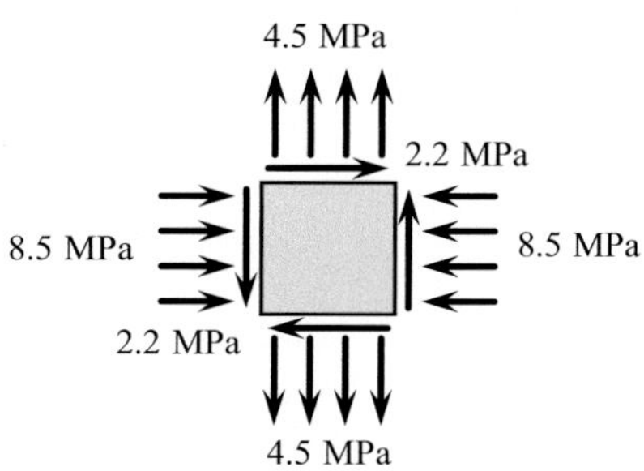

5.4 Consider a hollow sphere with inside radius r_i and outside radius r_o subjected to an internal pressure of p_o. A candidate solution for the stress field is

$$\mathbf{S}(\mathbf{x}) = \frac{p_o}{a}\left[\left(1 + \frac{b}{r^3}\right)\mathbf{I} - \frac{c}{r^3}[\mathbf{n} \otimes \mathbf{n}]\right]$$

where $\mathbf{x}$ is the position measured relative to the center of the sphere, $r = \|\mathbf{x}\|$ is the radial distance to point $\mathbf{x}$, and $\mathbf{n} = \mathbf{x}/r$ is a unit vector pointing in the same direction as $\mathbf{x}$. Find the values of the constants a, b, and c, in terms of r_i and r_o, for which the stress state satisfies the surface traction conditions and the equations of equilibrium. Assume that the body forces are negligible.

Chapter 6
Constitutive Models

Through equilibrium we relate applied forces to stress, and through kinematics we relate displacement to strain. To close the loop and form a complete theory, we must relate stress to strain. A relationship between stress and strain is called a *constitutive model*. While the ultimate goal of constitutive theory is to write a universal law governing material behavior, multiple practical considerations make it infeasible to describe all materials with one law. Hence, unlike equilibrium and kinematics, constitutive relationships are empirical. Thus, it is more accurate to call them constitutive *models* rather than *laws*. In this chapter we explore one of the most common constitutive models: linear elastic response.

Generally speaking, in constitutive theory we seek mathematical functions with empirically determined parameters that imitate the response of a material over a broad range of excitation scenarios. These functions generally model the mechanical behavior well within certain classes of materials (e.g., metals or plastics). A model that works for one material may not work well for another one. Once the mathematical form of the model is selected, the values of the parameters, called *material constants*, are established through laboratory testing and curve fitting.

Consider a simple example to illustrate the context. The equation

$$y = ax + b$$

describes a line in the (x, y) plane. The constants a and b are the parameters. The equation is *linear*, no matter what the parameter values are. As such, it represents a *class* of models. Specific values of a and b describe specific lines. The values $a = 1$ and $b = 2$ yield the specific function $y = x + 2$, a line with slope 1 and y-intercept 2. Once the parameters are set, we can use the model to predict values of y given values of x (or vice versa).

For many materials, the values of the material constants are stable enough that they are nearly independent of the particular circumstances of material processing. In such a case, the material properties can be cataloged, and engineers can safely use

K. D. Hjelmstad, *Mechanics of Deformable Solids: An Introduction*,
https://doi.org/10.1007/978-3-031-89873-0_6

the published values in the design process. For example, Young's modulus for steel is approximately 200 GPa across the many different types of steel and from one batch to the next. Hence, an engineer can use a tabulated value in the design process with the expectation that the material will have an actual value very close to the tabulated one when the system is built. Other materials are not as predictable. For example, the properties of concrete are usually verified in each application by taking samples at the construction site and testing them to make sure the material properties used in design adequately represent the material used in construction.

To model the behavior of materials, it turns out that a very useful assumption is that stress at a point is related only to the strain at that point. This assumption is called the *axiom of locality*. The implication of this assumption is that the constitutive model will be an algebraic relationship, not a differential equation.

6.1 Hooke's Law

Hooke's law is a constitutive model for isotropic linear elasticity.[1] We call it a *law* in keeping with tradition. It is really just a model of material behavior, but if a model has been around and successfully used for centuries, perhaps it can be elevated to the status of law (like Newton's *laws* of motion). The term *isotropic* means that the material behaves the same in all directions. A counterexample is wood, which has an oriented grain structure. The properties along the grain are different from those across the grain. Steel is generally assumed to be isotropic. The term *linear* simply means loading and unloading happens along a straight line and if we double the strain, we will double the stress. In this section we will write the equations of Hooke's law in various forms, e.g., stress as a function of strain or strain as a function of stress, and show some equivalent but different representations of the material parameters.

Hooke's law is a relationship between stress and strain. The constitutive relationship can be written in simplest form with stress as a function of strain. To wit,

$$\mathbf{S} = \lambda\,\mathrm{tr}(\mathbf{E})\,\mathbf{I} + 2\mu\,\mathbf{E} \tag{6.1}$$

where $\mathbf{S}$ is the stress tensor, $\mathbf{E}$ is the strain tensor, and $\mathbf{I}$ is the identity. The material parameters λ and μ are called the *Lamé parameters*. One of the most important observations about Hooke's law is that it has *two* independent material constants. In Chap. 3 we were introduced to only one parameter (Young's modulus). We will soon see why that is appropriate for the uniaxial state of stress of the axial bar, why we actually need two material constants in general, and how Young's modulus relates to the Lamé parameters.

[1] Robert Hooke (1635–1703) was a contemporary of Isaac Newton and one of the leading scientists of the time. He worked mostly with helical springs and posited the linear response model through a statement *ut tensio, sic vis* which translates to "as the extension, so the force."

First, let us examine how a tensor equation like Eq. 6.1 works. In components, it is a matrix relationship, and each one of the nine elements actually gives an individual scalar relationship. Thus, the tensor equation is actually nine scalar equations, each of which must be satisfied for the tensor equation to be valid. Since both the stress and strain tensors are symmetric, we only have six independent scalar equations. Let us write out Eq. 6.1 in components as

$$\begin{bmatrix} \sigma_{xx} & \sigma_{xy} & \sigma_{xz} \\ \sigma_{xy} & \sigma_{yy} & \sigma_{yz} \\ \sigma_{xz} & \sigma_{yz} & \sigma_{zz} \end{bmatrix} = \lambda \begin{bmatrix} e & 0 & 0 \\ 0 & e & 0 \\ 0 & 0 & e \end{bmatrix} + 2\mu \begin{bmatrix} \varepsilon_{xx} & \varepsilon_{xy} & \varepsilon_{xz} \\ \varepsilon_{xy} & \varepsilon_{yy} & \varepsilon_{yz} \\ \varepsilon_{xz} & \varepsilon_{yz} & \varepsilon_{zz} \end{bmatrix}$$

where $e = \text{tr}(\mathbf{E}) = \varepsilon_{xx} + \varepsilon_{yy} + \varepsilon_{zz}$. Recall from Sect. 4.5.4 that e is an approximate measure of volume change for small deformations.

We can select individual equations from the matrix equation. For example, taking the first row and column of each matrix, i.e., the (1, 1) element, gives the equation

$$\sigma_{xx} = \lambda\big(\varepsilon_{xx} + \varepsilon_{yy} + \varepsilon_{zz}\big) + 2\mu\,\varepsilon_{xx}$$

Observe that in the equation for σ_{xx}, all the normal strains are present. This feature provides coupling of the normal stresses and strains, a key feature of Hooke's law that captures the *Poisson effect*,[2] wherein a bar pulled in uniaxial tension will contract laterally and a bar in compression will expand laterally.

Taking the first row and second column gives the equation

$$\sigma_{xy} = 2\mu\,\varepsilon_{xy}$$

In the equation for σ_{xy}, there is no coupling effect. Therefore, if $\varepsilon_{xy} = 0$, then $\sigma_{xy} = 0$ and vice versa. Some authors refer to the *shear modulus* as

$$G = \mu \tag{6.2}$$

and write $\sigma_{xy} = 2G\varepsilon_{xy}$. Shear stress is proportional to twice the shear strain, and the constant of proportionality is G.

We can put Hooke's law into a form that gives strain as a function of stress. The easiest way to do that is to take the trace of both sides of Eq. 6.1 to get an expression for tr($\mathbf{E}$) in terms of tr($\mathbf{S}$). To wit,

$$\text{tr}(\mathbf{S}) = 3\lambda\,\text{tr}(\mathbf{E}) + 2\mu\,\text{tr}(\mathbf{E}) = \big(3\lambda + 2\mu\big)\text{tr}(\mathbf{E})$$

where we have made use of the fact that $\text{tr}(\mathbf{I}) = 3$. Now, solve for tr($\mathbf{E}$) and substitute the result into Eq. 6.1 to get

---

[2] Siméon Denis Poisson (1781–1840) was a French mathematician and physicist who made numerous contributions to mechanics.

$$\mathbf{S} = \frac{\lambda}{3\lambda + 2\mu} \operatorname{tr}(\mathbf{S})\,\mathbf{I} + 2\mu\,\mathbf{E}$$

Finally, solving this equation for **E** gives

$$\mathbf{E} = -\frac{\lambda}{2\mu(3\lambda + 2\mu)} \operatorname{tr}(\mathbf{S})\,\mathbf{I} + \frac{1}{2\mu}\,\mathbf{S} \tag{6.3}$$

Equation 6.3 is exactly the same constitutive model as Eq. 6.1, just in the form of strain as a function of stress.

6.2 Young's Modulus and Poisson's Ratio

Hooke's law, as formulated in Eqs. 6.1 and 6.3, is expressed in terms of the Lamé parameters λ and μ. In the latter case, one might argue that the appearance of the material constants is unnecessarily complicated, whereas the appearance of the constants in the former version of the equations seems canonically simple. It is reasonable to wonder why the material constants appear in the form that they do and to ask if this is the only possible way to express Hooke's law.

The answer to the question is no. In fact, one can write Hooke's law as

$$\mathbf{S} = \frac{E\nu}{(1+\nu)(1-2\nu)} \operatorname{tr}(\mathbf{E})\mathbf{I} + \frac{E}{1+\nu}\mathbf{E} \tag{6.4}$$

for stress in terms of strain and

$$\mathbf{E} = -\frac{\nu}{E} \operatorname{tr}(\mathbf{S})\mathbf{I} + \frac{1+\nu}{E}\mathbf{S} \tag{6.5}$$

for strain in terms of stress. We call E Young's modulus and ν Poisson's ratio. By comparison with Eq. 6.1, it is evident that

$$\lambda = \frac{E\nu}{(1+\nu)(1-2\nu)}, \qquad 2\mu = \frac{E}{1+\nu} \tag{6.6}$$

By comparison with Eq. 6.2, we can see that the shear modulus G is

$$G = \frac{E}{2(1+\nu)} \tag{6.7}$$

If the values of E and ν are known, then the values of λ and μ can be computed from Eq. 6.6. Therefore, these parameters are not additional material constants, but simply a different representation of the two constants that characterize Hooke's law.

Because the equations are equivalent, the key phenomena embedded in Hooke's law will manifest in either form of the equations.

Example 6.1
Find the constants E and ν in terms of the Lamé parameters λ and μ.

Solution. Substituting the second of Eq. 6.6 into the first, we can write

$$\lambda = \frac{2\mu\nu}{1-2\nu}$$

Multiplying through by $1-2\nu$ and rearranging terms, we get $\lambda = 2\nu(\lambda+\mu)$. Solving for ν gives Poisson's ratio in terms of the Lamé parameters:

$$\nu = \frac{\lambda}{2(\lambda+\mu)}$$

Now, solve the second of Eq. 6.6 for E and substitute the expression for ν just found to get

$$E = 2\mu(1+\nu) = 2\mu\left[1+\frac{\lambda}{2(\lambda+\mu)}\right] = \mu\left(\frac{3\lambda+2\mu}{\lambda+\mu}\right)$$

Thus, given the Lamé parameters, we can compute Young's modulus and Poisson's ratio.

There is one other material constant worth mentioning. Take the trace of Eq. 6.5 to get

$$\text{tr}(\mathbf{E}) = \frac{1-2\nu}{E}\,\text{tr}(\mathbf{S})$$

We know from Eq. 4.39 that $e = \text{tr}(\mathbf{E})$ is, for small strains, a measure of volume change. In a state of hydrostatic stress, the stress tensor has the form $\mathbf{S} = p\mathbf{I}$, where p is the pressure. For this case, $\text{tr}(\mathbf{S}) = 3p$. We can, then, relate the pressure to change in volume as

$$p = Ke$$

where K is called the *bulk modulus* and is related to Young's modulus and Poisson's ratio as

$$K = \frac{E}{3(1-2\nu)}$$

The bulk modulus measures a material's resistance to volume change under pressure. This form of Hooke's law makes it clearer why the upper limit on Poisson's ratio is one half. As $\nu \to 0.5$, the bulk modulus approaches infinity, i.e., $K \to \infty$. For values of ν greater than one half, the bulk modulus is negative, implying that under constant compressive pressure the volume of the material compressed would increase, which is not physically reasonable. To be more rigorous, one can show that values of $\nu > 0.5$ violate the laws of thermodynamics.

Example 6.2
The *Poisson effect* is where a body in uniaxial tension contracts laterally, as shown in Fig. 6.1. Show that Hooke's law incorporates this behavior for a condition of uniaxial tension.

Solution. For the state of uniaxial tension, we have

$$\sigma_{xx} = \sigma, \qquad \sigma_{yy} = \sigma_{zz} = \sigma_{xy} = \sigma_{xz} = \sigma_{yz} = 0$$

Using the equations for σ_{yy} and σ_{zz} in Eq. 6.4 gives

$$\begin{aligned} \sigma_{yy} &= \frac{E\nu}{(1+\nu)(1-2\nu)}\left(\varepsilon_{xx}+\varepsilon_{yy}+\varepsilon_{zz}\right)+\frac{E}{1+\nu}\varepsilon_{yy}=0 \\ \sigma_{zz} &= \frac{E\nu}{(1+\nu)(1-2\nu)}\left(\varepsilon_{xx}+\varepsilon_{yy}+\varepsilon_{zz}\right)+\frac{E}{1+\nu}\varepsilon_{zz}=0 \end{aligned} \tag{6.8}$$

Adding these two equations together and solving for $\varepsilon_{yy} + \varepsilon_{zz}$ gives

$$\varepsilon_{yy} + \varepsilon_{zz} = -2\nu\,\varepsilon_{xx}$$

Substituting this expression back into Eq. 6.8 yields

$$\varepsilon_{yy} = -\nu\,\varepsilon_{xx}, \qquad \varepsilon_{zz} = -\nu\,\varepsilon_{xx}$$

which proves that in a uniaxial tension test, an axial strain of ε_{xx} is accompanied by a lateral strain of $-\nu\varepsilon_{xx}$. Now, substitute this result into the equation for σ_{xx} to find a specialized relationship between axial stress and strain for the uniaxial tension scenario. To wit,

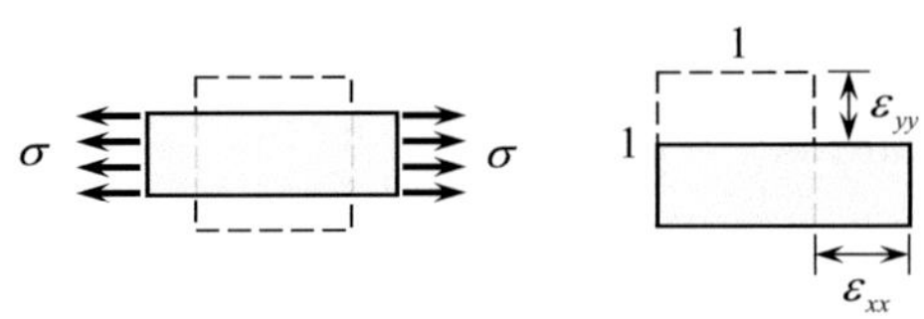

Fig. 6.1 *Example.* Poisson's effect for uniaxial tension

$$
\begin{aligned}
\sigma_{xx} &= \frac{E\nu}{(1+\nu)(1-2\nu)}\left(\varepsilon_{xx}+\varepsilon_{yy}+\varepsilon_{zz}\right)+\frac{E}{1+\nu}\varepsilon_{xx} \\
&= \frac{E\nu}{(1+\nu)(1-2\nu)}\left(\varepsilon_{xx}-2\nu\varepsilon_{xx}\right)+\frac{E}{1+\nu}\varepsilon_{xx} \\
&= \frac{E\nu}{(1+\nu)(1-2\nu)}(1-2\nu)\varepsilon_{xx}+\frac{E}{1+\nu}\varepsilon_{xx} \\
&= E\,\varepsilon_{xx}
\end{aligned}
$$

Hence, for a bar in uniaxial tension, the constitutive equation reduces to the simple form $\sigma = E\varepsilon$, as we assumed in Chap. 3 for the axial bar problem.

This example shows why a good test to determine Young's modulus E is a uniaxial tension test. If you measure the applied force and divide it by the area, you get σ. If you measure the change in length over a gauge length of the test piece and divide it by the gauge length, you get strain. Plotting the stress values against the strain values will plot as a straight line (if the material is truly linear). The slope of that line is E.

The upshot of this section is that the representation of Hooke's law, in terms of the material constants, is not unique, but all forms of the equation are equivalent. The existence of multiple definitions of the material constants relates more to the tests one might do to determine those constants. We have already shown that the uniaxial tension test provides a way to directly measure Young's modulus. A torsion test is a good way to directly measure the shear modulus (as we will appreciate in Chap. 11). For some materials, most notably granular materials, it is not possible to execute a tension test or a torsion test. One might measure the change in the volume of a sample under constant pressure to provide a direct measure of the bulk modulus. The so-called *triaxial test* places a confining pressure around a sample to hold it together and then compresses the sample longitudinally with a different pressure. If the changes in length and diameter of the specimen are measured, this test can directly measure the shear modulus of the granular material.

6.3 Properties of Common Materials

The values of Young's modulus, Poisson's ratio, and density for some common materials are given in Table 6.1, organized from the highest value of Young's modulus to the lowest. Some of the values are given as ranges because certain factors influence them. For example, the properties of concrete vary depending upon factors like the water-to-cement ratio at the time of creation. The properties of concrete change over time with the official values being specified at some agreed-upon time (e.g., 28

Table 6.1 *Properties of common materials.* Young's modulus E, Poisson's ratio ν, and density ρ for common materials

Material	E (GPa)	ν	ρ (kg/m^3)
Steel (mild)	200	0.30	7.82
Titantium	120	0.32	4.5
Copper	117	0.36	8.79
Glass (structural)	70	0.22	2.5
Aluminum	69	0.33	2.7
Concrete (medium)	17	0.1–0.2	1.3–1.7
Wood (Douglas fir)	13	–	0.53
Ice	9	0.33	0.92
Rubber	0.1	0.48–0.499	1.1
Cork	0.02	≈ 0	0.20–0.25

days). Additives are often used to change certain properties of concrete, e.g., its flow characteristics in its fluid state or the porosity in the hardened state.

Many of the materials in Table 6.1 have values of Poisson's ratio in the neighborhood of 0.30–0.36. The theoretical limits on Poisson's ratio are between -1.0 and 0.5, with negative values being quite rare. Rubber has a Poisson's ratio very near to the 0.5 limit, which means that when it deforms, the volume of material changes very little. We call such a material *incompressible*. Cork is included in the table because its Poisson's ratio is approximately zero, which is the practical lower limit. This unusual property, coupled with its low Young's modulus, makes the material a good choice for the stoppers in wine bottles.

Wood is included in the table even though it is not a material that meets the criterion of Hooke's law of being isotropic. The properties of wood along the grain versus perpendicular to the grain can be quite different. Wood is an example of what we might characterize as an *orthotropic* material where there are certain preferred directions. A model of this kind of material would have more than two material constants required to fully describe it. The Young's modulus value given in the table for Douglas fir is actually the ratio of axial stress to axial strain in a uniaxial tension test. The properties of wood are affected by the moisture content, among other factors.

The properties E and ν are the constants that represent the *stiffness* of the materials. Engineers are also quite interested in other properties of materials, chief among those being the properties that define the *strength* of the material. Each material exhibits a different type of failure mode, so it is difficult to summarize strength properties in a single value in a table. Steel, for example, is a metal with a crystal lattice structure. At a certain level of stress, the imperfections in the lattice start to slip, giving the macroscopic effect we call *yielding*. This property allows mild steel to deform at the same stress level for a fairly large range of strains. Eventually, the material hardens and the stress increases again under deformation, albeit at a much smaller rate than

E. Finally, steel fractures to give ultimate failure. Yielding is generally the end of linear elastic behavior.

Concrete cracks in tension and crushes in compression. Because concrete is composed of aggregates (sand or gravel) embedded in a cement matrix, the etiology of failure is very complex. Concrete is generally idealized as having a range of behavior that can be modeled with linear elasticity, but there is no clearly discernible level of stress or strain that delineates the end of the elastic range, like there is in mild steel. The strength of concrete depends on many factors, including the nature of the aggregates, the water-to-cement ratio at the time of mixing, chemical additives, the environment during the time of curing, and others.

6.4 Plane Stress and Plane Strain

Plane stress and plane strain are special states of stress and strain that can be treated as two-dimensional problems. In each case, some of the stress or strain components are known (or assumed) to be zero. In these cases, we can use the original three-dimensional equations to reduce, say, Eq. 6.4 to two dimensions. In this section we show how to do that.

Plane stress is a condition in which $\sigma_{xz} = \sigma_{yz} = \sigma_{zz} = 0$. In this case, the stress tensor in component form reduces to

$$\mathbf{S} = \begin{bmatrix} \sigma_{xx} & \sigma_{xy} & 0 \\ \sigma_{xy} & \sigma_{yy} & 0 \\ 0 & 0 & 0 \end{bmatrix}$$

Because of the coupling in Hooke's law, if the normal stress component is zero, then the normal strain component *cannot* be zero. We can compute it from the (3, 3) component of Eq. 6.4 as

$$\begin{aligned} \sigma_{zz} &= \frac{E\nu}{(1+\nu)(1-2\nu)}\big(\varepsilon_{xx} + \varepsilon_{yy} + \varepsilon_{zz}\big) + \frac{E}{1+\nu}\varepsilon_{zz} \\ &= \frac{E\nu}{(1+\nu)(1-2\nu)}\big(\varepsilon_{xx} + \varepsilon_{yy}\big) + \frac{E(1-\nu)}{(1+\nu)(1-2\nu)}\varepsilon_{zz} \end{aligned}$$

Set $\sigma_{zz} = 0$ and solve this equation for ε_{zz} to get

$$\varepsilon_{zz} = -\frac{\nu}{1-\nu}\big(\varepsilon_{xx} + \varepsilon_{yy}\big) \tag{6.9}$$

Thus, the strain in the z direction is dictated by the strains in the (x, y) plane and the value of Poisson's ratio. This equation also gives some insight into the practical bounds on Poisson's ratio. If it is equal to zero, then the straining in the plane does not increase or decrease the thickness. For a negative value of Poisson's ratio, stretching in the plane would *increase* the thickness. Most materials get thinner upon stretching, which explains why negative values of Poisson's ratio are rare.

Now, we can substitute Eq. 6.9 into the original three-dimensional version of Hooke's law to get a two-dimensional version that respects the condition of zero stress in the z direction and eliminates ε_{zz} from the equations for the components in the plane. The plane stress equations are

$$\begin{aligned} \sigma_{xx} &= \frac{E}{1-\nu^2}\left(\varepsilon_{xx} + \nu\,\varepsilon_{yy}\right) \\ \sigma_{yy} &= \frac{E}{1-\nu^2}\left(\nu\,\varepsilon_{xx} + \varepsilon_{yy}\right) \\ \sigma_{xy} &= \frac{E}{1+\nu}\varepsilon_{xy} \end{aligned} \tag{6.10}$$

Keep in mind that these equations apply only to the *plane stress* condition, as the nonzero strain ε_{zz} has been directly incorporated in the equations.

Plane strain is a condition in which $\varepsilon_{xz} = \varepsilon_{yz} = \varepsilon_{zz} = 0$. We can find a two-dimensional form of Hooke's law in the same way we did for plane stress. First, note that from Eq. 6.5 we can write the (3, 3) equation as

$$\begin{aligned} \varepsilon_{zz} &= -\frac{\nu}{E}\left(\sigma_{xx} + \sigma_{yy} + \sigma_{zz}\right) + \frac{1+\nu}{E}\sigma_{zz} \\ &= -\frac{\nu}{E}\left(\sigma_{xx} + \sigma_{yy}\right) + \frac{1}{E}\sigma_{zz} \end{aligned}$$

Now, set $\varepsilon_{zz} = 0$ and solve this equation for σ_{zz} to get

$$\sigma_{zz} = \nu\left(\sigma_{xx} + \sigma_{yy}\right)$$

We can substitute this expression into the original three-dimensional version of Hooke's law to get a two-dimensional version that respects the condition of zero strain in the z direction and eliminates σ_{zz} from the equations for the strain components in the plane. The plane strain equations are

$$\begin{aligned} \varepsilon_{xx} &= \frac{1+\nu}{E}\left[(1-\nu)\sigma_{xx} - \nu\,\sigma_{yy}\right] \\ \varepsilon_{yy} &= \frac{1+\nu}{E}\left[-\nu\,\sigma_{xx} + (1-\nu)\sigma_{yy}\right] \\ \varepsilon_{xy} &= \frac{1+\nu}{E}\sigma_{xy} \end{aligned} \tag{6.11}$$

Keep in mind that these equations apply only to the *plane strain* condition, $\varepsilon_{zz} = 0$, with the nonzero stress component σ_{zz} directly incorporated in the equations.

6.5 Principal Directions of Stress and Strain

Hooke's law has implications about the relationship between the principal directions and principal values of stress and strain. In fact, we can show that the principal directions are the *same* for stress and strain if they are related by Hooke's law. Let us start with the principal values and principal directions for the strain tensor **E**. We have three pairs $(\varepsilon_i, \mathbf{n}_i)$ that satisfy the eigenvalue problem

$$\mathbf{E}\mathbf{n}_i = \varepsilon_i \mathbf{n}_i$$

The values ε_i are the principal values, and the unit vectors $\mathbf{n}_i$ represent the principal directions of the strain tensor. Next, note that stress is related to strain by Eq. 6.1, i.e.,

$$\mathbf{S} = \lambda\,\mathrm{tr}\big(\mathbf{E}\big)\mathbf{I} + 2\mu\,\mathbf{E}$$

Let us multiply both sides by the vector $\mathbf{n}_i$ to see what happens:

$$\begin{aligned}\mathbf{S}\mathbf{n}_i &= \lambda\,\mathrm{tr}\big(\mathbf{E}\big)\mathbf{I}\mathbf{n}_i + 2\mu\,\mathbf{E}\mathbf{n}_i \\ &= \lambda\,\mathrm{tr}\big(\mathbf{E}\big)\mathbf{n}_i + 2\mu\,\varepsilon_i\mathbf{n}_i \\ &= \big[\lambda\,\mathrm{tr}\big(\mathbf{E}\big) + 2\mu\,\varepsilon_i\big]\mathbf{n}_i\end{aligned}$$

In the second line of this derivation, we used the fact that the identity multiplying any vector gives back the same vector. We also used the eigenvalue property of **E**. In the last line, we grouped together all terms multiplying $\mathbf{n}_i$. The term in square brackets is a scalar. Hence, we can write the equation as

$$\mathbf{S}\mathbf{n}_i = \sigma_i \mathbf{n}_i$$

where

$$\sigma_i = \lambda\,\mathrm{tr}(\mathbf{E}) + 2\mu\,\varepsilon_i \tag{6.12}$$

Therefore, $\mathbf{n}_i$ is also an eigenvector of the stress tensor. The principal directions of **E** and **S** are the same. The principal values are not the same, though. However, the principal values of stress, σ_i, are related to the principal values of strain ε_i by Eq. 6.12.

6.6 Combining Kinematics, Equilibrium, and Constitution

The three legs of the mechanics stool are now in place for multiaxial problems, and it is possible to look at problems that include the connection between the motion through the mapping function and its relationship to strain, the applied loads through

Cauchy's principle and its relationship to stress, and the constitutive behavior through Hooke's law, which relates stress and strain.

Example 6.3

Reconsider the mapping shown in Fig. 6.2 from the example on Sect. 4.2. Recall that we found the strain tensor associated with the mapping in the example on Sect. 4.3 to be (noting that the triangle is in a state of plane strain)

$$\mathbf{E} = \tfrac{1}{8}\begin{bmatrix} 4 & -6 & 0 \\ -6 & 5 & 0 \\ 0 & 0 & 0 \end{bmatrix}$$

Find the stress tensor and the forces required on the edges of the triangle required to accomplish the deformation. Young's modulus is 1000 MPa and Poisson's ratio is 0.3.

Solution. The stress can be found using Hooke's law. First, compute

$$\lambda = \frac{E\nu}{(1+\nu)(1-2\nu)} = \frac{(1000)(0.3)}{(1+0.3)(1-2(0.3))} = 576.92\,\text{MPa}$$

$$2\mu = \frac{E}{1+\nu} = \frac{1000}{1+0.3} = 769.23\,\text{MPa}$$

The trace of the strain tensor (i.e., the sum of the diagonal elements) is $\text{tr}(\mathbf{E}) = 9/8$. Substituting these values into Eq. 6.1 gives

$$\mathbf{S} = (576.92)\left(\tfrac{9}{8}\right)\begin{bmatrix} 1 & 0 & 0 \\ 0 & 1 & 0 \\ 0 & 0 & 1 \end{bmatrix} + (769.23)\left(\tfrac{1}{8}\right)\begin{bmatrix} 4 & -6 & 0 \\ -6 & 5 & 0 \\ 0 & 0 & 0 \end{bmatrix}$$

$$= \begin{bmatrix} 1033.7 & -576.9 & 0 \\ -576.9 & 1129.8 & 0 \\ 0 & 0 & 649.0 \end{bmatrix}$$

The tractions that must be applied to each face can be determined by Cauchy's formula $\mathbf{t}(\mathbf{n}) = \mathbf{S}\mathbf{n}$. We need the unit normal vector for each face in the deformed configuration. The three normal vectors have components

$$\mathbf{n}_1 = \begin{Bmatrix} -1 \\ 0 \\ 0 \end{Bmatrix}, \quad \mathbf{n}_2 = \begin{Bmatrix} 0 \\ 1 \\ 0 \end{Bmatrix}, \quad \mathbf{n}_3 = \begin{Bmatrix} 0.7071 \\ -0.7071 \\ 0 \end{Bmatrix}$$

Applying Cauchy's formula to each face gives

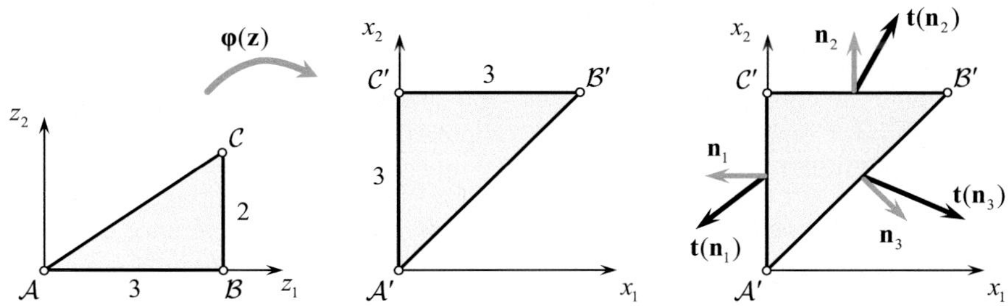

Fig. 6.2 *Example.* Mapping function from the example on Sect. 4.2. The sketch on the left is the reference configuration. The sketch in the center is the deformed configuration. The sketch on the right shows the tractions on the surface in the deformed configuration

$$\mathbf{t}(\mathbf{n}_1) = \begin{Bmatrix} -1033.7 \\ 576.9 \\ 0 \end{Bmatrix}, \quad \mathbf{t}(\mathbf{n}_2) = \begin{Bmatrix} -576.9 \\ 1129.8 \\ 0 \end{Bmatrix}, \quad \mathbf{t}(\mathbf{n}_3) = \begin{Bmatrix} 1138.8 \\ -1206.8 \\ 0 \end{Bmatrix}$$

All terms have units of MPa. Observe that the sum of the tractions multiplied by their respective areas is equal to zero, as guaranteed by equilibrium.

For some problems we might have partial information on the mapping, partial information on the state of strain (e.g., through strain gauges), and partial information on the stress. The following example shows how all of those ingredients interact in a single problem.

Example 6.4
Consider the deformation of the initially square region with dimension 1 in the plane, and thickness 0.1 out of the plane, shown in Fig. 6.3. The two strain gauges shown read $\varepsilon_n = 0.2$ (horizontal gauge) and $\varepsilon_m = 0.3$ (inclined gauge) after deformation. We know that $\sigma_{xx} = 500$ MPa, but σ_{yy} and σ_{xy} are not known. Assume that the plate is in a state of *plane stress* and that the material

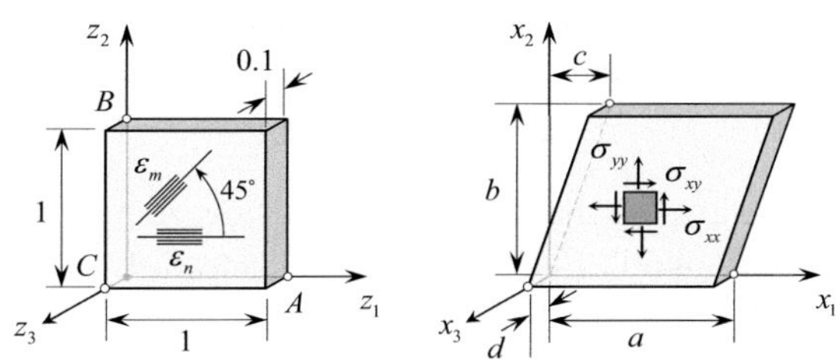

Fig. 6.3 *Example.* Problem with mixed information. The sketch on left is the undeformed configuration, showing the orientation of two strain gauges, one reading ε_n and one reading ε_m. The sketch on the right shows the deformed configuration

is elastic with Young's modulus $E = 2000$ MPa and Poisson's ratio $\nu = 0.4$. Compute the components of the stress and strain tensors and the values of a, b, c, and d. Compute the strain ε_{zz}. Compute the principal stresses and draw the Mohr circle for the stress.

Solution. The position vectors for points A, B, and C in the reference configuration are

$$\mathbf{z}_A = \begin{Bmatrix} 1 \\ 0 \\ 0 \end{Bmatrix}, \quad \mathbf{z}_B = \begin{Bmatrix} 0 \\ 1 \\ 0 \end{Bmatrix}, \quad \mathbf{z}_C = \begin{Bmatrix} 0 \\ 0 \\ 0.1 \end{Bmatrix}$$

The position vectors for those points in the deformed configuration are

$$\mathbf{x}_A = \begin{Bmatrix} a \\ 0 \\ 0 \end{Bmatrix}, \quad \mathbf{x}_B = \begin{Bmatrix} c \\ b \\ 0 \end{Bmatrix}, \quad \mathbf{x}_C = \begin{Bmatrix} 0 \\ 0 \\ d \end{Bmatrix}$$

These vectors comprise the columns of the matrices $\mathbf{Z}$ and $\mathbf{X}$, respectively. Thus,

$$\mathbf{Z} = \begin{bmatrix} 1 & 0 & 0 \\ 0 & 1 & 0 \\ 0 & 0 & 0.1 \end{bmatrix}, \quad \mathbf{X} = \begin{bmatrix} a & c & 0 \\ 0 & b & 0 \\ 0 & 0 & d \end{bmatrix}$$

Compute the deformation gradient $\mathbf{F}$ from Eq. 4.4:

$$\mathbf{F} = \mathbf{X}\mathbf{Z}^{-1} = \begin{bmatrix} a & c & 0 \\ 0 & b & 0 \\ 0 & 0 & d \end{bmatrix} \begin{bmatrix} 1 & 0 & 0 \\ 0 & 1 & 0 \\ 0 & 0 & 10 \end{bmatrix} = \begin{bmatrix} a & c & 0 \\ 0 & b & 0 \\ 0 & 0 & 10d \end{bmatrix}$$

The Green deformation tensor can be computed as

$$\mathbf{C} = \mathbf{F}^T\mathbf{F} = \begin{bmatrix} a & 0 & 0 \\ c & b & 0 \\ 0 & 0 & 10d \end{bmatrix} \begin{bmatrix} a & c & 0 \\ 0 & b & 0 \\ 0 & 0 & 10d \end{bmatrix} = \begin{bmatrix} a^2 & ac & 0 \\ ac & c^2+b^2 & 0 \\ 0 & 0 & 100d^2 \end{bmatrix}$$

and the Lagrangian strain tensor is

$$\mathbf{E} = \tfrac{1}{2}[\mathbf{C} - \mathbf{I}] = \tfrac{1}{2} \begin{bmatrix} a^2-1 & ac & 0 \\ ac & c^2+b^2-1 & 0 \\ 0 & 0 & 100d^2-1 \end{bmatrix}$$

We can use the strain gauge readings to get relationships among the constants a, b, and c. For the horizontal gauge, $\mathbf{n} = \mathbf{e}_1$. Thus, substituting values into $\varepsilon_n = \mathbf{n}^T\mathbf{E}\mathbf{n}$ gives

$$\varepsilon_n = \begin{bmatrix} 1 & 0 & 0 \end{bmatrix} \frac{1}{2} \begin{bmatrix} a^2-1 & ac & 0 \\ ac & c^2+b^2-1 & 0 \\ 0 & 0 & 100d^2-1 \end{bmatrix} \begin{Bmatrix} 1 \\ 0 \\ 0 \end{Bmatrix} = \tfrac{1}{2}\big(a^2-1\big)$$

We know the value of $\varepsilon_n = 0.2$. Therefore, we can solve for a as

$$\varepsilon_n = \tfrac{1}{2}\big(a^2-1\big) = 0.2 \qquad \rightarrow \qquad a = 1.1832$$

The inclined gauge has $\mathbf{m} = \frac{\sqrt{2}}{2}\big(\mathbf{e}_1 + \mathbf{e}_2\big)$. Thus, we can compute $\varepsilon_m = \mathbf{m}^T\mathbf{E}\mathbf{m}$ as

$$\varepsilon_m = \begin{bmatrix} \frac{\sqrt{2}}{2} & \frac{\sqrt{2}}{2} & 0 \end{bmatrix} \frac{1}{2} \begin{bmatrix} a^2-1 & ac & 0 \\ ac & c^2+b^2-1 & 0 \\ 0 & 0 & 100d^2-1 \end{bmatrix} \begin{Bmatrix} \frac{\sqrt{2}}{2} \\ \frac{\sqrt{2}}{2} \\ 0 \end{Bmatrix}$$

Carrying out the matrix multiplication gives

$$\varepsilon_m = \tfrac{1}{4}\big(a^2 + 2ac + c^2 + b^2 - 2\big)$$

We know $\varepsilon_m = 0.3$ and we know the value of a. Hence, we can plug in those values to get

$$0.3 = \tfrac{1}{4}\Big(\big(1.1832\big)^2 + 2\big(1.1832\big)c + c^2 + b^2 - 2\Big)$$

which reduces to

$$2.3664\,c + c^2 + b^2 = 1.8$$

Finally, we know $\sigma_{xx} = 500$ MPa. We can use the plane stress version of Hooke's law, Eq. 6.10, noting that $\varepsilon_{xx} = \varepsilon_n = 0.2$, to compute

$$\begin{aligned} \sigma_{xx} &= \frac{E}{1-\nu^2}\Big(\varepsilon_{xx} + \nu\,\varepsilon_{yy}\Big) \\ 500 &= \frac{2000}{1-(0.4)^2}\Big(0.2 + (0.4)\,\tfrac{1}{2}\big(c^2+b^2-1\big)\Big) \\ c^2 + b^2 &= \frac{(500)(0.84)}{(0.2)(2000)} = 1.05 \end{aligned}$$

Substituting this value into the previous equation gives

$$2.3664\,c + 1.05 = 1.8 \qquad \rightarrow \qquad c = 0.31694$$

Solving for b from $b^2 + c^2 = 1.05$ gives $b = 0.97445$. These values can be used to compute

$$\varepsilon_{yy} = \tfrac{1}{2}\big(c^2 + b^2 - 1\big) = 0.025$$

$$\varepsilon_{xy} = \tfrac{1}{2}\big(ac\big) = 0.1875$$

We can find d from the plane stress condition Eq. 6.9 that

$$\begin{aligned}\varepsilon_{zz} &= -\frac{\nu}{1-\nu}\big(\varepsilon_{xx} + \varepsilon_{yy}\big)\\ &= -\frac{0.4}{1-0.4}(0.2 + 0.025) = -0.150 = \tfrac{1}{2}\big(100d^2 - 1\big)\end{aligned}$$

Solving for d gives $d = 0.0837$, which is the thickness of the deformed plate. Now, all the constants a, b, c, and d are known.

We can complete the computation of the stresses using the plane stress version of Hooke's law. To wit,

$$\begin{aligned}\sigma_{yy} &= \frac{E}{1-\nu^2}\Big(\varepsilon_{yy} + \nu\,\varepsilon_{xx}\Big)\\ &= \frac{2000}{1-(0.4)^2}\Big(0.025 + (0.4)\,(0.2)\Big) = 250\,\text{MPa}\end{aligned}$$

and

$$\sigma_{xy} = \frac{E}{1+\nu}\varepsilon_{xy} = \frac{2000}{1+0.4}\Big(0.1875\Big) = 267.9\,\text{MPa}$$

The center of the Mohr circle is

$$\hat{c} = \tfrac{1}{2}\big(\sigma_{xx} + \sigma_{yy}\big) = \tfrac{1}{2}\big(500 + 250\big) = 375$$

where we have used the symbol $\hat{c}$ to avoid conflict with the constant c of the mapping defined earlier in Fig. 6.3. The radius of the Mohr circle is

$$\begin{aligned}R &= \tfrac{1}{2}\sqrt{(\sigma_{xx} - \sigma_{yy})^2 + 4\sigma_{xy}^2}\\ &= \tfrac{1}{2}\sqrt{(500 - 250)^2 + 4(267.9)^2} = 295.6\end{aligned}$$

The Mohr circle is plotted in Fig. 6.4. The principal stresses are

$$\begin{aligned}\sigma_1 &= \hat{c} + R = 375 + 295.6 = 670.6\\ \sigma_2 &= \hat{c} - R = 375 - 295.6 = 79.4\end{aligned}$$

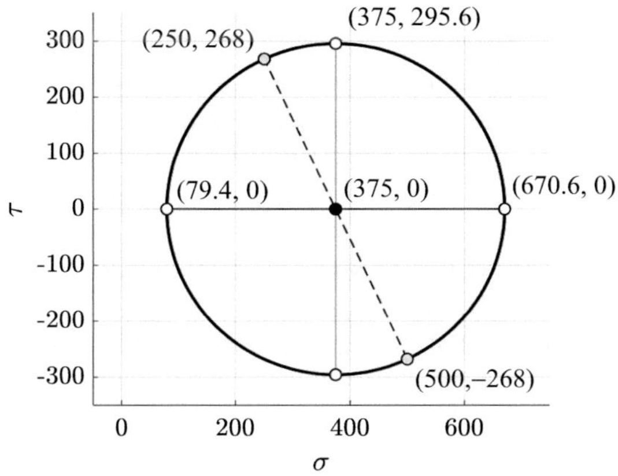

Fig. 6.4 *Example.* The Mohr circle for the example. The gray dots are the stresses in the (x, y) coordinate system. The open dots on the axis are the principal stresses. The open dots on top and bottom are the maximum shear stress

Of course, $\sigma_3 = 0$ is also a principal value. Note that the maximum shear stress would be the radius of the Mohr circle between σ_1 and σ_3. Hence, $\tau_{\text{max}} = 335.3$ MPa (see Fig. 5.12) rather than the value 295.6 MPa, which is the radius of the Mohr circle shown in Fig. 6.4.

6.7 Other Models

The behavior of materials is a huge and important part of solid mechanics. All sorts of behaviors have been observed in material testing. The linear elastic model we have presented in this chapter is just one very important example. Some other models that have been developed to describe the response of common materials include:

- *Nonlinear elasticity*. There are many models of *nonlinear* elasticity, where the body loads and unloads along the same path, but that path is not linear. For all elastic models, the body returns to its original shape upon unloading. Materials like rubber exhibit nonlinear elastic response.
- *Anisotropic elasticity*. Hooke's law can be extended to cases that are linear but *not* isotropic. Materials like wood exhibit anisotropic elastic behavior for at least some range of response.
- *Inelasticity*. Some materials suffer damage that prevents the body from returning to its original state upon unloading. Elastoplasticity is a model wherein the behavior is elastic if the stresses are below some threshold value. Once the stress state reaches the yield point, inelastic strains accrue in a way prescribed by the model. Metals generally exhibit inelastic response.

- *Viscoelasticity* or *viscoplasticity*. The stress-strain response of some materials has time-dependent (or viscous) effects. For a viscoelastic material, a body under constant load will continue to deform with time, with the amount of additional deformation getting smaller and smaller with time. If the load is removed, then the body will return to its original configuration, but the process again takes time to complete. For a viscoplastic material, deformation again accrues with time. However, when the loads are removed, the body does not return to its original shape, but rather retains some residual unrecoverable deformation. Wood and concrete exhibit viscous response over long timescales. The phenomenon is called *creep* when referring to deformation with time under constant loads and *relaxation* when referring to reduction in stress with time for fixed deformations.

There are many other material models.

6.8 Summary

This chapter has presented only the simple case of isotropic linear elastic material behavior, or Hooke's law. With Hooke's law in three dimensions, we can relate multiaxial states of stress to multiaxial states of strain. The model incorporates the important observation that a bar stretched in tension will contract laterally (Poisson's effect) as well as the normal straining caused by normal stresses and shear straining caused by shear stresses. This model was used for the axial bar problem and will be used in the following chapters for beams and torsion.

6.9 Problems

6.1 A thin rectangular plate is in a state of *plane stress* and is subjected to stresses σ_{xx} and σ_{yy}, as shown in the figure. The plate has (x, y, z) dimensions of $550 \times 400 \times 20$ mm and is made of steel with Young's modulus $E = 200$ GPa and Poisson's ratio $\nu = 0.30$. Find the stress and strain tensors, the angle of maximum shear strain in the plane of the plate, the maximum shear stress, the change in thickness of the plate, and the change in volume of the plate.

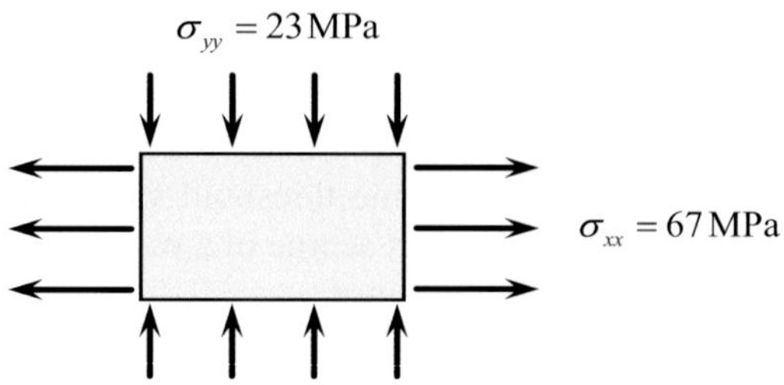

6.2 Consider the deformation of the initially square region $\mathcal{ABCD}$ shown. The distances a, b, and c are unknown and need to be determined. The two strain gauges read $\varepsilon_n = 0.1$ (horizontal gauge) and $\varepsilon_m = 0.2$ (45° gauge) after deformation. We know that $\sigma_{xx} = 200$ MPa, but σ_{yy} and σ_{xy} are not known. Assume that the plate is in a state of plane stress and that the material is elastic with Young's modulus $E = 3000$ MPa and Poisson's ratio $\nu = 0.2$. Compute the components of the stress and strain tensors and the values of a, b, and c. Compute the strain ε_{zz}, the principal stresses, and draw a Mohr circle for the stress.

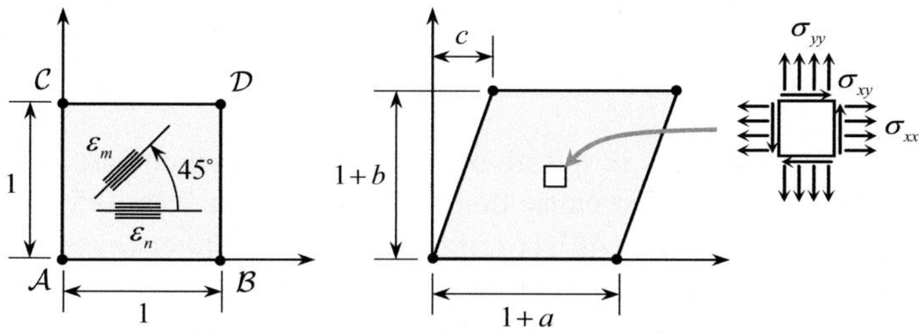

Reference Configuration Deformed Configuration

6.3 Consider the deformation of the initially square region shown. Compute the components of the stress and strain tensors resulting from the deformation. Compute the principal stresses. Determine the tractions that would need to be applied to the edges of the region to accomplish the deformation. Assume that the plate is in a state of plane stress and that the material is elastic with Young's modulus $E = 3000$ MPa and Poisson's ratio $\nu = 0.2$. Draw a Mohr circle for the stress.

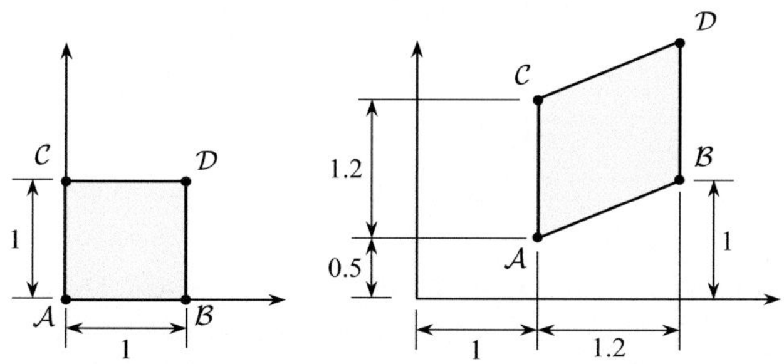

Reference Configuration Deformed Configuration

6.4 Consider the deformation of the triangular region shown. Compute the components of the stress tensor in the deformed configuration. Compute the principal strains and stresses. Determine the tractions that would need to be applied to the edges of the triangular region to accomplish the deformation. Assume that the plate is in a state of plane stress and that the material has Young's modulus $E = 1.0$ MPa and Poisson's ratio $\nu = 0.25$.

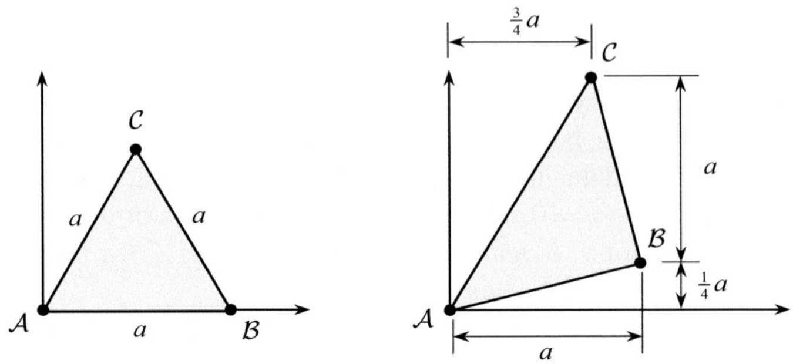

Reference Configuration Deformed Configuration

6.5 A thin rhombic plate is in a state of plane stress. Tractions are applied to the perimeter of the rhombus that cause diagonal $\mathcal{BD}$ to stretch by 0.5%, line $\mathcal{CA}$ to shrink by 0.2%, and the angle $\mathcal{AOB}$ (a right angle in the reference configuration) to reduce by 0.0092 rad. Find the principal stresses and the maximum shearing stress. Assume that the material is elastic and has Young's modulus $E = 25{,}000$ kPa and Poisson's ratio $\nu = 0.15$. Draw a Mohr circle for the stress. Compute the tractions that would need to be applied to the edges of the region to accomplish the deformation. The reference configuration is shown in the sketch.

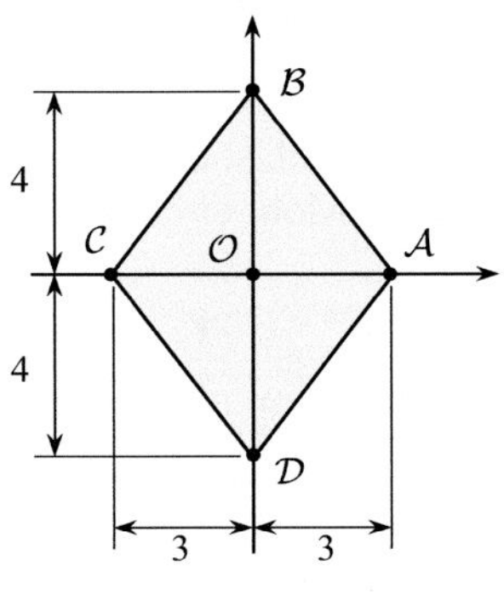

6.6 Consider the deformation of the triangular region shown. The triangle is isosceles in the reference configuration. Compute the components of the stress and strain tensors resulting from the deformation. Compute the principal stresses. Compute the tractions that would need to be applied to the edges of the region to accomplish the deformation. Assume that the plate is in a state of plane stress and that the material is elastic with Young's modulus $E = 2500$ kPa and Poisson's ratio $\nu = 0.15$. Draw a Mohr circle for the stress.

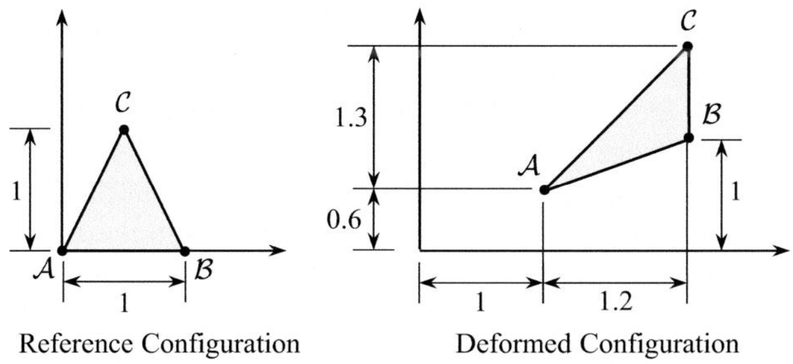

6.7 The thin square plate is in a state of uniform plane stress. Tractions are applied to the perimeter of the region that cause edge $\mathcal{AB}$ to stretch by 0.10%, line $\mathcal{AD}$ to shrink by 0.25%, and the angle $\mathcal{DAB}$ (a right angle in the reference configuration) to reduce by 0.01 rad. Find the principal stresses and the maximum shearing stress. Assume that the material is elastic and has Young's modulus $E = 25{,}000$ psi and Poisson's ratio $\nu = 0.25$. Draw a Mohr circle for the stress. Compute the tractions that would need to be applied to the edges of the region to accomplish the deformation. The reference configuration is shown in the sketch.

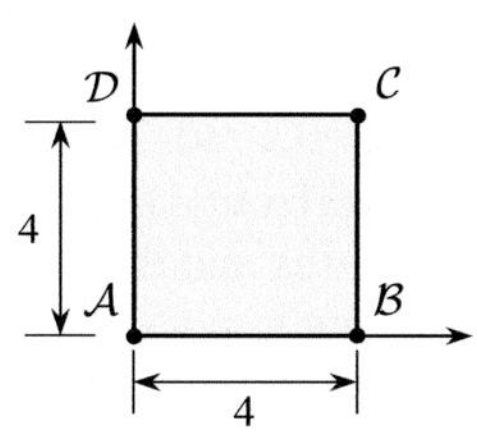

6.8 Consider the deformation of the initially square region $\mathcal{ABCD}$ shown. When subjected to a uniaxial state of tension, the square deforms into a rectangle of dimensions a by b, which are unknown and need to be determined. The strain gauge shown reads $\varepsilon_n = 0.25$ (45° gauge) after deformation. We know that $p_o = 500$ psi, and that the top and bottom surface, as well as the two faces, are traction free (the plate is in a state of plane stress). The material is elastic with Young's modulus $E = 4000$ psi and Poisson's ratio $\nu = 0.4$. Compute the components of the stress and strain tensors and the values of a and b. Compute the principal stresses and draw a Mohr circle for the stress.

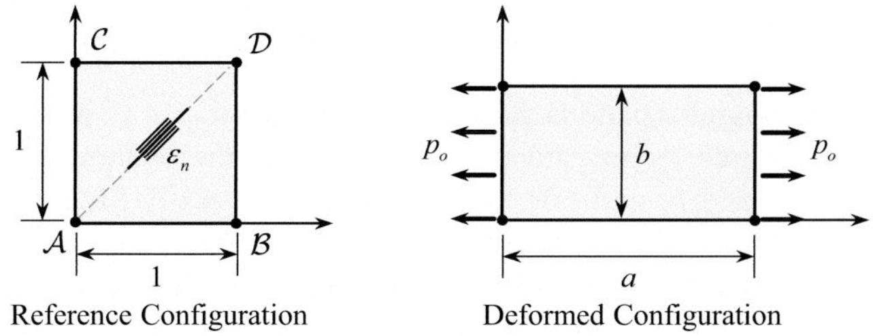

6.9 The square plate is in a state of homogeneous deformation. It is subjected to the stress shown (the shear stress is unknown). The strain gauge, oriented at 45° from the horizontal, reads $\varepsilon_n = 0.15$ after deformation. Find the strains, the shear stress, and the stress on the front and back faces of the plate. Find the principal stresses, principal angle, and the maximum shear stress. Assume that the material is elastic and has Young's modulus $E = 2500$ psi and Poisson's ratio $\nu = 0.25$. Draw a Mohr circle for the stress.

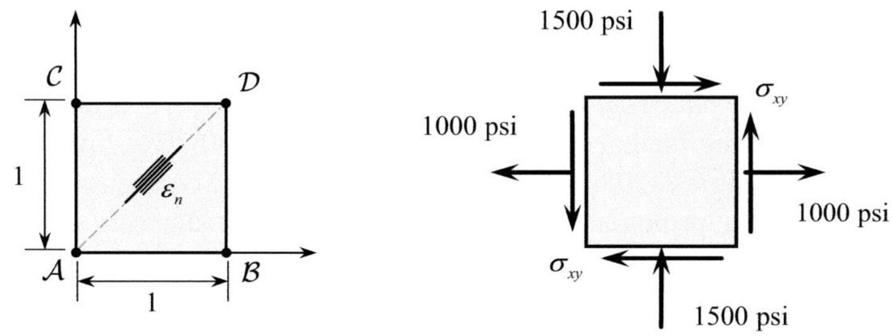

Reference Configuration

6.10 The square plate $\mathcal{ABCD}$ is in a state of homogeneous deformation. Tractions are applied to the perimeter of the region that cause the angle $\mathcal{CAB}$ (a right angle in the reference configuration) to reduce by 0.1 rad. The sides also change length. The two strain gauges shown read $\varepsilon_n = 0.1$ (horizontal gauge) and $\varepsilon_m = 0.15$ (30° gauge) after deformation. Find the principal stresses and the maximum shearing stress. Assume that the material is elastic with Young's modulus $E = 3000$ psi and Poisson's ratio $\nu = 0.25$. Draw a Mohr circle for the stress. Compute the tractions that would need to be applied to the edges of the region to accomplish the deformation.

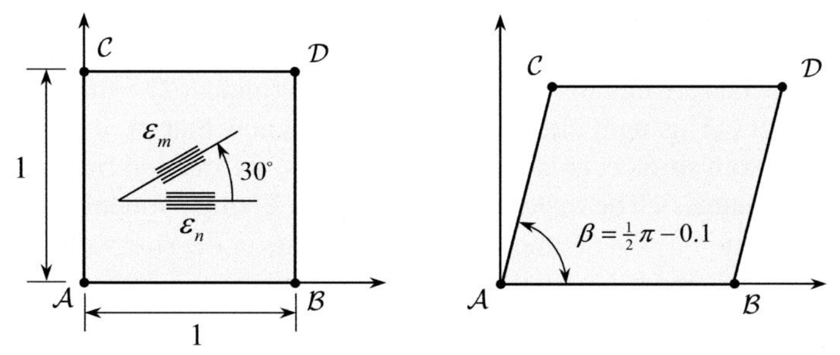

Reference Configuration Deformed Configuration

6.11 A thin rectangular plate is subjected to stresses σ_{xx}, σ_{xy}, and σ_{yy} in a plane stress condition. Three strain gauges are arranged in the pattern shown with 60° between gauges a and b, and 60° between gauges b and c. Measurements show that the strains in the three directions are $\varepsilon_a = 40 \times 10^{-6}$, $\varepsilon_b = 980 \times 10^{-6}$, and $\varepsilon_c = 330 \times 10^{-6}$. Young's modulus is $E = 30 \times 10^6$ psi and Poisson's ratio is $\nu = 0.3$. Find the components of the strain tensor and the associated stresses. Find the principal strains and the maximum shear strain. Draw a Mohr circle for the strains.

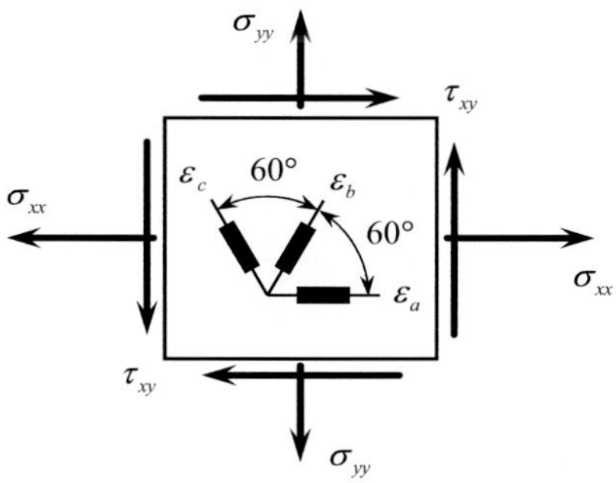

6.12 A thin right-triangular sheet with side lengths of 1.2 and 1.0 in. is deformed as shown. Point $\mathcal{C}$ is at the midpoint of the hypotenuse in the reference configuration. The coordinates of two points in the deformed configuration are given. Young's modulus is $E = 300$ psi and Poisson's ratio is $\nu = 0.30$. Find the deformation map, the location of the deformed point $\mathcal{B}$, and the components of the stress and strain tensors associated with the deformation. If the allowable normal stress of the material is 35 psi, will it withstand this deformation? Draw a Mohr circle for the state of strain.

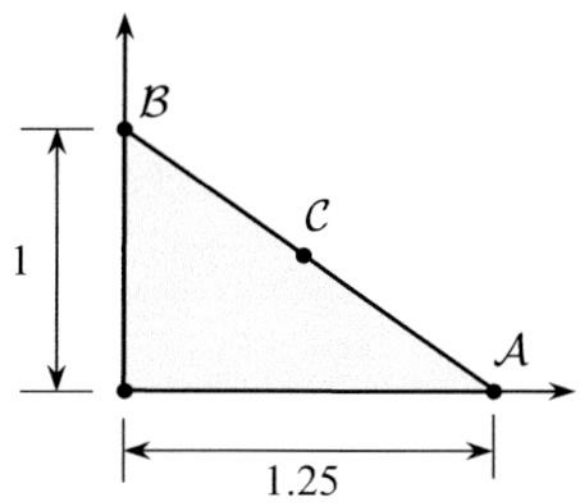

Reference Configuration

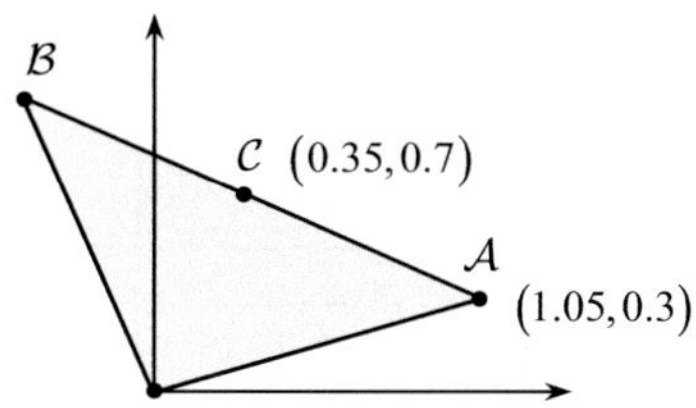

Deformed Configuration

6.13 A thin sheet of width 2.0 and height 1.5 in. has the given deformation map **F**

$$\mathbf{F} = \begin{bmatrix} 1.25 & 0.5 \\ 0.3 & 1.1 \end{bmatrix}$$

Draw the deformed rectangle. Find the components of the strain tensor associated with the deformation. Find the principal strains. Find the stresses associated with the strains assuming a state of plane stress. Draw a Mohr circle for the state of strain. Use the values of $E = 1400$ psi and $\nu = 0.30$ for Young's modulus and Poisson's ratio, respectively.

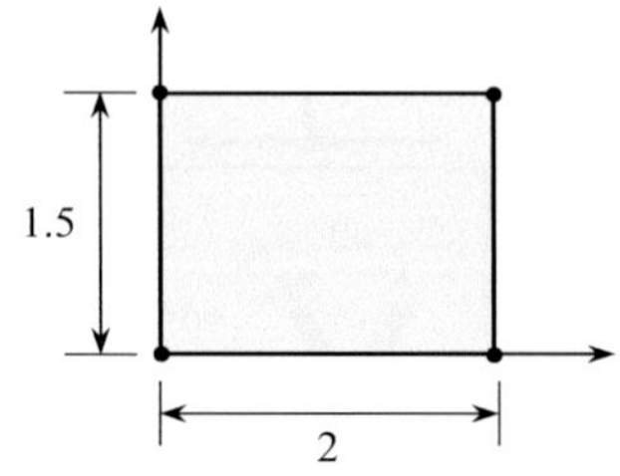

6.14 A thin rectangular sheet of width 1.5 and height 1.0 in. is subjected to a homogeneous deformation, as shown. The coordinates of two points in the deformed configuration are given. Young's modulus is $E = 300$ psi and Poisson's ratio is $\nu = 0.30$. Find the deformation map, the location of the deformed point C, and the components of the strain tensor associated with the deformation. Find the principal strains. Find the stresses, noting that the sheet is in a state of plane stress. Draw a Mohr circle for the state of strain.

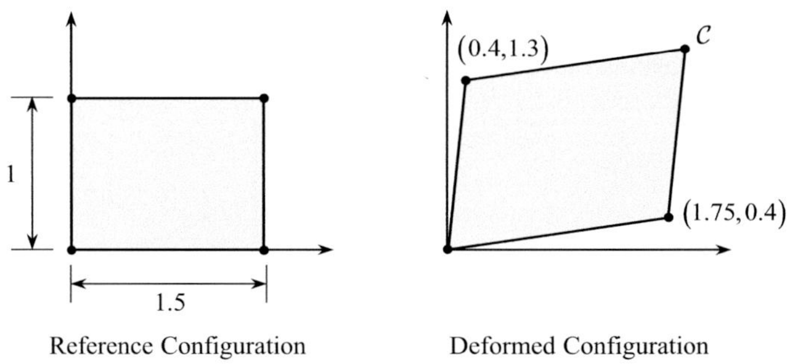

6.15 *Computation.* Write a code that creates a (two-dimensional) deformation gradient **F** and computes the Lagrangian strain tensor from it. Compute the stress tensor from strain using Hooke's law, with a provision for either plane strain or plane stress. Analyze the state of stress and strain by finding the principal angles, maximum shear angles, and principal values. Plot the results on Mohr's circle (one for stress and one for strain). Include the original components in the (x, y) system, the center of the circle, and the principal values (see Fig. 4.17 for an example of what the Mohr circle should look like).

Include in the code the vertices of a particular region (e.g., a unit square, a pentagon, or a circle). The vertices should be in sequential counterclockwise order and stored in an array z that has 2 rows and N columns. Each column contains the (x, y) coordinates of a vertex, with the first row being the x value and the second row being the y value. Make sure the region is closed by returning to the first point. With this input, you can plot the region with the MATLAB code

```
plot(z(1,:),z(2,:),'-','Color','b','LineWidth',2)
```

Create the deformed region by multiplying the coordinates by **F**. Create new vertices, rotating the original ones through an angle θ by multiplying the coordinates by the rotation matrix

$$\mathbf{R} = \begin{bmatrix} \cos\theta & -\sin\theta \\ \sin\theta & \cos\theta \end{bmatrix}$$

Find the deformed configuration for the rotated figure. Rotate the region through the principal angle to show that there is no shear deformation at that angle.

6.16 *Computation.* Write a code that extends the one from Problem 6.15 to three dimensions. Create a 3×3 deformation gradient **F**. Find the principal values and principal directions of stress and strain using the MATLAB `eig` function, for example,

```
[Evec,Evals] = eig(E)
```

Create three Mohr circles from the principal values. Create the vertices of a unit cube in a 3×8 array `verts` and an index array `VisitOrder` that gives the visit order to draw all the edges without lifting the pen.

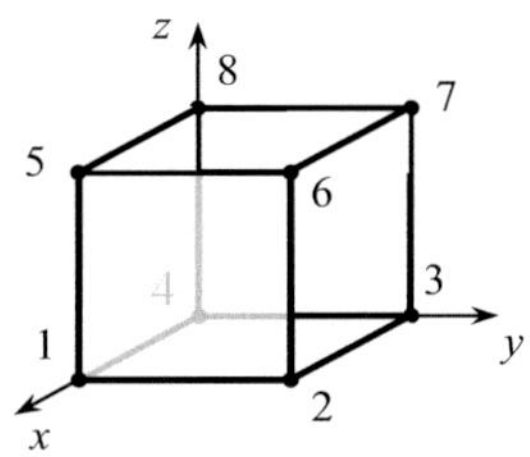

For the vertex numbering scheme here, the visit order can be specified as

```
VisitOrder = [1,2,3,4,1,5,6,7,8,5,6,2,3,7,8,4]
```

Notice that some lines get drawn twice. Now, create the coordinates `z` as

```
z = verts(:,VisitOrder)
```

which generates (x, y, z) values so that the `plot` statement

```
plot3(z(1,:),z(2,:),z(3,:),'-',LineWidth',2)
```

will plot the region. The rotation matrix can be set equal to the eigenvalue matrix `Evec` generated from the `eig` function. This rotation will orient the edges of the original cube in the principal directions.

Chapter 7
Beam Theory

Beam theory is one of the most successful and widely used theories in engineering, having been refined and validated over the course of hundreds of years. It forms the basis of the subject of structural analysis and is at the heart of all commercial structural analysis codes today. This chapter shows how a simple kinematic hypothesis leads to the formulation of a complete mechanical theory governed by an ordinary differential equation that can be readily solved to generate practical results. From the theory, we can estimate the deflections of a beam under load and the stresses and strains that accrue in the process. In this chapter we establish the theory of beams, confining our attention to planar motion of beams. The developments of this chapter provide the theoretical foundation needed for most of the subsequent chapters.

Figure 7.1 illustrates the problem of a planar beam of length L, subjected to applied loads $\mathbf{q}(x)$. The loads have a component transverse to the beam axis, which causes the beam to bend. We also embrace the possibility of applied axial loads and applied couple loads that are not shown in the sketch. As a result of the loads, the beam displaces *transverse* to the longitudinal axis of the beam, and cross sections of the beam rotate in the plane of motion. Recall from Fig. 3.1 that a *cross section* is a planar region perpendicular to the axis of the beam. The relevant geometric properties of the cross section for beam theory are its area and the first and second moments of the area relative to the centroid. The derivation of the theory will show why these properties are important, and in Chap. 9 we will explore ways to compute them. The relevant material property is Young's modulus E. We will take the line connecting the centroids of the cross sections to be the x axis. The standard base vectors $\{\mathbf{e}_1, \mathbf{e}_2, \mathbf{e}_3\}$ are shown in the sketch. We take the base vector $\mathbf{e}_3$ to be downward because most of the forces in our problems will act in that direction. Each beam will have certain boundary conditions at its ends. For example, the beam in Fig. 7.1 is *fixed* at the left end and *simply supported* at the right end.

K. D. Hjelmstad, *Mechanics of Deformable Solids: An Introduction*,
https://doi.org/10.1007/978-3-031-89873-0_7

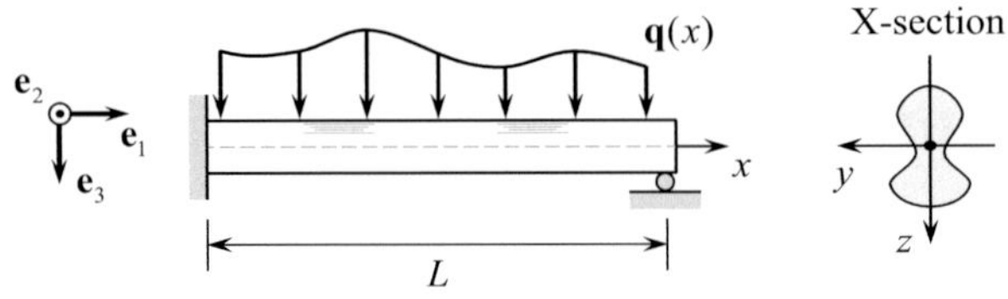

Fig. 7.1 *Beam.* A beam subjected to transverse forces $\mathbf{q}(x)$ responds by bending and deflecting transverse to the original x axis. The cross section is perpendicular to the original axis of the beam, in the (y, z) plane

7.1 Beam Kinematics

Beam theory is based upon the kinematic hypothesis that plane sections remain plane after deformation. As such, a cross section moves like a rigid body, but adjacent cross sections can rotate and displace relative to each other, as illustrated in Fig. 7.2. This hypothesis reduces the three-dimensional partial differential equations of mechanics (e.g., Eqs. 4.47 and 5.29) to ordinary differential equations of the variable x. In this section we derive the equations of kinematics for the planar beam. The resulting equations of kinematics will be *nonlinear*. In the next chapter, we will linearize the equations in service of finding analytical approaches to solving beam problems.

7.1.1 Motion of a Beam Cross Section

If we assume that plane cross sections in the reference configuration remain plane after deformation, then the motion of the beam can be completely characterized by the displacement of the centroid $\mathbf{w}(x)$ and the rotation of the cross section $\theta(x)$, as illustrated in Fig. 7.3. The position vector for the typical particle $\mathcal{B}$ in cross section $\mathcal{A}$, which is originally located a distance x from the left end, is given by

$$\mathbf{x} = x\mathbf{e}_1 + \mathbf{w}(x) + z\,\mathbf{m}(x) \tag{7.1}$$

Fig. 7.2 *Plane sections hypothesis.* The kinematic hypothesis of beam theory is that cross sections in the beam remain plane after deformation. The cross sections are the black lines, which remain straight in the deformed configuration

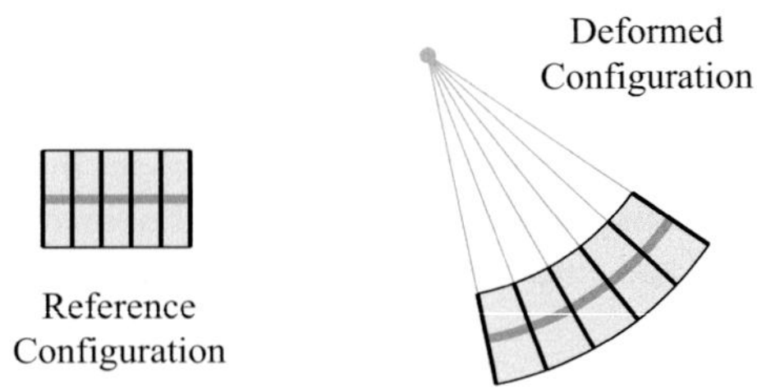

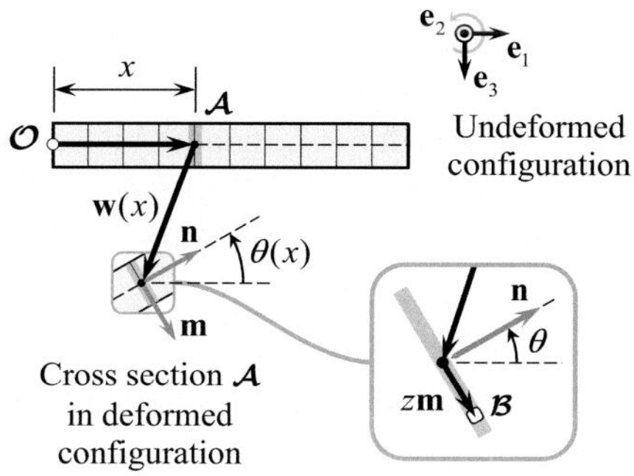

Fig. 7.3 *Beam kinematics.* The kinematic consequences of the *plane sections* hypothesis. The magnified view of the cross section at $\mathcal{A}$ shows the typical particle $\mathcal{B}$ at distance z from the centroidal axis

where x is the distance along the axis of the undeformed beam and z is the distance from the axis to the typical particle. Note that Eq. 7.1 describes the deformation map $\mathbf{x} = \boldsymbol{\varphi}(\mathbf{z})$ for planar beam theory (see Eq. 4.1). In this case, the components of $\mathbf{z}$ are called (x, y, z). Since the motion is restricted to the (x, z) plane, the motion does not depend explicitly upon the coordinate y. The deformation map in Eq. 7.1 tracks the motion of each particle in the beam and respects the kinematic hypothesis that plane sections remain plane after deformation.

The displacement vector can be expressed in terms of components relative to the standard base vectors as

$$\mathbf{w}(x) = u(x)\mathbf{e}_1 + w(x)\mathbf{e}_3 \tag{7.2}$$

where $u(x)$ is the displacement along, and $w(x)$ is the displacement transverse to, the *original* axis of the beam. The angle of rotation $\theta(x)$ of the cross section is measured relative to the orientation of the undeformed axis of the beam, as shown in Fig. 7.3, and is positive if the rotation is counterclockwise. The unit vectors $\mathbf{n}(x)$ and $\mathbf{m}(x)$ track the rotation of the cross section at x, and can be expressed in terms of components relative to the standard base vectors as

$$\begin{aligned} \mathbf{n}(x) &= \cos\theta(x)\,\mathbf{e}_1 - \sin\theta(x)\,\mathbf{e}_3 \\ \mathbf{m}(x) &= \sin\theta(x)\,\mathbf{e}_1 + \cos\theta(x)\,\mathbf{e}_3 \end{aligned} \tag{7.3}$$

Observe that $\mathbf{m}(x)$ is in the plane of the rotated cross section and $\mathbf{n}(x)$ is normal to it. Both vectors depend upon x implicitly through the function $\theta(x)$. The derivatives of these vectors with respect to x are

$$\mathbf{n}'(x) = -\theta'(x)\mathbf{m}(x), \qquad \mathbf{m}'(x) = \theta'(x)\mathbf{n}(x) \tag{7.4}$$

To prove these relationships, just do the indicated derivative using the chain rule, factor out θ', and recognize the remaining terms as the other unit vector (or the negative of it). These derivatives will come in handy in some of the later derivations. Note that we are using a prime to represent derivatives with respect to x. For example,

$$\mathbf{w}' = \frac{d\mathbf{w}}{dx}, \quad \mathbf{w}'' = \frac{d^2\mathbf{w}}{dx^2}$$

This notation, due to Lagrange,[1] is possible because there is only one spatial variable, x. To keep the notation compact, we will often suppress the dependence on x, e.g., write $\mathbf{w}$ rather than $\mathbf{w}(x)$ in many of the equations, unless doing so creates ambiguity.

7.1.2 Strain-Displacement Relationships

To see what happens as a result of the plane sections hypothesis, consider the sketch in Fig. 7.4. There are two neighboring cross sections shown, $\mathcal{A}_1$ and $\mathcal{A}_2$, the former initially located a distance x and the latter a distance $x + \Delta x$ from the left end of the beam (i.e., at the origin $\mathcal{O}$). Both cross sections translate and rotate relative to their initial positions. The translation of the centroid of the cross section at x is $\mathbf{w}(x)$, a vector going from the open circle at x in the undeformed configuration to the associated open circle in the deformed configuration. We can construct that vector in two pieces by vector addition, as shown. Similarly, the centroid of the cross section on the right translates by $\mathbf{w}(x+\Delta x)$. We can construct that vector in three pieces. The two cross sections rotate by $\theta(x)$ and $\theta(x+\Delta x)$, respectively. Note that the unit vectors $\mathbf{n}$ and $\mathbf{m}$ also have different values at x and $x + \Delta x$.

The *plane sections* kinematic hypothesis allows neighboring cross sections to rotate relative to each other. If they do, then the lines extending along the two cross sections must intersect at some point $\mathcal{C}$. The deformed axis of the beam is the circular

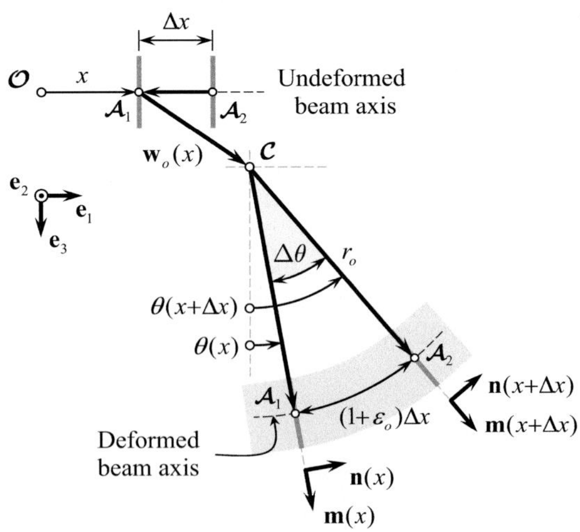

Fig. 7.4 *Beam motion.* Two cross sections, $\mathcal{A}_1$ and $\mathcal{A}_2$, are initially a distance Δx apart. The centroid of each cross section is at the open circle. The motion includes translation and rotation of the cross sections, and the deformation includes stretching of the axis

[1] Joseph-Louis Lagrange (1736–1813), an Italian mathematician, made many contributions to mathematics and mechanics.

arc that connects the two open circles (the centroids of the cross sections) in the deformed configuration. Call the arc length between the cross sections $(1+\varepsilon_o)\Delta x$. We will eventually identify the quantity ε_o as a *strain resultant*, specifically the axial strain which measures the change in length of the beam axis. Note that if $\varepsilon_o = 0$, then the length of the deformed axis is the same as the original length.

Let us designate the radial distance from $\mathcal{C}$ to the centroidal circular arc as r_o, the *radius of curvature* of the beam. We can relate the radius of curvature to the relative rotations of the adjacent cross sections. The difference in angles, i.e., $\Delta\theta = \theta(x+\Delta x) - \theta(x)$, subtends the arc. Therefore,

$$r_o\big[\,\theta(x+\Delta x) - \theta(x)\,\big] = \big(1+\varepsilon_o\big)\Delta x \tag{7.5}$$

In other words, the arc length is equal to the radius multiplied by the angle. Dividing through by $r_o\Delta x$ and taking the limit as Δx goes to zero, we get

$$\theta'(x) = \lim_{\Delta x \to 0} \frac{\theta(x+\Delta x) - \theta(x)}{\Delta x} = \frac{1+\varepsilon_o}{r_o} \tag{7.6}$$

Let us define the *curvature* of the beam to be

$$\kappa_o = \frac{1+\varepsilon_o}{r_o} \tag{7.7}$$

The curvature is a strain-like quantity that measures the amount of bending of the beam. If r_o is large, then the beam is bent only a little. In fact, as $r_o \to \infty$, the beam is straight. The advantage of using curvature rather than radius of curvature to characterize bending is that the curvature is zero when the beam is straight. On the other hand, if r_o is small, then the beam is bent a lot. High curvature (i.e., large κ_o) implies lots of bending. While κ_o and r_o measure the same physical phenomenon, it is more convenient to use κ_o to derive the theory. Using Eqs. 7.6 and 7.7, we can express the *rotation-curvature* relationship as

$$\kappa_o(x) = \theta'(x) \tag{7.8}$$

This is the first key kinematic relationship of beam theory. We will often use the generic term *strain-displacement relationship* to classify such an equation because it connects motion variables like θ with strain variables like κ_o. Rotation is not a displacement per se, but it is displacement-like because it relates to the motion, causing points away from the centroidal axis to displace. Curvature is not a strain per se, but it is strain-like because it relates to the deformation and will show up in the beam strain equation. Strain-displacement relationships are always differential equations. Strain, roughly speaking, is related to the rate of change of displacement.

The displacement from the centroid in the undeformed configuration to the centroid in the deformed configuration for the cross section originally located at x is, by vector addition,

$$\mathbf{w}(x) = \mathbf{w}_o(x) + r_o\,\mathbf{m}(x)$$

where $\mathbf{w}_o(x)$ is the vector from the centroid of the cross section in the undeformed configuration at x to the point $\mathcal{C}$, as shown in Fig. 7.4. Similarly, the displacement from the undeformed to deformed centroid of the cross section originally located at $x + \Delta x$ is

$$\mathbf{w}(x+\Delta x) = -\Delta x\,\mathbf{e}_1 + \mathbf{w}_o(x) + r_o\,\mathbf{m}(x+\Delta x)$$

Ultimately, we do not intend to keep track of point $\mathcal{C}$ when we solve beam problems. It turns out that the vector $\mathbf{w}_o(x)$ will naturally drop out of our equations. However, it is the key to connecting the displacements of the two centroids.

The derivative of $\mathbf{w}(x)$ with respect to x is, by definition,

$$\mathbf{w}'(x) = \lim_{\Delta x \to 0} \frac{\mathbf{w}(x+\Delta x) - \mathbf{w}(x)}{\Delta x}$$

Substituting the expressions for the displacements of the two cross sections and taking the limit as $\Delta x \to 0$ gives

$$\mathbf{w}'(x) = \lim_{\Delta x \to 0} \left\{ r_o \frac{\mathbf{m}(x+\Delta x) - \mathbf{m}(x)}{\Delta x} - \mathbf{e}_1 \right\} = r_o \mathbf{m}' - \mathbf{e}_1$$

Noting from Eq. 7.4 that $\mathbf{m}' = \theta'\mathbf{n}$, and substituting Eq. 7.6, we find that

$$\mathbf{w}' = \big(1+\varepsilon_o\big)\mathbf{n} - \mathbf{e}_1 \tag{7.9}$$

This is the second *strain-displacement* relationship for beam theory. It relates the strain ε_o to the displacement $\mathbf{w}$ and the rotation θ, implicitly through the unit vector $\mathbf{n}$.

We can write component expressions for Eq. 7.9, either as strain in terms of displacement or as displacement in terms of strain. First, dot Eq. 7.9 with $\mathbf{n}$ to get

$$\varepsilon_o = \big(1 + u'\big)\cos\theta - w'\sin\theta - 1 \tag{7.10}$$

To get the displacement derivatives in terms of the strains, dot Eq. 7.9 with $\mathbf{e}_1$ and $\mathbf{e}_3$, respectively, to get

$$\begin{aligned} u' &= \big(1 + \varepsilon_o\big)\cos\theta - 1 \\ w' &= -\big(1 + \varepsilon_o\big)\sin\theta \end{aligned} \tag{7.11}$$

Equations 7.9, 7.10, and 7.11 are equivalent. We can also dot Eq. 7.9 with $\mathbf{m}$ to get

$$\big(1 + u'\big)\sin\theta - w'\cos\theta = 0 \tag{7.12}$$

which is useful for expressing the relationship between θ and the derivatives of displacement, especially for the linearized theory. This is an equation of constraint among the three fields $u(x)$, $w(x)$, and $\theta(x)$ that results from the assumption that the shear deformation is zero, which is implicit in the kinematic hypothesis. Beam theory with this constraint is called Bernoulli-Euler beam theory. For static beams, the effect of shear deformation is typically small. Timoshenko[2] developed a beam theory that lifts this constraint. We will not pursue Timoshenko beam theory in this book but observe that it can be important for static response of short beams and in dynamic response of beams in general.

Equations 7.8 and 7.9 constitute the complete *strain-displacement* equations for the beam. The generalized strain resultants ε_o and κ_o are adequate to completely characterize the deformation of the beam within the context of the kinematic hypothesis. Observe that the equations of kinematics are nonlinear, mostly due to the presence of the sine and cosine of θ.

7.1.3 Normal Strain

Figure 7.5 shows the deformed positions of the two cross sections originally located at x and $x + \Delta x$. This picture will aid us in deriving expressions for the normal strains at other points in the beam cross section. Recall that z measures position in the cross section relative to its centroid. The length of the circular arc at z is

$$d(z) = \big(r_o + z\big)\Delta\theta = \big(r_o + z\big)\left(\frac{1+\varepsilon_o}{r_o}\right)\Delta x$$

where $\Delta\theta = \theta(x + \Delta x) - \theta(x)$, and we have substituted Eq. 7.5 to eliminate $\Delta\theta$ in favor of Δx. The engineering strain of a fiber in the beam is the change in length divided by the original length, which is Δx for all fibers. Thus,

$$\varepsilon(x, y, z) = \frac{d(z) - \Delta x}{\Delta x} = \varepsilon_o + z\left(\frac{1+\varepsilon_o}{r_o}\right)$$

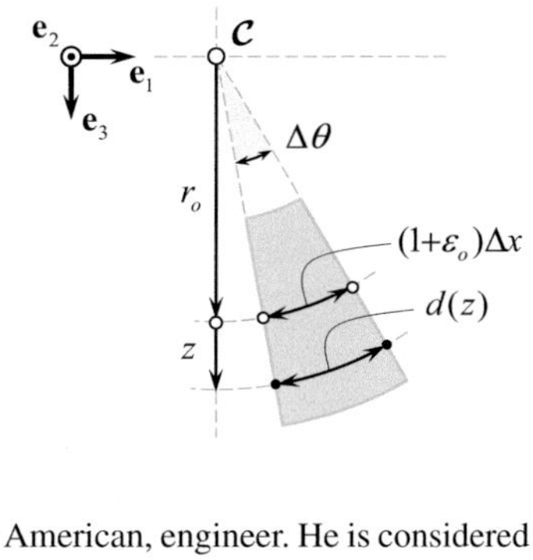

Fig. 7.5 *Strain and curvature.* Longitudinal strain comes from bending of the beam. Some fibers get longer, some get shorter. The quantity ε_o measures the strain of the centroidal fiber. The length of the fiber at location z is $d(z)$

[2] Stephen P. Timoshenko (1878–1972) was a Ukrainian, later American, engineer. He is considered by many to be the father of modern engineering mechanics.

Noting the definition of curvature from Eq. 7.7, we can simplify this expression for the normal strain at a cross section to

$$\varepsilon(x, y, z) = \varepsilon_o(x) + z\,\kappa_o(x) \tag{7.13}$$

Hence, the normal strain varies *linearly* over the cross section. The strain resultant ε_o is the axial strain at the centroid. The amount of strain experienced by a fiber depends upon the curvature and how far that fiber is from the centroid. Because z takes on both positive and negative values, the strains can be positive or negative. If the axial strain at the centroid is zero, then positive curvature results in compressive strains for fibers above the centroid and tensile strains for fibers below the centroidal axis. If the curvature is zero, then all fibers in the cross section strain by the same amount, which was the case for the axial bar problem of Chap. 3.

7.2 Stress Resultants and Equilibrium

Let us now turn our attention to the force side of the picture. The deformed beam is in a state of stress. We can take a cut for a free body diagram at the cross section located at position x, as illustrated in Fig. 7.6a. The traction field $\mathbf{t}(x, y, z)$ acts on the face of the cut. We can write the traction vector in terms of stress components as

$$\mathbf{t} = \sigma\mathbf{n} + \tau\mathbf{m}$$

where σ is the normal component and τ is the shear component of stress. The unit vectors $\mathbf{n}$ and $\mathbf{m}$ point normal to the cross section and in the plane of the cross section, respectively.

The *stress resultants* $N(x)$, $V(x)$, and $M(x)$, shown in Fig. 7.6d, represent the *net* effect of the tractions over the cross section. These resultants are computed by integrating the normal and shear stresses over the cross section, as shown in Fig. 7.6b, c. To wit,

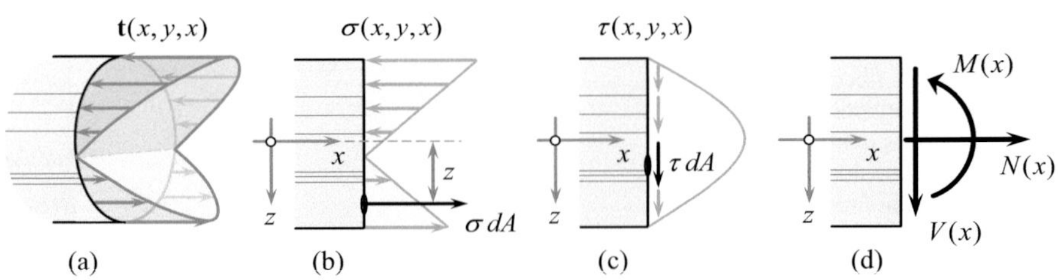

Fig. 7.6 *Planar beam stress resultants.* A cut through a cross section at x exposes a traction field $\mathbf{t}$ that can be characterized by the stress resultants N, V, and M

$$
\begin{aligned}
N(x) &= \int_A \sigma(x, y, z)\, dA \\
V(x) &= \int_A \tau(x, y, z)\, dA \\
M(x) &= \int_A z\,\sigma(x, y, z)\, dA
\end{aligned}
\tag{7.14}
$$

We call N the *axial force*, V the *shear force*, and M the *bending moment*. The resultant moment M is the moment caused by the normal tractions relative to the centroid of the cross section. Note that the vector direction of the moment is often represented with a double-headed arrow, which is defined by the right-hand rule. The bending moment vector points in the $\mathbf{e}_2$ direction, i.e., out of the page. To see why, compute the moment vector as

$$
\mathbf{M}(x) = \int_A z\mathbf{m} \times \sigma(x, y, z)\mathbf{n}\, dA = \int_A z\,\sigma(x, y, z)\, dA\, \mathbf{e}_2 \tag{7.15}
$$

where we have noted that $\mathbf{m} \times \mathbf{n} = \mathbf{e}_2$. Hence, the moment vector has component form $\mathbf{M}(x) = M(x)\mathbf{e}_2$.

Ultimately, we will sum forces and moments. Since the integral is a sum, that means we would execute the integrals represented in Eq. 7.14 every time we establish equilibrium of a beam with a cut. For this reason, working directly with N, V, and M represents no loss of generality in writing equations of equilibrium.

Equations of Equilibrium To establish the differential equations of equilibrium, we isolate a free body diagram of a segment of beam between x and $x + \Delta x$. The forces and moments acting on the deformed free body diagram are shown in Fig. 7.7. The externally applied loads are $\mathbf{q}(x)$ and the applied moment is $m(x)$ about the $\mathbf{e}_2$ axis. Sometimes it is convenient to break the applied load into components as

$$
\mathbf{q}(x) = p(x)\,\mathbf{e}_1 + q(x)\,\mathbf{e}_3
$$

where $p(x)$ is the load in the direction of the original axis of the beam and $q(x)$ is the load transverse to the original axis.

Assume that the applied force does not change with elongation, i.e., the total force on the beam segment is $\mathbf{q}\,\Delta x$, even though the beam axis stretches to $(1 + \varepsilon_o)\Delta x$. For example, the load might be caused by gravity, but stretching does not increase the number of particles contributing to the weight. Thus, the force associated with weight does not change, but just spreads out a little. A counterexample would be forces caused by fluid pressure, like wind on a surface. If the surface gets bigger while the pressure stays the same, the net force increases.[3]

[3] The assumption we are making here about the load is a good example of why going back to the fundamental derivations is important when using a theory to do practical calculations. Always verify that your theory does what you think it does.

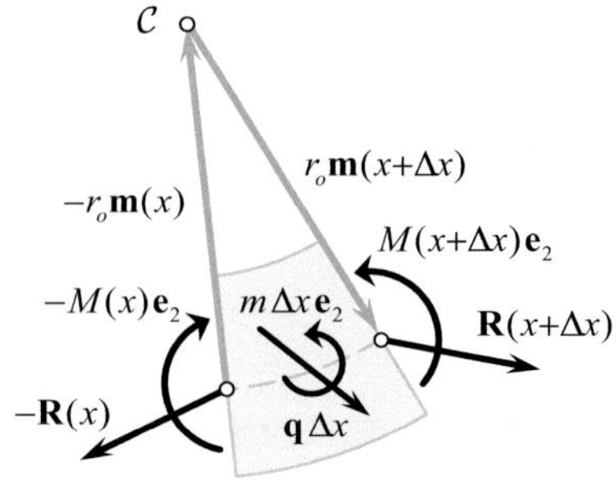

Fig. 7.7 *Free body diagram.* The free body diagram of a piece of beam between x and $x + \Delta x$. The applied loads are $\mathbf{q}(x)$ and the applied moment is $m(x)$

The net force acting on the cross section is designated by the vector $\mathbf{R}(x)$. This force can be resolved into components as

$$\mathbf{R} = N\mathbf{n} + V\mathbf{m} = H\mathbf{e}_1 + Q\mathbf{e}_3 \tag{7.16}$$

as shown in Fig. 7.8, where N and V are the axial force and shear force, respectively, and H and Q are simply the components of force relative to the fixed basis $\{\mathbf{e}_1, \mathbf{e}_2, \mathbf{e}_3\}$. The vectors $\mathbf{n}$ and $\mathbf{m}$ are defined in Eq. 7.3. The relationships among the force components are

$$\begin{aligned} H &= N\cos\theta + V\sin\theta \qquad & N &= H\cos\theta - Q\sin\theta \\ Q &= -N\sin\theta + V\cos\theta \qquad & V &= H\sin\theta + Q\cos\theta \end{aligned} \tag{7.17}$$

Dot Eq. 7.16 with $\mathbf{e}_1$ and $\mathbf{e}_3$, respectively, to get the equations on the left and dot with $\mathbf{n}$ and $\mathbf{m}$, respectively, to get the equations on the right. There are a few reasons to distinguish between the two sets of components that represent the internal resultant force $\mathbf{R}$. First, when we do constitutive equations, it is more natural to refer to the resultant axial and shear forces. For some problems, it is more convenient to refer to the components relative to the standard basis when those components are constant. For problems with small rotations, the difference is negligible.

Balance of Forces Using the free body diagram in Fig. 7.7, establish equilibrium by setting the sum of forces equal to zero. To wit,

$$\mathbf{R}(x + \Delta x) - \mathbf{R}(x) + \mathbf{q}\,\Delta x = \mathbf{0}$$

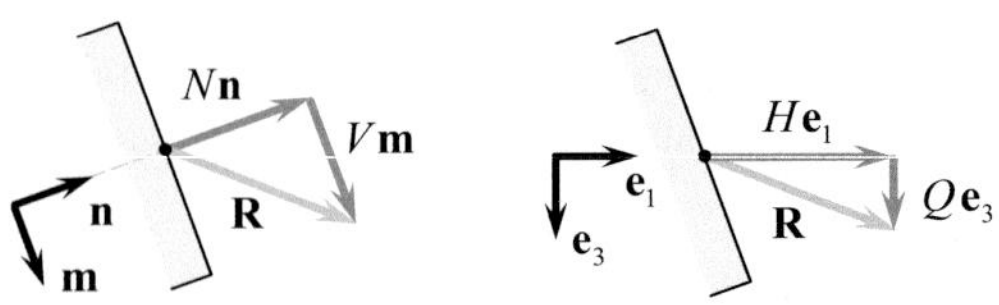

Fig. 7.8 *Resultant force components.* The resultant force $\mathbf{R}$ can be broken into components (N, V) or (H, Q)

Divide this equation through by Δx and take the limit as $\Delta x \to 0$ to show that forces balance if

$$\mathbf{R}' + \mathbf{q} = \mathbf{0} \tag{7.18}$$

It is interesting to note that we do not need to specify the components of $\mathbf{R}(x)$ for force equilibrium at this point. This equation encompasses both the axial part and the transverse part.

Balance of Moments Balance of moments can be done by taking moments about the centroid on the left side (i.e., at x) of the free body diagram in Fig. 7.7. The moment arm for the stress resultant at $x + \Delta x$ is the vector from the centroid at x to the centroid at $x + \Delta x$, which is most easily defined using the two radial vectors from the respective centroids to the point where they meet at C (see also Fig. 7.4). To wit,

$$\begin{aligned}\big[\, r_o\mathbf{m}(x+\Delta x) - r_o\mathbf{m}(x)\,\big] &\times \mathbf{R}(x+\Delta x) \\ &+ \big[M(x+\Delta x) - M(x) + m\Delta x\,\big]\,\mathbf{e}_2 = \mathbf{0}\end{aligned}$$

The moment caused by $\mathbf{q}$ will vanish in the limit, so it has been omitted. Again, divide the equation through by Δx and take the limit as $\Delta x \to 0$. Note that

$$\lim_{\Delta x \to 0} \frac{r_o\mathbf{m}(x+\Delta x) - r_o\mathbf{m}(x)}{\Delta x} = r_o\mathbf{m}' = \big(1 + \varepsilon_o\big)\mathbf{n}$$

because Eq. 7.4 gives $\mathbf{m}' = \theta'\mathbf{n}$ and Eq. 7.6 gives $\theta' = (1 + \varepsilon_o)/r_o$. Thus, balance of moments holds if

$$\big(M' + m\big)\mathbf{e}_2 + (1+\varepsilon_o)\,\mathbf{n} \times \mathbf{R} = \mathbf{0}$$

This is a vector equation, but every term points in the $\mathbf{e}_2$ direction because it is a planar problem. Therefore, we can get a scalar equation by dotting the equation with $\mathbf{e}_2$. For the term that includes the cross product, we will make use of the cyclic nature of the triple scalar product in Eq. 1.10. Specifically,

$$(\mathbf{n} \times \mathbf{R}) \cdot \mathbf{e}_2 = (\mathbf{e}_2 \times \mathbf{n}) \cdot \mathbf{R} = -\mathbf{m} \cdot \mathbf{R}$$

because $\mathbf{e}_2 \times \mathbf{n} = -\mathbf{m}$ from Eq. 7.3. Putting all these facts together, we can express the balance of moments for the beam as

$$M' - (1+\varepsilon_o)\,\mathbf{m} \cdot \mathbf{R} + m = 0 \tag{7.19}$$

This is the most basic form of the moment equation, as it does not commit to any particular set of components of the internal force vector $\mathbf{R}$. However, recognizing that $\mathbf{R} = N\mathbf{n} + V\mathbf{m}$, we can express Eq. 7.19 in the form

$$M' - (1+\varepsilon_o)\,V + m = 0 \tag{7.20}$$

Alternatively, substituting the expressions for N and V from Eq. 7.17 and using Eq. 7.10, we can express Eq. 7.19 as

$$M' + Hw' - Q\left(1+u'\right) + m = 0 \tag{7.21}$$

which will be useful when we discuss buckling in Chap. 12. We can use any of these forms of the equation of balance of moments (i.e., Eqs. 7.19, 7.20, and 7.21) interchangeably.

7.3 Constitutive Equations for Resultants

We will assume that the fibers of the beam are in a uniaxial state of stress. While that assumption is not precisely consistent with the kinematic constraint associated with the cross sections being rigid, it provides the best estimate of the stresses in the beam. If we assume that there are no confining normal stresses in the cross section (i.e., the stress components $\sigma_{yy}=0$ and $\sigma_{zz}=0$), then Hooke's law reduces to $\sigma = E\varepsilon$, where E is Young's modulus (see the example on Sect. 6.2).

A relationship between the axial force N and the strain resultants can be found as follows. Start with the definition of resultant axial force, Eq. 7.14. Substitute Hooke's law $\sigma = E\varepsilon$ and $\varepsilon = \varepsilon_o + z\kappa_o$ from Eq. 7.13 to get

$$N = \int_A \sigma\, dA = \int_A E\left(\varepsilon_o + z\,\kappa_o\right) dA$$

Carrying out the integral over the cross section gives

$$N = EA\,\varepsilon_o \tag{7.22}$$

Note that the term with curvature vanishes if we take the x axis to be at the *centroid* of the cross section because, by definition of the centroid,

$$\int_A z\, dA = 0$$

This simplification makes it clear why the centroid is a good choice for the location of the x axis. In Chap. 9 we will discuss how to compute the location of the centroid of a cross section.

Similarly, for the bending moment, substitute Hooke's law and Eq. 7.13 into the definition of resultant moment from Eq. 7.14 to get

$$M = \int_A z\,\sigma\,dA = \int_A zE\big(\varepsilon_o + z\,\kappa_o\big)dA$$

Again, carrying out the integral gives

$$M = EI\kappa_o \tag{7.23}$$

where I is the second moment of the area relative to the centroid, i.e.,

$$I = \int_A z^2\,dA$$

The second moment of the area is often called the *moment of inertia*,[4] and the quantity EI is called the *effective flexural stiffness* of the beam. Chapter 9 will explore the details of how to compute this essential property of the cross section. Qualitatively, the moment of inertia measures how much material is in the cross section and where it is located, weighting each bit by the square of the distance from the centroid. Hence, the parts of the cross section farther from the centroid contribute much more to the moment of inertia than do the parts close to the centroid. That fact explains why an I-beam is much more efficient at resisting bending than a solid square cross section of the same area and depth. The two flanges of an I-beam contain a large fraction of the material in the cross section, and they are located distant from the centroid. In a solid rectangular cross section, much of the material is located close to the centroid. In both cases, the centroid is in the middle of the cross section.

The elastic constitutive equations in Eqs. 7.22 and 7.23 are the resultant equivalent of Hooke's law. Notice that both involve Young's modulus E, which is a material property. However, they each also involve some property of the cross-sectional geometry, specifically, the area A and the second moment of the area relative to the centroid I. These properties play a role in the evaluation of the deformation of the beam and the stresses induced by the applied forces.

7.4 Beam Sign Conventions

Sign conventions are an important part of every theory. The field variables of the theory, specifically the motions $(u,\ w,\ \theta)$, the applied loads $(p,\ q,\ m)$, the stress resultants $(N,\ V,\ M)$, and the strain resultants $(\varepsilon_o,\ \kappa_o)$ can take on positive and negative values. The sign convention gives a way to interpret what, for example, a negative value of M means. The positive sign convention on the stress resultants, displacements, and rotation are shown in Fig. 7.9. The sign conventions play a key role in the derivations and solution processes.

---

[4] The name *moment of inertia* is a bit of a misnomer in this context because inertia is a dynamic quantity. However, there are deep similarities between the second moment of area and the *mass moment of inertia* that we encounter in rigid-body dynamics.

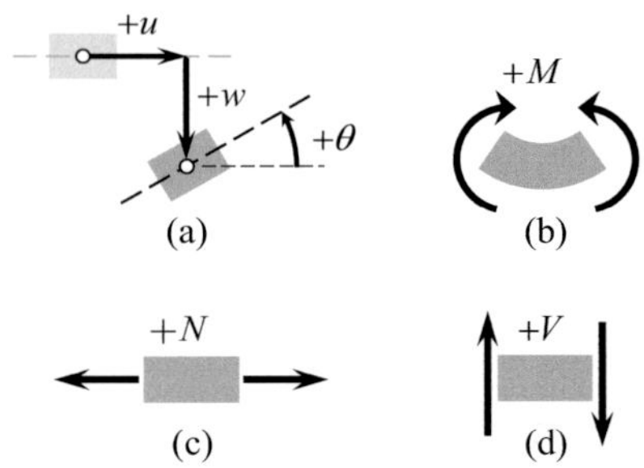

Fig. 7.9 *Sign convention for beams.* **a** Positive directions for displacement and rotation. Applied loads and moment have positive directions corresponding to their associated displacement. **b** Positive moment (and also κ_o). **c** Positive axial force (and also ε_o). **d** Positive shear force

The displacements u and w are positive in the positive directions of $\mathbf{e}_1$ and $\mathbf{e}_3$, respectively. The positive directions of the applied loads p and q are in these same directions. The rotation is positive in the counterclockwise direction. The axial strain is positive if the axis gets longer, and the curvature is positive if the beam bends in a way that looks like Fig. 7.9b, which looks like a smile. Negative curvature is the opposite of that, which looks like a frown. Positive axial force results in positive axial strain and positive moment results in positive curvature.

7.5 Stress Formulas

The stress resultants $N(x)$, $V(x)$, and $M(x)$ represent the aggregate effects of the tractions on the cross section located at x. As we have seen, these resultants are sufficient for establishing equilibrium and for computing the deflections of a beam. In this section we derive a formula for the normal stress based upon Hooke's law and the kinematic hypothesis. We also derive a formula for shear stress based upon equilibrium.

7.5.1 Normal Stress Formula

The normal stress at a cross section can be determined from Hooke's law in conjunction with the kinematic hypothesis. Consider the following derivation:

$$\begin{aligned}\sigma(x, y, z) &= E\,\varepsilon(x, y, z)\\ &= E\Big(\varepsilon_o(x) + z\,\kappa_o(x)\Big)\\ &= E\Big(\frac{N(x)}{EA} + z\frac{M(x)}{EI}\Big)\end{aligned}$$

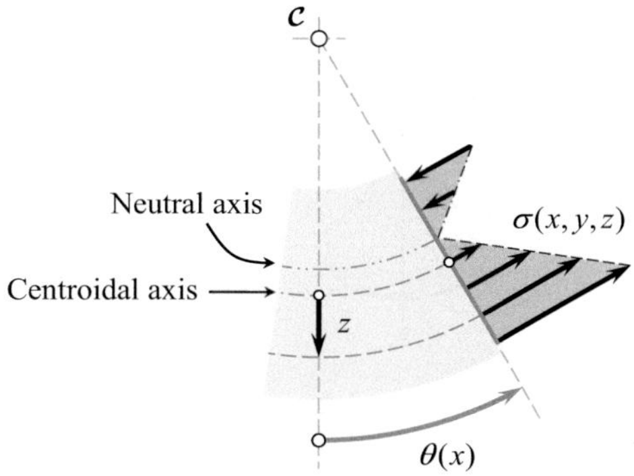

Fig. 7.10 *Beam normal stress*. The normal stress on a cross section is distributed linearly. The *neutral axis* is the location of zero normal stress

Going from the first to second line of this derivation uses Eq. 7.13, which reflects the kinematic hypothesis that plane sections remain plane. Going from the second to third line uses the results from Eqs. 7.22 and 7.23, which also depend crucially on the kinematic hypothesis. If E is constant throughout the cross section, then the planar beam *normal stress formula* reduces to

$$\sigma(x, y, z) = \frac{N(x)}{A} + \frac{M(x)z}{I} \tag{7.24}$$

The normal stresses are distributed linearly over the cross section, as shown in Fig. 7.10. This equation assumes that the beam is prismatic. A non-prismatic beam would simply replace A with $A(x)$ and I with $I(x)$. If there is no bending moment, the normal stress equation reduces to the stress formula for the axial bar.

Equation 7.24 can be used to find the location of the so-called *neutral axis*, which is the place in the cross section where the normal stress is equal to zero, as indicated in Fig. 7.10. Setting the normal stress equal to zero gives the location z_o of the neutral axis. To wit,

$$\frac{N(x)}{A} + \frac{M(x)z_o}{I} = 0 \quad \rightarrow \quad z_o = -\frac{N(x)I}{M(x)A} \tag{7.25}$$

Observe that the location of the neutral axis varies in accord with the stress resultants $N(x)$ and $M(x)$. If the axial force is zero, then $z_o = 0$, which means that the neutral axis is located at the centroid for pure bending. It is possible for the neutral axis not to lie in the domain of the cross section. For example, for pure axial force without bending moment, the location is $z_o \rightarrow -\infty$.

7.5.2 Shear Stress Formula

Unlike the normal stresses, we cannot develop a formula for shear stress at a cross section using Hooke's law because our kinematic hypothesis assumes that there is

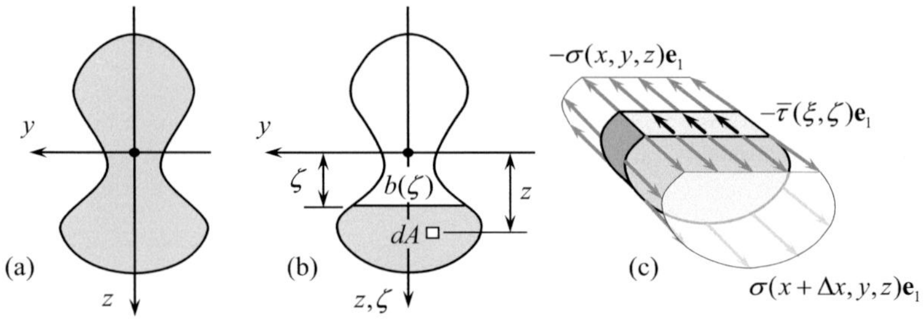

Fig. 7.11 *Shear stress formula.* Derivation of the shear stress formula. **a** The cross section at x. **b** Location of a horizontal cut at ζ in the cross section. **c** Free body diagram of a beam segment between x and $x + \Delta x$, cut at ζ. The y axis is at the centroid of the cross section

no shear deformation and, therefore, $\varepsilon_{xz} = 0$. However, we *can* develop a shear stress formula using an equilibrium argument.

Consider the beam whose cross section at x is shown in Fig. 7.11a. Let us take a free body diagram with cuts perpendicular to the beam axis at locations x and $x + \Delta x$ and another cut through the cross section at a distance ζ from the centroid. The width of the cross section at the location of the cut is $b(\zeta)$. The free body diagram shown in Fig. 7.11c shows the normal stresses on the faces at x and $x + \Delta x$ along with the shear stresses $\bar{\tau}(\xi, \zeta)$, acting at the cut, where ξ is some value between x and $x + \Delta x$. Note that there are shear stresses on the front and back faces, too, but those point in the z direction and will not affect the sum of forces in the x direction. Therefore, they have been omitted from the sketch.

Force equilibrium of the free body diagram in Fig. 7.11c in the $\mathbf{e}_1$ direction can be written as

$$\int_{A_f} \sigma(x + \Delta x, y, z)\, dA - \int_{A_f} \sigma(x, y, z)\, dA \;-\; \bar{\tau}(\xi, \zeta)\, b(\zeta)\Delta x = 0$$

where A_f is the shaded area in Fig. 7.11b, and $\bar{\tau}(\xi, \zeta)$ is the average shear stress over the rectangular horizontal cut, which has area $b(\zeta)\Delta x$. All these quantities depend on the cut location ζ. The region of integration of the normal stress is the same for both faces. Divide through by Δx and take the limit as $\Delta x \to 0$ to get

$$\lim_{\Delta x \to 0} \int_{A_f} \frac{\sigma(x + \Delta x, y, z) - \sigma(x, y, z)}{\Delta x}\, dA - \bar{\tau}(x, \zeta)\, b(\zeta) = 0 \tag{7.26}$$

where we have noted that $\xi \to x$ in the limit. The limit can be moved inside the integral, and we can note that

$$\lim_{\Delta x \to 0} \frac{\sigma(x + \Delta x, y, z) - \sigma(x, y, z)}{\Delta x} = \frac{\partial \sigma}{\partial x}$$

Now we can use the normal stress formula, Eq. 7.24, in conjunction with the moment equilibrium equation, Eq. 7.20, which linearizes to $M' - V + m = 0$. For simplicity, let us only consider the case of a prismatic beam with constant axial force and no distributed applied moments $m(x)$. In that case,

$$\frac{\partial \sigma}{\partial x} = \frac{M'(x)z}{I} = \frac{V(x)z}{I}$$

Substituting this result into Eq. 7.26, we can write the average shear stress as

$$\bar{\tau}(x, \zeta) = \frac{V(x)Q(\zeta)}{I\, b(\zeta)} \tag{7.27}$$

where $V(x)$ is the net shear acting on the cross section, I is the moment of inertia relative to the centroid, $b(\zeta)$ is the width of the cross section at location ζ, and the cross-sectional property $Q(\zeta)$ is defined as

$$Q(\zeta) = \int_{A_f} z\, dA \tag{7.28}$$

Notice that $Q(\zeta)$ is the *first moment of the area* relative to the centroid and the integral is only over the portion A_f of the cross section, which depends upon ζ. If ζ is at the location of the bottom of the cross section, then $Q = 0$ because the area A_f goes to zero. If ζ is at the top of the cross section, then $Q = 0$ because the centroid is defined to be the location where the first moment of the area of the entire cross section is zero. Hence, the shear stress vanishes at the top and bottom fibers of the cross section.

The largest value of Q occurs at the centroid. The reason is as follows. The integrand z takes on positive values below the centroid and negative values above the centroid. The integral is a sum. Hence, for all parts of the cross section below the centroid, we are adding positive values. The more cross section we include, the greater is the sum. If we add values above the centroid, the negative values take away from the sum. The more of the cross section we include, the more we take away (until Q finally vanishes at the top). Therefore, the maximum of Q must be at the centroid, i.e.,

$$Q_{\max} = Q(0)$$

In many cases, we seek the maximum shear stress. The maximum for a prismatic beam will be at the location x where the net shear force V is largest and at the centroid where Q is largest. There are some exceptions, though, because the shear stress depends upon $b(\zeta)$. Hence, the maximum shear stress occurs where $Q(\zeta)/b(\zeta)$ is largest.

The derivation for shear stress finds the horizontal shear stress on the free body diagram in Fig. 7.11c. However, we know that the transverse shear stress acting in the plane of the cross section is equal, as illustrated in Fig. 7.12. In Chap. 5 we

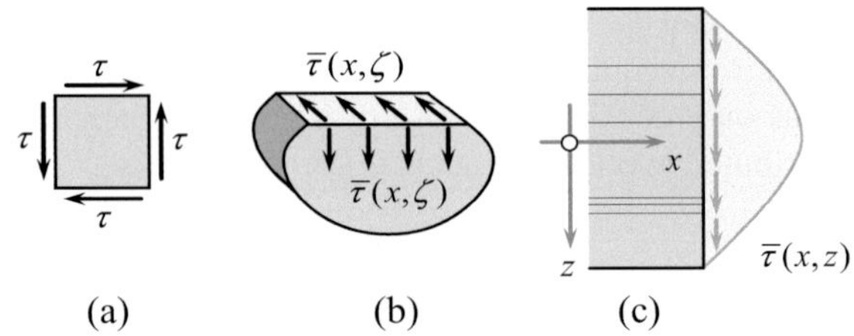

Fig. 7.12 *Transverse shear.* The transverse shear in the cross section is equal to the horizontal shear. **a** Shear stresses on an infinitesimal cube. **b** Shear stresses on the horizontal cut from Fig. 7.11. **c** Shear stress distribution on the cross section

learned that the shear stresses on an infinitesimal cube must be equal on all sides for moments to balance. Hence, the values of $\bar{\tau}(x, \zeta)$ on the horizontal cut in Fig. 7.12b are the same as the shear stresses in the cross section. Figure 7.12c shows the shear stress distribution over the cross section. The shear stress is zero at the top and bottom of the cross section and achieves its maximum value at the centroidal axis.

7.6 Summary

This chapter has laid the theoretical foundations of planar beam theory. It is all about derivation. To fully understand any theory, one must know how the equations come about. The derivations define the language of the theory, identify the assumptions, and tell us about the limitations. Students who are new to mechanics generally benefit from solving problems to cement their understanding of theory. This chapter does not have any chapter end problems because the next three chapters are devoted to the details of solving beam problems. Look to the end of those chapters to find the exercises of the concepts of this one. The following chapter, which focuses on *solving* linear beam problems, is accessible with only rudimentary mastery of this one. Mastery of any theory is a process of going back and forth between following through derivations and crunching numbers with specific problems. So, learn to crunch the numbers in the next chapter, but return to this one not only to recall formulas and sign conventions, but also to build mastery of the concepts.

Chapter 8
Linear Theory of Beams

The equations governing the planar beam derived in the previous chapter are non-linear. In order to make progress with analytical solutions to beam problems, we will linearize the equations. The linear equations will be amenable to classical solution strategies, and through these solutions we can gain insight into the behavior of beams. Of course, the linear theories are applicable only if the deformations are small relative to the physical dimensions of the beam, an assumption that can be verified once the solution is found.

8.1 Linearization of the Beam Equations

A good way to think about linearization of equations is through the lens of the Taylor series of an ordinary function. If we have a function $g(x)$, we can find its Taylor series at the point $x = 0$ as

$$g(x) = g(0) + g'(0)x + \tfrac{1}{2}g''(0)x^2 + \tfrac{1}{6}g'''(0)x^3 + \cdots$$

where a prime denotes derivative with respect to x. The quantities $g(0)$, $g'(0)$, $g''(0)$, and $g'''(0)$ are the values of the function and its first few derivatives, evaluated at zero. Therefore, they are constants. The first two terms in this infinite series constitute the *linear* part. The next term is quadratic, etc. When we linearize, we neglect all the terms after the first two. Whether or not such a simplification is warranted depends upon how much those neglected terms contribute to the function. If they contribute a lot, then the approximation is not a good one. If they contribute little, then the approximation provides a good representation of what the function does. For values of x close to zero, x^2 is much smaller than x and x^3 is much smaller than x^2. Thus, in the neighborhood of $x = 0$, one can argue that the linear approximation is

K. D. Hjelmstad, *Mechanics of Deformable Solids: An Introduction*,
https://doi.org/10.1007/978-3-031-89873-0_8

reasonable. For larger values of x, the deviation of the original function from the linear approximation grows.

A *linear differential equation* is one that involves only the functions and their derivatives in linear form. The assumption required to make our beam equations linear is that the displacements and rotations are small in comparison with the dimensions of the beam. If, for example, w and w' are small, then products and powers of those functions, e.g., w^2, $(w')^2$, or $w\,w'$, will be much smaller. When we neglect those higher order terms, the resulting equations are linear.

Linearized Kinematics Let us assume that the rotation θ is small enough to warrant the approximations $\sin\theta \approx \theta$ and $\cos\theta \approx 1$, and that the axial strains ε_o are small compared to one, i.e., $1+\varepsilon_o \approx 1$. With these approximations, the strain-displacement equations (Eqs. 7.8 and 7.11) reduce to the linear form

$$\kappa_o(x) = \theta'(x), \qquad \varepsilon_o(x) = u'(x), \qquad w'(x) = -\theta(x) \tag{8.1}$$

We can algebraically eliminate θ from the equations by differentiating the third relationship once and substituting it into the first equation to get

$$\kappa_o(x) = -w''(x) \tag{8.2}$$

In this approximation, the curvature is directly related to the transverse displacement. The beam cross sections still rotate, and the rotation is equal to the negative of the slope of the deflected axis, i.e., $\theta = -w'$.

Linearized Equations of Equilibrium If we assume that the rotations and strains are small, then the equations of equilibrium, Eqs. 7.18 and 7.20, reduce to

$$\begin{aligned} N'(x) + p(x) &= 0 \\ V'(x) + q(x) &= 0 \\ M'(x) - V(x) + m(x) &= 0 \end{aligned} \tag{8.3}$$

The first of these three equations is the axial bar problem, which we considered in great detail in Chap. 3. Note that it is not coupled with the other two equations in the linear theory. The second and third equations are coupled through the shear force. We can reduce the equations of equilibrium involving shear and moment to a single second-order equation by algebraically eliminating shear from the equations. Take the derivative of Eq. 8.3(c) to get

$$M''(x) - V'(x) + m'(x) = 0$$

Now, substitute the shear equation $V'(x) = -q(x)$ from Eq. 8.3(b) to get

$$M''(x) + q(x) + m'(x) = 0 \tag{8.4}$$

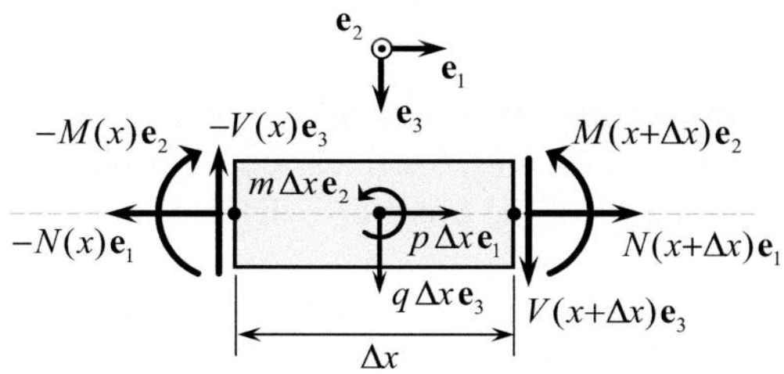

Fig. 8.1 *Equilibrium in the reference configuration*. The linear equations of equilibrium for the beam can be derived by establishing equilibrium in the undeformed configuration

While this equation does not directly involve the shear $V(x)$, it can always be computed from Eq. 8.3(c), once the moment is known. Also, Eq. 8.3(c) can be used to establish boundary conditions that involve shear.

It is worth noting that Eq. 8.3 are the equilibrium equations that would result from establishing equilibrium in the *reference configuration*. In rigid-body statics the free body diagrams we use to establish equilibrium are in the reference configuration because there is no deformation. For deformable bodies it is an approximation.

To clarify this point, consider the segment of beam of length Δx, shown in Fig. 8.1. Unlike the case shown in Fig. 7.7, the free body diagram places the forces on the undeformed (or reference) configuration of the beam. Force equilibrium implies that

$$\Big[N(x+\Delta x) - N(x) + p\Delta x\Big]\mathbf{e}_1 + \Big[V(x+\Delta x) - V(x) + q\Delta x\Big]\mathbf{e}_3 = \mathbf{0}$$

Divide through by Δx, take the limit as $\Delta x \to 0$, and dot the equation successively with $\mathbf{e}_1$ and $\mathbf{e}_3$ to get the equations

$$N'(x) + p(x) = 0, \qquad V'(x) + q(x) = 0$$

which are identical to the first two of Eq. 8.3. Taking moments about the left end of the segment (noting that the moments of p and q will vanish in the limit), we get

$$\big[M(x+\Delta x) - M(x) + m\Delta x\big]\mathbf{e}_2 + \Delta x\,\mathbf{e}_1 \times V(x+\Delta x)\mathbf{e}_3 = \mathbf{0}$$

Again, dividing through by Δx, taking the limit as $\Delta x \to 0$, and taking the dot product with $\mathbf{e}_2$, we get

$$M'(x) - V(x) + m(x) = 0$$

which is identical to the third of Eq. 8.3.

This observation has a practical implication. For many of the problems we face, it will be convenient to find $N(x)$, $V(x)$, and $M(x)$ using finite free body diagrams rather than integrating the differential equations of equilibrium. The preceding derivation shows that carrying out equilibrium computations on the reference configuration will yield exactly the same result as solving the linearized differential equations.

Equations in terms of displacement Using the linear strain-displacement equations in conjunction with the linear constitutive equation $M = EI\kappa_o$, we can express Eq. 8.4 in terms of the transverse displacement $w(x)$ alone as

$$\left[EIw''(x) \right]'' = q(x) + m'(x) \tag{8.5}$$

This equation is the most common form of *Bernoulli-Euler beam theory*.[1] It is a linear, fourth-order, ordinary differential equation. Because the loads $q(x)$ and $m(x)$ are known functions, this differential equation is classified as a *separable differential equation*, for which classical solution methods are well known. The equation is valid for nonprismatic beams in which $EI(x)$ varies along the length of the beam. If EI is constant, then it is a fourth-order, ordinary differential equation with constant coefficients. Assuming that there are no applied distributed moments $m(x)$, we can write the equation in the form

$$EIw''''(x) = q(x) \tag{8.6}$$

for which successive integration of both sides of the equation is a route to solution. This form of the beam equation is the version found in most elementary textbooks on structural analysis.

An alternative solution route recognizes that we can generally find the internal moment $M(x)$ from equilibrium of finite free body diagrams of the beam. In that case, we only need to solve the second-order differential equation

$$w''(x) = -\frac{M(x)}{EI} \tag{8.7}$$

which is the linearized version of Eq. 7.23. Again, this differential equation is separable and can be solved by successive integration. The effects of the applied loads $q(x)$ and applied moments $m(x)$ find their way into this equation through the equilibrium equations that define $M(x)$. The two approaches are equivalent, as we will demonstrate through a series of examples.

8.2 Boundary Conditions

The deflection of a beam is governed by a differential equation. Hence, we must specify boundary conditions to uniquely determine the solution. Boundary conditions are conditions that are true at the two ends of the beam. We can have any one of four

---

[1] Jacob Bernoulli (1654–1705) made some of the initial contributions to beam theory, including the kinematic hypothesis, but he made some mistakes. The theory was put on sound footing by Daniel Bernoulli (1700–1782) and Leonhard Euler (1707–1783) around 1750.

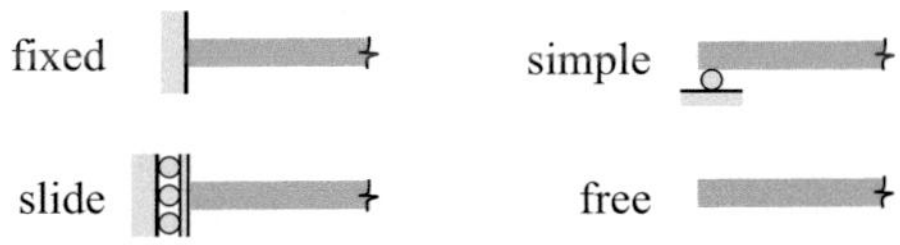

Fig. 8.2 *Beam boundary conditions.* Each end of the beam has one of the four boundary conditions shown here

Table 8.1 *Beam boundary conditions.* There are two conditions at each end of the beam. The various combinations are illustrated in Fig. 8.2. Note that • stands for either $x = 0$ or $x = L$

BC 1	BC 2	Designation in Fig. 8.2
$w(\bullet) = 0$	$w'(\bullet) = 0$	Fixed at end •
$w'(\bullet) = 0$	$V(\bullet) = 0$	Slide at end •
$w(\bullet) = 0$	$M(\bullet) = 0$	Simple at end •
$M(\bullet) = 0$	$V(\bullet) = 0$	Free at end •

conditions: fixed, slide, simple, or free at either end, as illustrated in Fig. 8.2, with the characteristics listed in Table 8.1. Note that both the roller support (pictured in the figure) and pin support fit the conditions of the *simple* boundary condition. The difference between the two concerns the axial displacement $u(x)$, not the transverse displacement and rotation.

Noting the constitutive relationships and the definition of the stress resultant M, we can write the moment conditions (simple and free end) in terms of w as

$$M(\bullet) = -EIw''(\bullet)$$

Hence, if the boundary condition is $M(0) = 0$, we can substitute the equivalent condition $w''(0) = 0$. For the Bernoulli-Euler beam, the shear is actually a reaction force because of the kinematic constraint that forces the shear deformation to be zero. Hence, the shear boundary condition in these theories must be obtained from balance of moments as

$$V(\bullet) = M'(\bullet) + m(\bullet)$$

Substituting the constitutive equation for moment and the displacement derivative for the rotation, the shear boundary conditions can be written for the linearized theory in terms of w as

$$V(\bullet) = -EIw'''(\bullet) + m(\bullet) \tag{8.8}$$

We will often write the boundary conditions in terms of the displacement variable in the solution approaches that follow.

It is possible to have a concentrated force or moment applied at the end of a beam. If a downward concentrated transverse force P is applied at a free (or slide) boundary condition, then the shear boundary condition would be

$$V(L) = P$$

A similar modification would be made for a concentrated moment applied at a free end or simple support. These conditions can be found from equilibrium of an infinitesimal free body diagram at the end of the beam. We need this free body diagram because it exposes the internal stress resultants. Even in the infinitesimal limit, the concentrated force or moment at the end will be present. If there are distributed loads $q(x)$, then as the length of the free body diagram goes to zero, so does the effect of the distributed load. Therefore, the boundary conditions on shear and moment are not influenced by the distributed load.

8.3 Continuity Conditions

Beams that have intermediate supports, point loads, or abrupt changes in EI have discontinuities in some of the state variables. To solve this type of problem, we will break the beam into segments at the points of discontinuity and establish *continuity conditions* for the state variables that must be continuous at those points. For example, consider the two-span beam shown in Fig. 8.3. It is fixed at the left end and simply supported at the right end. The presence of the roller at $x = L$ means that there will be a reaction force R at that point. Therefore, the shear will be discontinuous, jumping by the amount R, while the moment, rotation, and displacement will be continuous at that point.

To analyze this beam, let us break it into two segments, labeled as 1 and 2 in the figure. Each piece will have its own state variables, i.e., w_1, θ_1, M_1, and V_1 for piece 1 and w_2, θ_2, M_2, and V_2 for piece 2. Both of these sets of state variables characterize the transverse displacement, rotation, moment, and shear, respectively, and both sets are functions of x (which is measured from the left end of the beam). As such, the state variables are defined *piecewise*. For example, the transverse displacement is

$$w(x) = \begin{cases} w_1(x) & 0 \leq x \leq L \\ w_2(x) & L \leq x \leq 2L \end{cases}$$

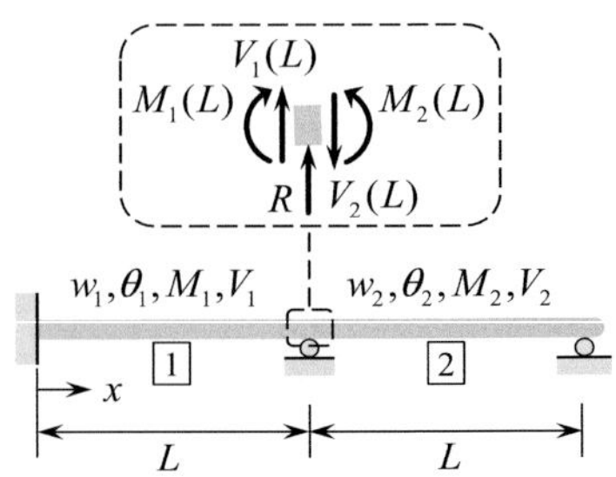

Fig. 8.3 *Beam continuity conditions.* A two-span beam with a roller support in the middle. Displacement, rotation, and moment are continuous at the support. The free body diagram of the joint establishes moment continuity

with similar expressions for the other state variables.

Each of the quantities in the two segments of the beam will be determined independently by integrating the governing equations in that region. The only way to guarantee that they match where the two regions join is to enforce the continuity conditions. For example, if a variable, say $\theta(x)$, must be continuous at $x = L$, then the continuity condition is $\theta_1(L) = \theta_2(L)$.

There are two kinds of continuity conditions: kinematic (conditions on w and θ) and kinetic (conditions on M and V). The simplest way to derive the kinetic continuity conditions is to take a tiny free body diagram of the neighborhood of the region where the two segments join, as shown in Fig. 8.3. Now, equilibrium of the free body diagram requires that

$$M_1(L) = M_2(L), \qquad V_1(L) + R = V_2(L)$$

Usually, when the equilibrium conditions involve an unknown reaction force, as does the shear in this case, we call it a *discontinuity* condition. It can be used to find the reaction once the shears are completely determined. Moment equilibrium in this case shows that the moment is continuous at $x = L$. If the approach to equilibrium is to take finite free body diagrams, then M_1 can be established by taking a cut at x in piece 1, and M_2 can be established by taking a cut at x in piece 2. If these calculations are done correctly, then the continuity of moment should be automatically satisfied. The continuity condition would then be a verification of the equilibrium calculations.

Kinematic continuity conditions are evident directly from the sketch showing the support conditions. If the beam is continuous over the support, then the rotation on the left side of the roller at $x = L$ must be the same as the rotation on the right side of the roller. If that is not the case, then the beam has a kink, which is inconsistent with the physical condition of the beam. If the beam has an *internal hinge*, then the rotation will be discontinuous and the value of $\theta(x)$ will be different on the two sides of the hinge.

The roller support in Fig. 8.3 prevents transverse displacement. Therefore,

$$w_1(L) = 0, \qquad w_2(L) = 0$$

Thus, there are *two* independent displacement conditions (they are independent because they show up in separate equations). Both displacements are equal to zero, therefore, they are equal to each other. Hence, the continuity condition $w_1(L) = w_2(L)$ is automatically satisfied and does not represent an additional independent condition.

The combination of boundary and continuity conditions always provides exactly the amount of information needed to uniquely solve a problem. Consider that each beam segment must satisfy Eq. 8.6, which is a fourth-order differential equation and will, therefore, yield four constants of integration. With two beams, there will be a total of eight constants of integration. There are four boundary conditions, two at each end, and four continuity conditions at the location where the regions join.

Example 8.1
Consider the two-span beam of Fig. 8.3. What are the boundary and continuity conditions for this system?

Solution. The beam is fixed at the left end and simply supported at the right end. The boundary conditions at the left end are

$$w_1(0) = 0, \qquad \theta_1(0) = 0$$

The boundary conditions at the right end are

$$w_2(2L) = 0, \qquad M_2(2L) = 0$$

The continuity conditions at the roller support in the middle are

$$w_1(L) = 0, \qquad w_2(L) = 0, \qquad \theta_1(L) = \theta_2(L), \qquad M_1(L) = M_2(L)$$

Note that for the linear beam, we can substitute $\theta = -w'$ in the boundary and continuity conditions. Hence, we can write the rotation condition at the fixed end as $w_1'(0) = 0$, and we can write the rotation continuity condition as $w_1'(L) = w_2'(L)$. In that same vein, we can write the moment continuity condition as $w_1''(L)/EI_1 = w_2''(L)/EI_2$.

8.4 Approaches to Solving Linear Beam Problems

The linear theory of beams is defined by the differential equations relating shear, moment, rotation, and transverse displacement. These relationships have been described in various forms including (a) four first-order differential equations: Eq. 8.3 for static equilibrium and Eqs. 8.1 for kinematics, (b) two second-order differential equations: Eq. 8.4 for equilibrium and Eq. 8.6 for kinematics, and (c) one fourth-order differential equation, Eq. 8.5, that combines all the differential equations for the beam together. These equations are nothing more than different ways to express the same Bernoulli-Euler beam theory.

The various ways of writing the differential equations for beam theory allow different strategies in the solution process for particular problems. Each strategy has its own advantages, and each offers slightly different insights into beam response and the solution process. For example, viewing the equations as a system of four first-order differential equations clearly shows the hierarchical and sequential nature of the equations. The shear can be determined from the applied load. The moment can be determined from the shear. The rotation can be determined from the moment. And the transverse displacement can be determined from the rotation. These equations also provide insight about how one can qualitatively sketch the response functions (i.e., shear, moment, rotation, and deflection diagrams). The slope of the shear diagram is equal to the negative of the loading diagram at each point. The slope of the moment diagram is equal to the shear, etc.

The version of the equations with two second-order differential equations affords the possibility of finding the moment $M(x)$ by solving a statics problem using a finite free body diagram with cuts to expose the internal moment and then integrating the second-order kinematic equations to find displacement and its first derivative, which is the negative of the rotation. In this approach, we do not explicitly introduce $\theta(x)$ into the equations but acknowledge its presence through the function $w'(x)$. If the internal moment is determined from a free body diagram, then the boundary conditions on moment and shear are automatically satisfied.

The version of the equation with a single fourth-order differential equation allows us to express everything in terms of the transverse displacement $w(x)$. All boundary conditions must be translated into conditions on w, w' (for rotation), w'' (for moment), or w''' (for shear).

We will discuss the framework of four first-order equations extensively in Chap. 10, with the aim of writing a code to solve the prismatic beam for any applied loading and any boundary conditions. In this chapter we examine the remaining approaches through a series of examples, starting with a couple of examples that illustrate different approaches to establishing equilibrium. Next, we look at the computations of deflections and rotation for statically determinate beams. We end with a couple of examples that deal with statically indeterminate beams and nonprismatic beams. These examples should allow the reader to see the advantages and disadvantages of the various approaches. Along the way we will run into the *internal hinge*, and we will see how to handle point loads.

8.4.1 Beam Statics

Equilibrium is one of the key aspects of every beam problem. We need to determine the internal stress resultants that arise due to applied loads. The internal stress resultants will be useful both for computing deflections and stresses in the beam. In this section we consider two approaches: (1) using free body diagrams, and (2) integrating the differential equation of equilibrium to find the internal state.

Example 8.2
Consider a simply supported beam of length L subjected to a uniformly distributed load of magnitude q_o, shown in Fig. 8.4. Find the internal shear $V(x)$ and bending moment $M(x)$, first using a finite free body diagram and second by integrating the differential equations of equilibrium.

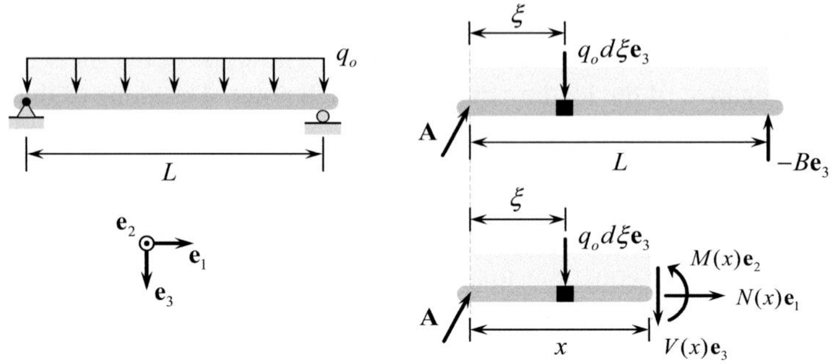

Fig. 8.4 *Example*. Simply supported beam of length L, subjected to a uniformly distributed load of magnitude q_o. The free body diagrams are on the right, the top being the entire beam and the bottom being the beam with a cut at x. The typical particle is located a distance ξ from the left end

Finite free body diagram approach. The finite free body diagrams are shown on the right side of Fig. 8.4, the top being a free body diagram of the entire beam, which we will use to find the reactions, and the bottom being a free body diagram with a cut at x, exposing the internal stress resultants $M(x)$, $V(x)$, and $N(x)$.

Moment equilibrium of the free body diagram of the entire system, taking moments at the left end, gives

$$
\begin{aligned}
\int_0^L (\xi \mathbf{e}_1) \times (q_o\, d\xi \mathbf{e}_3) + (L\mathbf{e}_1) \times (-B\mathbf{e}_3) &= \mathbf{0} \\
-q_o \mathbf{e}_2 \int_0^L \xi\, d\xi + BL\mathbf{e}_2 &= \mathbf{0} \\
-\tfrac{1}{2} q_o L^2 \mathbf{e}_2 + BL\mathbf{e}_2 &= \mathbf{0}
\end{aligned}
$$

where we have noted that $\mathbf{e}_1 \times \mathbf{e}_3 = -\mathbf{e}_2$. The moment equation can be dotted with $\mathbf{e}_2$ and solved for B. Force equilibrium gives

$$
\begin{aligned}
\mathbf{A} + \int_0^L q_o\, d\xi\, \mathbf{e}_3 - B\mathbf{e}_3 &= \mathbf{0} \\
\mathbf{A} + q_o L\mathbf{e}_3 - B\mathbf{e}_3 &= \mathbf{0}
\end{aligned}
$$

Solving the moment equilibrium equation for B and substituting it into the force equilibrium equation yields the reactions

$$
B = \tfrac{1}{2} q_o L, \qquad \mathbf{A} = -\tfrac{1}{2} q_o L\mathbf{e}_3
$$

Observe that both reactions are upward, equilibrating the downward load, and that the reaction **A** does not have a component in the $\mathbf{e}_1$ direction. The reactions are equal, as one would expect from the symmetry of the loading.

Now we can establish equilibrium of the free body diagram with a cut at x. Taking moments at the cut gives

$$
\begin{aligned}
-x\mathbf{e}_1 \times \mathbf{A} + \int_0^x -(x-\xi)\mathbf{e}_1 \times q_o\mathbf{e}_3\,d\xi + M(x)\mathbf{e}_2 &= \mathbf{0} \\
-x\mathbf{e}_1 \times -\tfrac{1}{2}q_oL\mathbf{e}_3 + q_o\mathbf{e}_2\int_0^x (x-\xi)\,d\xi + M(x)\mathbf{e}_2 &= \mathbf{0} \\
-\tfrac{1}{2}q_oLx\mathbf{e}_2 + q_o\mathbf{e}_2\Big[x\xi - \tfrac{1}{2}\xi^2\Big]_0^x + M(x)\mathbf{e}_2 &= \mathbf{0} \\
-\tfrac{1}{2}q_oLx\mathbf{e}_2 + \tfrac{1}{2}q_ox^2\mathbf{e}_2 + M(x)\mathbf{e}_2 &= \mathbf{0}
\end{aligned}
$$

The first equation makes clear why we need both x and ξ to measure distance along the beam axis. The distance from the cut to the typical particle is $x-\xi$, and the moment arm points in the $-\mathbf{e}_1$ direction. The cross product $\mathbf{e}_1 \times \mathbf{e}_3 = -\mathbf{e}_2$ appears frequently in these problems. Dotting the last equation with $\mathbf{e}_2$ and solving for $M(x)$ gives

$$
M(x) = \tfrac{1}{2}q_o\big(Lx - x^2\big)
$$

Observe that the moment satisfies the boundary conditions $M(0)=0$ and $M(L)=0$ automatically. We can find $V(x)$ and $N(x)$ from the free body diagram with a cut at x. Force equilibrium requires

$$
\begin{aligned}
\mathbf{A} + \int_0^x q_o\mathbf{e}_3\,d\xi + N(x)\mathbf{e}_1 + V(x)\mathbf{e}_3 &= \mathbf{0} \\
-\tfrac{1}{2}q_oL\mathbf{e}_3 + q_ox\mathbf{e}_3 + N(x)\mathbf{e}_1 + V(x)\mathbf{e}_3 &= \mathbf{0}
\end{aligned}
$$

Dotting this equation, respectively, with $\mathbf{e}_1$ and $\mathbf{e}_3$ gives

$$
N(x) = 0, \qquad V(x) = \tfrac{1}{2}q_o\big(L - 2x\big)
$$

Note that the shear satisfies the differential equation $V = M'$. The shear at $x=0$ is equal to $-\mathbf{A}\cdot\mathbf{e}_3$ and the shear at $x=L$ is equal to $-B$, as equilibrium requires.

Differential equation approach. Integrate the shear equation first, i.e.,

$$
\begin{aligned}
V'(x) &= -q_o \\
V(x) &= -q_ox + C_1
\end{aligned}
$$

Now, integrate the moment equation $M' = V$, i.e.,

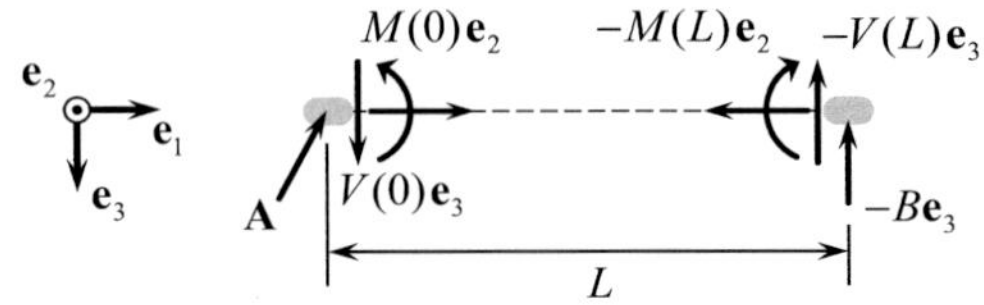

Fig. 8.5 *Example (cont'd).* Infinitesimal free body diagrams at the ends of the beam can be used to relate the reaction forces to the internal stress resultants

$$M'(x) = -q_o x + C_1$$
$$M(x) = -\tfrac{1}{2}q_o x^2 + C_1 x + C_2$$

To find the constants C_1 and C_2, we use the boundary conditions $M(0) = 0$ and $M(L) = 0$. To wit,

$$M(0) = -\tfrac{1}{2}q_o(0)^2 + C_1(0) + C_2 = 0$$
$$M(L) = -\tfrac{1}{2}q_o(L)^2 + C_1(L) + C_2 = 0$$

Solving these equations gives

$$C_1 = \tfrac{1}{2}q_o L, \qquad C_2 = 0$$

Thus, the final expression for shear and moment is

$$V(x) = -q_o x + \tfrac{1}{2}q_o L, \qquad M(x) = -\tfrac{1}{2}q_o x^2 + \tfrac{1}{2}q_o L x$$

which are the same as the ones found by the finite free body diagram method.

In contrast to the previous approach, the differential equation approach does not require the evaluation of the reaction forces to get the internal stress resultants. To find the reaction forces, we can take infinitesimal free body diagrams in the neighborhood of $x = 0$ and $x = L$, as shown in Fig. 8.5. Because the neighborhood is infinitesimal, the distributed load causes a vanishingly small contribution. Summing forces gives $\mathbf{A} + V(0)\mathbf{e}_3 = \mathbf{0}$ (noting that the axial force is zero). Similarly, an infinitesimal free body diagram in the neighborhood of $x = L$ gives $-B\mathbf{e}_3 - V(L)\mathbf{e}_3 = \mathbf{0}$. Simplifying these equations gives

$$\mathbf{A} = -V(0)\mathbf{e}_3 = -\tfrac{1}{2}q_o L\mathbf{e}_3, \qquad B = -V(L) = \tfrac{1}{2}q_o L$$

Of course, these are the same results found previously using the finite free body diagram.

This example is instructive because it illustrates that there are multiple ways to set up the equilibrium calculations. While they appear, at first glance, to be different methods, they are flip sides of the same coin. It appears that we do not need a free body diagram for the differential equations approach, but remember that we used a free body diagram of length Δx to *derive* the differential equations of equilibrium. The summation of forces and moment in the finite free body diagram approach includes integrals of the distributed loads. In the differential equations approach, we do those same integrals by finding the antiderivatives (and adding the obligatory constants of integration). For the finite free body diagram approach, we do not use the moment boundary conditions, whereas for the differential equations approach, we do. For the finite free body diagram approach, we compute the reactions first. For the differential equations approach, we compute them last.

It appears from this example that the differential equations approach requires less computation than the finite free body diagram approach. That may not always be the case, but recall that when we use the differential equation, we already benefit from having done equilibrium on a free body diagram to generate the starting point for that method.

8.4.2 The Internal Hinge

One device that deserves some special consideration is the *internal hinge*. The hinge allows the beam segments on either side of it to rotate relative to each other but forces the two segments to translate together at the hinge. The hinge can transmit shear and axial force, but the moment is zero at the hinge. To see what is at stake, consider the following example. We will again do the problem with both methods.

Example 8.3
Consider the propped cantilever beam of length L with an internal hinge, subjected to a point load of magnitude P, as shown in Fig. 8.6. Find the internal shear $V(x)$ and bending moment $M(x)$, first using a finite free body diagram and second by integrating the differential equations of equilibrium.

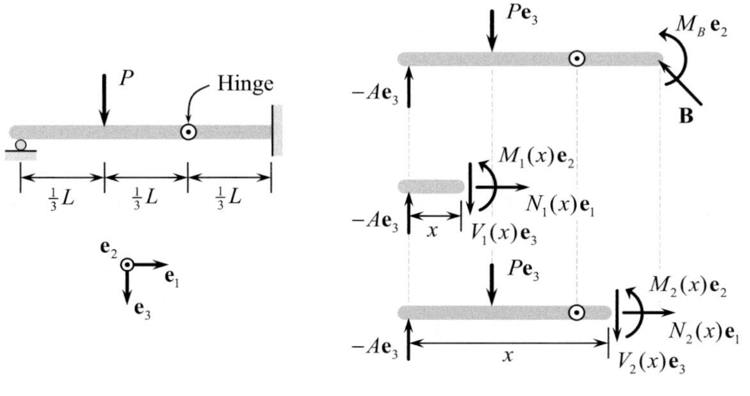

Fig. 8.6 *Example*. Propped cantilever beam of length L with an internal hinge, subjected to a point load of magnitude P. The free body diagrams are on the right, the top being the entire beam and the middle being the beam with a cut at x to the left of the load, and the bottom being the beam with a cut at x to the right of the load

Solution. The point load causes a discontinuity in the shear and a kink in the moment. Therefore, we will define the internal stress resultants piecewise. For example, the moment is defined as

$$M(x) = \begin{cases} M_1(x) & 0 \leq x \leq \frac{1}{3}L \\ M_2(x) & \frac{1}{3}L \leq x \leq L \end{cases}$$

The shear and axial forces are defined in a similar manner. The need to define the internal stress resultants as piecewise functions is common to any approach to solving the equilibrium problem with a discontinuity caused by a point load.

Finite free body diagram approach. The finite free body diagrams are shown on the right side of Fig. 8.6, the top being a free body diagram of the entire beam, which we will use to find the reactions, and the bottom two being free body diagrams with cuts at x, exposing the internal stress resultants $M(x)$, $V(x)$, and $N(x)$. Force equilibrium of the free body diagram of the entire system gives

$$-A\mathbf{e}_3 + P\mathbf{e}_3 + \mathbf{B} = \mathbf{0}$$

Moment equilibrium, taking moments at the right end, gives

$$\begin{aligned} \left(-L\mathbf{e}_1\right) \times \left(-A\mathbf{e}_3\right) + \left(-\tfrac{2}{3}L\mathbf{e}_1\right) \times P\mathbf{e}_3 + M_B\mathbf{e}_2 &= \mathbf{0} \\ -AL + \tfrac{2}{3}PL + M_B &= 0 \end{aligned}$$

where we have noted that $\mathbf{e}_1 \times \mathbf{e}_3 = -\mathbf{e}_2$ and dotted the equation with $\mathbf{e}_2$. These equations relate the reactions A, $\mathbf{B}$, and M_B to each other and to the applied load P.

Now we can establish equilibrium of the free body diagram with a cut at x left of the load P. Taking moments at the cut gives

$$\begin{aligned} -x\mathbf{e}_1 \times -A\mathbf{e}_3 + M_1(x)\mathbf{e}_2 &= \mathbf{0} \\ -Ax\mathbf{e}_2 + M_1(x)\mathbf{e}_2 &= \mathbf{0} \end{aligned}$$

The moment arm $-x\mathbf{e}_1$ is a vector from the cut to the force. The value x is positive for any location along the beam, the minus sign is because the direction of the vector is $-\mathbf{e}_1$. Dotting the last equation with $\mathbf{e}_2$ and solving for $M_1(x)$ gives

$$M_1(x) = Ax$$

Force equilibrium requires

$$-A\mathbf{e}_3 + N_1(x)\mathbf{e}_1 + V_1(x)\mathbf{e}_3 = \mathbf{0}$$

Dotting this equation, respectively, with $\mathbf{e}_1$ and $\mathbf{e}_3$ gives

$$N_1(x) = 0, \qquad V_1(x) = A$$

The functions $M_1(x)$, $V_1(x)$, and $N_1(x)$ apply only for values of x between zero and $L/3$.

Repeat those same calculations for the free body diagram with the cut at x to the right of the load P. Moment equilibrium, with moments taken at the cut, requires

$$-x\mathbf{e}_1 \times -A\mathbf{e}_3 - \big(x - \tfrac{1}{3}L\big)\mathbf{e}_1 \times P\mathbf{e}_3 + M_2(x)\mathbf{e}_2 = \mathbf{0}$$

$$-Ax\mathbf{e}_2 + \big(x - \tfrac{1}{3}L\big)P\mathbf{e}_2 + M_2(x)\mathbf{e}_2 = \mathbf{0}$$

Dotting the last equation with $\mathbf{e}_2$ and solving for $M_2(x)$ gives

$$M_2(x) = Ax - \big(x - \tfrac{1}{3}L\big)P$$

Force equilibrium requires

$$-A\mathbf{e}_3 + P\mathbf{e}_3 + N_2(x)\mathbf{e}_1 + V_2(x)\mathbf{e}_3 = \mathbf{0}$$

Dotting this equation, respectively, with $\mathbf{e}_1$ and $\mathbf{e}_3$ gives

$$N_2(x) = 0, \qquad V_2(x) = A - P$$

To solve for the reaction A, we can use the condition that the moment at the hinge must be zero, specifically, $M_2(\frac{2}{3}L) = 0$. Substituting the value $x = \frac{2}{3}L$ into $M_2(x)$ gives

$$M_2(\tfrac{2}{3}L) = A\big(\tfrac{2}{3}L\big) - \big((\tfrac{2}{3}L) - \tfrac{1}{3}L\big)P = 0$$

Solving for A gives

$$A = \tfrac{1}{2}P$$

Finally, we can summarize the shears and moments, substituting the value of A just found. They are

$$V_1(x) = \tfrac{1}{2}P \qquad V_2(x) = -\tfrac{1}{2}P$$

$$M_1(x) = \tfrac{1}{2}Px \qquad M_2(x) = \tfrac{1}{6}P\big(2L - 3x\big)$$

The reaction forces at the fixed end can also be computed from the relationships derived previously. They are

$$\mathbf{B} = \big(A - P\big)\mathbf{e}_3 = -\tfrac{1}{2}P\mathbf{e}_3, \qquad M_B = AL - \tfrac{2}{3}PL = -\tfrac{1}{6}PL$$

Once the reaction force A is found, all the other force quantities—reactions and internal stress resultants—are completely and uniquely determined. This observation highlights the critical importance of the hinge condition in this approach.

Differential equation approach. There is no distributed load on either segment. Therefore, the rate of change of shear is zero ($V' = 0$) and, thus, the shear is constant in those regions. The moment is the antiderivative of shear ($M' = V$) and is therefore linear. Hence, we can write

$$\begin{aligned} V_1(x) &= C_1 & V_2(x) &= C_3 \\ M_1(x) &= C_1 x + C_2 & M_2(x) &= C_3 x + C_4 \end{aligned}$$

The conditions that are available to solve for the constants of integration include the boundary condition on moment at the left end, the continuity condition on moment at the point of load, the zero-moment condition at the hinge, and the discontinuity condition on the shear at the point of load, i.e.,

$$\begin{aligned} M_1\big(0\big) &= 0 & M_1(\tfrac{1}{3}L) &= M_2(\tfrac{1}{3}L) \\ M_2(\tfrac{2}{3}L) &= 0 & V_1(\tfrac{1}{3}L) &= P + V_2(\tfrac{1}{3}L) \end{aligned}$$

Implementing these conditions gives, in the same order,

$$\begin{aligned} C_1(0) + C_2 &= 0 & C_1\big(\tfrac{1}{3}L\big) + C_2 &= C_3\big(\tfrac{1}{3}L\big) + C_4 \\ C_3\big(\tfrac{2}{3}L\big) + C_4 &= 0 & C_1 &= P + C_3 \end{aligned}$$

This is a system of four equations in the four unknowns C_1, C_2, C_3, and C_4. There is often a route through the equations to expedite solving them by substitution. These equations can also be put into matrix form. Multiplying the second and third equations by 3 to clear the fractions, we can write

$$\begin{bmatrix} 0 & 1 & 0 & 0 \\ L & 3 & -L & -3 \\ 0 & 0 & 2L & 3 \\ 1 & 0 & -1 & 0 \end{bmatrix} \begin{Bmatrix} C_1 \\ C_2 \\ C_3 \\ C_4 \end{Bmatrix} = \begin{Bmatrix} 0 \\ 0 \\ 0 \\ P \end{Bmatrix}$$

Solving the equations gives

$$C_1 = \tfrac{1}{2}P, \qquad C_2 = 0, \qquad C_3 = -\tfrac{1}{2}P, \qquad C_4 = \tfrac{1}{3}PL$$

Substituting these values back into the equations for shear and moment gives the same results as the method using the finite free body diagram. The reaction forces can be found by taking infinitesimal free body diagrams at the ends to relate the reactions to the internal stress resultants at the ends, as we did in the previous example. The details are left to the reader.

This example demonstrates why we need to use piecewise functions for the shear and moment fields. The function $V(x)$ is constant in each segment, but it is not the same constant (i.e., $C_1 \neq C_3$). By that same token, $M(x)$ is linear in each segment, but it is not the same linear function. The discontinuity in shear causes a kink in the moment diagram.

It is also important to realize that the internal hinge does not cause a discontinuity in the shear or moment fields. That is why there was no need to break the second segment of the beam into two parts divided at the hinge location. The hinge condition is essential to finding the solution to the problem because it adds a kinetic condition, i.e., $M(\tfrac{2}{3}L) = 0$. From the perspective of the approach using the differential equations, the four constants of integration require four equations to determine them. The hinge condition was one of those.

The hinge *does* cause a discontinuity, but that discontinuity is in $\theta(x)$, which is a kinematic variable. The linearized equilibrium equations do not involve the kinematic variables. We will see in an example in the next section how an internal hinge affects the solution for kinematic variables.

8.4.3 Beam Deflections and Rotations

To find the deflection and rotation functions $w(x)$ and $\theta(x)$, we must integrate the kinematic equations. In this section, we will show through some examples what is at stake in these calculations. We will consider two approaches: (1) integrating the second-order kinematic equations using the moment obtained from a free body diagram, and (2) integrating the fourth-order kinematic equations.

Example 8.4
Consider the simply supported beam of length L and bending modulus EI, subjected to a uniformly distributed load of magnitude q_o, shown in Fig. 8.4. In a previous example, we determined the moment to be

$$M(x) = \tfrac{1}{2}q_o\big(Lx - x^2\big)$$

Use this moment in conjunction with Eq. 8.7 to find the transverse displacement $w(x)$. Solve the problem again starting from Eq. 8.6.

Compute deflection from moment. First integrate the equation $w'' = -M/EI$ (Eq. 8.7), relating the second derivative of w to the moment. To wit,

$$\begin{aligned} w''(x) &= -\frac{q_o}{2EI}\left(Lx - x^2\right) \\ w'(x) &= -\frac{q_o}{2EI}\left(\tfrac{1}{2}Lx^2 - \tfrac{1}{3}x^3 + C_1\right) \\ w(x) &= -\frac{q_o}{2EI}\left(\tfrac{1}{6}Lx^3 - \tfrac{1}{12}x^4 + C_1 x + C_2\right) \end{aligned}$$

The boundary conditions for the simple beam are

$$M(0) = 0, \qquad w(0) = 0, \qquad M(L) = 0, \qquad w(L) = 0$$

The boundary conditions on moment are already satisfied by the $M(x)$ function. Thus, we have two conditions on displacement $w(x)$ available to determine the constants of integration C_1 and C_2. Plugging the values $x = 0$ and $x = L$ into the expression for $w(x)$, we get

$$\begin{aligned} w(0) &= -\frac{q_o}{2EI}\left(\tfrac{1}{6}L(0)^3 - \tfrac{1}{12}(0)^4 + C_1(0) + C_2\right) = 0 \\ w(L) &= -\frac{q_o}{2EI}\left(\tfrac{1}{6}L(L)^3 - \tfrac{1}{12}(L)^4 + C_1(L) + C_2\right) = 0 \end{aligned}$$

The first equation gives $C_2 = 0$ and the second equation gives $C_1 = -\frac{1}{12}L^3$. Substituting these values back into the equation for $w(x)$ gives the final expression

$$w(x) = -\frac{q_o}{2EI}\left(\tfrac{1}{6}Lx^3 - \tfrac{1}{12}x^4 - \tfrac{1}{12}L^3 x\right)$$

While this solution is complete, it is useful to simplify it with a change of variable $\zeta = x/L$ to the form

$$w(\zeta) = \frac{q_o L^4}{24EI}\left(\zeta^4 - 2\zeta^3 + \zeta\right)$$

The *shape* of the deflection curve is the term in parentheses, which involves only dimensionless quantities. The magnitude of deflection is the leading coefficient. In this form it is easy to see what influences the magnitude of deflection (i.e., it is proportional to q_o and L^4 but inversely proportional to EI).

Integrate the fourth-order differential equation. Let us solve the problem again, starting from Eq. 8.6. This approach does not assume that we know $M(x)$. Integrating the differential equation gives

$$
\begin{aligned}
EIw''''(x) &= q_o \\
EIw'''(x) &= q_o\big(x + C_1\big) \\
EIw''(x) &= q_o\big(\tfrac{1}{2}x^2 + C_1x + C_2\big) \\
EIw'(x) &= q_o\big(\tfrac{1}{6}x^3 + \tfrac{1}{2}C_1x^2 + C_2x + C_3\big) \\
EIw(x) &= q_o\big(\tfrac{1}{24}x^4 + \tfrac{1}{6}C_1x^3 + \tfrac{1}{2}C_2x^2 + C_3x + C_4\big)
\end{aligned}
$$

Since $M(x) = EIw''(x)$, the moment boundary conditions can be translated to conditions on w''. Specifically, $EIw''(0) = 0$ and $EIw''(L) = 0$. The displacement boundary conditions are the same as before. Implementing the boundary conditions gives

$$
\begin{aligned}
EIw''(0) &= q_o\big(\tfrac{1}{2}(0)^2 + C_1(0) + C_2\big) = 0 \\
EIw(0) &= q_o\big(\tfrac{1}{24}(0)^4 + \tfrac{1}{6}C_1(0)^3 + \tfrac{1}{2}C_2(0)^2 + C_3(0) + C_4\big) = 0 \\
EIw''(L) &= q_o\big(\tfrac{1}{2}(L)^2 + C_1(L) + C_2\big) = 0 \\
EIw(L) &= q_o\big(\tfrac{1}{24}(L)^4 + \tfrac{1}{6}C_1(L)^3 + \tfrac{1}{2}C_2(L)^2 + C_3(L) + C_4\big) = 0
\end{aligned}
$$

The first two equations give $C_2 = 0$ and $C_4 = 0$. The third equation gives $C_1 = -\frac{1}{2}L$. Substituting these values into the fourth equation and dividing out q_o gives

$$\tfrac{1}{24}(L)^4 + \tfrac{1}{6}\big(-\tfrac{1}{2}L\big)(L)^3 + C_3(L) = 0$$

which can be solved to get $C_3 = \frac{1}{24}L^3$. Plugging the constants back into the equation for w gives

$$w(x) = \frac{q_o}{EI}\big(\tfrac{1}{24}x^4 + \tfrac{1}{6}\big(-\tfrac{1}{2}L\big)x^3 + \tfrac{1}{24}L^3x\big)$$

which simplifies to

$$w(x) = \frac{q_o}{EI}\big(\tfrac{1}{24}x^4 - \tfrac{1}{12}Lx^3 + \tfrac{1}{24}L^3x\big)$$

The solution is, of course, the same as the one we got with the other method. Note that we can compute the rotation, shear, and moment from these equations by differentiation as

$$
\begin{aligned}
\theta(x) &= -w'(x) = -\frac{q_o}{EI}\big(\tfrac{1}{6}x^3 - \tfrac{1}{4}Lx^2 + \tfrac{1}{24}L^3\big) \\
M(x) &= -EIw''(x) = -q_o\big(\tfrac{1}{2}x^2 - \tfrac{1}{2}Lx\big) \\
V(x) &= -EIw'''(x) = -q_o\big(x - \tfrac{1}{2}L\big)
\end{aligned}
$$

The shear, moment, and rotation get computed along the way for this approach. In fact, the sequential integration of the fourth-order differential equation is the same as the successive integration of the four first-order equations.

Example 8.5
Consider the propped cantilever beam of length L and bending modulus EI with a hinge at midspan, subjected to a uniformly distributed load of magnitude q_o, shown in Fig. 8.7. Compute the deflection $w(x)$ and find the rotation of the hinge.

Solution. To determine the moment $M(x)$, we only need one free body diagram with a cut at x, as shown in Fig. 8.7. The hinge does *not* cause a discontinuity in the moment or shear. Summing moments at the cut gives

$$
\begin{aligned}
-x\mathbf{e}_1 \times -A\mathbf{e}_3 + \int_0^x -(x-\xi)\mathbf{e}_1 \times q_o\mathbf{e}_3\, d\xi + M(x)\mathbf{e}_2 &= \mathbf{0} \\
-Ax\,\mathbf{e}_2 + q_o\Big[x\xi - \tfrac{1}{2}\xi^2\Big]_0^x \mathbf{e}_2 + M(x)\mathbf{e}_2 &= \mathbf{0} \\
-Ax + \tfrac{1}{2}q_o x^2 + M(x) &= 0
\end{aligned}
$$

The hinge condition is $M(\tfrac{1}{2}L) = 0$. Hence,

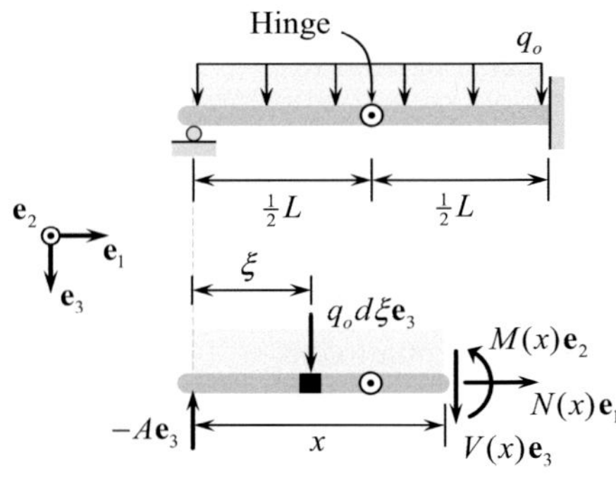

Fig. 8.7 *Example.* Propped cantilever beam of length L with an internal hinge, subjected to a uniformly distributed load of magnitude q_o. The free body diagram of the beam with a cut at x is shown

$$-A\left(\tfrac{1}{2}L\right) + \tfrac{1}{2}q_o\left(\tfrac{1}{2}L\right)^2 + 0 = 0 \qquad \rightarrow \qquad A = \tfrac{1}{4}q_oL$$

Substituting this result back into the moment equation and solving for $M(x)$ gives

$$M(x) = \tfrac{1}{4}q_o\left(Lx - 2x^2\right)$$

The hinge creates a *kinematic* discontinuity. Therefore, we will break the beam into two segments and define the deflection piecewise as

$$w(x) = \begin{cases} w_1(x) & 0 \le x \le \tfrac{1}{2}L \\ w_2(x) & \tfrac{1}{2}L \le x \le L \end{cases}$$

The equation $w'' = -M/EI$ must be integrated for both segments. Note that the expression for the moment is the same for both. Hence,

$$\begin{aligned} w_1''(x) &= -\frac{q_o}{4EI}\left(Lx - 2x^2\right) \\ w_1'(x) &= -\frac{q_o}{4EI}\left(\tfrac{1}{2}Lx^2 - \tfrac{2}{3}x^3 + C_1\right) \\ w_1(x) &= -\frac{q_o}{4EI}\left(\tfrac{1}{6}Lx^3 - \tfrac{1}{6}x^4 + C_1x + C_2\right) \end{aligned}$$

and

$$\begin{aligned} w_2''(x) &= -\frac{q_o}{4EI}\left(Lx - 2x^2\right) \\ w_2'(x) &= -\frac{q_o}{4EI}\left(\tfrac{1}{2}Lx^2 - \tfrac{2}{3}x^3 + C_3\right) \\ w_2(x) &= -\frac{q_o}{4EI}\left(\tfrac{1}{6}Lx^3 - \tfrac{1}{6}x^4 + C_3x + C_4\right) \end{aligned}$$

The difference between the two displacement expressions lies in the constants. The boundary and continuity conditions for the beam are

$$w_1(0) = 0, \qquad w_2(L) = 0, \qquad w_2'(L) = 0, \qquad w_1\left(\tfrac{1}{2}L\right) = w_2\left(\tfrac{1}{2}L\right)$$

Implementing the boundary conditions gives

$$\begin{aligned} w_1(0) &= -\frac{q_o}{4EI}\left(\tfrac{1}{6}L(0)^3 - \tfrac{1}{6}(0)^4 + C_1(0) + C_2\right) = 0 \\ w_2(L) &= -\frac{q_o}{4EI}\left(\tfrac{1}{6}L(L)^3 - \tfrac{1}{6}(L)^4 + C_3(L) + C_4\right) = 0 \\ w_2'(L) &= -\frac{q_o}{4EI}\left(\tfrac{1}{2}L(L)^2 - \tfrac{2}{3}(L)^3 + C_3\right) = 0 \end{aligned}$$

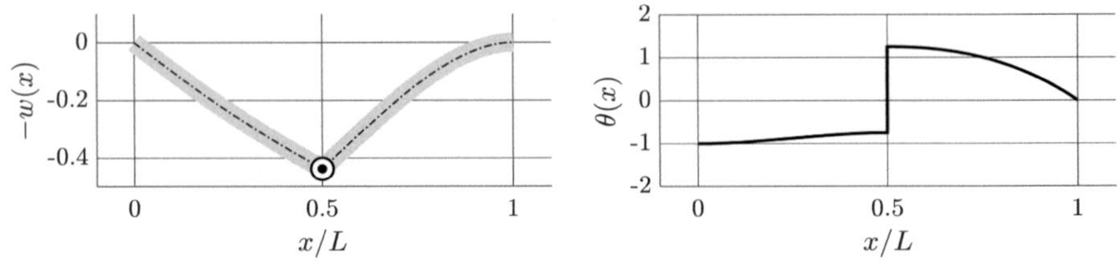

Fig. 8.8 *Example results*. Rotation and deflection results for the propped cantilever beam with an internal hinge of Fig. 8.7. The rotation is scaled by $q_o L^3/24EI$ and the deflection is scaled by $q_o L^4/24EI$

The first equation gives $C_2 = 0$. The third equation gives $C_3 = \frac{1}{6}L^3$, which can be substituted into the second equation to give $C_4 = -\frac{1}{6}L^4$. Using these values, and noting that the first two terms in the expressions for $w_1(x)$ and $w_2(x)$ and dividing out the leading coefficient, the continuity condition takes the form

$$\begin{aligned} C_1(\tfrac{1}{2}L) + C_2 &= C_3(\tfrac{1}{2}L) + C_4 \\ C_1(\tfrac{1}{2}L) &= (\tfrac{1}{6}L^3)(\tfrac{1}{2}L) + (-\tfrac{1}{6}L^4) \\ C_1 &= -\tfrac{1}{6}L^3 \end{aligned}$$

Substituting the constants, the deflection is

$$w(x) = -\frac{q_o}{24EI}\begin{cases} Lx^3 - x^4 - L^3x & 0 \le x \le \frac{1}{2}L \\ Lx^3 - x^4 + L^3x - L^4 & \frac{1}{2}L \le x \le L \end{cases}$$

and the rotation is $\theta = -w'$. Hence,

$$\theta(x) = \frac{q_o}{24EI}\begin{cases} 3Lx^2 - 4x^3 - L^3 & 0 \le x \le \frac{1}{2}L \\ 3Lx^2 - 4x^3 + L^3 & \frac{1}{2}L \le x \le L \end{cases}$$

The relative angle of rotation of the hinge is

$$\theta_2(\tfrac{1}{2}L) - \theta_1(\tfrac{1}{2}L) = \frac{q_o L^3}{12EI}$$

The rotation and deflection functions are plotted in Fig. 8.8. The discontinuity in the rotation at the hinge is evident and manifests in the deflection plot as a kink at the hinge.

8.4.4 Statically Indeterminate Beams

A *statically indeterminate* beam is one for which we cannot completely determine the reaction forces and internal stress resultants using only the equations of equilibrium. For planar beam problems, we have three equations of equilibrium (two component equations from balance of forces and one from balance of moments). Thus, if we take a free body diagram of the entire system, then we can use these three equations to find three unknown reaction forces. A problem that has more than three reactions is, therefore, statically indeterminate. The number of reaction forces in excess of three is called the *degree of static indeterminacy*. In essence, it is the number of additional equations that would be needed to solve for all the reactions.

Looking back at the problems we have solved using the finite free body diagram approach, it is evident that when we take a cut at x, we introduce $N(x)$, $V(x)$, and $M(x)$. In other words, the free body diagrams with cuts introduce three new unknowns. However, each free body diagram gives three new equations of equilibrium. Therefore, there is no way to take a free body diagram with a cut that will enable finding additional unknown reaction forces (with the exception of cutting exactly at a hinge location, but that would then use up the zero-moment condition at the hinge). We can, therefore, determine the degree of static indeterminacy from the free body diagram of the entire system.

The process of solution for problems with static indeterminacy is exactly the same as problems that are statically determinate, except that we must carry forward the excess unknown reaction forces in the solution of the differential equation $w'' = -M/EI$. While we do not know these forces, they are not functions of x, so they do not cause any difficulty in the integration of the differential equation.

Reaction forces are created by support conditions. Having additional reaction forces implies that there are additional supports. Each one of these represents an additional boundary condition on either w or θ. The number of constants that appear when we integrate the differential equation will not change (i.e., a second-order differential equation will have two constants, and a fourth-order differential equation will have four). Therefore, the additional boundary conditions will enable the solution of the remaining reaction forces, as the following example illustrates.

Example 8.6
Consider a propped cantilever beam of length L and bending modulus EI, fixed at the right end and propped on a roller at the left end, shown in Fig. 8.9. The beam is subjected to a uniformly distributed load of magnitude q_o. Find the shear, moment, rotation, and deflection. Find the maximum deflection.

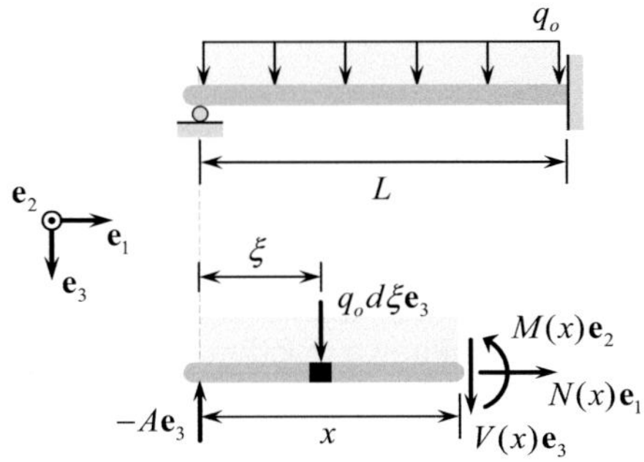

Fig. 8.9 *Example.* Propped cantilever beam of length L and bending modulus EI, subjected to a uniformly distributed load of magnitude q_o. The free body diagram of the beam with a cut at x is shown

Solution. This problem is statically indeterminate to the first degree. There are three reactions at the fixed end (i.e., the moment and two components of force) and one reaction A at the roller. Four unknown reactions minus three equations of equilibrium gives one degree of static indeterminacy. We know that we will need to carry one of these reactions through to the differential equation. We can select any of them (except for the horizontal component of the reaction at the fixed end, as that one can be completely determined from statics). In this case, we will take the reaction A at the roller. That influences our selection of the free body diagram to be the one shown. It has A in it but not the other reactions. With this selection, we can proceed with the solution without first finding the other reaction forces, which is the reason for not having a free body diagram of the overall system.

We need the moment $M(x)$ to advance the differential equation. Establishing moment equilibrium, taking moments at the cut, gives

$$
\begin{aligned}
-x\mathbf{e}_1 \times -A\mathbf{e}_3 + \int_0^x -(x-\xi)\mathbf{e}_1 \times q_o\mathbf{e}_3\, d\xi + M(x)\mathbf{e}_2 &= \mathbf{0} \\
-Ax\,\mathbf{e}_2 + q_o\Big[x\xi - \tfrac{1}{2}\xi^2\Big]_0^x \mathbf{e}_2 + M(x)\mathbf{e}_2 &= \mathbf{0} \\
-Ax + \tfrac{1}{2}q_o x^2 + M(x) &= 0
\end{aligned}
$$

Therefore, the moment is

$$
M(x) = Ax - \tfrac{1}{2}q_o x^2
$$

The shear is simply the derivative of the moment, i.e., $V = M'$. Thus,

$$
V(x) = A - q_o x
$$

We can solve the kinematic equation $w'' = -M/EI$ as follows:

$$
\begin{aligned}
w''(x) &= -\frac{1}{EI}\big(Ax - \tfrac{1}{2}q_o x^2\big)\\
w'(x) &= -\frac{1}{EI}\big(\tfrac{1}{2}Ax^2 - \tfrac{1}{6}q_o x^3 + C_1\big)\\
w(x) &= -\frac{1}{EI}\big(\tfrac{1}{6}Ax^3 - \tfrac{1}{24}q_o x^4 + C_1 x + C_2\big)
\end{aligned}
$$

We have the following boundary conditions

$$
w(0) = 0, \qquad M(0) = 0, \qquad w(L) = 0, \qquad w'(L) = 0
$$

The moment condition is already satisfied by $M(x)$. That leaves three conditions on the deflection or rotation. Implementing these conditions gives

$$
\begin{aligned}
w(0) &= -\frac{1}{EI}\big(\tfrac{1}{6}A(0)^3 - \tfrac{1}{24}q_o(0)^4 + C_1(0) + C_2\big) = 0\\
w'(L) &= -\frac{1}{EI}\big(\tfrac{1}{2}A(L)^2 - \tfrac{1}{6}q_o(L)^3 + C_1\big) = 0\\
w(L) &= -\frac{1}{EI}\big(\tfrac{1}{6}A(L)^3 - \tfrac{1}{24}q_o(L)^4 + C_1(L) + C_2\big) = 0
\end{aligned}
$$

The first equation gives $C_2 = 0$. The second equation gives

$$
C_1 = \tfrac{1}{6}q_o L^3 - \tfrac{1}{2}AL^2
$$

Substituting these into the third equation gives

$$
\tfrac{1}{6}AL^3 - \tfrac{1}{24}q_o L^4 + \big(\tfrac{1}{6}q_o L^3 - \tfrac{1}{2}AL^2\big)L = 0
$$

which can be solved to give

$$
A = \tfrac{3}{8}q_o L \qquad \rightarrow \qquad C_1 = \tfrac{1}{6}q_o L^3 - \tfrac{1}{2}\big(\tfrac{3}{8}q_o L\big)L^2 = -\tfrac{1}{48}q_o L^3
$$

Substituting these values, we can get the final expressions for the deflection. To wit,

$$
\begin{aligned}
w(x) &= -\frac{1}{EI}\big(\tfrac{1}{6}\big(\tfrac{3}{8}q_o L\big)x^3 - \tfrac{1}{24}q_o x^4 + \big(-\tfrac{1}{48}q_o L^3\big)x\big)\\
&= -\frac{q_o}{48EI}\big(3Lx^3 - 2x^4 - L^3 x\big)
\end{aligned}
$$

Again, it is instructive to implement a change of variable $\zeta = x/L$ to put the deflection in the form

$$
w(\zeta) = \frac{q_o L^4}{48EI}\big(2\zeta^4 - 3\zeta^3 + \zeta\big)
$$

The rotation can be found by taking the derivative of this expression. Note that the derivative is with respect to x. By the chain rule

$$\frac{d(\bullet)}{dx} = \frac{d(\bullet)}{d\zeta}\frac{d\zeta}{dx} = \frac{d(\bullet)}{d\zeta}\frac{1}{L}$$

Hence,

$$\theta(\zeta) = -\frac{q_o L^3}{48EI}\left(8\zeta^3 - 9\zeta^2 + 1\right)$$

Similarly, we can compute the moment and shear from $M = EI\theta'$ and $V = M'$. To wit,

$$M(\zeta) = -\frac{q_o L^2}{48}\left(24\zeta^2 - 18\zeta\right), \qquad V(\zeta) = -\frac{q_o L}{48}\left(48\zeta - 18\right)$$

Of course, the moment and shear can also be found by simply substituting A into the original equations derived from the free body diagram.

The maximum deflection occurs at the location where the rotation is zero. Setting $\theta = 0$ yields the equation

$$8\zeta_o^3 - 9\zeta_o^2 + 1 = 0$$

where ζ_o represents the location of the maximum deflection. Solving the cubic equation (e.g., by Newton's method) gives $\zeta_o = 0.412535$. The maximum deflection at that point is

$$w(\zeta_o) = \frac{q_o L^4}{48EI}\left(2\zeta_o^4 - 3\zeta_o^3 + \zeta_o\right) = 0.005416\frac{q_o L^4}{EI} = w_{\max}$$

The reaction forces at the fixed end can be determined from an infinitesimal free body diagram of the beam in the neighborhood of $x = L$, shown in Fig. 8.10. Equilibrium then requires that

$$M_B = M(L) = -\tfrac{1}{8}q_o L^2, \qquad \mathbf{B} = V(L)\mathbf{e}_3 = -\tfrac{5}{8}q_o L\mathbf{e}_3$$

since $N(x) = 0$ throughout the beam. The shear, moment, rotation, and transverse deflection are plotted in Fig. 8.11.

This example illustrates the fact that the beam problem can be solved for a statically indeterminate problem in the same way that we solve a statically determinate problem. The only difference was that we used the boundary conditions to find not only the constants of integration, but also the value of the additional reaction force.

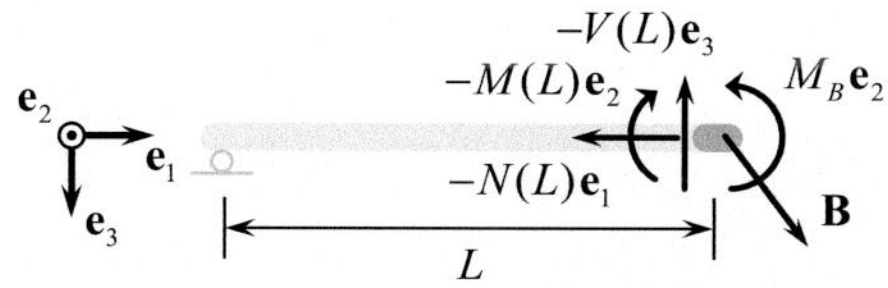

Fig. 8.10 *Example results.* Computation of the reaction force and moment at the fixed end from the internal stress resultants at $x = L$

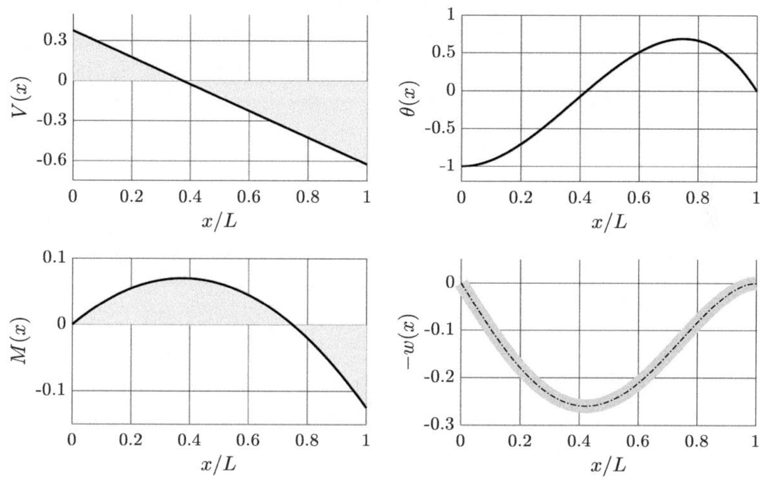

Fig. 8.11 *Example results.* Rotation and deflection results for the propped cantilever beam of Fig. 8.9. The shear is scaled by $q_o L$, the moment is scaled by $q_o L^2$, the rotation is scaled by $q_o L^3/48EI$, and the deflection is scaled by $q_o L^4/48EI$

8.4.5 Nonprismatic Beams

A *nonprismatic beam* has a cross-section geometry that varies with x. For computing the rotation and deflection of the beam, the most important feature of that variation is that the bending modulus $EI(x)$ depends upon x. The variable modulus does not affect the equilibrium calculations, but it does affect the kinematic computations, as the following example illustrates.

Example 8.7
Consider a cantilever beam of length L, free at the left end and fixed at the right end. The beam is subjected to a downward tip load of magnitude P at the free end, as shown in Fig. 8.12. The beam cross section is rectangular with constant depth and linearly varying width such that the variable bending modulus can be described as

$$EI(x) = EI_o\left(1 + \frac{x}{L}\right)$$

where EI_o is the bending stiffness at the left end and $2EI_o$ is the bending stiffness at the right end. Find the shear, moment, rotation, and deflection of the beam as a function of x.

Solution. We can determine the moment $M(x)$ by establishing moment equilibrium using the free body diagram with a cut at x. Summing moments at the cut gives

$$-x\mathbf{e}_1 \times P\mathbf{e}_3 + M(x)\mathbf{e}_2 = \mathbf{0}$$
$$Px\,\mathbf{e}_2 + M(x)\mathbf{e}_2 = \mathbf{0}$$

Therefore, the moment is

$$M(x) = -Px$$

We can write the kinematic equation $w'' = -M/EI$ as follows:

$$w''(x) = \frac{Px}{EI(x)} = \frac{PL}{EI_o}\left(\frac{x/L}{1 + x/L}\right) = \frac{PL}{EI_o}\left(\frac{\zeta}{1 + \zeta}\right)$$

where we have substituted the explicit expression for $EI(x)$ and used a change of variable $\zeta = x/L$. The challenge in integrating this differential equation is that the right side is a rational fraction, not a simple polynomial as previous examples have been. Note the following antiderivative

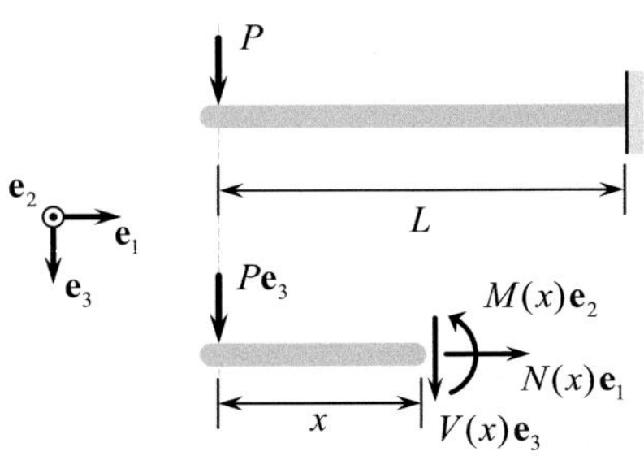

Fig. 8.12 *Example.* Cantilever beam of length L and variable bending modulus $EI(x)$, subjected to a tip load of magnitude P at the free end. The free body diagram of the beam with a cut at x is shown

$$\int \frac{\zeta}{1+\zeta}\, d\zeta = \zeta - \ln(1+\zeta) + C$$

where $\ln(\bullet)$ is the natural log of $(\bullet)$. Using this fact, we can integrate the kinematic equation once to get

$$w'(\zeta) = \frac{PL^2}{EI_o}\Big(\zeta - \ln(1+\zeta) + C_1\Big)$$

Notice that an additional L appeared in the leading term because the integration is with respect to x. Now, noting the antiderivative

$$\int \ln(1+\zeta) = (1+\zeta)\ln(1+\zeta) - \zeta + C$$

we can integrate one more time to get

$$w(\zeta) = \frac{PL^3}{EI_o}\Big(\tfrac{1}{2}\zeta^2 - \big[(1+\zeta)\ln(1+\zeta) - \zeta\big] + C_1\zeta + C_2\Big)$$

We have the following boundary conditions

$$V(0) = -P, \qquad M(0) = 0, \qquad w(L) = 0, \qquad w'(L) = 0$$

The shear and moment boundary conditions are already satisfied because we used a free body diagram to get the moment. Implementing the fourth boundary condition first gives

$$w'(1) = \frac{PL^2}{EI_o}\Big(1 - \ln 2 + C_1\Big) = 0$$

from which we can solve for C_1 as

$$C_1 = \ln 2 - 1$$

Implementing the boundary condition on deflection at the fixed end gives

$$w(1) = \frac{PL^3}{EI_o}\Big(\tfrac{1}{2} - \big[2\ln 2 - 1\big] + C_1(1) + C_2\Big) = 0$$

Substituting C_1 and solving for C_2 gives

$$C_2 = -\tfrac{3}{2} + 2\ln 2 - \ln 2 + 1 = \ln 2 - \tfrac{1}{2}$$

Substituting the constants, the expressions for the displacement and rotation come out as

$$w(\zeta) = \frac{PL^3}{EI_o}\Big(\tfrac{1}{2}(\zeta^2 - 1) - 2(1+\zeta)\big[\ln(1+\zeta) - \ln 2\big]\Big)$$

$$\theta(\zeta) = -\frac{PL^2}{EI_o}\Big(\zeta - \ln(1+\zeta) + \ln 2 - 1\Big)$$

It is simple to verify that the deflection and rotation are zero at $\zeta = 1$ from these expressions. The deflection and rotation at the tip can be found by substituting $\zeta = 0$ into the above relationships, giving

$$w(0) = \big(\ln 2 - \tfrac{1}{2}\big)\frac{PL^3}{EI_o} = 0.1931\frac{PL^3}{EI_o}$$

$$\theta(0) = \big(1 - \ln 2\big)\frac{PL^2}{EI_o} = 0.3069\frac{PL^2}{EI_o}$$

For comparison, a prismatic beam with bending stiffness EI_o has a tip deflection of $0.333PL^3/EI_o$ and a prismatic beam with bending stiffness $2EI_o$ has a tip deflection of $0.167PL^3/EI_o$. Thus, the nonprismatic beam of this example fits between those bounds.

This example highlights the challenges in integrating the differential equation when the bending modulus is variable, the linear variation probably being the simplest continuously nonprismatic case. Another form of nonprismatic beam is a stepped beam where EI is constant over each segment, but different segments have different EI values. The way to approach such a problem is to define the state variable piecewise and treat each piece as a prismatic beam. Then use continuity equations to relate the state variables of the two pieces at the place where they join.

8.4.6 Structures

Frame structures are nothing more than an assembly of beams working together to resist applied loads. The theory of beams applies to each element, and the main challenge is in specifying what it means mathematically for the elements to work together. There are many ways to accomplish this goal, and that is the primary focus of methods of structural analysis. One way is to establish continuity conditions at the location where the elements join, as we have done for cases where we needed to use piecewise functions. We will take a deeper look at this issue in Chap. 10 but illustrate the concepts in the following example.

Example 8.8

Consider the right-angle bent structure shown in Fig. 8.13a with two beams of length L and bending modulus EI and axial modulus EA, fixed at $\mathcal{A}$, on a roller at $\mathcal{B}$, and free at $\mathcal{C}$. The horizontal beam is subjected to a load of magnitude P at the free end. Find the stress resultants, rotations, and deflections. Find the displacement at point $\mathcal{C}$. Assume that the relationship between the axial stiffness and bending stiffness for this system is $EA = 100EI/L^2$.

Solution. The vertical element is designated as beam 1 and the horizontal element as beam 2. The coordinate axes for each beam are shown in the figure, with z measuring distance along beam 1 and x measuring distance along beam 2. The system is statically indeterminate to the first degree. We will use the reaction at $\mathcal{B}$ as the unknown force carried forward through the kinematic calculations.

Equilibrium. To find the internal stress resultants in the two beams, we will use the free body diagrams shown in Fig. 8.13e, f. For beam 1, taking moments at the cut gives

$$(L - z)\mathbf{e}_3 \times B\mathbf{e}_1 + \big((L - z)\mathbf{e}_3 + L\mathbf{e}_1\big) \times P\mathbf{e}_3 - M_1(z)\mathbf{e}_2 = \mathbf{0}$$

Carry out the cross products, dot with $\mathbf{e}_2$, and solve for $M_1(z)$ to get

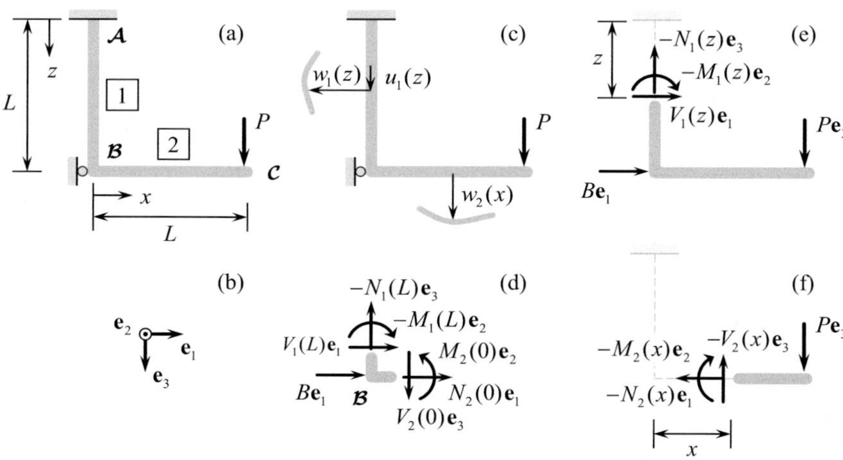

Fig. 8.13 *Example.* Structure with two beams of length L, bending modulus EI, and axial modulus EA, fixed at $\mathcal{A}$, on a roller at $\mathcal{B}$, and free at $\mathcal{C}$. Beam 2 is subjected to a load of magnitude P at the free end. **a** Definition of structure, **b** base vectors, **c** positive deflection, **d** free body diagram of joint at $\mathcal{B}$, **e** free body diagram with cut in beam 1 at z, and **f** free body diagram with cut in beam 2 at x

$$M_1(z) = B(L - z) - PL$$

Summing forces, we get

$$B\mathbf{e}_1 + P\mathbf{e}_3 - N_1(z)\mathbf{e}_3 + V_1(z)\mathbf{e}_1 = \mathbf{0}$$

Dotting this equation with $\mathbf{e}_1$ and $\mathbf{e}_3$, respectively, gives

$$V_1(z) = -B, \qquad N_1(z) = P$$

Follow the same steps for beam 2. Summing moments at the cut gives

$$(L - x)\mathbf{e}_1 \times P\mathbf{e}_3 - M_2(x)\mathbf{e}_2 = \mathbf{0}$$

Solving for $M_2(x)$ gives

$$M_2(x) = -P(L - x)$$

Summing forces results in

$$V_2(x) = P, \qquad N_2(x) = 0$$

All the internal stress resultants are known in terms of the reaction B or are known explicitly.

Boundary and continuity conditions. At point $\mathcal{A}$, all the motion quantities are known. Specifically,

$$u_1(0) = 0, \qquad w_1(0) = 0, \qquad \theta_1(0) = 0$$

At point $\mathcal{C}$, all the stress resultants are known. Specifically,

$$N_2(L) = 0, \qquad V_2(L) = P, \qquad M_2(L) = 0$$

Note that all the force conditions are automatically satisfied because we derived the stress resultants from finite free body diagrams. The kinematic conditions at point $\mathcal{B}$ are

$$w_1(L) = 0, \qquad w_2(0) = u_1(L), \qquad \theta_1(L) = \theta_2(0)$$

The kinetic continuity conditions are

$$B + V_1(L) + N_2(0) = 0, \qquad N_1(L) = V_2(0), \qquad M_1(L) = M_2(0)$$

all of which are automatically satisfied because the internal forces are based upon finite free body diagrams. These conditions can, of course, be checked using the derived expressions for the internal stress resultants.

Integrate to get displacements. There are three tasks to carry out with respect to integrating the strain-displacement relationships. One is the axial bar problem in beam 1 (since there is a nonzero axial force). The other two are the transverse deflection calculations for both beams. Let us start with the axial bar problem. Integrate $u_1' = N_1/EA$ to get

$$u_1'(z) = \frac{P}{EA} \qquad \rightarrow \qquad u_1(z) = \frac{P}{EA}\big(z + C_1\big)$$

Using the boundary condition $u_1(0) = 0$ gives $C_1 = 0$. We need the value at $z = L$ for one of the boundary conditions on beam 2. Thus, we can note for future reference that $u_1(L) = PL/EA$.

Integrating $w_1'' = -M_1/EI$ in beam 1 gives

$$\begin{aligned}
w_1''(z) &= -\frac{1}{EI}\Big(B\big(L - z\big) - PL\Big) \\
w_1'(z) &= -\frac{1}{EI}\Big(B\big(Lz - \tfrac{1}{2}z^2\big) - PLz + C_2\Big) \\
w_1(z) &= -\frac{1}{EI}\Big(B\big(\tfrac{1}{2}Lz^2 - \tfrac{1}{6}z^3\big) - \tfrac{1}{2}PLz^2 + C_2 z + C_3\Big)
\end{aligned}$$

Similarly, for beam 2, we get

$$\begin{aligned}
w_2''(x) &= \frac{1}{EI}\Big(P\big(L - x\big)\Big) \\
w_2'(x) &= \frac{1}{EI}\Big(P\big(Lx - \tfrac{1}{2}x^2\big) + C_4\Big) \\
w_2(x) &= \frac{1}{EI}\Big(P\big(\tfrac{1}{2}Lx^2 - \tfrac{1}{6}x^3\big) + C_4 x + C_5\Big)
\end{aligned}$$

The constants in these expressions can be determined by the boundary and continuity conditions.

Implement kinematic boundary and continuity conditions. There are five constants of integration and one unknown reaction force. Therefore, we must have six kinematic boundary or continuity conditions. We have already used one of those to find C_1 for the axial bar problem in beam 1. At point $\mathcal{A}$ we have

$w_1(0) = 0$ and $w_1'(0) = 0$. Therefore,

$$w_1'(0) = -\frac{1}{EI}\Big(B\big(L(0) - \tfrac{1}{2}(0)^2\big) - PL(0) + C_2\Big) = 0$$
$$w_1(0) = -\frac{1}{EI}\Big(B\big(\tfrac{1}{2}L(0)^2 - \tfrac{1}{6}(0)^3\big) - \tfrac{1}{2}PL(0)^2 + C_2(0) + C_3\Big) = 0$$

From these equations, we get

$$C_2 = 0, \qquad C_3 = 0$$

Implementing $w_1(L) = 0$, noting the values of the constants already determined, gives

$$w_1(L) = -\frac{1}{EI}\Big(B\big(\tfrac{1}{2}L(L)^2 - \tfrac{1}{6}(L)^3\big) - \tfrac{1}{2}PL(L)^2\Big)$$
$$0 = -\frac{1}{EI}\Big(\tfrac{1}{3}BL^3 - \tfrac{1}{2}PL^3\Big)$$

This equation allows us to find the reaction B as

$$B = \tfrac{3}{2}P$$

The continuity condition $w_2(0) = u_1(L)$ gives

$$\frac{1}{EI}\Big(P\big(\tfrac{1}{2}L(0)^2 - \tfrac{1}{6}(0)^3\big) + C_4(0) + C_5\Big) = \frac{PL}{EA}$$

from which we can solve for C_5 as

$$C_5 = PL^3\Big(\frac{EI}{EAL^2}\Big) = \tfrac{1}{100}PL^3$$

where we have noted the given relationship $EA = 100EI/L^2$. Finally, implementing $w_1'(L) = w_2'(0)$, noting constants already determined, gives

$$-\frac{1}{EI}\Big(\big(\tfrac{3}{2}P\big)\big(L(L) - \tfrac{1}{2}(L)^2\big) - PL(L)\Big) = \frac{1}{EI}\Big(P\big(L(0) - \tfrac{1}{2}(0)^2\big) + C_4\Big)$$

This equation allows us to solve for C_4 as

$$C_4 = \tfrac{1}{4}PL^2$$

We can now get the final expressions for all the state variables by substituting the value of the reaction B and the constants of integration. The internal stress resultants can be summarized as

$$N_1(z) = P \quad V_1(z) = -\tfrac{3}{2}P \quad M_1(z) = \tfrac{1}{2}P(L - 3z)$$

$$N_2(x) = 0 \quad V_2(x) = P \quad M_2(x) = -P(L - x)$$

The displacement fields take the final form

$$w_1(z) = \frac{P}{EI}\left(\tfrac{1}{4}z^3 - \tfrac{1}{4}Lz^2\right)$$

$$w_2(x) = \frac{P}{EI}\left(-\tfrac{1}{6}x^3 + \tfrac{1}{2}Lx^2 + \tfrac{1}{4}L^2x + \tfrac{1}{100}L^3\right)$$

The displacement at the point of load is

$$w_2(L) = \frac{P}{EI}\left(-\tfrac{1}{6}L^3 + \tfrac{1}{2}L^3 + \tfrac{1}{4}L^3 + \tfrac{1}{100}L^3\right) = \tfrac{89}{150}\frac{PL^3}{EI}$$

The axial, shear, and bending moment diagrams are plotted in Fig. 8.14. It is customary to draw the diagrams directly on the backbone of the structure itself, showing the sign convention with sketches of positive or negative axial force, shear force, and bending moment.

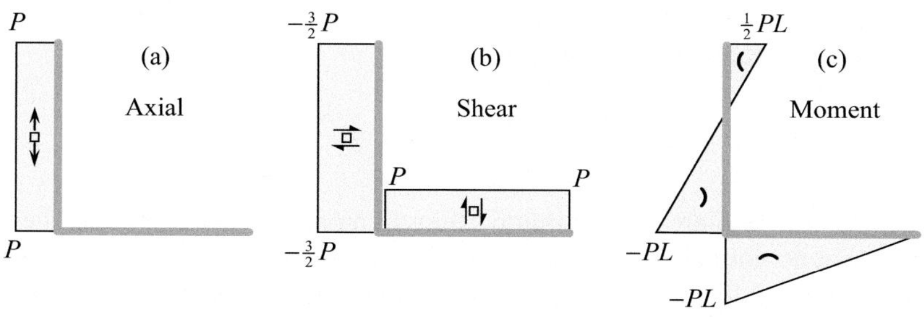

Fig. 8.14 *Example results.* Internal stress resultant diagrams for the structure shown in Fig. 8.13: **a** axial force, **b** shear force, and **c** bending moment

8.4.7 Normal and Shear Stresses

In the examples up to this point in the chapter, the displacement and rotation fields were computed in terms of the bending stiffness EI, where I is the moment of inertia of the cross section relative to the centroid. The calculations can be completed without getting into the details of the geometry of the cross section. To get specific results, of course, the moment of inertia must be evaluated. The same is true for the calculation of normal and shear stresses in the beam, but those depend more intimately on the geometry of the cross section.

The normal and shear stress formulas for beam bending (from Eqs. 7.24 and 7.27) are, respectively,

$$\sigma(x,z) = \frac{N(x)}{A} + \frac{M(x)z}{I}, \qquad \tau(x,z) = \frac{V(x)Q(z)}{I\,b(z)} \tag{8.9}$$

where $N(x)$, $V(x)$, and $M(x)$ are the internal axial force, shear force, and bending moment. The properties of the cross section include the area, A, the first moment of area, $Q(z)$, the second moment of area, I, and the width of the cross section, $b(z)$ at z, which is measured relative to the centroid of the cross section.

For most of the beam problems in this chapter, the axial force is zero, eliminating the first term in the normal stress formula. For the purposes of the design of beams, we are most interested in the maximum values of the stresses. The maximum normal stress occurs at the fiber with the smallest value of $S = I/z_{\max}$, where $z_{\max}$ is the distance to the point in the cross section that is farthest from the centroid. We call that point the *extreme fiber*. The property S is often referred to as the *section modulus* of the cross section. The maximum shear stress occurs at the value of z that maximizes $Q(z)/b(z)$. If b is constant in the neighborhood of the centroid, then the maximum value will be $Q(0)$, i.e., the value of the first moment at the centroid. In such cases, the maximum normal and shear stresses are

$$\sigma_{\max} = \frac{M_{\max} z_{\max}}{I}, \qquad \tau_{\max} = \frac{V_{\max} Q(0)}{I\,b(0)}$$

where $M_{\max}$ and $V_{\max}$ are the maximum values of moment and shear along the span. These values do not necessarily happen at the same x location. The normal stress, in absence of axial force, will be tensile at one extreme fiber and compressive at the other one. If the material tolerates compression differently than tension, then both the tensile and compressive normal stresses might be important.

The next chapter is completely devoted to the calculation of area, centroid, and the various moments of the area. Therefore, the ability to complete the calculation of normal and shear stresses in beams will be strengthened by mastering the concepts in that chapter. The biggest challenge in finding the stresses lies in calculating the properties of the cross section, once the moment and shear diagrams are available. We end this chapter with a simple example of computing the normal and shear stresses in a beam.

Example 8.9
Reconsider the propped cantilever beam of length L, fixed at the right end, supported by a roller on the left end, subjected to a uniform transverse load of magnitude q_o (see Fig. 8.9). Assume that the beam has a solid rectangular cross section with height h and width $b = 0.4h$. Find the maximum normal and shear stress in the beam.

Solution. From Example 8.6, which starts on Sect. 8.4, we know that the moment and shear in the beam are

$$M(\zeta) = -\frac{q_o L^2}{48}\left(24\zeta^2 - 18\zeta\right), \qquad V(\zeta) = -\frac{q_o L}{48}\left(48\zeta - 18\right)$$

where $\zeta = x/L$, with x measured from the left end. Figure 8.11 shows that the shear and moment both have their maximum values at $\zeta = 1$. The values are

$$V_{\max} = -\tfrac{5}{8} q_o L, \qquad M_{\max} = -\tfrac{1}{8} q_o L^2$$

For a rectangular cross section, the centroid is in the middle, the moment of inertia relative to the centroid is $I = bh^3/12$ and the first moment of area relative to the centroid and at the centroid is $Q(0) = bh^2/8$. The details of the calculation of these values can be found in Example 9.1 in Sect. 9.1. Using these values in the normal stress formula gives

$$\sigma_{\max} = \frac{M_{\max} z_{\max}}{I} = \frac{\left(q_o L^2/8\right)\left(h/2\right)}{bh^3/12} = \frac{12}{16}\frac{q_o L^2 h}{(0.4h)h^3} = 1.875\frac{q_o L^2}{h^3}$$

The shear stress formula gives

$$\tau_{\max} = \frac{V_{\max}\, Q(0)}{I\, b(0)} = \frac{\left(5q_o L/8\right)\left(bh^2/8\right)}{\left(bh^3/12\right)\left(b\right)} = \frac{15}{16}\frac{\left(q_o L\right)\left(0.4h\right)h^2}{\left(0.4h\right)h^3\left(0.4h\right)} = 2.344\frac{q_o L}{h^2}$$

Both stresses are proportional to the load intensity q_o. Thus, if you double the load, the stresses will double. The normal stress is proportional to L^2/h^3 and the shear stresses are proportional to L/h^2. Thus, the normal stress increases more rapidly with length than the shear stress does.

8.5 Summary

Linear beam theory is a mathematical model for which the governing differential equations can be solved by classical methods. In this chapter, we showed first how the linear equations come from the nonlinear equations of the previous chapter. Then, through a series of examples, we demonstrated how to solve beam problems, covering statically determinate and indeterminate beams, prismatic and nonprismatic beams, beams with internal hinges and point loads, and structures comprised of beams. In all these cases, the solution process is identical with the variety manifesting in the form of the applied loads and the support conditions (and for nonprismatic beams, the variation of the bending modulus).

The response of a beam to loads depends upon the material used, through the value of Young's modulus, the geometric features of the cross section, through the area and the first and second moments of area, the nature of the applied loads, and the boundary conditions. In the next chapter, we will focus on the cross-sectional properties. In the following chapter, we will organize the computations done by hand in this chapter into a computational algorithm and code.

8.6 Problems

8.1 The cantilever beam $\mathcal{AB}$ is carrying a uniform load over half its length. Derive the expressions for shear force and bending moment along the beam axis. Draw the shear and moment diagrams for the beam. Neglect the self-weight of the beam.

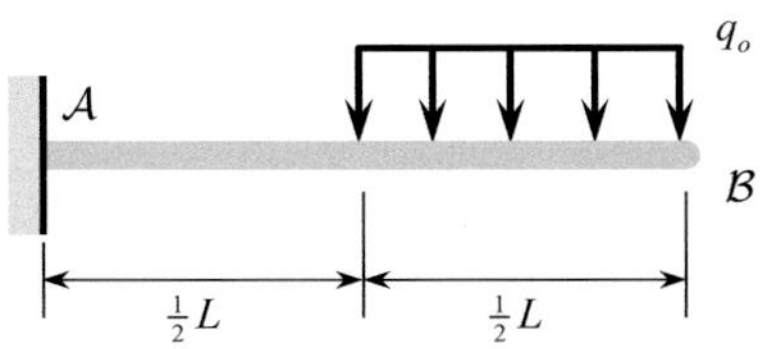

8.2 A beam of length L is subjected to a uniformly distributed load of magnitude q_o over its entire length. Determine the distance a to the simple supports so that the maximum internal moment in the beam is the smallest possible value. Find the maximum moment for this position of the supports. Plot the shear and moment diagrams. Neglect the self-weight of the beam.

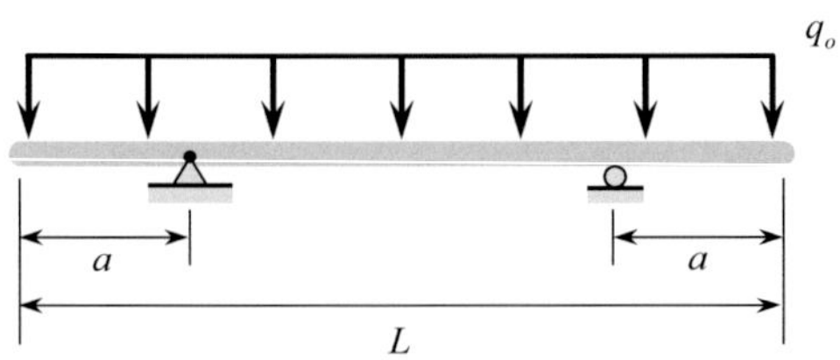

8.3 Beam $\mathcal{ABCD}$ has an overhang at one end, supports a uniform load of intensity q_o over one-third of its length, and has a concentrated moment of magnitude M_o at $\mathcal{D}$. The self-weight of the beam can be neglected. Derive the expressions for shear force and bending moment along the beam axis and draw the diagrams for the beam. Find the location and value for the maximum bending moment in the beam. Evaluate for $M_o = 3$ kN m, $q_o = 12$ kN/m, and $L = 4.8$ m.

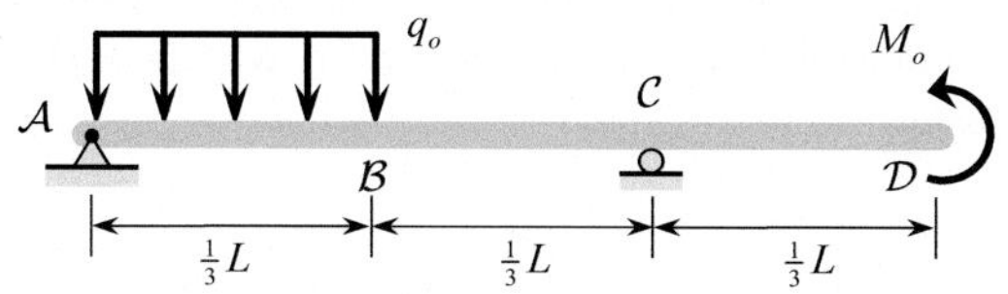

8.4 Beam $\mathcal{ABCD}$ is fixed at the right end and has a roller support at the left end. There is an internal hinge at $\mathcal{B}$. The distributed load varies linearly from zero at $\mathcal{A}$ to an intensity of q_o at $\mathcal{C}$ and is constant from $\mathcal{C}$ to $\mathcal{D}$. Derive the expressions for shear force and bending moment along the axis of the beam and draw the shear and moment diagrams for the beam. Neglect the self-weight of the beam. Evaluate for $q_o = 7.5$ kN/m and $L = 6$ m.

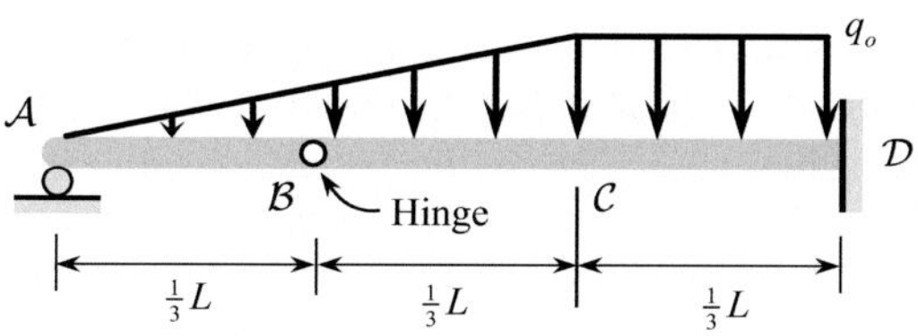

8.5 The continuous beam assembly $\mathcal{ABC}$ is made up of two straight segments connected at right angles and continuously welded together. The system is subjected to a downward distributed load of magnitude q_o over the segment $\mathcal{BC}$ and a linearly varying distributed load over segment $\mathcal{AB}$ with maximum magnitude of q_o at $\mathcal{B}$. The beam is pinned at $\mathcal{A}$ and has a roller support at $\mathcal{C}$. Find the reaction forces. Derive the expressions for the axial force, shear force, and bending moment in both segments of the beam. Plot the axial, shear, and bending moment diagrams. Determine the maximum moment and the maximum shear. Neglect the self-weight of the beams.

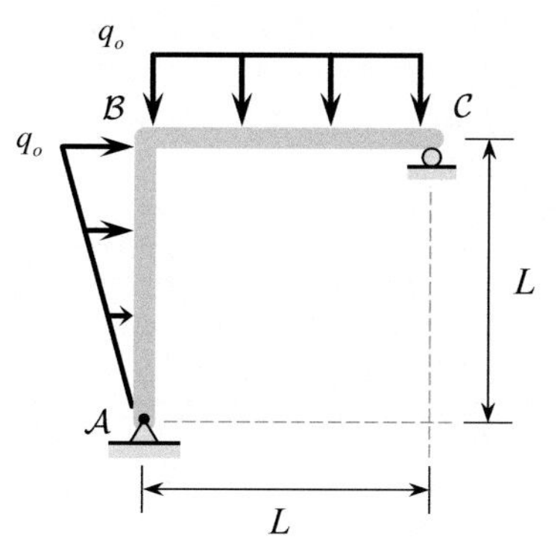

8.6 The continuous beam assembly $\mathcal{ABCD}$ is made up of three straight segments connected at right angles and continuously welded together. The system is subjected to a downward distributed load of magnitude q_o over the segment $\mathcal{CD}$ and a downward point load of magnitude P at $\mathcal{B}$. The beam is pinned at $\mathcal{A}$ and has a roller support at $\mathcal{D}$. Derive the expressions for the axial force, shear force, and bending moment in all three segments of the beam. Find the expressions for the reaction forces. Plot the axial, shear, and bending moment diagrams for the specific case where $P = q_o L$ and determine the maximum moment and the maximum shear for that case. Neglect the self-weight of the beam.

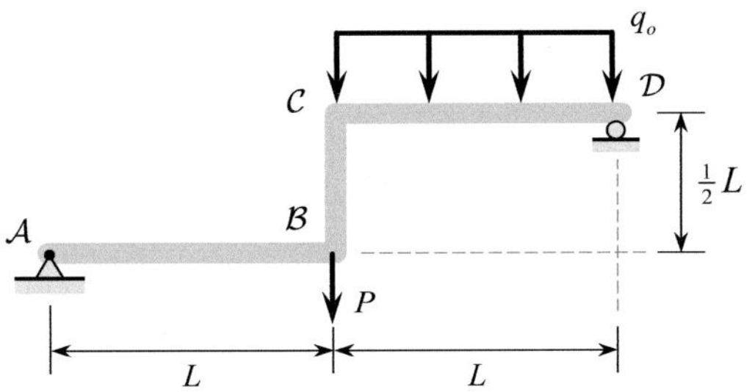

8.7 The beam $\mathcal{ACB}$ is subjected to a triangular distributed load over half its length. The load has a magnitude of q_o at $\mathcal{C}$ and decreases to 0 at end $\mathcal{B}$. The beam is fixed at $\mathcal{A}$ and has a roller support at $\mathcal{C}$. Derive the expressions for axial force, shear force, and bending moment. Find the expressions for the reaction forces. Determine the maximum moment and the maximum shear. Neglect the self-weight of the beam.

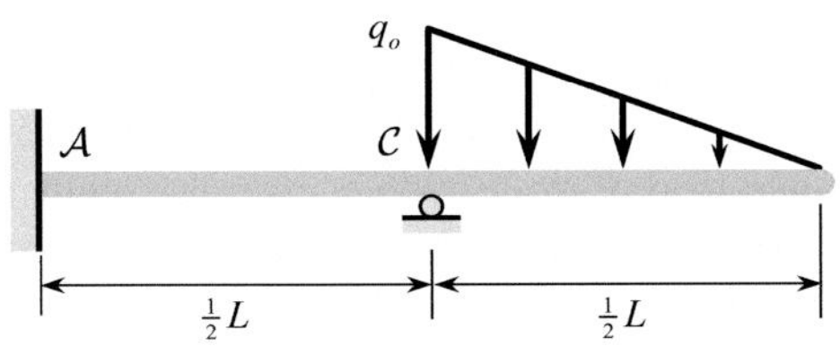

8.8 Beam $\mathcal{ABCD}$ of length L is pinned at $\mathcal{A}$, has roller supports at $\mathcal{B}$ and $\mathcal{D}$, and an internal hinge at $\mathcal{C}$. There is a constant distributed load with intensity of q_o acting downward between $\mathcal{B}$ and $\mathcal{D}$. Derive the expressions for axial force, shear force, and bending moment along the beam axis and draw the diagrams for the beam. Find the maximum shear and moment and the locations at which those maxima occur. Neglect the self-weight of the beam.

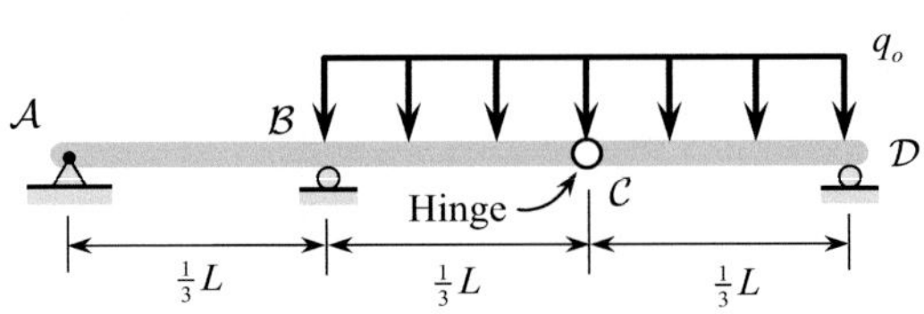

8.9 The continuous beam assembly $\mathcal{ABCD}$ is made up of two straight segments connected at right angles and continuously welded together. The system is subjected to a downward distributed load of magnitude q_o over the segment $\mathcal{ABC}$. The beam is fixed at $\mathcal{D}$ and has a roller support at $\mathcal{A}$. There is an internal hinge at $\mathcal{B}$ that can transmit axial and shear force, but not moment. Find the reaction forces. Derive the expressions for the axial force, shear force, and bending moment in both segments of the beam. Plot the axial, shear, and bending moment diagrams. Determine the maximum moment and the maximum shear. Neglect the self-weight of the beam.

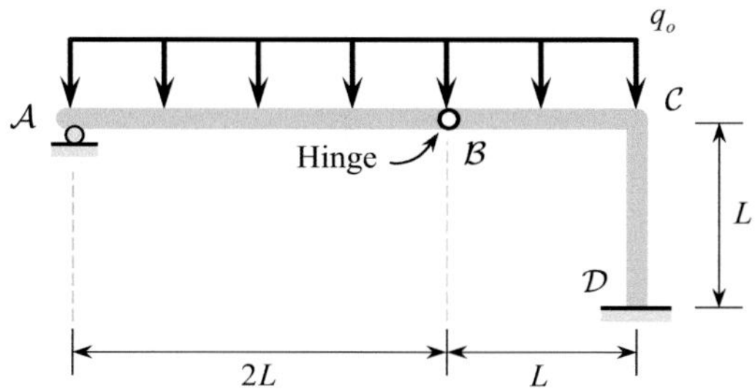

8.10 The continuous beam assembly $\mathcal{ABC}$ is made up of two straight segments connected at right angles and continuously welded together. The system is subjected to a load to the right of magnitude P at $\mathcal{B}$. The beam is pinned at $\mathcal{A}$ and has a roller support at $\mathcal{C}$. Find the reaction forces. Derive expressions for the axial force, shear force, and bending moment in both segments of the beam. Plot the axial, shear, and bending moment diagrams. Determine the maximum moment and the maximum shear. Neglect the self-weight of the beams.

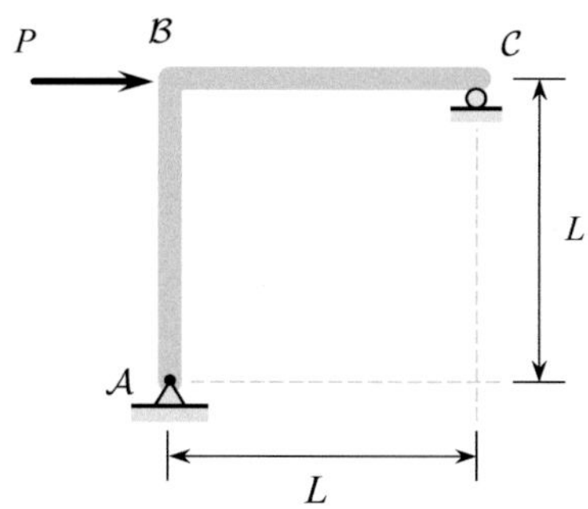

8.11 Beam $\mathcal{ABCD}$ of length $3L$ is pinned at $\mathcal{A}$, has roller supports at $\mathcal{B}$ and $\mathcal{D}$, and an internal hinge at $\mathcal{C}$. There is a triangular distributed load with increasing intensity from 0 at $\mathcal{A}$ to q_o at $\mathcal{D}$, acting downward over the entire beam. Derive the expressions for shear force and bending moment and draw the shear and moment diagrams for the beam. Find the maximum shear and moment and the locations at which those maxima occur. Neglect the self-weight of the beam.

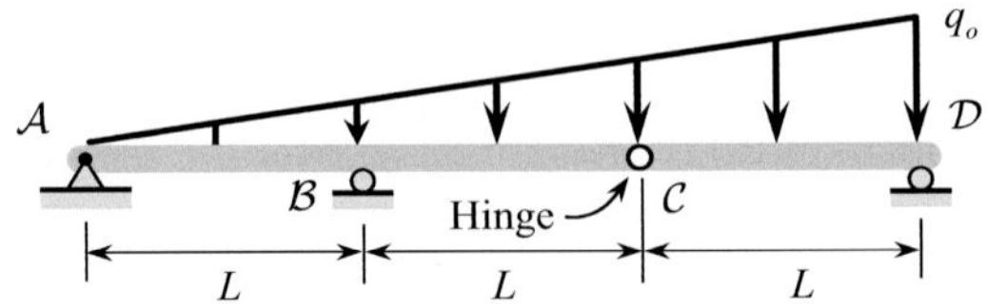

8.12 The simply supported beam $\mathcal{AB}$ has a point load a distance a away from the pin support at $\mathcal{A}$. The beam has a rectangular cross section with width b and depth h, as shown. Find the maximum normal stress and shear stress. Find the deflection at the point of load and the maximum deflection.

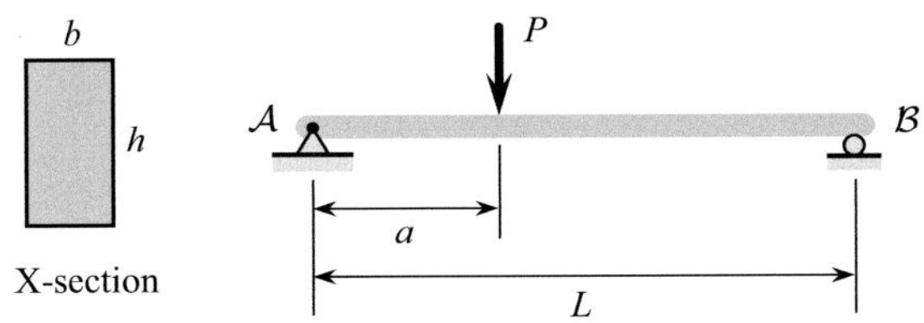

8.13 Beam $\mathcal{ABC}$ has an overhang at one end and supports a uniform load of intensity q_o on the cantilever portion. The beam has a square hollow tube cross section. Find the deflection of the beam. Find the maximum normal and shear stress. The thickness of the tube is $t = 0.1h$. Neglect the self-weight of the beam.

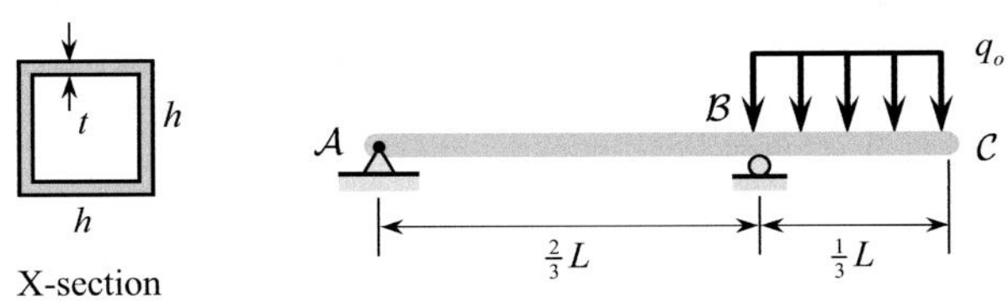

8.14 The beam $\mathcal{AB}$ is carrying a constant distributed load over its length of magnitude q_o. The beam is simply supported at end $\mathcal{B}$ and has a roller support at end $\mathcal{A}$. The beam has a nonprismatic rectangular cross section with constant depth h and linearly varying width $b(x)$. Find the expressions for the deflection and rotation along the beam. Find the maximum normal and shear stress in the beam. Neglect the self-weight of the beam.

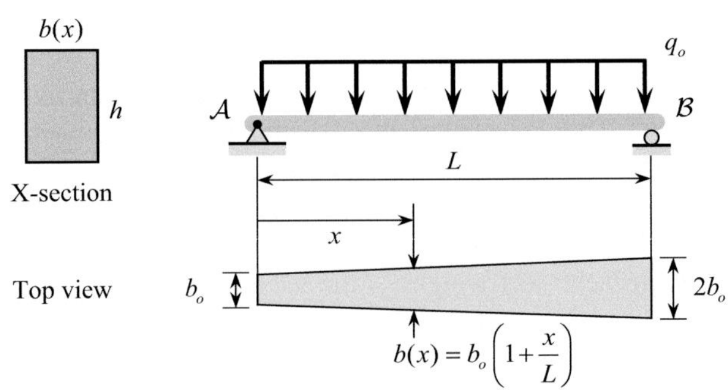

8.15 The beam $\mathcal{ABC}$ carries a point load of magnitude P that is acting at the end of the beam at $\mathcal{C}$. It is pinned at $\mathcal{A}$ and has a roller support at $\mathcal{B}$. The beam has a hollow rectangular cross section with dimensions $t = 0.1h$ and $b = 0.5h$. Find the expressions for the deflection and rotation along the beam. Find the maximum normal and shear stress in the beam. Neglect the self-weight of the beam.

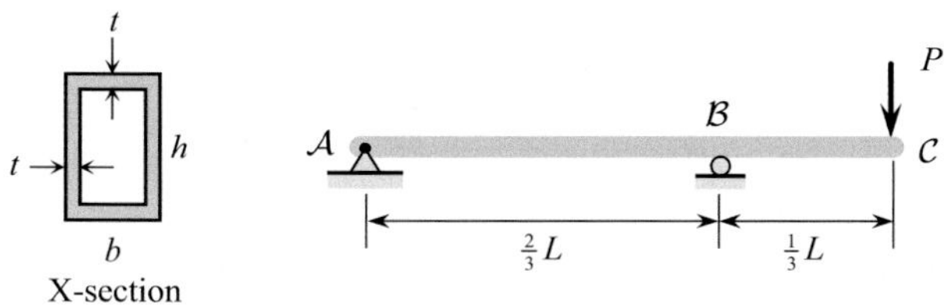

8.16 The overhanging beam $\mathcal{ABC}$ of length L is subjected to a continuous distributed load, q_o, and has an I-shaped cross section. The beam is pinned at point $\mathcal{A}$, has a roller at point $\mathcal{B}$, and is free at point $\mathcal{C}$. Derive expressions for the rotation and deflection. Find the maximum normal and shear stresses in the beam. If the maximum allowable stress in the beam is 20 ksi, determine the flange width b_f assuming that $L = 20$ ft, $q_o = 40$ lb/in., $d = 8$ in., $t_w = 0.1$ in., and $t_f = 0.25$ in. Neglect the self-weight of the beam.

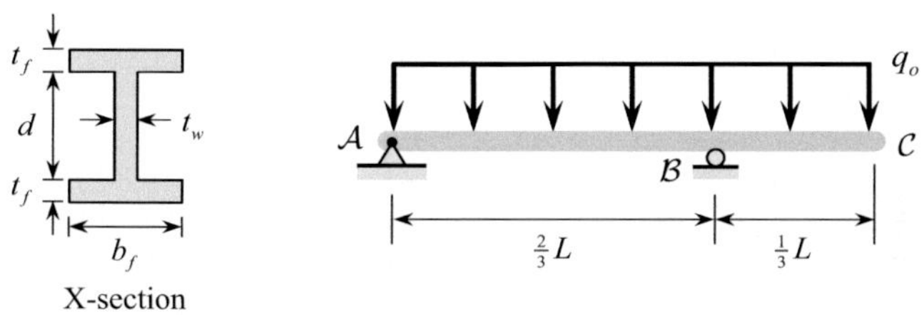

8.17 The continuous beam assembly $\mathcal{ABC}$ is made up of two straight segments of length L and bending stiffness EI, connected at right angles and continuously welded together. The system is subjected to a point load of magnitude P at $\mathcal{C}$. The beam is pinned at $\mathcal{A}$ and has a roller support at $\mathcal{B}$. Find the reaction forces. Derive the expressions for the axial force, shear force, and bending moment in both beams and plot them. Compute the rotation and deflection of both beams as a function of their axial coordinate. Find the maximum deflection and the rotation at the junction at $\mathcal{B}$. Neglect the self-weight of the beams and neglect the axial elongation of the beam in tension.

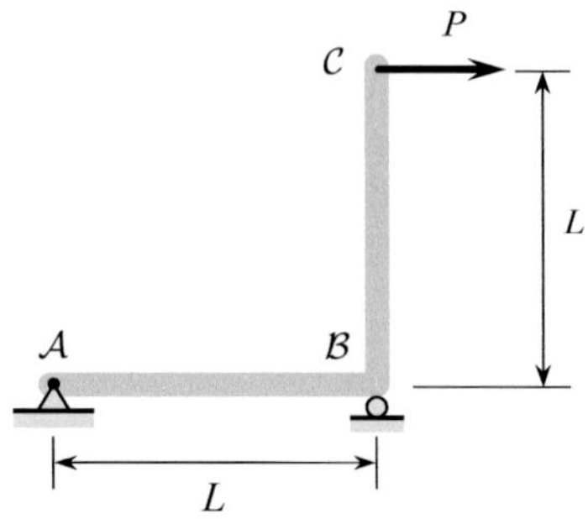

8.18 The cantilever beam $\mathcal{AB}$ is fixed at $\mathcal{A}$ and is subjected to a constant vertical distributed load of magnitude q_o. The beam has a rectangular cross section with height h and width $0.5h$. Find expressions for the shear, moment, deflection, and rotation along the length of the beam. Plot the shear and moment diagrams. Sketch the deflected shape. Find the maximum normal and shear stress in the beam and the maximum deflection. Neglect the self-weight of the beam.

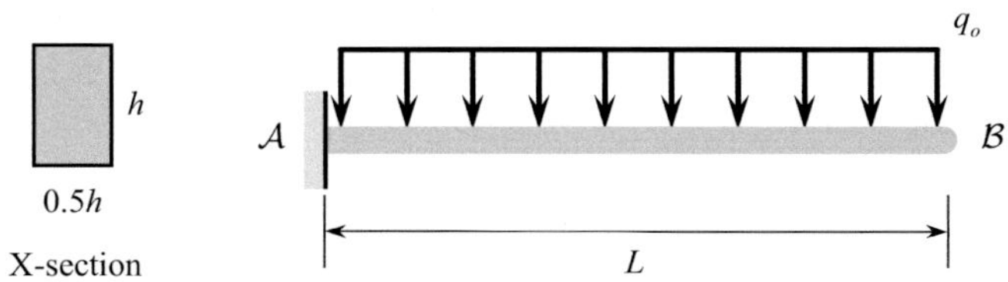

8.19 The overhanging beam $\mathcal{ABC}$ of length $2L$ and Young's modulus E is subjected to a linearly increasing distributed load with magnitude $2q_o$ at the right end. The beam has an I-shaped cross section. The beam is pinned at point $\mathcal{A}$, has a roller at point $\mathcal{B}$, and is free at point $\mathcal{C}$. Derive expressions for moment, shear, rotation, and deflection. Find the maximum normal stress and shear stress in the beam. Find the displacement at $\mathcal{C}$. Express answers in terms of q_o, E, L, and h. Neglect the self-weight of the beam.

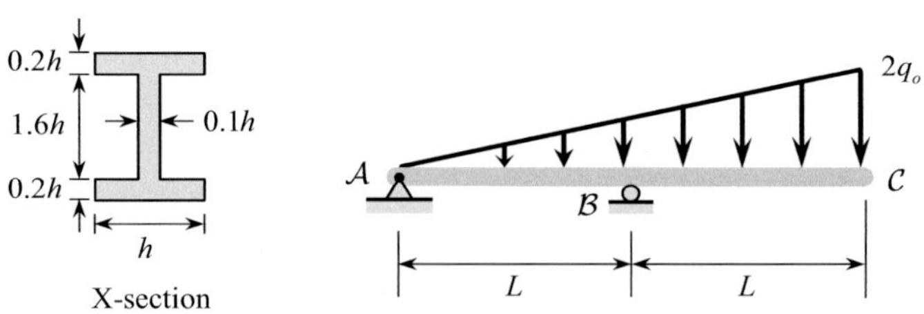

8.20 The beam $\mathcal{ABC}$ is carrying a concentrated clockwise moment of magnitude M_o that is acting at the end of the beam at $\mathcal{C}$. It is pinned at $\mathcal{A}$ and has a roller support at $\mathcal{B}$. The beam has a hollow rectangular cross section. Find the shear force and bending moment along the beam. Find the expressions for the deflection and rotation along the beam. Find the maximum normal and shear stress in the beam. Find the maximum displacement. Neglect the self-weight of the beam.

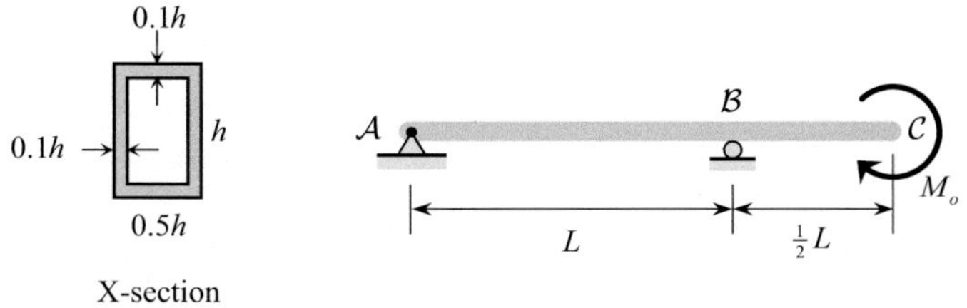

8.21 The continuous beam assembly $\mathcal{ABC}$ is made up of two straight segments of length L and bending stiffness EI, connected at right angles and continuously welded together. The system is subjected to a uniformly distributed load of magnitude q_o along bar $\mathcal{BC}$. The beam is pinned at $\mathcal{B}$ and has a roller support at $\mathcal{A}$. Find the reaction forces. Derive the expressions for axial force, shear force, and bending moment in both beams and plot them. Compute the rotation and deflection of both beams as a function of their axial coordinate. Find the maximum deflection and the rotation of the beams at $\mathcal{B}$. Neglect the self-weight of the beams.

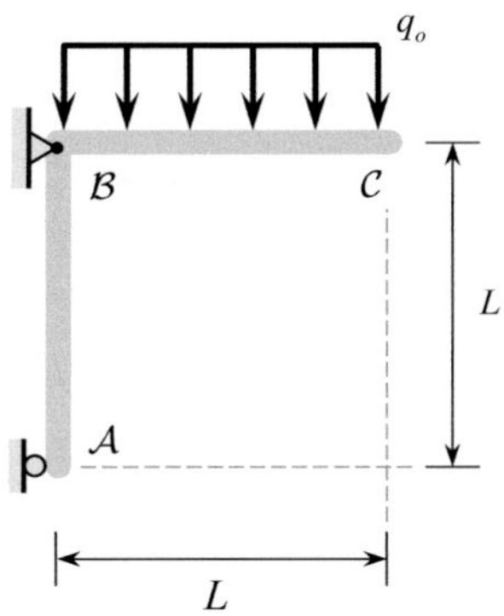

8.22 The beam $\mathcal{AB}$ is carrying a uniform distributed load over half its length that has magnitude q_o. The beam is fixed at the left end, has a roller support at the right end, and has a rectangular cross section. Find the reaction forces, the internal forces (i.e., N, V, and M) and expressions for the deflection and rotation along the beam. Find the maximum normal and shear stress in the beam. Neglect the self-weight of the beam.

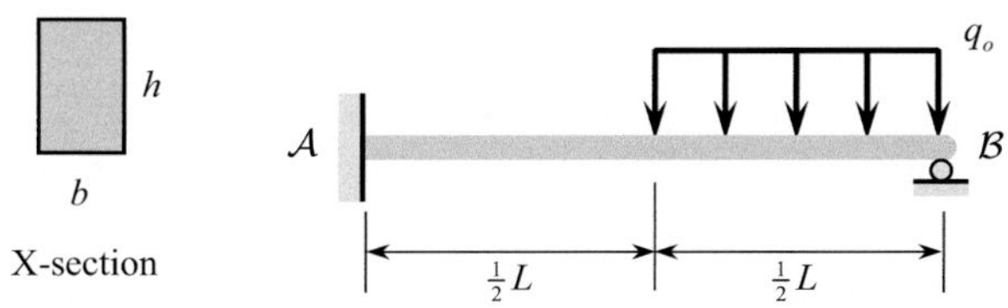

8.23 The beam $\mathcal{AB}$ is carrying a sinusoidally varying load over its length that goes from 0 to a magnitude q_o at the middle. The beam is fixed at end $\mathcal{A}$ and has a roller support at $\mathcal{B}$. The cross section is a symmetric diamond shape with a width half the height, h. Find the reaction force at $\mathcal{B}$. Find the expressions for the deflection and rotation along the beam. Find the maximum normal stress and shear stress (Hint: it is not at the centroid). Neglect the self-weight of the beam.

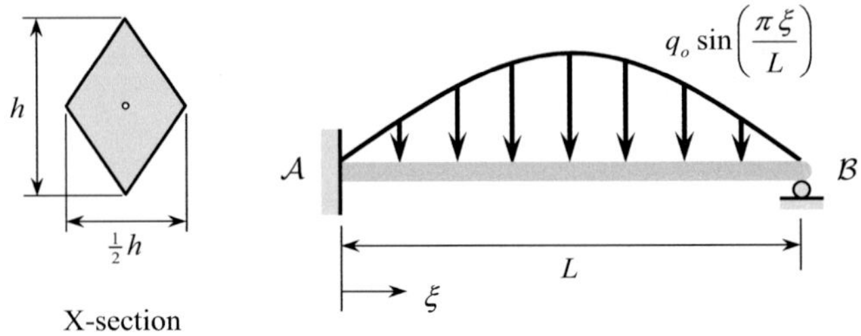

8.24 The beam $\mathcal{AB}$ is carrying a linearly varying load over its length that goes from 0 at $\mathcal{A}$ to a magnitude q_o at $\mathcal{B}$. The beam is fixed at $\mathcal{B}$ and has a roller support at $\mathcal{A}$. The beam is I-shaped with the cross-sectional dimensions shown. Find the expressions for the moment, shear, rotation, and deflection along the beam. Find the maximum normal and shear stress in the beam. Neglect the self-weight of the beam.

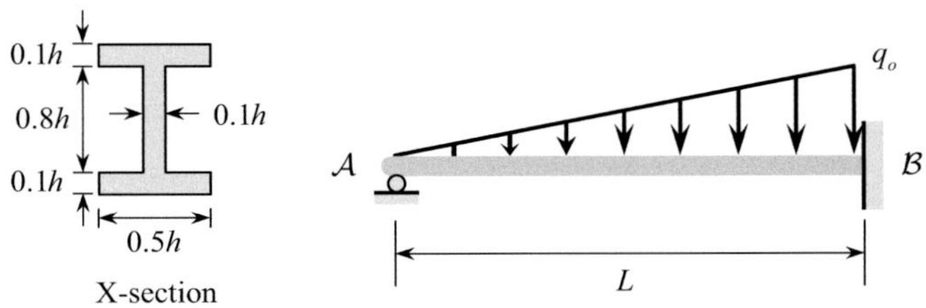

8.25 The beam $\mathcal{ABC}$ is carrying a uniform load over its length of magnitude q_o. The beam is pinned at $\mathcal{C}$ and on rollers at $\mathcal{A}$ and $\mathcal{B}$. Span $\mathcal{AB}$ has bending stiffness EI, and span $\mathcal{BC}$ has bending stiffness $2EI$. Find the reaction forces. Find the expressions for the deflection and rotation along the beam. The self-weight of the beam can be neglected.

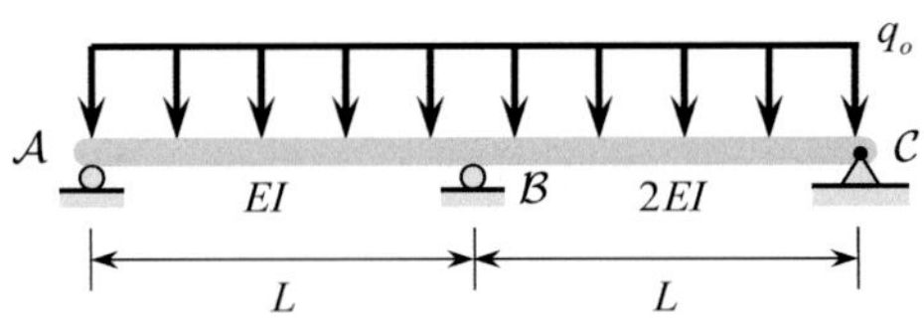

8.26 The propped cantilever beam with overhang $\mathcal{ABC}$ has length $1.5L$ and is subjected to a point load P at $\mathcal{C}$ and has the cross section shown. The beam is fixed at point $\mathcal{A}$, has a roller at point $\mathcal{B}$, and is free at point $\mathcal{C}$. Derive expressions for the shear, moment, rotation, and deflection. Find the support reactions. Find the maximum normal and shear stresses in the beam. Draw the shear and moment diagrams and the deflected shape. Use $L = 12$ ft, $P = 20$ k, $d = 12$ in., $t_w = 0.2$ in., $b_f = 10$ in, and $t_f = 0.3$ in. to compute the stresses. The self-weight of the beam can be neglected.

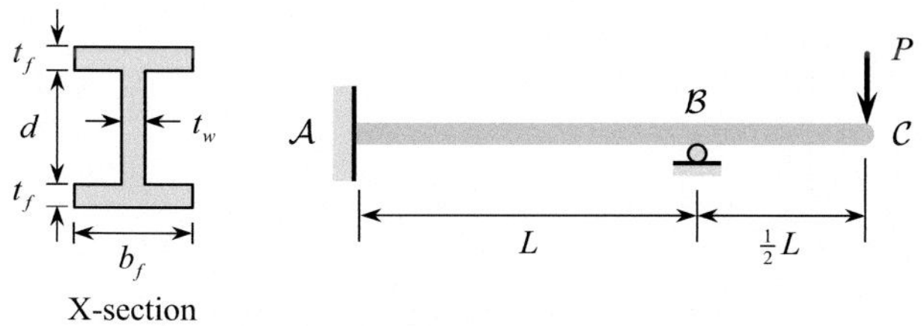

8.27 A fixed-fixed beam of length L and Young's modulus E is subjected to a linearly increasing distributed load with magnitude q_o at the right end. The moment expression (with M_A and M_B in the directions shown) is

$$M(x) = -\frac{M_A}{L}(L - x) - \frac{M_B}{L}x + \tfrac{1}{6}\frac{q_o}{L}(L^2x - x^3)$$

The beam has the T-shaped cross section shown. Derive expressions for shear, rotation, and deflection. Find the reaction moments M_A and M_B. Find the maximum normal and shear stresses in the beam. Express answers in terms of q_o, E, L, and h. Draw the shear and moment diagrams and the deflected shape. The self-weight of the beam can be neglected.

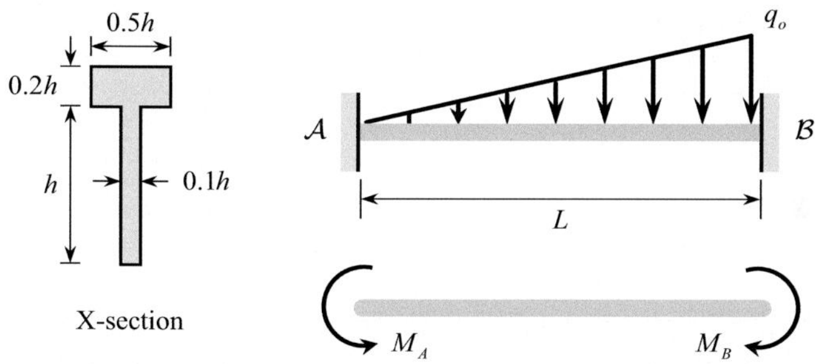

8.28 The continuous beam assembly $\mathcal{ABC}$ is made up of two straight segments of length $3L$ and $5L$ with Young's Modulus E and the cross section shown. The two beams are connected at the angle shown and continuously welded together. The system is subjected to a point load of magnitude P acting downward at $\mathcal{C}$. The beam is fixed at $\mathcal{A}$ and has a roller support at $\mathcal{B}$. Find the reaction forces. Derive the expressions for axial force, shear force, and bending moment in both beams and plot them. Compute the rotation and deflection of both beams as a function of their own axial coordinate. Find the deflection at $\mathcal{C}$ and the rotation of the beams at $\mathcal{B}$. Compute the maximum normal and shear stresses. Express results in terms of P, L, E, and h. Neglect the self-weight of the beams and neglect axial deformation.

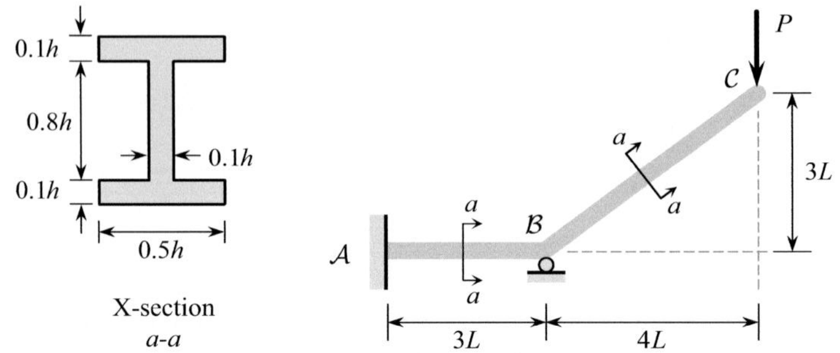

8.29 The beam $\mathcal{ABC}$ of length $2L$ is carrying a linearly increasing distributed load over the left half of its length with magnitude q_o at $\mathcal{B}$. The beam is pinned at $\mathcal{C}$ and on rollers at $\mathcal{A}$ and $\mathcal{B}$. The beam has the hollow tube cross section shown with all wall thicknesses the same. The cutouts are the same size. Find the reaction forces. Find the expressions for the moment, shear, deflection, and rotation along the beam. Plot the moment and shear diagrams. Sketch the deflected shape. Find the maximum normal and shear stress. Express answers in terms of q_o, E, L, and h. The self-weight of the beam can be neglected.

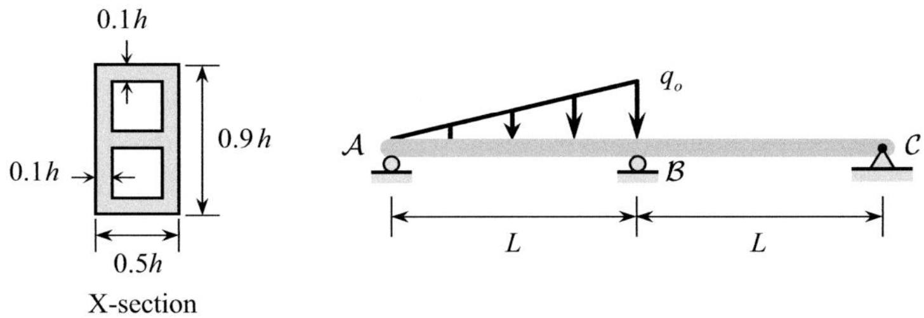

8.30 The beam $\mathcal{ABC}$ is subjected to a triangular distributed load over half its length. The load has a magnitude of q_o at the middle of the beam and decreases to 0 at end $\mathcal{C}$. The beam is fixed at $\mathcal{A}$ and has a roller support at the midpoint of the beam. The beam has a square cross section of dimension b. Find the expressions for the reaction forces, moment, shear, rotation, and transverse deflection along the beam. Find the maximum normal and shear stress in the beam. Neglect the self-weight of the beam.

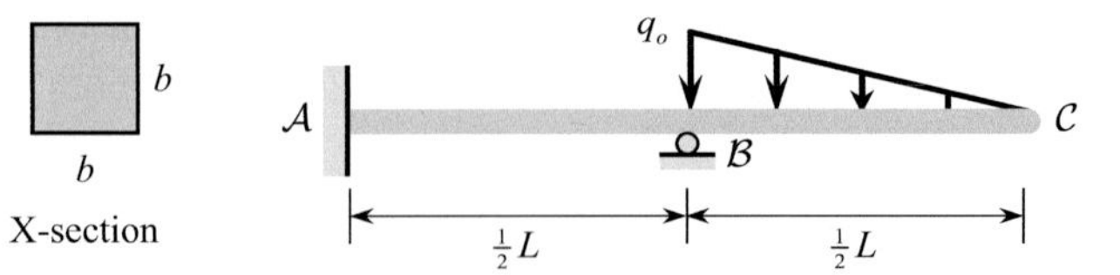

8.31 The beam $\mathcal{ABC}$ is fixed at both ends (with a slide at the right end), subjected to a linearly distributed load over half its length, and a uniform load over the other half. The load has a maximum magnitude of q_o. The beam has a rectangular cross section with the given dimensions. Find the expressions for the reaction forces, shear force, bending moment, rotation, and transverse deflection along the beam. Find the maximum normal and shear stress in the beam. Neglect the self-weight of the beam.

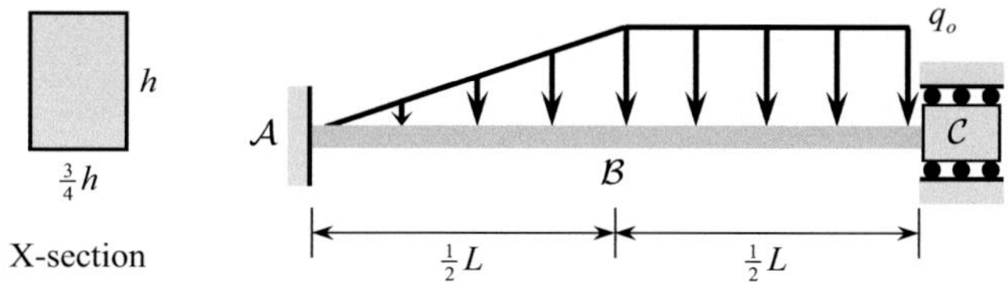

Chapter 9
Properties of Cross Sections

One important observation from the derivations in Chap. 7 is that the geometric properties of the cross section play a fundamental role in beam theory. The area, centroid, second moment of the area relative to the centroid (commonly called the *moment of inertia*), and the first moment of the area relative to the centroid show up in the resultant constitutive Eq. (7.24), the transverse deflection Eqs. (8.6 and 8.7), and the stress formulas (Eqs. 7.24 and 7.27). The geometric properties appear in these equations due to the kinematic hypothesis that plane cross sections remain plane after deformation. To understand why they appear, it is essential to retrace the steps of the derivation of the beam equations in the previous two chapters.

The properties of the cross section tell us a lot about how beams carry load, and they give us a way to think about how to design beams to do that task efficiently. In this chapter we delve into the computation of these geometric properties, first by setting up the relevant integrals and solving them directly, second by introducing a couple of integration tricks that can simplify the computation in certain cases, and finally by creating an algorithm and code that will compute the properties for any cross section with a polygonal shape (or a shape that can be approximated by a polygon).

9.1 Basic Approach

When deriving the equations of beam theory, we assumed that the x axis of the beam is at the *centroid* of the cross section. This choice uncouples the axial part of the deflection equations from the bending part. While the cross-sectional area is not affected by this choice, the first and second moments of the area are defined relative to the centroid. Thus, we must first compute the area, followed by the location of the centroid, which depends on the area, followed by the moments of the area, which depend upon both the area and centroid. To illustrate what is at stake, consider the

K. D. Hjelmstad, *Mechanics of Deformable Solids: An Introduction*,
https://doi.org/10.1007/978-3-031-89873-0_9

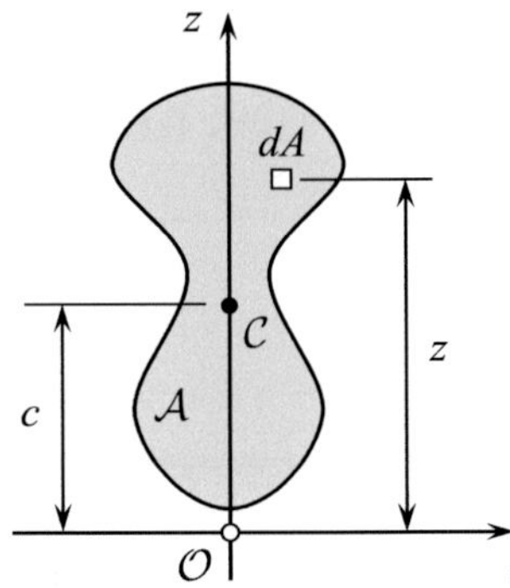

Fig. 9.1 *Properties of cross section $\mathcal{A}$.* The origin $\mathcal{O}$ defines the (y, z) coordinates. The centroid is at point $\mathcal{C}$, located a distance c from the origin. The element of integration dA is located at distance z from the origin

general cross section shown in Fig. 9.1. Let us refer to the region of the cross section as $\mathcal{A}$ in what follows. We will focus first on planar bending, so for simplicity we can consider the cross section to be symmetric about the z axis. In this case, the centroid $\mathcal{C}$ lies on the z axis. We will later lift that restriction.

The *area* of the cross section can be written as an integral in the form

$$A = \int_{\mathcal{A}} dA \tag{9.1}$$

What this equation means is that we will compute the total area by summing the infinitesimal areas dA over the region $\mathcal{A}$. The notation we use to express this concept is a single integral sign, with the subscript indicating the region over which integration takes place. In calculus, and in some of our examples, that process is often denoted with a double integral sign, and the element of integration is written as $dA = dx\,dy$ if rectangular Cartesian coordinates are used. The notation of Eq. 9.1 is designed to emphasize *what* an integral is rather than *how* to execute the calculation. Thinking of the area as an integral stands apart from *mensuration formulas* for area that are available for common shapes like circles ($A = \pi r^2$), rectangles ($A = bh$), and triangles ($A = bh/2$). Many of those formulas were discovered long before calculus was invented. All of them can be found using calculus.

A simple way of thinking about the integral is to draw the cross section on a grid, as illustrated in Fig. 9.2. The region $\mathcal{A}$ lives in the (y, z) plane. If we superpose a grid on top of the region, then we can think of the dark gray square as an element dA of known area. Now, count the squares. The total area of the region is the number of squares counted times the area of the elemental square.[1] Integration, then, is simply adding up the squares, and we can distinguish that action from the *techniques of integration* commonly taught in integral calculus courses, which involve finding the antiderivative of the integrand. If the integral over the region has some function

[1] Note that the elemental piece does not have to be a square. It could be any shape that allows tiling of the region (i.e., fitting parts together with no gaps), e.g., triangles or parallelograms. In calculus it is common to use rectangles where $dA = dx\,dy$ or polar coordinates where $dA = r\,dr\,d\theta$, which are circular segments with thickness dr.

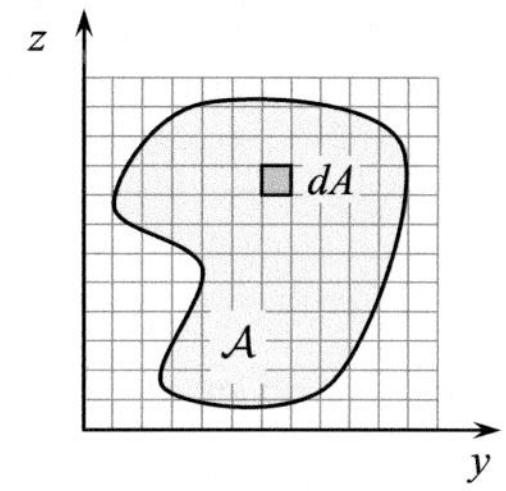

Fig. 9.2 *Concept of integration.* Integration over a region $\mathcal{A}$ is simply adding up contributions from each elemental region dA

$g(y, z)$ as an integrand, i.e.,

$$\int_{\mathcal{A}} g(y, z)\, dA$$

then the integration process simply adds up the values of the function at the center of each square multiplied by the elemental area dA. For the calculation of the area or region $\mathcal{A}$, the function is $g(y, z) = 1$.

At the edges of the region in the figure, it is apparent that only parts of some squares are covered. To remedy this situation, we can take a finer grid, thereby reducing the portion of the region associated with partial squares. Integral calculus lives in the limit as the squares reduce to infinitesimal size, eliminating the partial square issue. Ultimately, we do not count squares, but rather use various techniques of integration to support the calculation. Specific methods to carry out the computation of cross-sectional properties will be shown through examples. Later we will apply the divergence theorem of calculus to derive yet another way to compute the integrals over the region. The key is to visualize the integrals over the region $\mathcal{A}$ as a summing process, and to use calculus techniques to carry out the details of the calculation in specific instances.

After computing the area, the next challenge is to find the location of the *centroid* of the cross section. In the derivations of beam theory, we took the origin of the (y, z) coordinates at the centroid (e.g., as shown in Fig. 7.1). However, we do not yet know where that is. So, we just put the origin $\mathcal{O}$ of the (y, z) axes at a convenient point in space, as shown in Fig. 9.1. The choice will usually be evident for a particular cross section. For example, for a rectangular cross section, the origin might be located at the bottom edge of the region. For a circle, the center is a good choice because it enables the use of polar coordinates. The choice of the origin is arbitrary. It does not affect beam theory, only the intermediate calculations of the properties of the cross section. In fact, the centroid itself does not appear directly in beam theory, but rather implicitly through the first and second moments of the area. Hence, we can always reposition the (y, z) axes once the location of the centroid is determined.

The *centroid* is defined to be the location where the first moment of the area is equal to zero. In the current coordinate system, that means

$$\int_{\mathcal{A}} \big(z - c\big)\, dA = 0 \tag{9.2}$$

Even though we do not know the value of c, we know that it is constant and can, therefore, be pulled out of the integral. Thus, we can rewrite Eq. 9.2 as

$$\int_{\mathcal{A}} z\,dA - c\int_{\mathcal{A}} dA = 0$$

The second integral is simply the total area A, and we can solve this equation for c as

$$c = \frac{1}{A}\int_{\mathcal{A}} z\,dA \tag{9.3}$$

This equation defines the location of the centroid. Notice that the integral in this equation is carried out in the coordinate system defined by the origin $\mathcal{O}$. Referring back to Fig. 9.2, that means the z values in the integrand are the distances of each square relative to the y axis. If z is relative to $\mathcal{O}$, then so is c.

The *second moment of the area*, or moment of inertia, relative to the centroid $\mathcal{C}$ is defined as[2]

$$I_{\mathcal{C}} = \int_{\mathcal{A}} (z-c)^2\,dA \tag{9.4}$$

It is possible to simplify this calculation by first distributing the integrand. To wit,

$$I_{\mathcal{C}} = \int_{\mathcal{A}} \left(z^2 - 2cz + c^2\right) dA$$

The integral of a sum is the sum of the individual integrals. Therefore, each term in the integral can be evaluated separately. Again, noting that c is constant, we can write

$$I_{\mathcal{C}} = \int_{\mathcal{A}} z^2\,dA - 2c\int_{\mathcal{A}} z\,dA + c^2\int_{\mathcal{A}} dA$$

The integral in the second term, by Eq. 9.3, is simply cA, and the integral in the third term is simply A. Thus, the second moment of the area reduces to

$$I_{\mathcal{C}} = \int_{\mathcal{A}} z^2\,dA - c^2 A \tag{9.5}$$

This form of the equation is often called the *parallel axis theorem* because it allows us to compute the second moment of area about point $\mathcal{O}$ (the first term) and then transform it with the adjusting term c^2A, which is tantamount to a parallel shift of the axis from $\mathcal{O}$ to $\mathcal{C}$.

---

[2] We will use the subscript on I to indicate what the second moment is relative to. In beam theory we simply use the notation I without a subscript, with the implicit understanding that it is relative to the centroid of the cross section.

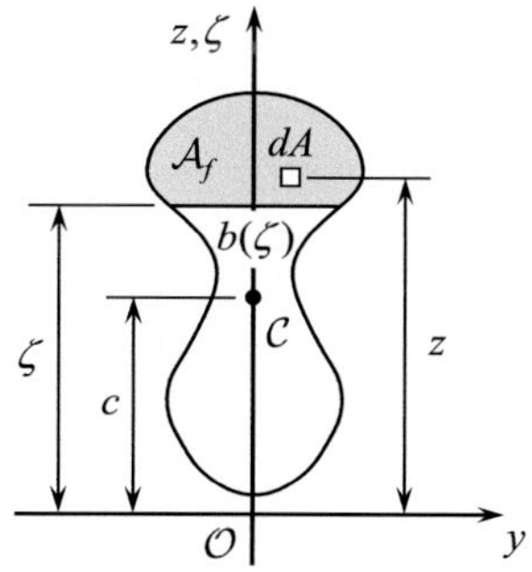

Fig. 9.3 *First moment of area.* A cross section with centroid at point $\mathcal{C}$, located a distance c from the origin. The element of integration is dA. The subregion $\mathcal{A}_f$ is defined by the cut at ζ. The width of the cross section at the cut is $b(\zeta)$

We can get some insight about the second moment of area from Eq. 9.5. First, note that there are an infinite number of second moments of area

$$I_{\mathcal{O}} = \int_{\mathcal{A}} z^2 \, dA$$

each one depending on where we put the origin $\mathcal{O}$. The area A is always positive and c^2 is always positive. Therefore, the moment of inertia relative to $\mathcal{C}$ is always smaller than the moment of inertia relative to $\mathcal{O}$, because we are taking away a positive value, i.e.,

$$I_{\mathcal{C}} = I_{\mathcal{O}} - c^2 A \tag{9.6}$$

The smallest reduction happens when $c = 0$, i.e., when point $\mathcal{O}$ is coincident with $\mathcal{C}$ or, said a different way, when the origin of coordinates for z is located at the centroid.

We can follow a similar path for the computation of the *first moment of the area* relative to the centroid, Q. This property requires us to identify a subregion $\mathcal{A}_f$ for the integral, the shaded region shown in Fig. 9.3. The first moment of area relative to the centroid is defined as

$$Q_{\mathcal{C}}(\zeta) = \int_{\mathcal{A}_f} (z - c) \, dA \tag{9.7}$$

where ζ is the location of the cut that defines the region $\mathcal{A}_f$. Again, noting that c is constant, we can write

$$Q_{\mathcal{C}}(\zeta) = \int_{\mathcal{A}_f} z \, dA - c \int_{\mathcal{A}_f} dA$$

or

$$Q_{\mathcal{C}}(\zeta) = \int_{\mathcal{A}_f} z \, dA - c A_f \tag{9.8}$$

where A_f is the area of the region $\mathcal{A}_f$. The first term on the right side is $Q_{\mathcal{O}}(\zeta)$, the first moment of the area $\mathcal{A}_f$ relative to $\mathcal{O}$. Thus, we can write Eq. 9.8 in the form

$$Q_C = Q_{\mathcal{O}} - cA_f \tag{9.9}$$

This equation is a sort of parallel axis theorem for Q, allowing us to compute Q relative to the origin $\mathcal{O}$ and then adjust it to the centroid by subtracting cA_f. If the origin $\mathcal{O}$ is above the centroid, then c will be negative. Of course, $\mathcal{A}_f$ is always positive.

There are a few important observations that we can make about $Q_C(\zeta)$. First, if the region $\mathcal{A}_f$ is taken to be the entire region $\mathcal{A}$ (i.e., ζ is at the bottom fiber of the cross section), then the value of Q_C is zero due to the definition of the location of the centroid in Eq. 9.2. Second, if ζ is taken at the top edge of the cross section, Q_C is also zero because the region of integration $\mathcal{A}_f$ goes to zero. Hence, Q_C vanishes at the top and bottom edges of the cross section, implying that it must have some maximum value somewhere between the top and bottom edges. For locations of the cut ζ between c and the top, we can observe that the integrand $z-c$ in Eq. 9.7 is positive at every point. Hence, Q_C must be positive for any cut above the centroid. The largest value would be a cut at the centroid because that is the largest region for which the integrand is entirely positive. For a cut below the centroid, some of the $z-c$ values are negative, taking away from the positive values accumulated above the centroid. Therefore, Q_C for cuts below the centroid are smaller than the value of Q_C at the centroid.

This argument proves that $Q_C(c)$ is the maximum value of $Q_C(\zeta)$. The reason this observation is important is because Q_C appears in the shear stress formula and when designing beams, we will be most interested in the maximum shear stress. For most cross sections, the shear stress will be greatest at the centroid. Because the average shear stress depends upon $b(\zeta)$, it is possible that the maximum shear stress will occur at some other location.

Example 9.1
Compute the first and second moments of the area relative to the centroid for a rectangular cross section of width b and height h, shown in Fig. 9.4, by directly integrating over the domain of the region.

Solution. One can show that the centroid must lie on an axis of symmetry. The rectangular region has two axes of symmetry. Therefore, the centroid must lie

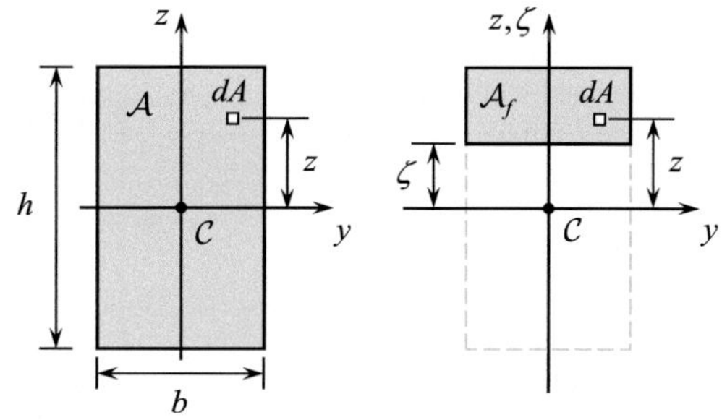

Fig. 9.4 *Example*. A rectangular cross section of width b and height h. The centroid of the cross section is in the middle at point $\mathcal{C}$, which is taken as the origin for the (y, z) coordinate system. The element of integration is dA. The subregion $\mathcal{A}_f$ is defined by the cut at ζ

on the intersection of those lines, i.e., at the center of the region. Thus, we can take the (y, z) coordinates at the centroid, and there is no need to compute the location of the centroid (it is $c = 0$). The area of the region is, obviously, $A = bh$. The second moment of the area can be computed as

$$I_{\mathcal{C}} = \int_{\mathcal{A}} z^2\, dA = \int_{-\frac{h}{2}}^{\frac{h}{2}} \int_{-\frac{b}{2}}^{\frac{b}{2}} z^2\, dy\, dz = \Big[\, y \,\Big]_{-\frac{b}{2}}^{\frac{b}{2}} \Big[\tfrac{1}{3} z^3\Big]_{-\frac{h}{2}}^{\frac{h}{2}} = \tfrac{1}{12} bh^3$$

Similarly, $Q_{\mathcal{C}}(\zeta)$ can be computed as

$$Q_{\mathcal{C}}(\zeta) = \int_{\mathcal{A}_f} z\, dA = \int_{\zeta}^{\frac{h}{2}} \int_{-\frac{b}{2}}^{\frac{b}{2}} z\, dy\, dz = \tfrac{1}{8} bh^2 \Big[1 - \frac{4\zeta^2}{h^2}\Big]$$

Notice that $Q_{\mathcal{C}}(\zeta)$ is zero at the top ($\zeta = h/2$) and bottom ($\zeta = -h/2$) edges and is maximum at the centroid, as expected. To summarize these results for the *rectangle*, we have

$$I_{\mathcal{C}} = \tfrac{1}{12} bh^3, \qquad Q_{\mathcal{C}}(0) = \tfrac{1}{8} bh^2 \tag{9.10}$$

We will use these results in several of our subsequent examples.

One important variation on the approach to integration in this example is the so-called *strip method*. Noting that the integrand is purely a function of z, we can take an elemental strip rather than an elemental particle. The strip extends from the left side to the right with height dz to give $dA = b\, dz$. With this choice, the integral can be written as

$$I_{\mathcal{C}} = \int_{-\frac{h}{2}}^{\frac{h}{2}} z^2\, b\, dz$$

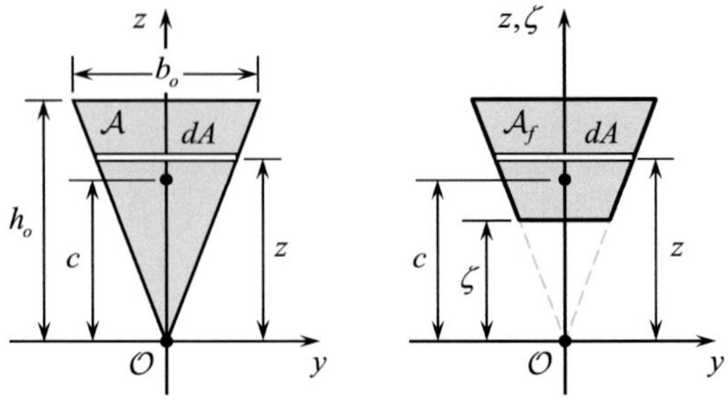

Fig. 9.5 *Example.* A triangular cross section of height h_o and width b_o at the top. The origin for the (y, z) coordinate system is taken at the tip of the triangle. The element of integration is $dA = b(z)\,dz$

resulting in only a single integral in the computation. This simplification is useful in many cases, even when b is not constant. To use the method for variable width, the function $b(z)$ must be known. The following example illustrates this approach.

Example 9.2

Compute the area and the location of the centroid of the triangular cross section shown in Fig. 9.5. Compute the first and second moments of the area relative to the centroid. Use the strip method to carry out the calculations.

Solution. The area of a triangle can be easily computed by a mensuration formula, but let us compute it by integration to show how to set up the problem using the strip method. Then we can use the same setup for the other properties. The width of a strip at height z is $b(z) = b_o z / h_o$, which goes from zero at the base to b_o at $z = h_o$ and varies linearly. Thus,

$$A = \int_A dA = \int_0^{h_o} b(z)\,dz = \int_0^{h_o} \frac{b_o z}{h_o}\,dz = \frac{b_o}{h_o}\Big[\tfrac{1}{2}z^2\Big]_0^{h_o} = \tfrac{1}{2}b_o h_o$$

as expected. To compute the centroid, we need to compute the first moment of the entire area. The first moment of the total area about the origin $\mathcal{O}$ is

$$\int_0^{h_o} z\,b(z)\,dz = \int_0^{h_o} \frac{b_o}{h_o} z^2\,dz = \frac{b_o}{h_o}\Big[\tfrac{1}{3}z^3\Big]_0^{h_o} = \tfrac{1}{3}b_o h_o^2$$

Now, compute the location of the centroid c as

$$c = \frac{1}{A}\int_0^{h_o} z\,b(z)\,dz = \frac{\tfrac{1}{3}b_o h_o^2}{\tfrac{1}{2}b_o h_o} = \tfrac{2}{3}h_o$$

We can compute the second moment of the area by first computing it about the origin and then adjusting it with the parallel axis theorem. The second moment of the area about $\mathcal{O}$ is

$$I_O = \int_0^{h_o} z^2 b(z)\,dz = \int_0^{h_o} \frac{b_o}{h_o} z^3\,dz = \frac{b_o}{h_o}\Big[\tfrac{1}{4}z^4\Big]_0^{h_o} = \tfrac{1}{4}b_o h_o^3$$

Using the parallel axis theorem, we can compute I_C relative to the centroid as

$$I_C = I_O - c^2 A = \tfrac{1}{4}b_o h_o^3 - \big(\tfrac{2}{3}h_o\big)^2\big(\tfrac{1}{2}b_o h_o\big) = \tfrac{1}{36}b_o h_o^3$$

Now, calculate the first moment of the area for a cut at ζ, which is defined in Eq. 9.8 as

$$Q_C(\zeta) = \int_{\mathcal{A}_f} z\,dA - cA_f$$

To complete this calculation, we need to compute both A_f and the first moment of the area above the cut relative to $\mathcal{O}$. The area above the cut is

$$A_f(\zeta) = \int_\zeta^{h_o} b(z)\,dz = \int_\zeta^{h_o} \frac{b_o}{h_o} z\,dz = \frac{b_o}{h_o}\Big[\tfrac{1}{2}z^2\Big]_\zeta^{h_o} = \frac{1}{2}\frac{b_o}{h_o}\Big[h_o^2 - \zeta^2\Big]$$

Observe that at $\zeta = 0$ the area is equal to the total area of the triangle and that at $\zeta = h_o$ the area is equal to zero, as expected. The first moment of the area above the cut is

$$Q_O(\zeta) = \int_\zeta^{h_o} z\,b(z)\,dz = \int_\zeta^{h_o} \frac{b_o}{h_o} z^2\,dz = \frac{b_o}{h_o}\Big[\tfrac{1}{3}z^3\Big]_\zeta^{h_o} = \frac{1}{3}\frac{b_o}{h_o}\Big[h_o^3 - \zeta^3\Big]$$

Therefore, $Q_C(\zeta)$ is

$$Q_C(\zeta) = \frac{1}{3}\frac{b_o}{h_o}\Big[h_o^3 - \zeta^3\Big] - \Big(\tfrac{2}{3}h_o\Big)\frac{1}{2}\frac{b_o}{h_o}\Big[h_o^2 - \zeta^2\Big] = \frac{1}{3}\frac{b_o}{h_o}\Big[h_o\zeta^2 - \zeta^3\Big]$$

It is clear that this function is equal to zero at $\zeta = 0$ and at $\zeta = h_o$. It achieves its maximum value at the centroid. Hence,

$$Q_C(c) = \frac{1}{3}\frac{b_o}{h_o}\Big[h_o\big(\tfrac{2}{3}h_o\big)^2 - \big(\tfrac{2}{3}h_o\big)^3\Big] = \tfrac{4}{81}b_o h_o^2$$

The maximum value of Q_C can also be determined by setting $dQ_C/d\zeta = 0$ and solving for the value of ζ that gives the maximum (which, of course, will be the centroid).

For many cross sections, the maximum shear stress happens at the centroid. For cross sections with variable width, that is not necessarily true. The triangular cross section of the previous example is a case in point. Recall from Eq. 7.27 that the shear

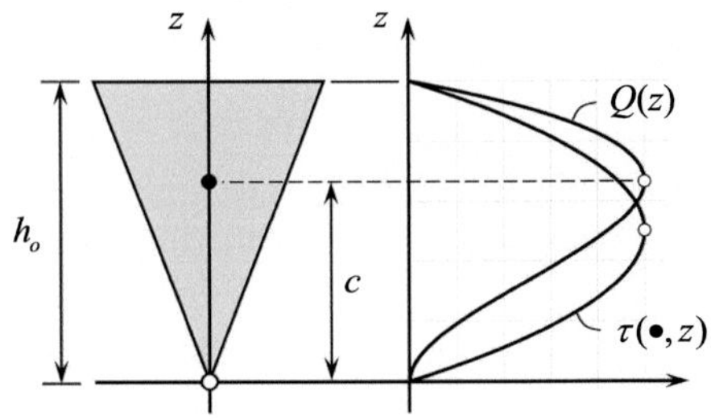

Fig. 9.6 *Triangle cross section*. Distribution of $Q(z)$ and $\tau(\bullet, z)$ for a triangular cross section. The functions are normalized to have the same maximum value

stress formula is

$$\bar{\tau}(x, z) = \frac{V(x)}{I} \frac{Q(z)}{b(z)}$$

where both I and Q are relative to the centroid. The maximum value will occur at the x location where the shear force is largest. In the cross section at that location, the shear stress varies as $Q(z)/b(z)$. For the triangular cross section of Fig. 9.5, that gives

$$\bar{\tau}(x, z) = \frac{V(x)}{I} \frac{1}{3}\left[h_o z - z^2 \right]$$

The maximum value of this function with respect to z occurs at $z = h_o/2$, which is below the centroid. While $Q(z)$ achieves its maximum value at the centroid, the width decreases linearly from top to bottom. Hence, the average shear stress is relative to a smaller and smaller area. The reduction in $b(z)$ is faster in the neighborhood of the centroid than the reduction in $Q(z)$. The reduction in $Q(z)$ starts dominating the reduction in $b(z)$ at $z = h_o/2$, as shown in Fig. 9.6.

For problems with circular geometries, the evaluation of integrals over the region are best done in polar coordinates, as the following example shows.

Example 9.3
Compute the first and second moment of area for the hollow circular tube cross section with inside radius R_i and outside radius R_o, shown in Fig. 9.7. Evaluate only the maximum value of the first moment of area.

Solution. The centroid of a circle is at the center. Hence, we can take our coordinate axes at point C so that $c = 0$. To compute the properties of the cross section for a circular geometry, it is convenient to use polar coordinates (r, θ), as shown in Fig. 9.7. Hence, the element of integration over the area is $dA = r\,dr\,d\theta$. The value of the vertical distance z can be converted to polar coordinates as $z = r \sin\theta$. Using these definitions, the second moment of area can be computed as

$$I_C = \int_A z^2\, dA = \int_0^{2\pi}\int_{R_i}^{R_o} \big(r \sin\theta\big)^2\, r\, dr\, d\theta = \int_{R_i}^{R_o} r^3\, dr \int_0^{2\pi} \sin^2\theta\, d\theta$$

Using the trigonometric identity $\sin^2\theta = \frac{1}{2}(1 - \cos 2\theta)$, we can write

$$I_C = \int_{R_i}^{R_o} r^3\, dr \int_0^{2\pi} \tfrac{1}{2}\big(1 - \cos 2\theta\big)\, d\theta = \Big[\tfrac{1}{4}r^4\Big]_{R_i}^{R_o}\Big[\tfrac{1}{2}\theta - \tfrac{1}{4}\sin 2\theta\Big]_0^{2\pi}$$

Substituting the limits of integration gives the final result:

$$I_C = \tfrac{1}{4}\pi\big(R_o^4 - R_i^4\big)$$

Similarly, $Q_C(0)$ can be computed as

$$Q_C(0) = \int_{A_f} z\, dA = \int_0^{\pi}\int_{R_i}^{R_o} \big(r \sin\theta\big)\, r\, dr\, d\theta = \int_{R_i}^{R_o} r^2\, dr \int_0^{\pi} \sin\theta\, d\theta$$

Carrying out the integrals and substituting the limits of integration gives

$$Q_C(0) = \tfrac{2}{3}\big(R_o^3 - R_i^3\big)$$

We can recover the values for the solid circular cross section by taking $R_i = 0$.

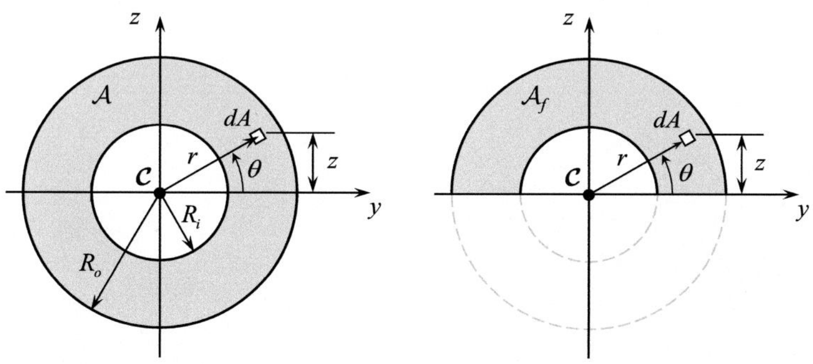

Fig. 9.7 *Example.* A hollow circular tube cross section of inside radius R_i and outside radius R_o. The centroid of the cross section is at the center of the circles, and is the origin for the (r, θ) coordinate system. The element of integration is $dA = r\, dr\, d\theta$

9.2 Computations for Composite Areas

An integral is nothing more than a sum, and a sum can be organized in multiple different ways. We can use this observation to develop a simplified approach to computing the properties of a cross section when the cross section can be divided up into simple geometric shapes (e.g., triangles or rectangles). We will look at two basic approaches: *Trick* 1, wherein the cross section is broken up into subregions and the properties are computed as the sum of the contributions of the pieces, and *Trick* 2, wherein the cross section is created by including parts that do not exist and then subtracting those contributions back out. Both methods are effective in organizing the hand calculations of the geometric properties of cross sections.

Trick 1 Let us first consider the decomposition of a region $\mathcal{A}$ into subregions $\mathcal{A}_i$, as shown in Fig. 9.8. Any integral over the entire region can be expressed as the sum of integrals over the subregions. To wit,

$$\int_{\mathcal{A}} (\bullet)\, dA = \sum_{i=1}^{N} \int_{\mathcal{A}_i} (\bullet)\, dA$$

where N is the number of subregions and $\mathcal{A}_i$ is the ith subregion. This property holds for *any* integrand $(\bullet)$. Each individual subregion $\mathcal{A}_i$ can be viewed as a geometric figure with properties like area, centroid, and moment of inertia on its own, as shown in Fig. 9.8 on the right. Thus, we can compute the zeroth, first, and second moment of the area of region $\mathcal{A}_i$ relative to the y axis as

$$A_i = \int_{\mathcal{A}_i} dA, \qquad Q_i = \int_{\mathcal{A}_i} z\, dA, \qquad R_i = \int_{\mathcal{A}_i} z^2\, dA$$

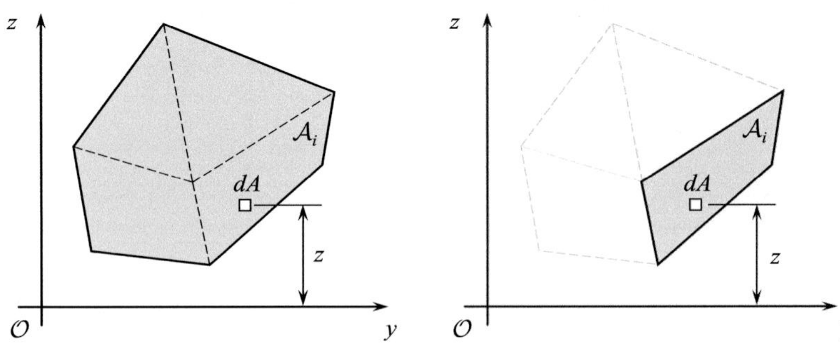

Fig. 9.8 *Trick 1*. A cross section can be broken into subregions. The shaded subregion $\mathcal{A}_i$ can also be viewed as a geometric region in its own right, with its own area, A_i, centroid location, c_i, and second moment of area about its own centroid, I_i

Note that we are using the notation Q and R to represent the first and second moment relative to the arbitrarily selected origin $\mathcal{O}$, which is the common origin for all the subregions. The location of the centroid of region i and the second moment of the area relative to the centroid are then

$$c_i = \frac{Q_i}{A_i}, \qquad I_i = R_i - c_i^2 A_i$$

where the notation I_i in this section is the same as I_C in the previous section, i.e., the second moment of piece i relative to its own centroid, and R_i is the same as $I_{\mathcal{O}}$ in the previous section, i.e., the second moment of subregion i relative to $\mathcal{O}$. We can turn these equations around and use them to compute the integrals Q_i and R_i. To wit,

$$Q_i = \int_{\mathcal{A}_i} z\,dA = c_i A_i, \qquad R_i = \int_{\mathcal{A}_i} z^2\,dA = I_i + c_i^2 A_i$$

The integrals over the entire region $\mathcal{A}$ can be computed as the sum over all the subregions. To wit,

$$\int_{\mathcal{A}} dA = \sum_{i=1}^{N} A_i, \qquad \int_{\mathcal{A}} z\,dA = \sum_{i=1}^{N} c_i A_i, \qquad \int_{\mathcal{A}} z^2\,dA = \sum_{i=1}^{N} \left(I_i + c_i^2 A_i\right)$$

These relationships can, in turn, be used to compute the properties of the overall cross section as

$$\begin{aligned} A &= \sum_{i=1}^{N} A_i \\ c &= \frac{1}{A} \sum_{i=1}^{N} c_i A_i \\ I_C &= \sum_{i=1}^{N} \left(I_i + c_i^2 A_i\right) - c^2 A \end{aligned} \tag{9.11}$$

These equations can be very useful if the cross section can be broken down into subregions for which the areas, centroids, and moments of inertia are known. For example, any cross section that can be broken down into rectangles of width b and height h have area $A = bh$ and second moment of area $bh^3/12$ relative to the center of the rectangle where the centroid of that piece is located. With these items easily computed, the computation of the overall cross section can be carried out with Eq. 9.11, as the following example illustrates.

Example 9.4
Consider the T-beam cross section shown in Fig. 9.9. Compute the area, location of the centroid, and the second moment of the area relative to the centroid using *Trick* 1.

Solution. The cross section can be subdivided into two rectangular pieces, $\mathcal{A}_1$ and $\mathcal{A}_2$, as shown in Fig. 9.9 on the right. The first rectangle has dimensions $0.2h$ by $0.8h$, and the second has dimensions $0.2h$ and h. The individual areas are $A_1 = 0.16h^2$ and $A_2 = 0.20h^2$. The total area is the sum of the areas of the two regions,

$$A = 0.16h^2 + 0.20h^2 = 0.36h^2$$

The centroid of $\mathcal{A}_1$ is located at a distance $c_1 = 0.4h$ from the axis and the centroid of $\mathcal{A}_2$ is located at a distance $c_2 = 0.9h$ from the base of the cross section (i.e., the origin $\mathcal{O}$), as shown in Fig. 9.9. Using Eq. 9.11(b), we can compute the location of the centroid as

$$\begin{aligned} c &= \frac{1}{A}\Big[c_1 A_1 + c_2 A_2\Big] \\ &= \frac{1}{0.36h^2}\Big[(0.4h)(0.16h^2) + (0.9h)(0.20h^2)\Big] = 0.6778\,h \end{aligned}$$

The second moment of a rectangle about its own centroid is $I = bh^3/12$. Therefore, by the parallel axis theorem, the second moments of area of the two pieces relative to the origin $\mathcal{O}$ at the base are

$$\begin{aligned} I_1 + c_1^2 A_1 &= \tfrac{1}{12}(0.2h)(0.8h)^3 + (0.4h)^2(0.16h^2) = 0.03413h^4 \\ I_2 + c_2^2 A_2 &= \tfrac{1}{12}(1.0h)(0.2h)^3 + (0.9h)^2(0.20h^2) = 0.16267h^4 \end{aligned}$$

Observe that we are using Eq. 9.6 here to compute the moments of inertia of the pieces relative to the origin $\mathcal{O}$ at the base from the moments of inertia relative to their individual centroids. Finally, the second moment of the entire cross section can be computed from Eq. 9.11(c)as

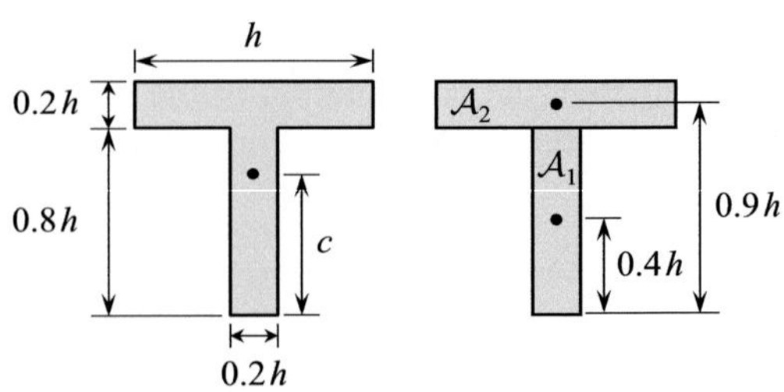

Fig. 9.9 *Example.* T-shaped cross section with height h, flange width h, and thickness $0.2h$. The locations of the individual centroids of pieces $\mathcal{A}_1$ and $\mathcal{A}_2$ are shown in the sketch on the right

$$
\begin{aligned}
I_C &= \sum_{i=1}^{2} \left(I_i + c_i^2 A_i \right) - c^2 A \\
&= 0.03413\, h^4 + 0.16267\, h^4 - (0.6778\, h)^2 (0.36\, h^2) \\
&= 0.03142\, h^4
\end{aligned}
$$

Observe that I_C has units of length to the fourth power, which is a good check on the integrity of the calculation.

The upshot of the preceding example is that the properties of the region were computed without explicitly carrying out integration over the domain. More precisely, the integration was done, once and for all, for a rectangular region, and *Trick* 1 shows how to make use of those results to compute the properties of cross sections that are not rectangular but can be subdivided into rectangular regions.

Trick 2 The second integration trick is based upon computing the geometric properties by including regions that are not part of the original cross section, which is justified if you subtract those contributions back out. Consider the situation shown in Fig. 9.10. We can write the integral with *any* integrand (•) over the region $\mathcal{A}$ as

$$
\int_{\mathcal{A}} (\bullet)\, dA = \int_{\mathcal{A}} (\bullet)\, dA + \int_{\mathcal{A}_2} \big[(\bullet) - (\bullet) \big]\, dA
$$

Because the integrand in the second integral is zero (the quantity minus itself), the value of the integral is zero. Thus, adding this term does not change the result. Next, we can observe that

$$
\mathcal{A}_1 = \mathcal{A} \cup \mathcal{A}_2
$$

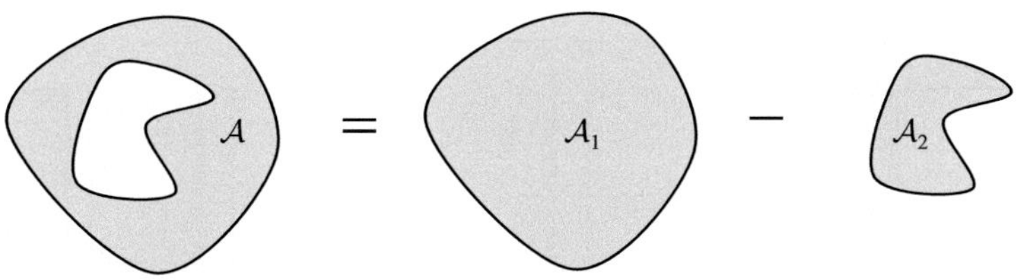

Fig. 9.10 *Trick 2.* A cross section $\mathcal{A}$ with a cutout. The integration can be accomplished as an integral over region $\mathcal{A}_1$ minus an integral over region $\mathcal{A}_2$. The process is the same for any number of cutouts

In other words, the region $\mathcal{A}_2$ is exactly the same region as the cutout in the original cross section. Therefore, the sum of integrals over $\mathcal{A}$ and $\mathcal{A}_2$ is the same as an integral over $\mathcal{A}_1$. Hence, we can write

$$\int_{\mathcal{A}} (\bullet)\, dA = \int_{\mathcal{A}_1} (\bullet)\, dA - \int_{\mathcal{A}_2} (\bullet)\, dA$$

This is *Trick* 2, and it is useful when one can identify regions that simplify the individual calculations, i.e., if the integrals over the regions $\mathcal{A}_1$ and $\mathcal{A}_2$ are simple to carry out or if those regions are simple geometries like rectangles, for which the integrals are already known. The following example illustrates how to use this integration trick.

Example 9.5
Compute the second moment of the area and the value of Q at the centroid of the I-shaped cross section shown in Fig. 9.11 using Trick 2.

Solution. The beam has web depth h, flange width h, flange thickness $0.2h$, and web thickness $0.1h$. Region $\mathcal{A}_1$ is the rectangle of width h and height $1.4h$. Region $\mathcal{A}_2$ is the two small rectangles of width $0.45h$ and height h.

The second moment of area can be computed using the formula for a rectangle given in Eq. 9.10. To wit,

$$I = \tfrac{1}{12}\big(h\big)\big(1.4h\big)^3 - (2)\tfrac{1}{12}\big(0.45h\big)\big(h\big)^3 = 0.1537h^4$$

The value $Q(0)$ can be computed similarly as

$$Q(0) = \tfrac{1}{8}\big(h\big)\big(1.4h\big)^2 - (2)\tfrac{1}{8}\big(0.45h\big)\big(h\big)^2 = 0.1325h^3$$

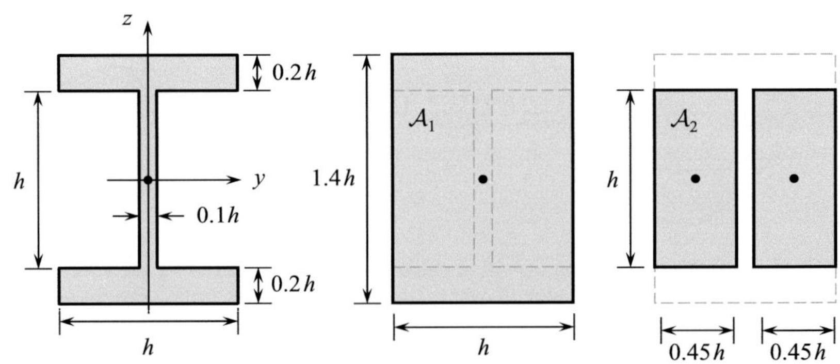

Fig. 9.11 *Example.* I-shaped cross section with web depth h, flange width h, flange thickness $0.2h$, and web thickness $0.1h$

The simplicity of this calculation is mainly due to the fact that the centroids of the three subregions are in the same location as the centroid of the overall cross section. Hence, the parallel axis theorem does not come into play (or, said another way, $c=0$).

9.3 Algorithm for Polygonal Cross Sections

Cross sections often have complex shapes for which the integrals associated with the area, centroid, and moment of inertia cannot be easily computed. In this section we lay out an algorithm for computing these properties for any cross section whose exterior surface can be represented as, or approximated by, a polygon.

Up to this point we have focused on cross-sectional properties associated with planar beam bending problems. Of course, for motion in three dimensions, a beam can bend about the z axis just as it does about the y axis. Thus, there are corresponding cross-sectional properties that go along with biaxial bending (e.g., moment of inertia about the z axis). In this section, we will generalize the problem of computing the properties of a cross section to incorporate all such possibilities.

Centroid Consider the general cross section shown in Fig. 9.12 with origin at $\mathcal{O}$. The *centroid*, $\mathcal{C}$, is defined to be the point that has the property

$$\int_{\mathcal{A}} (\mathbf{x}-\mathbf{c})\, dA = \mathbf{0}$$

This condition is the two-dimensional counterpart of Eq. 9.2. In fact, since the vectors have two components, i.e., $\mathbf{x} = y\mathbf{e}_2 + z\mathbf{e}_3$, the second component of this equation is exactly the same as Eq. 9.2. Noting that $\mathbf{c}$ is constant, we can write

$$\int_{\mathcal{A}} \mathbf{x}\, dA - \mathbf{c}\int_{\mathcal{A}} dA = \mathbf{0}$$

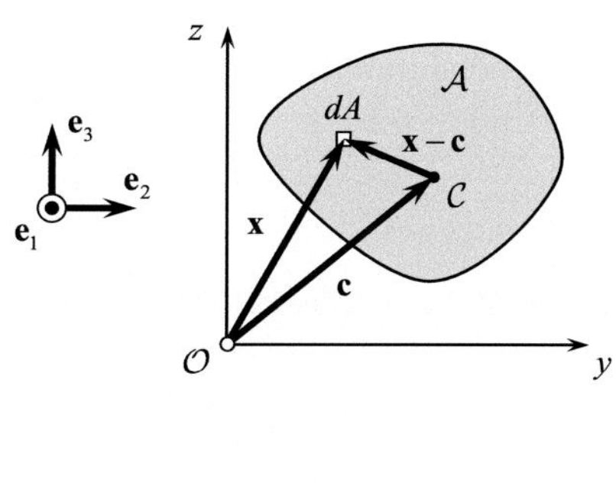

Fig. 9.12 *General cross section.* Cross-sectional region $\mathcal{A}$ with centroid $\mathcal{C}$. The coordinate system has origin $\mathcal{O}$. The typical particle is located at $\mathbf{x}$

and solve for vector **c**, the location of the centroid, as

$$\mathbf{c} = \frac{1}{A}\int_{\mathcal{A}} \mathbf{x}\,dA \tag{9.12}$$

Of course, this result also implies that

$$\int_{\mathcal{A}} \mathbf{x}\,dA = \mathbf{c}A \tag{9.13}$$

which will be useful in some of the subsequent derivations. It should be evident that the developments in this more general setting mirror those of the two-dimensional case, replacing z with $\mathbf{x}$ and c with $\mathbf{c}$.

Moment of inertia Let us define the second moment of area relative to the centroid as the tensor

$$\mathbf{J}_C = \int_{\mathcal{A}} (\mathbf{x}-\mathbf{c}) \otimes (\mathbf{x}-\mathbf{c})\,dA \tag{9.14}$$

The integrand is a tensor, so the result is, too. Since the vector **c** is constant, we can distribute the tensor product in the integrand to get

$$\begin{aligned}\mathbf{J}_C &= \int_{\mathcal{A}} (\mathbf{x}\otimes\mathbf{x} - \mathbf{x}\otimes\mathbf{c} - \mathbf{c}\otimes\mathbf{x} + \mathbf{c}\otimes\mathbf{c})\,dA \\ &= \int_{\mathcal{A}} (\mathbf{x}\otimes\mathbf{x})\,dA - \int_{\mathcal{A}} \mathbf{x}\,dA\otimes\mathbf{c} - \mathbf{c}\otimes\int_{\mathcal{A}} \mathbf{x}\,dA + (\mathbf{c}\otimes\mathbf{c})\int_{\mathcal{A}} dA\end{aligned}$$

Now, the integral in the last term is simply the total area A, and the integrals in the second and third terms, from Eq. 9.13, are equal to $\mathbf{c}A$. Thus, we have

$$\mathbf{J}_C = \int_{\mathcal{A}} (\mathbf{x}\otimes\mathbf{x})\,dA - \mathbf{c}A\otimes\mathbf{c} - \mathbf{c}\otimes\mathbf{c}A + (\mathbf{c}\otimes\mathbf{c})A$$

Gathering terms gives the *parallel axis theorem* for the general case:

$$\mathbf{J}_C = \int_{\mathcal{A}} (\mathbf{x}\otimes\mathbf{x})\,dA - (\mathbf{c}\otimes\mathbf{c})A \tag{9.15}$$

The second moment of the area relative to the origin $\mathcal{O}$ can be defined as

$$\mathbf{J}_O = \int_{\mathcal{A}} (\mathbf{x}\otimes\mathbf{x})\,dA \tag{9.16}$$

Thus, the parallel axis theorem can be written as

$$\mathbf{J}_C = \mathbf{J}_O - (\mathbf{c}\otimes\mathbf{c})A \tag{9.17}$$

The component expression for the position vector is $\mathbf{x} = y\mathbf{e}_2 + z\mathbf{e}_3$. The symbol $\otimes$ indicates the *tensor product* of vectors. As mentioned in Chap. 1, the matrix representation of the tensor product of $\mathbf{x}$ with itself can be written as

$$\{\mathbf{x} \otimes \mathbf{x}\} = \{\mathbf{x}\}\{\mathbf{x}\}^T = \begin{Bmatrix} y \\ z \end{Bmatrix} \begin{bmatrix} y & z \end{bmatrix} = \begin{bmatrix} y^2 & yz \\ yz & z^2 \end{bmatrix}$$

Therefore, we can write

$$\mathbf{J}_O = \int_A \begin{bmatrix} y^2 & yz \\ yz & z^2 \end{bmatrix} dA$$

Thus, the (2, 2) component of $\mathbf{J}_O$ is I_O, the second moment of area relative to the y axis from the previous section. The (1,1) component is the second moment of the area relative to the z axis. The off-diagonal terms, with integrands yz, are called *products of inertia*. They had no counterpart in the previous section.

The second moment of area is a symmetric *tensor* in two dimensions. Therefore, it enjoys all the properties and interpretations of any tensor (like stress and strain). In components, we can write the tensor as

$$\mathbf{J}_C = \begin{bmatrix} I_{yy} & I_{yz} \\ I_{yz} & I_{zz} \end{bmatrix} \tag{9.18}$$

Observe that I_{zz} is the moment of inertia about the y axis, i.e.,

$$I_{zz} = \int_A (z - c_z)^2 \, dA$$

which is the moment of inertia in Eq. 9.4 from earlier in the chapter and it is what we called I in the previous chapters on planar beam bending in the context of the deflection equation $w'' = -M/EI$ and the normal stress formula $\sigma = Mz/I$. The quantity I_{yy} is the moment of inertia about the z axis. This quantity represents resistance to bending about the other axis. The *product of inertia* I_{yz} is a bit harder to interpret physically. If the product of inertia is zero, then I_{zz} and I_{yy} are the principal moments of inertia.

Like stress and strain, the second moment of area can be represented with a Mohr circle, as shown in Fig. 9.13. The center and radius of the Mohr circle can be computed from the components of $\mathbf{J}_C$ as

$$c = \tfrac{1}{2}\big(I_{yy} + I_{zz}\big), \qquad R = \sqrt{\tfrac{1}{4}(I_{zz} - I_{yy})^2 - I_{yz}^2}$$

and the *principal angle* can be computed as

$$\tan 2\theta_p = \frac{2I_{yz}}{I_{yy} - I_{zz}}$$

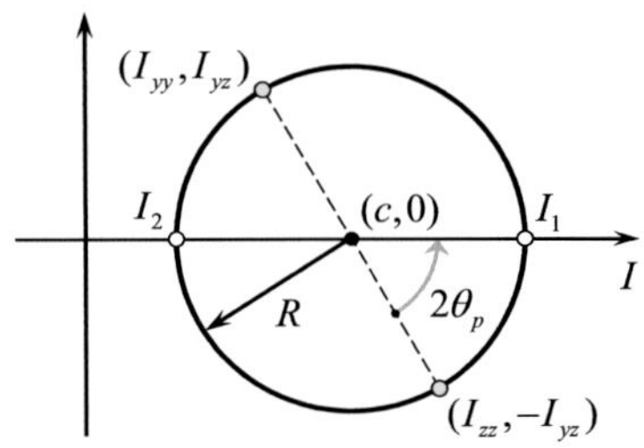

Fig. 9.13 *Mohr circle for* $\mathbf{J}_{\mathcal{C}}$. Since the second moment of area is a tensor, it also has a Mohr circle of radius R centered at $(c, 0)$

The principal moments of inertia are $I_1 = c + R$ and $I_2 = c - R$. One way to think about the principal moments of inertia is through the lens of planar bending of a beam. If the cross section is oriented so that the major principal direction is aligned with the plane of bending, then the moment of inertia is I_1 and that is the orientation that resists bending best, i.e., it yields the lowest deflections and stresses. On the other hand, if the cross section is oriented so that the minor principal direction is aligned with the plane of bending, then the moment of inertia is I_2, and that is the least effective orientation for resisting bending. An I-beam laid on its side is much more flexible than one oriented vertically. The product of inertia I_{yz} is zero if the principal axes align with the coordinate axes (y, z).

The principal values and principal directions of the moment of inertia tensor can also be computed from the matrix eigenvalue problem

$$\left[\mathbf{J}_{\mathcal{C}} - \mu\,\mathbf{I}\right]\mathbf{n} = \mathbf{0}$$

where $\mathbf{I}$ is the 2×2 identity matrix. This eigenvalue problem has two solutions $(\mu_1, \mathbf{n}_1)$ and $(\mu_2, \mathbf{n}_2)$, where $\mu_1 = I_1$ and $\mu_2 = I_2$ are the principal moments of inertia relative to the centroid. The unit vectors $\mathbf{n}_1$ and $\mathbf{n}_2$ point in the directions of the principal axes of the cross section. If a cross section has an axis of symmetry, then the principal directions lie along and perpendicular to that axis.

We can use the parallel axis theorem to translate second moments of area from one point to another, as the following example shows.

Example 9.6
Assume that the second moment of the area relative to point $\mathcal{B}$, which is not the centroid, is known. Compute the second moment of the area relative to point $\mathcal{D}$, also not the centroid, using the parallel axis theorem.

Solution. Let $\mathbf{c}_1$ be a vector from $\mathcal{B}$ to the location of the centroid $\mathcal{C}$, and let $\mathbf{c}_2$ be a vector from $\mathcal{D}$ to $\mathcal{C}$. Using the parallel axis theorem, noting that the centroid is the same for both, we can write

$$\mathbf{J}_{\mathcal{C}} = \mathbf{J}_{\mathcal{B}} - (\mathbf{c}_1 \otimes \mathbf{c}_1)A = \mathbf{J}_{\mathcal{D}} - (\mathbf{c}_2 \otimes \mathbf{c}_2)A$$

from which we can solve for $\mathbf{J}_{\mathcal{D}}$ as

$$\mathbf{J}_{\mathcal{D}} = \mathbf{J}_{\mathcal{B}} + \big[(\mathbf{c}_2 \otimes \mathbf{c}_2) - (\mathbf{c}_1 \otimes \mathbf{c}_1)\big]A$$

Observe that the location of the centroid $\mathcal{C}$ is needed to complete this calculation, but the second moment relative to the centroid is not.

First moment Q The first moment of the entire area of the cross section was used to define the location of the centroid. For the shear stress formula, we need the first moment of a region of the cross section defined by a cut, as shown in Fig. 9.14. To wit,

$$\mathbf{Q}_{\mathcal{C}} = \int_{\mathcal{A}_f} (\mathbf{x} - \mathbf{c})\, dA \tag{9.19}$$

where $\mathbf{c}$ is the location of the centroid of the region $\mathcal{A}$ defined in Eq. 9.12. Since $\mathbf{c}$ is constant, we can write this equation as

$$\mathbf{Q}_{\mathcal{C}} = \int_{\mathcal{A}_f} \mathbf{x}\, dA - A_f \mathbf{c} \tag{9.20}$$

where A_f is the area of region $\mathcal{A}_f$. The quantity $\mathbf{Q}_{\mathcal{C}}$ depends upon the cut, and there are an infinite number of ways to make such a cut, even to evaluate $\mathbf{Q}_{\mathcal{C}}$ at the centroid of the cross section. In essence, we must define the dot-dash line in Fig. 9.14, which we can do, for example, by specifying a point on the line and a unit vector along the line. The first moment of area is useful in the shear stress formula, Eq. 7.27.

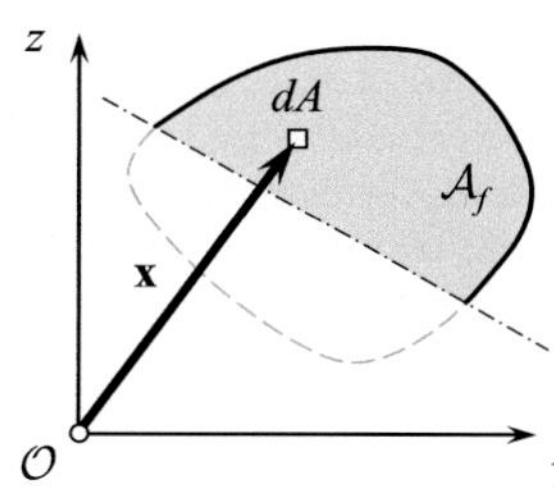

Fig. 9.14 *Cut to define* $\mathbf{Q}_{\mathcal{C}}$. A cross-sectional region $\mathcal{A}_f$, defined by the cut, is a portion of the total cross section $\mathcal{A}$

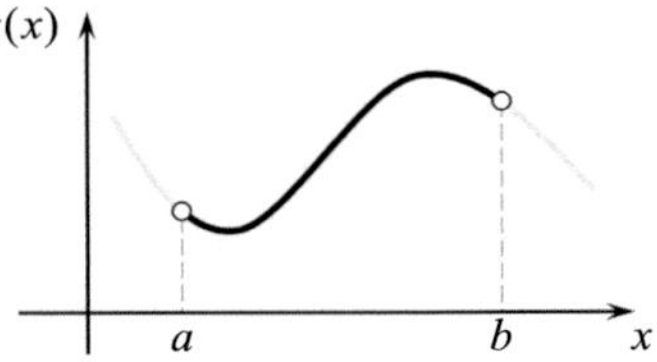

Fig. 9.15 *The fundamental theorem of calculus.* A function $g(x)$ between a to b

9.3.1 The Divergence Theorem

The algorithm for computing the geometric properties of a cross section relies on the *divergence theorem* of calculus in two dimensions. To warm up to the idea, it is worth reviewing the *fundamental theorem of calculus* for functions of a single variable, like $g(x)$ shown in Fig. 9.15. The theorem states that

$$\int_a^b g'(x)\,dx = g(b) - g(a) \tag{9.21}$$

where $g'(x)$ is the first derivative of $g(x)$. The theorem states that the integral of the derivative of $g(x)$ between the limits a and b is equal to the difference in the function values $g(b)$ and $g(a)$. This theorem connects integration of functions and differentiation of functions.

There is an analogous result for a scalar function $g(\mathbf{x})$, a vector function $\mathbf{w}(\mathbf{x})$, and a tensor function $\mathbf{T}(\mathbf{x})$ of the vector variable $\mathbf{x}$. These functions are defined on a region $\mathcal{A}$ with boundary $\partial\mathcal{A}$. The boundary has an outward unit normal vector field $\mathbf{n}$ at each point on the boundary, as shown in Fig. 9.16. In two dimensions, the vector variable has component representation $\mathbf{x} = x_1\mathbf{e}_1 + x_2\mathbf{e}_2$.[3]

There are two main questions that arise in going to multiple dimensions: (1) What is the analog to $g'(x)$ in Eq. 9.21? In multiple dimensions there are several kinds of derivatives. (2) What is the analog for evaluating the function at the end points of

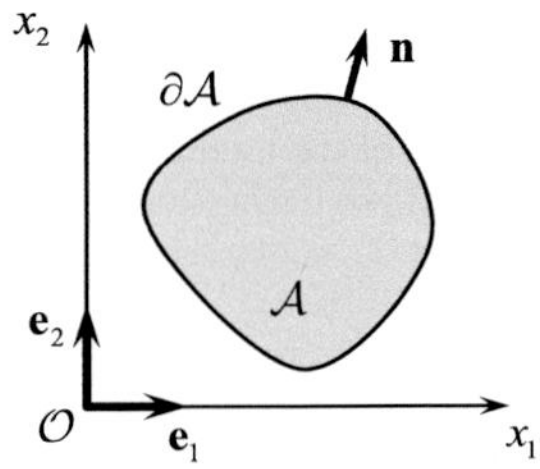

Fig. 9.16 *A cross section and its boundary.* A cross section $\mathcal{A}$ with boundary $\partial\mathcal{A}$ having outward unit normal vector field $\mathbf{n}$

[3] The development of this theorem is best done with *index notation*. So, for this section, we will use coordinates x_1 and x_2 (rather than y and z) and base vectors $\mathbf{e}_1$ and $\mathbf{e}_2$ (replacing $\mathbf{e}_2$ and $\mathbf{e}_3$, respectively). Once we have completed the derivations that rely on the divergence theorem, we can translate the results back to our usual coordinates.

the interval? The answer to the first question is the *divergence*, and the answer to the second is a line integral around the boundary of the region defining the cross section.

The divergence theorem allows us to relate the integral of the derivative of a function over the region $\mathcal{A}$ to an integral of a related function around the boundary $\partial\mathcal{A}$. The relationship can be stated as

$$\int_{\mathcal{A}} \operatorname{div}(\mathbf{w})\, dA = \int_{\partial\mathcal{A}} \mathbf{w} \cdot \mathbf{n}\, ds \tag{9.22}$$

where $\partial\mathcal{A}$ is the boundary of the region, $\mathbf{n}$ is the unit normal to the boundary, as shown in Fig. 9.16, and div($\bullet$) is the *divergence* of ($\bullet$). This theorem holds for any vector or tensor field $\mathbf{w}$ for which the divergence is well behaved (i.e., the integral exists). The divergence of a vector field $\mathbf{w}$ can be computed in Cartesian coordinates as

$$\operatorname{div}(\mathbf{w}) = \sum_{i=1}^{2} \frac{\partial w_i}{\partial x_i} \tag{9.23}$$

where w_i is the ith component of the vector field $\mathbf{w}(\mathbf{x})$ and x_i is the ith component of the vector $\mathbf{x}$. The following example shows how to compute the divergence of a vector function of a vector variable.

Example 9.7
Compute the divergence of the vector field $\mathbf{w}(\mathbf{x}) = x_1^3 x_2^2 \mathbf{e}_1 + x_1 x_2^3 \mathbf{e}_2$.

Solution. First, identify the components of $\mathbf{w}$ as

$$w_1 = x_1^3 x_2^2 \qquad w_2 = x_1 x_2^3$$

Now, apply the formula for the divergence from Eq. 9.23. To wit,

$$\begin{aligned}
\operatorname{div}(\mathbf{w}) &= \frac{\partial w_1}{\partial x_1} + \frac{\partial w_2}{\partial x_2} \\
&= \frac{\partial}{\partial x_1}\left(x_1^3 x_2^2\right) + \frac{\partial}{\partial x_2}\left(x_1 x_2^3\right) \\
&= 3x_1^2 x_2^2 + 3x_1 x_2^2
\end{aligned}$$

The divergence of the vector field $\mathbf{w}(\mathbf{x}) = \mathbf{x}$ can be computed as

$$\operatorname{div}(\mathbf{x}) = \sum_{i=1}^{2} \frac{\partial x_i}{\partial x_i} = \frac{\partial x_1}{\partial x_1} + \frac{\partial x_2}{\partial x_2} = 2 \tag{9.24}$$

The formula for the divergence of a tensor field $\mathbf{T}$ is

$$\operatorname{div}(\mathbf{T}) = \sum_{i=1}^{2}\sum_{j=1}^{2} \frac{\partial T_{ij}}{\partial x_j}\mathbf{e}_i$$

where T_{ij} are the components of the tensor relative to the basis $\mathbf{e}_i \otimes \mathbf{e}_j$. For example, the tensor $\mathbf{x}\otimes\mathbf{x}$ has components $x_i x_j$ and the divergence is

$$\operatorname{div}\big[\mathbf{x}\otimes\mathbf{x}\big] = \sum_{i=1}^{2}\sum_{j=1}^{2} \frac{\partial}{\partial x_j}(x_i x_j)\mathbf{e}_i = \sum_{i=1}^{2}\sum_{j=1}^{2}\left(\frac{\partial x_i}{\partial x_j}x_j + x_i\frac{\partial x_j}{\partial x_j}\right)\mathbf{e}_i$$

where we have used the product rule for differentiation. Now, we can note that the derivative of an independent variable with respect to itself, e.g., $\partial x_1/\partial x_1$, is one and the derivative of one independent variable with respect to another, e.g., $\partial x_1/\partial x_2$, is zero. To see what the sum implies, let us write out all the terms and simplify the result. To wit,

$$\begin{aligned}\operatorname{div}\big[\mathbf{x}\otimes\mathbf{x}\big] &= \left(\frac{\partial x_1}{\partial x_1}x_1 + x_1\frac{\partial x_1}{\partial x_1}\right)\mathbf{e}_1 + \left(\frac{\partial x_1}{\partial x_2}x_2 + x_1\frac{\partial x_2}{\partial x_2}\right)\mathbf{e}_1 \\ &+ \left(\frac{\partial x_2}{\partial x_1}x_1 + x_2\frac{\partial x_1}{\partial x_1}\right)\mathbf{e}_2 + \left(\frac{\partial x_2}{\partial x_2}x_2 + x_2\frac{\partial x_2}{\partial x_2}\right)\mathbf{e}_2\end{aligned}$$

where the first term has $i=1$ and $j=1$, the second term has $i=1$ and $j=2$, etc. Noting the derivatives that are one and those that are zero, we get

$$\begin{aligned}\operatorname{div}\big[\mathbf{x}\otimes\mathbf{x}\big] &= \Big((1)x_1 + x_1(1)\Big)\mathbf{e}_1 + \Big((0)x_2 + x_1(1)\Big)\mathbf{e}_1 \\ &+ \Big((0)x_1 + x_2(1)\Big)\mathbf{e}_2 + \Big((1)x_2 + x_2(1)\Big)\mathbf{e}_2\end{aligned}$$

Recognizing that $3x_1\mathbf{e}_1 + 3x_2\mathbf{e}_2 = 3\mathbf{x}$, the final result reduces to

$$\operatorname{div}\big[\mathbf{x}\otimes\mathbf{x}\big] = 3\mathbf{x}$$

We will apply the divergence theorem to three different vector or tensor functions to support our calculations of the properties of cross sections. The following table gives the function, the divergence of that function, and the dot product of the function with the unit normal vector $\mathbf{n}$:

$\mathbf{w}(\mathbf{x})$	$\operatorname{div}(\mathbf{w})$	$\mathbf{w}\cdot\mathbf{n}$
$\mathbf{x}$	2	$(\mathbf{x}\cdot\mathbf{n})$
$\mathbf{x}\otimes\mathbf{x}$	$3\,\mathbf{x}$	$(\mathbf{x}\cdot\mathbf{n})\mathbf{x}$
$\mathbf{x}\otimes\mathbf{x}\otimes\mathbf{x}$	$4\big[\mathbf{x}\otimes\mathbf{x}\big]$	$(\mathbf{x}\cdot\mathbf{n})\big[\mathbf{x}\otimes\mathbf{x}\big]$

The divergence of the third-order tensor $(\mathbf{x}\otimes\mathbf{x}\otimes\mathbf{x})$ follows the pattern set by the other two cases. Notice that the divergence of a vector is a scalar, the divergence of a second-order tensor is a vector, and the divergence of a third-order tensor is a second-order tensor. Also, the tensorial order of $\mathbf{w}\cdot\mathbf{n}$ matches the order of $\text{div}(\mathbf{w})$.

If we interpret the vector $\mathbf{x}$ as the position vector from the origin to the element of integration in the cross section, then we can convert the integrals associated with the properties of a cross section from area integrals to line integrals around the boundary. The advantage of making this conversion is that we will be able to compute the properties of the cross section from a description of the boundary of the region. Further, if we approximate the boundary with straight lines, then we will be able to compute those integrals explicitly and develop a simple formula for the properties. Using the divergence theorem, we can write the following integrals over the cross section as line integrals around the boundary of the cross section:

$$\begin{aligned}\int_{\mathcal{A}} dA &= \tfrac{1}{2}\int_{\partial\mathcal{A}}(\mathbf{x}\cdot\mathbf{n})\,ds\\ \int_{\mathcal{A}} \mathbf{x}\,dA &= \tfrac{1}{3}\int_{\partial\mathcal{A}}(\mathbf{x}\cdot\mathbf{n})\,\mathbf{x}\,ds\\ \int_{\mathcal{A}} \big[\mathbf{x}\otimes\mathbf{x}\big]\,dA &= \tfrac{1}{4}\int_{\partial\mathcal{A}}(\mathbf{x}\cdot\mathbf{n})\big[\mathbf{x}\otimes\mathbf{x}\big]\,ds\end{aligned}$$

These three integrals play a key role in computing the properties of a cross section, as we shall see in the next section.

9.3.2 Polygonal Cross Sections

If we approximate the boundary of region $\mathcal{A}$ with straight line segments (or if the region is already a polygon), then the integral over the entire boundary can be expressed as the sum of the integrals over the individual straight segments. We can compute those line integrals exactly. Define

$$\begin{aligned}A_n &= \tfrac{1}{2}\int_{\partial\mathcal{A}_n}(\mathbf{x}\cdot\mathbf{n})\,ds\\ \mathbf{r}_n &= \tfrac{1}{3}\int_{\partial\mathcal{A}_n}(\mathbf{x}\cdot\mathbf{n})\,\mathbf{x}\,ds\\ \mathbf{R}_n &= \tfrac{1}{4}\int_{\partial\mathcal{A}_n}(\mathbf{x}\cdot\mathbf{n})\big[\mathbf{x}\otimes\mathbf{x}\big]\,ds\end{aligned}$$

where $\partial\mathcal{A}_n$ is the part of the boundary associated with segment n.

Consider the representation of the boundary segment n shown in Fig. 9.17. The origin of coordinates is $\mathcal{O}$. The coordinates of the two adjacent vertices are $\mathbf{x}_i$ and $\mathbf{x}_j$. The unit vector $\mathbf{m}$ and the segment length h can be computed from the coordinates

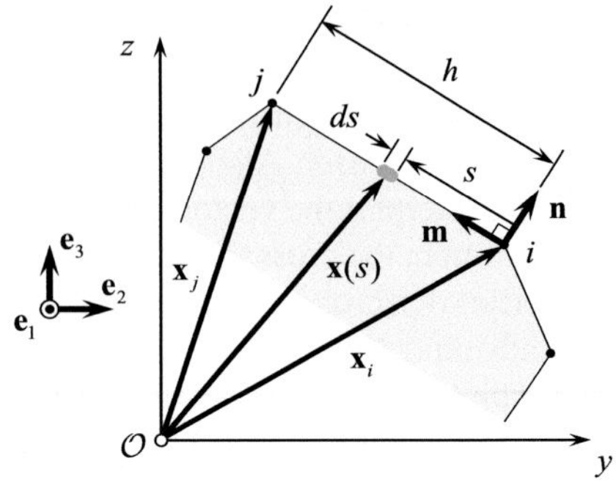

Fig. 9.17 *Geometry of boundary segment.* The boundary segment n between adjacent vertices i and j, with unit vectors $\mathbf{m}$ and $\mathbf{n}$. The coordinate system for $\mathbf{x}$ has origin at $\mathcal{O}$

of the vertices as

$$\mathbf{m} = \frac{\mathbf{x}_j - \mathbf{x}_i}{h}, \qquad h = \|\mathbf{x}_j - \mathbf{x}_i\|$$

The vector $\mathbf{x}(s)$ to the typical particle of length ds along the segment is

$$\mathbf{x}(s) = \mathbf{x}_i + s\mathbf{m}$$

where $s \in [0, h]$ measures distance along the segment from vertex i to vertex j. Observe that the dot product between $\mathbf{x}$ and $\mathbf{n}$ reduces to $\mathbf{x} \cdot \mathbf{n} = \mathbf{x}_i \cdot \mathbf{n}$, which is constant along the segment because $\mathbf{n}$ is orthogonal to $\mathbf{m}$. Substituting $\mathbf{x}(s)$ into the area integral for segment n gives

$$A_n = \tfrac{1}{2}\big(\mathbf{x}_i \cdot \mathbf{n}\big) \int_0^h ds = \tfrac{1}{2}\big(\mathbf{x}_i \cdot \mathbf{n}\big)h$$

This equation can be put into a form that eliminates h and $\mathbf{n}$ by noting that $\mathbf{n} = \mathbf{m} \times \mathbf{e}_1$ and $h\mathbf{m} = \mathbf{x}_j - \mathbf{x}_i$. Now, compute

$$\begin{aligned} A_n &= \tfrac{1}{2}\,\mathbf{x}_i \cdot \big(h\mathbf{m} \times \mathbf{e}_1\big) \\ &= \tfrac{1}{2}\,\mathbf{x}_i \cdot \big(\big(\mathbf{x}_j - \mathbf{x}_i\big) \times \mathbf{e}_1\big) \\ &= \tfrac{1}{2}\,\mathbf{e}_1 \cdot \big(\mathbf{x}_i \times \big(\mathbf{x}_j - \mathbf{x}_i\big)\big) = \tfrac{1}{2}\big(\mathbf{x}_i \times \mathbf{x}_j\big) \cdot \mathbf{e}_1 \end{aligned}$$

where we have used the cyclic property of the triple scalar product and noted that a vector crossed with itself is zero. The vectors $\mathbf{x}_i$ and $\mathbf{x}_j$ are

$$\mathbf{x}_i = y_i\mathbf{e}_2 + z_i\mathbf{e}_3, \qquad \mathbf{x}_j = y_j\mathbf{e}_2 + z_j\mathbf{e}_3$$

where (y_i, z_i) are the components of the vector $\mathbf{x}_i$, and (y_j, z_j) are the components of $\mathbf{x}_j$. Thus, the contribution to the area of segment n can be computed as

$$A_n = \tfrac{1}{2}\big(y_i z_j - y_j z_i\big) \tag{9.25}$$

The integral for $\mathbf{r}_n$ can be carried out as

$$\begin{aligned}\mathbf{r}_n &= \tfrac{1}{3}\int_0^h (\mathbf{x}_i \cdot \mathbf{n})(\mathbf{x}_i + s\mathbf{m})\,ds \\ &= \tfrac{1}{3}(\mathbf{x}_i \cdot \mathbf{n})\Big[\mathbf{x}_i s + \tfrac{1}{2}s^2\mathbf{m}\Big]_0^h \\ &= \tfrac{1}{3}(\mathbf{x}_i \cdot \mathbf{n})(\mathbf{x}_i h + \tfrac{1}{2}h^2\mathbf{m})\end{aligned}$$

Recognizing that $h\mathbf{m} = \mathbf{x}_j - \mathbf{x}_i$ and $(\mathbf{x}_i \cdot \mathbf{n})h = 2A_n$, the expression simplifies to

$$\mathbf{r}_n = \tfrac{1}{3}(\mathbf{x}_i + \mathbf{x}_j)A_n \tag{9.26}$$

A similar derivation holds for $\mathbf{R}_n$. After some calculation, we get

$$\mathbf{R}_n = \tfrac{1}{12}\Big(2[\,\mathbf{x}_i \otimes \mathbf{x}_i\,] + [\,\mathbf{x}_i \otimes \mathbf{x}_j\,] + [\,\mathbf{x}_j \otimes \mathbf{x}_i\,] + 2[\,\mathbf{x}_j \otimes \mathbf{x}_j\,]\Big)A_n \tag{9.27}$$

The details are left to the reader. The edge contributions can be summed to obtain the complete integrals. To wit,

$$A = \sum_{n=1}^{N} A_n, \qquad \mathbf{c} = \frac{1}{A}\sum_{n=1}^{N} \mathbf{r}_n, \qquad \mathbf{J}_{\mathcal{O}} = \sum_{n=1}^{N} \mathbf{R}_n \tag{9.28}$$

The moment of inertia relative to the centroid, $\mathbf{J}_{\mathcal{C}}$, can be computed from $\mathbf{J}_{\mathcal{O}}$ by the parallel axis theorem, i.e., Eq. 9.15, as

$$\mathbf{J}_{\mathcal{C}} = \mathbf{J}_{\mathcal{O}} - A\big[\,\mathbf{c} \otimes \mathbf{c}\,\big]$$

These calculations are implemented in code in the following section.

9.3.3 Code Implementation

Code 9.1 `ComputeProperties` is a function that computes the properties of the cross section A, $\mathbf{c}$, $\mathbf{Q}_{\mathcal{O}}$, and $\mathbf{J}_{\mathcal{C}}$ from the coordinates of the vertices defining a polygonal region, using the equations derived in the previous section. The vertex coordinates are defined relative to the origin $\mathcal{O}$. The value $\mathbf{Q}_{\mathcal{O}}$ is a bit redundant, as it could be computed as $A\mathbf{c}$, but it will be useful for computing various versions of $\mathbf{Q}$ later.

Code 9.1 MATLAB code `ComputeProperties` to compute the area, centroid, moment of inertia, and Q of a cross section whose boundary is a polygon

```
function [A,Q,c,J] = ComputeProperties(x)
%  Compute properties of polygonal cross section with vertices x

%. Extract number of sides
   nSides = size(x,1) - 1;

%. Initialize values
   A = 0;                          % Area
   Q = zeros(2,1);                 % First moment relative to origin
   R = zeros(2,2);                 % Second moment relative to origin

   for n=1:nSides

%... Fetch vectors at end points
     xi = x(n,:)';                 % x at vertex i
     xj = x(n+1,:)';               % x at vertex j

%... Compute contribution from side n and add to totals
     An = (xi(1)*xj(2) - xi(2)*xj(1))/2;
     A = A + An;
     Q = Q + An*(xi + xj)/3;
     R = R + An*(2*(xi*xi') + (xi*xj') + (xj*xi') + 2*(xj*xj'))/12;

   end

%. Finish the calculation of centroid and moment of inertia
   c = Q/A;
   J = R - A*(c*c');

end
```

This code takes in the vertex coordinates `x` and computes the cross-sectional area `A`, the first moment of area `Q` (relative to the origin), the location of the centroid `c`, and the moment of inertia tensor `J` (relative to the centroid). The code requires that the input vertex coordinates `x` be listed in sequential order in a *counterclockwise* sweep of the vertices. Any vertex can be the first. The list of vertices must include a repeat of the first vertex at the end to close the polygon.

If vertices are listed in clockwise order, then the result is the negative of the integral. This feature allows the creation of regions with cutouts. Traverse the exterior vertices in counterclockwise order, then move to one of the cutout vertices. Traverse the cutout vertices in clockwise order. Finally, list the first exterior vertex again to close the polygon. Any number of cutouts can be included. Think of the list of vertices as you would if you were going to connect the dots on a piece of paper without lifting the pen, including the move from the exterior to a cutout or from one cutout to the next.

The function `ComputeProperties` can be executed from a script to compute the basic properties of the overall cross section. Code 9.2 reads the input from `InputCrossSection`, computes the properties, computes the eigenvalues of $\mathbf{J}_C$, and then plots and prints the results.

Code 9.2 MATLAB code `XSectionProp` to compute the area, centroid, and moment of inertia relative to the centroid of a cross section whose surface can be represented as a polygon

```
   clear; clc;

%. Establish coordinates x of the polygon vertices
   [x,Section] = InputCrossSection;

%. Compute the properties of the x-section
   [A,Qo,c,Jc] = ComputeProperties(x);

%. Compute the principal values (eigenvalues) of Jc
   [eVecs,eVals] = eig(Jc);

%. Plot cross section and print results
   PlotResults(x,c,eVecs)
   PrintResults(Section,A,c,Jc,eVals,x)
```

The output functions `PlotResults` and `PrintResults` are not included here. The printing function just prints the input and computed values in a nicely formatted way. The plotting function renders the cross section, plots the location of the centroid, and draws lines in the directions of the principal axes. Code 9.2 `InputCrossSections` is a function that generates the vertex coordinates as input to Code 9.2 for various cross sections.

Code 9.3 MATLAB code `InputCrossSection` to specify the vertex coordinates of a cross section defined as a polygonal region

```
function [x,Section] = InputCrossSection
%  Create the vertex locations, x, for a polygonal region.
%  Vertices must be numbered sequentially with positive areas numbered
%  counterclockwise and negative areas numbered clockwise (for cutouts)

%. Names of section types for menu
   SectionNames = {'Rectangle'; 'Rectanglular Tube'; ...
                   'Triangle'; 'Triangle with cutout'; ...
                   'Circular wedge'; 'Circular arc';...
                   'Solid N-gon'; 'T-Beam (plain)';...
                   'I-Beam (plain)'; 'I-Beam (fillets)';...
                   'Channel (fillet)'; 'L-section (fillet)';...
                   'Constant thickness rose'};

%. Select cross section by menu
   msg = 'Select Cross Section';
   Section.Type = listdlg('PromptString',msg,'ListString',SectionNames);
   Section.Name = char(SectionNames(Section.Type));

%. Establish vertex coordinates for given type
   switch Section.Type

     case 1  % Rectangle
       b = 2; h = 6;
       x = [0,0; b,0; b,h; 0,h; 0,0];

     case 2  % Rectanglular Tube
       b = 2; h = 5; tf = 0.5; tw = 0.5;
       x = [0,0; b,0; b,h; 0,h; 0,0; tw,tf;...
            tw,h-tf; b-tw,h-tf; b-tw,tf; tw,tf; 0,0];

     case 3  % Triangle
       b = 6; h = 6;
       x = [0,0; b,0; b/2,h; 0,0];
```

```
case 4  % Triangle with cutout
  b = 6; h = 6; a = 1; w = 1; c = 3;
  x = [0,0; b,0; b/2,h; 0,0;...
       a,w; b/2,h-2*w; 2*b/3,c; a,w; 0,0];

case 5  % Circular wedge
  r = 2; ArcAngle = pi/6;  nPts = 31;
  phi = linspace(ArcAngle,pi-ArcAngle,nPts)';
  x = r*[cos(phi), sin(phi)];
  x = [0,0; x; 0,0];

case 6  % Circlular arc
  rOuter = 5; rInner = 3.7; angle = -pi/6; nPts = 41;
  phiO = linspace(angle,pi-angle,nPts)';
  phiI = flip(phiO);
  x = rOuter*[cos(phiO), sin(phiO)];
  y = rInner*[cos(phiI), sin(phiI)];
  z = rOuter*[cos(angle), sin(angle)];
  x = [x; y; z];

case 7  % Solid N-gon
  r = 1;  nPts = 51;
  phi = linspace(0,2*pi,nPts)';
  x = r*[cos(phi), sin(phi)];

case 8  % T-Beam
  b = 10; tf = 2; h = 10; tw = 2;
  B = b/2; Tw = tw/2; H = h-tf;
  x = [Tw,0; Tw,H; B,H; B,h; -B,h; -B,H; -Tw,H; -Tw,0 ; Tw,0];

case 9  % I-Beam (plain)
  b = 12; tf = 1.5; h = 18; tw = 0.8;
  B = b/2; Tw = tw/2;
  fx = [B;  B; Tw];
  fy = [0; tf; tf];
  x = [fx,fy; flip([fx,h-fy]); -fx,h-fy; flip([-fx,fy]); B,0];

case 10  % I-Beam (fillet)
  b = 12; tf = 1.5; h = 18; tw = 0.8; r = 1; nPts = 6;
  B = b/2; Tw = tw/2; R = min(tf/2,r);
  theta = linspace(0,pi/2,nPts)';
  ct = 1-cos(theta); st = 1-sin(theta);
  fx = [B;  B-R*ct; Tw+r*st];
  fy = [0; tf-R*st; tf+r*ct];
  x = [fx,fy; flip([fx,h-fy]); -fx,h-fy; flip([-fx,fy]); B,0];

case 11  % Channel (fillet)
  b = 6; tf = 1.2; h = 15; tw = 0.8; r = 1; nPts = 6;
  theta = linspace(0,pi/2,nPts)'; R = min(tf/2,r);
  ct = 1-cos(theta); st = 1-sin(theta);
  fx = [0; b;  b-R*ct; tw+r*st];
  fy = [0; 0; tf-R*st; tf+r*ct];
  x = [fx,fy; flip([fx,h-fy]); 0,0];

case 12  % L-shape (fillet)
  t = 12; b = 100; h = 150; nPts = 6; r = 8;
  theta = linspace(0,pi/2,nPts)'; R = min(t/2,r);
  ct = 1-cos(theta); st = 1-sin(theta);
  fx = [0; b; b-R*ct; t+r*st; t-R*ct; t-R; 0; 0; 0];
  fy = [0; 0; t-R*st; t+r*ct; h-R*st;   h; h; h; 0];
  x = [fx,fy];

case 13 % Constant thickness n-leaved rose
  t = 2; R = 15; a = 4; m = 4; nPts = 1000;
  xOuter = zeros(nPts,2); xInner = zeros(nPts,2);
  for i=1:nPts
```

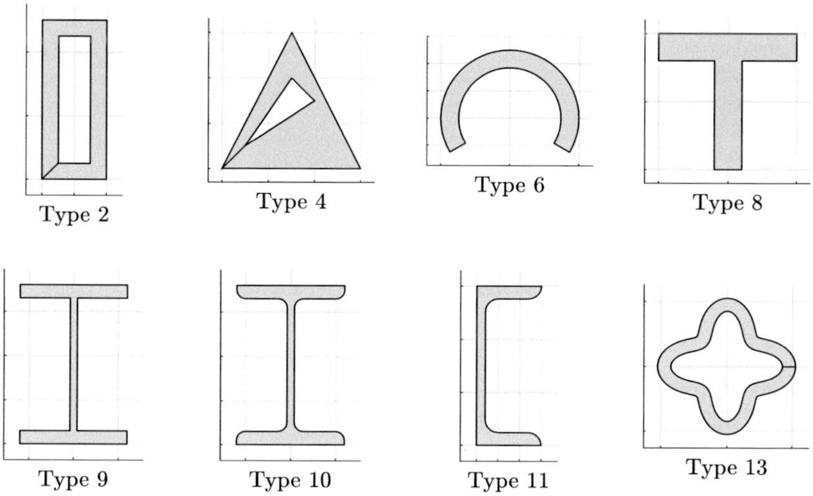

Fig. 9.18 *Sample cross sections.* Eight of the cross sections available in the function `InputCrossSections`

```
            th = 2*pi*(i-1)/(nPts-1);
            p = [ cos(th), sin(th)];
            q = [-sin(th), cos(th)];
            x0 = (R + a*cos(m*th))*p;
            N = m*a*sin(m*th)*q + (R + a*cos(m*th))*p;
            N = N/norm(N);
            xOuter(i,:) = x0 + t*N;
            xInner(i,:) = x0 - t*N;
        end
        x = [xOuter; flip(xInner); xOuter(1,:)];

    end % switch
end
```

This code is set up to have a menu selection of region type. The output of the function includes the vertex coordinates `x` and a 'struct' `Section` that holds the `Type` and `Name` of the cross section (for printing purposes). Most of the cross sections in Code 9.3 are parameterized so that the size can be changed easily. The library of regions can be expanded by adding new cases. A sampler of the cross sections available in the function is shown in Fig. 9.18.

9.3.4 Computing *Q*

The function `ComputeProperties` in Code 9.1 evaluates the first moment of the area relative to the origin as part of the task. For the entire cross section, that quantity is needed to compute the location of the centroid. The first moment of area is also

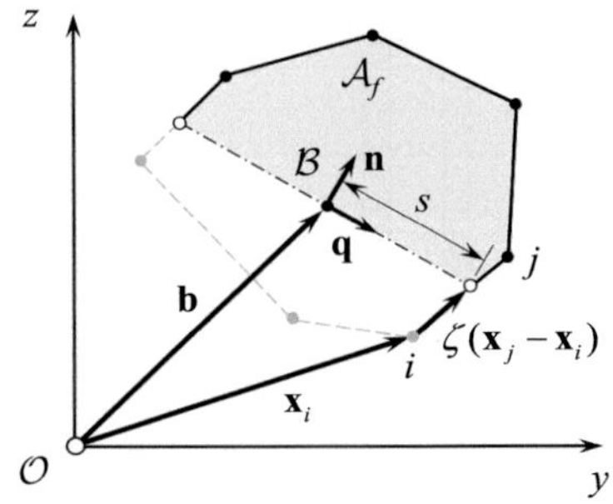

Fig. 9.19 *Cuts for Q*. The cut is defined by the location **b** of the point $\mathcal{B}$ and the direction perpendicular to the cut **n**. The black dots are the vertices of the original cross section. The open circles are new points needed to create the region $\mathcal{A}_f$

needed to compute **Q** associated with the region $\mathcal{A}_f$, which is only part of the cross section, as illustrated in Fig. 9.19. To compute first moments with cuts, we will need to first figure out how to make those cuts and second, how to modify the vertex array to produce a new polygon that only includes the region $\mathcal{A}_f$. If we can accomplish those two tasks, then we can use our function `ComputeProperties` to do the remaining computations.

Because we are considering a polygon in the code to define the cross section, we can define the subregion $\mathcal{A}_f$ as shown in Fig. 9.19. To define the cut, we must specify the location **b** of point $\mathcal{B}$ and a unit vector **n** perpendicular to the cut. The cut line is in the direction of the unit vector **q**, which is rotated 90° from **n**. The main problem we face is that the intersection of the cut with the edge of the region is not necessarily a vertex of the original cross section. For some problems it might be possible to include these points in the original list of vertices, but that would require a new definition of the coordinates for each value of **Q**. To solve this problem, we will define new points (the open circles at the ends of the cut line in Fig. 9.19) and then use them to establish a new set of vertices for the region $\mathcal{A}_f$.

Consider the case where the cut intersects the edge between vertex i and vertex j, shown in Fig. 9.19, which are the two vertices associated with side n. The intersection point lies a distance s from point $\mathcal{B}$ and a fraction ζ of the distance from vertex i to vertex j. Thus,

$$\mathbf{b} + s\mathbf{q} = \mathbf{x}_i + \zeta\left(\mathbf{x}_j - \mathbf{x}_i\right)$$

where $\mathbf{x}_j - \mathbf{x}_i$ is the vector from vertex i to vertex j. We do not need to find s, so we can dot the equation with **n** and solve for ζ. To wit,

$$\zeta = \frac{\left(\mathbf{b} - \mathbf{x}_i\right) \cdot \mathbf{n}}{\left(\mathbf{x}_j - \mathbf{x}_i\right) \cdot \mathbf{n}} \tag{9.29}$$

If $0 < \zeta < 1$, then the cut line intersects the edge segment between vertex i and vertex j. In this case, we can create a new vertex point at the intersection point and add it to the original set of vertices. Because the vertex numbering must be sequential, we have to move the index numbers of all the vertices greater than i up by one to fit the new point into the list.

Finally, to define the region, we must eliminate all the vertices below the cut line. To accomplish this task, we compute the vector from $\mathcal{B}$ to the vertex in question, i.e., $\mathbf{x}_i - \mathbf{b}$. If the dot product of that vector with $\mathbf{n}$ is greater than zero, then the vertex is above the cut. If not, then it is below the cut and needs to be eliminated. Code 9.4 does this task.

Code 9.4 MATLAB code `CutsForQ` to create the vertex coordinates for the region $\mathcal{A}_f$ from the vertices of the original cross section using a cut at $\mathcal{B}$ with normal direction $\mathbf{n}$

```
function [z] = CutsForQ(x,b,n)
%   Find new set of points to compute Q from vertices x
%      x = Coordinates defining entire cross section
%      b = Coordinates of reference point for cut
%      n = Vector defining direction perpendicular to cut

%. Number of sides of original cross section
   nSides = size(x,1) - 1;
   eps = 1.e-8;

%. Search for segments that intersect cut line
   ii = 0;
   for i=1:nSides
     ii = ii + 1;
     xi = x(ii,:)';
     xj = x(ii+1,:)';
     zeta = dot(b-xi,n)/dot(xj-xi,n);

%... Line intersects cut, add point
     if zeta>0 && zeta<1
        jj = ii+1;
        NewPoint = xi + zeta*(xj-xi);
        x = [x(1:ii,:); NewPoint'; x(jj:end,:)];
        ii = ii + 1;
     end
   end

%. Trim off vertices below the cut line
   nVerts = size(x,1);
   keep = zeros(nVerts,1);
   for i=1:nVerts
     d = x(i,:)' - b;
     proj = dot(d,n) + eps;
     if proj >= 0; keep(i) = 1; end
   end
   z = x(keep==1,:);         % trim off points below cut line
   z = [z; z(1,:)];          % close last point

end
```

This function takes in the array of cross section vertices `x`, the vector `b`, which defines the position of point $\mathcal{B}$ on the cut line, and the vector `n`, which defines the direction normal to the cut line and passes back the array of vertices `z` that defines the region $\mathcal{A}_f$. The algorithm loops over all sides of the polygon, computing ζ from

Eq. 9.29. If the cut fraction is between zero and one, then a new point is added to the array of vertices.

Once all sides have been investigated, the vertices below the cut line are eliminated. If the projection of the vector $\mathbf{x}_i - \mathbf{b}$ along $\mathbf{n}$ is positive, then the point lies above the cut. In order to avoid eliminating the points exactly on the cut, which we have just computed by finding those intersections, we add a small value `eps` to the projection before we test it for positivity. The array `keep` tracks which vertices are above (`keep` = 1) and which are below (`keep` = 0) the cut line.

Now, the first moment of the region $\mathcal{A}_f$ can be computed in exactly the same way as the original region defining the entire cross section, i.e., by calling the function `ComputeProperties`. Code 9.5 executes the computation of the cross-sectional properties, including the values of **Q**.

Code 9.5 MATLAB code `XSectionProp` (Q version) to compute the area, centroid, moment of inertia, and **Q** of a cross section represented as a polygon

```
   clear; clc;

%. Output control
   N = [1,0; 0,1];                  % Cut normal directions
   b = [0; 0];                      % Center for cut location
   nChoice = 1;                     % n: =0 use N given, =1 use eigenvectors
   bChoice = 1;                     % b: =0 use b given, =1 use centroid

%. Establish coordinates x of the polygon vertices
   [x,Section] = InputCrossSection;

%. Compute the properties of the x-section
   [A,~,c,Jc] = ComputeProperties(x);

%. Compute the principal values (eigenvalues) of Jc
   [eVecs,eVals] = eig(Jc);

%. Revised cut axes and cut location
   if nChoice==1; N = eVecs; end
   if bChoice==1; b = c; end

%. Compute first moment Qz for cut
   n = N(:,1);                          % Select cut normal direction
   [z] = CutsForQ(x,b,n);               % Compute vertices of region Af
   [Af,Qo,~,~] = ComputeProperties(z);  % Compute area and Q
   Qz = Qo - Af*c;                      % Apply parallel axis theorem

%. Compute first moment Qy for cut
   n = N(:,2);                          % Select cut normal direction
   [y] = CutsForQ(x,b,n);               % Compute vertices of region Af
   [Af,Qo,~,~] = ComputeProperties(y);  % Compute area and Q
   Qy = Qo - Af*c;                      % Apply parallel axis theorem

%. Output results
   PrintResultsQ(Section,A,c,Qy,Qz,Jc,eVals,x)
   PlotResultsQ(x,y,z,c,eVecs,b,N)
```

This code first sets up the quantities needed to take the cut in the cross section to define the region $\mathcal{A}_f$. Two options are provided for the cut point: either with the specified input cut reference point `b` or with the reference point taken at the centroid. Two options are provided for the cut direction: either with the normal directions specified as input `N` or with the cut directions in the principal directions of the moment of inertia tensor. These choices are activated by setting `nChoice` and `bChoice`. The cross section is defined by the vertex coordinates obtained from the function `InputCrossSection`.

Observe that in Code 9.5 we take two cuts, first with normal direction defined by the first column of `N` and subsequently with the normal direction defined by the second column of `N`. What that means is that the columns of the array `N` align with the directions normal to the cut lines. If `nChoice` is set to one, then the principal directions are used for the cut normal directions instead of those initially defined. Similarly, if `bChoice` is set to one, the centroid is substituted for reference point for the cut.

The two first moments of the area, i.e., `Qy` and `Qz`, are computed. First, the first moment of the region $\mathcal{A}_f$ is computed relative to the origin. Then the value is adjusted using the parallel axis theorem for $\mathbf{Q}$ by subtracting off the area times the distance from the origin to the centroid.

Example 9.8

To illustrate and verify the computation of area, centroid, moment of inertia, and $Q(0)$, consider the T-beam cross section shown in Fig. 9.20. Verify the results generated by the code through hand calculation of the properties.

Solution. Code 9.5 computes the location of the centroid `c = 6.7778`, the area as `A = 36`, and the moment of inertia of `Izz = 314.22`, and `Qy = 45.938`. To verify these properties, recall Example 9.4 in Sect. 9.2, which gave (using $h = 10$)

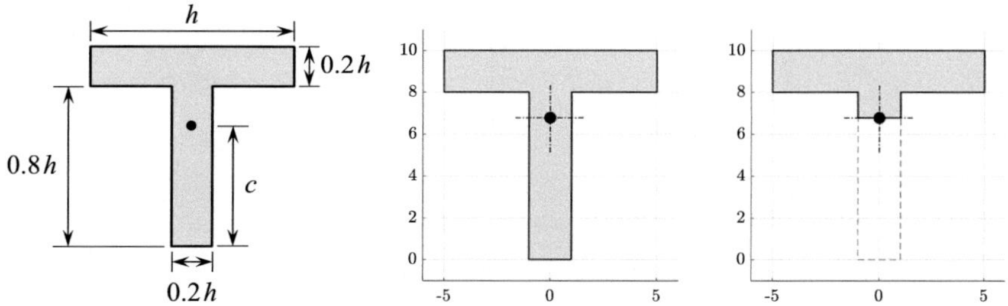

Fig. 9.20 *Example.* T-shaped cross section from Fig. 9.9 with $h = 10$. Compute the area and centroid. Compute the moment of inertia and $Q(0)$, relative to the centroid. The two plots on the right are generated by the code

$$I_C = 0.03142\, h^4 = 0.03142\,(10)^4 = 314.2$$

Thus, the value of the moment of inertia computed by the code is correct.

The example did not include the computation of $Q_C(c)$. The value relative to the origin can be computed as

$$\begin{aligned} Q_O(c) &= \int_c^8 2z\,dz + \int_8^{10} 10z\,dz \\ &= 2\big[\tfrac{1}{2}z^2\big]_c^8 + 10\big[\tfrac{1}{2}z^2\big]_8^{10} \\ &= (8^2 - c^2) + 5(10^2 - 8^2) = 198.062 \end{aligned}$$

where we have substituted $c = 6.7778$. The area of the region of the cross section above the centroid is

$$A_f = (2)(8 - 6.7778) + (10)(2) = 22.444$$

Adjusting Q to the centroid with Eq. 9.8 gives

$$Q_C(c) = Q_O(c) - cA_f = 198.06 - (6.7778)(22.444) = 45.938$$

thereby verifying the result produced by the code.

The advantage of having the code is in application to cross sections that are not readily amenable to hand calculations. Some examples are shown in Fig. 9.21. The cross section in Fig. 9.21a is `Case 13` in Code 9.3. Its middle line is generated by the function

$$\mathbf{z}(\theta) = \big[R + a\cos\big(m\theta\big)\big]\mathbf{p}(\theta)$$

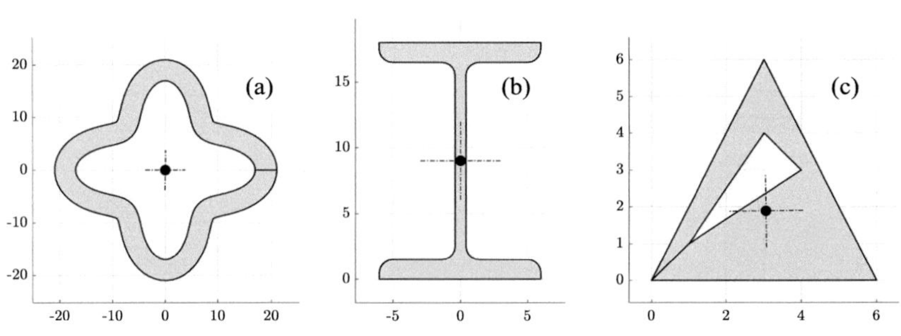

Fig. 9.21 *Examples of complex cross sections.* **a** Curved section of constant thickness. **b** I-beam with rounded filets. **c** Triangle with triangular cutout

where $\mathbf{p}(\theta) = \cos\theta\,\mathbf{e}_2 + \sin\theta\,\mathbf{e}_3$. The value R determines the average radius, the value a is the amplitude of deviations from the average radius, and m (an integer) is the frequency of those variations. The region is created by tracing lines a distance $\pm t$ in the direction normal to the middle line, assuring that the shape has a constant thickness. The case shown has $R = 15$, $a = 4$, $m = 4$, and $t = 2$.

Real I-beam shapes have rounded filets, as shown in Fig. 9.21b. This shape is `Case 10` in Code 9.3. The filets are created with quarter circles of radius r at all the interior edges. This cross section can be compared with the plain I-beam of `Case 9` in the code. The differences in the cross-sectional properties are generally small.

The triangular shape shown in Fig. 9.21c illustrates the ability of the code to capture cutouts in cross sections. The cutout in this case is awkwardly placed, making the hand calculation of the properties somewhat more tedious. To create a cross section with a cutout, the vertices that define the exterior of the cross section are input first in counterclockwise order. Next, the vertices that define the cutout are input in clockwise order. Finally, the cross section is closed by adding the initial point again at the end. Vertices in counterclockwise order produce positive contributions to the properties. Vertices in clockwise order produce negative contributions to the properties. This process is the algorithmic equivalent of *Trick* 2.

9.4 Summary

The ability of a beam to carry load effectively is governed by the geometric features of its cross section. The equations for deflection and stress all depend on the area and the first and second moments of the area relative to the centroid. Thus, one of the most important tasks of solving beam problems is the calculation of these properties. This chapter has provided a detailed outline of how to carry out those calculations by hand and how to organize them in a computational algorithm that works for a wide range of cross sections.

9.5 Problems

9.1 For the circular segment of radius R and angle 2α, find the location of the centroid, the moments of inertia relative to the centroid about the horizontal and vertical axes. Evaluate for the specific case of a semicircular region (i.e., $\alpha = \pi/2$). Find the first moment of area Q at the centroid for a semicircular region.

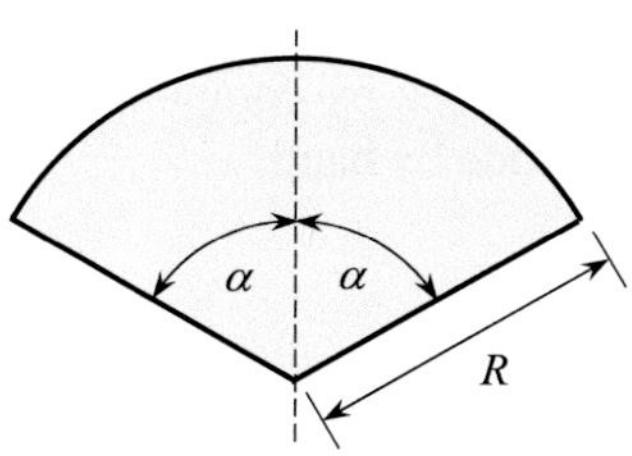

9.2 For the L-shaped cross section shown, find the centroid, the moment of inertia tensor, the principal moments of inertia relative to the centroid, and the first moment of area Q at points $\mathcal{A}$ and $\mathcal{B}$ (with horizontal cuts). Evaluate for $t = 25$ mm and $b = 200$ mm.

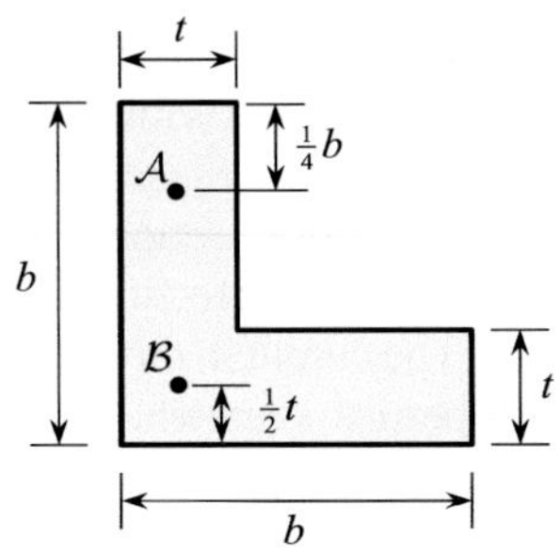

9.3 For the cross section shown, find the area, centroid, second moment of area, and the first moment of the area Q at the centroid for bending in the vertical plane. Carry out the computation with direct integration, *Trick 1*, and *Trick 2*.

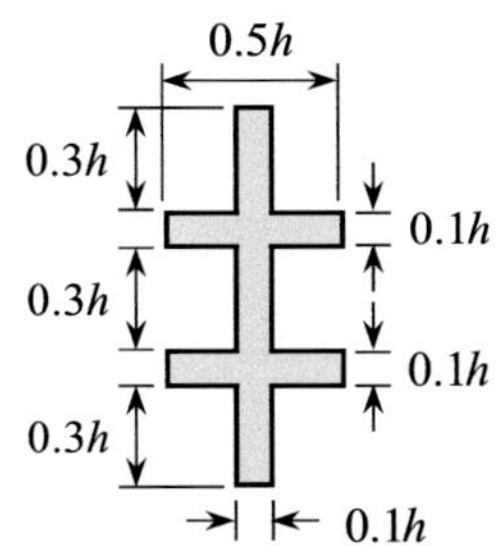

9.4 For the cross section shown, find the area, centroid, second moment of area, and the first moment of the area Q at the centroid for bending in the vertical plane. Carry out the computation with direct integration, *Trick 1*, and *Trick 2*.

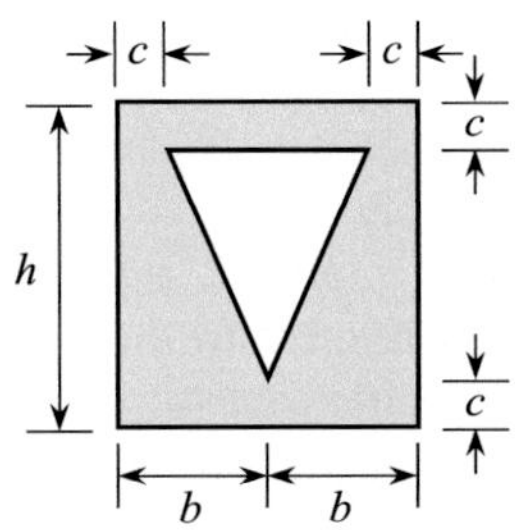

9.5 *Computation.* Verify Code 9.2 by computing the properties of some of the cross sections included in the code by hand.

9.6 *Computation.* The cross section shown is for a box girder beam for a highway bridge. Modify the function `InputCrossSection` to add the cross section with the parameters shown. Investigate the properties of the cross section by looking at various combinations of the parameters. Start without the filets for the rounded corners of the cutout sections. Add the filets (take the radius of the filet to be half of t_2) and compare the difference in the properties with and without them.

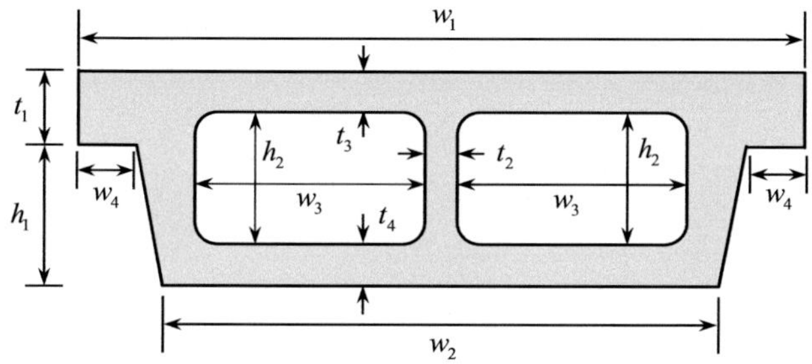

9.7 *Computation.* Use Code 9.2 to: (a) show that a circular tube has a larger polar moment of area, $J = I_{yy} + I_{zz}$, than a solid circle if both have the same amount of material. (b) Explore the tradeoffs for maximizing the moment of inertia for an I-beam. For a fixed amount of total material, is it better to increase the depth at the expense of the flange area or increase the flange area at the expense of the depth? (c) Find a set of dimensions for an I-beam that has the same moment of inertia relative to both axes. (d) Demonstrate that the principal axes of a symmetric cross section lie along the lines of symmetry.

9.8 *Computation.* Write a code based on Code 9.2 that supports having regions with different materials. Use the code to find the moment of inertia of a rectangular reinforced concrete beam with reinforcing bars in the top and bottom of the cross section. As a first approximation, treat the steel bars with area A_2 and modulus E_2 as *points* in the input and calculation. Then include them as solid regions. Remember to cut out the region of the bars from the rectangular region (i.e., do not have two materials occupying the same points in space). Show that you can use Trick 2 and *not* cut the circles out of the rectangular region but add in the material of the bars using an E value of $E_2 - E_1$.

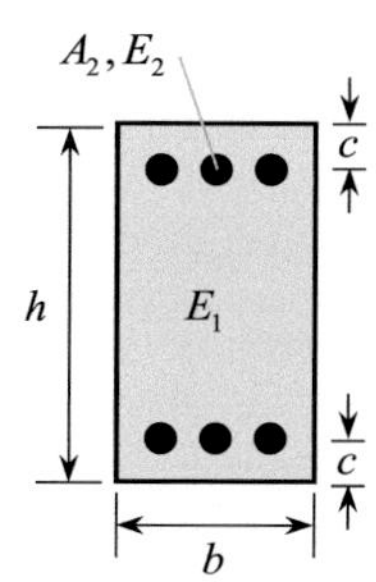

Chapter 10
A Computational Approach to Beams

Beam theory is one of the most popular and successful mathematical models in engineering, constituting a fundamental tool for structural engineering. It is no surprise, then, that many approaches to automatic computation for beam problems have been developed. Some of these approaches are the basis of commercial software for structural analysis. In this chapter, we will formulate an approach to computation with beam theory that will culminate in a code that will solve the problem of a beam with any boundary conditions and any form of distributed loading. The approach uses the tools of numerical integration to accomplish the primary tasks associated with solving the differential equation. We will then extend the code to incorporate point loads and end by implementing a code that solves the multispan beam problem. The main goal of the chapter is to illuminate the theory by organizing the solution process for beams into a tidy algorithm. A primary benefit of the developments of this chapter is a computational framework that can quickly solve a broad array of problems with different loading and support conditions, enabling systematic exploration of the behavior of beams.

10.1 Integrating the Governing Differential Equations

To set up the computational approach, we start by writing the governing differential equations of linear beam theory as a set of first-order differential equations. From Eq. 8.3, we have the equations of equilibrium, simplified by assuming that the distributed applied moment $m(x) = 0$. From Eq. 7.23, we have the constitutive equation relating bending moment to curvature. And from Eq. 8.1 we have the linearized kinematic relationships. These equations can be summarized as

K. D. Hjelmstad, *Mechanics of Deformable Solids: An Introduction*,
https://doi.org/10.1007/978-3-031-89873-0_10

$$
\begin{aligned}
V'(x) &= -q(x) \\
M'(x) &= V(x) \\
\theta'(x) &= M(x)/EI \\
w'(x) &= -\theta(x)
\end{aligned}
\tag{10.1}
$$

Note that these equations have already combined the constitutive relationship $\kappa_o = M/EI$, which is not a differential equation, with the kinematic equation $\kappa_o = \theta'$. These four equations can be integrated successively, as each one depends only on the previous one.

First, integrate the shear equation, Eq. 10.1(a), to get

$$
V(x) = V_o - \int_0^x q(\xi)\, d\xi \tag{10.2}
$$

where $V_o = V(0)$ is the value of the shear force at $x = 0$. In order to evaluate the integral that appears in this equation, we would need to have a specific function like $q(x) = q_o x/L$ or $q(x) = q_o \sin(\pi x/L)$ in mind. We can, however, make some more analytical progress without yet specifying the exact form of the loading function.

Now that we know $V(x)$, we can substitute it into the moment equation, Eq. 10.1(b), and integrate it to get

$$
\begin{aligned}
M(x) &= M_o + \int_0^x V(\xi)\, d\xi \\
&= M_o + \int_0^x \Big(V_o - \int_0^\xi q(\eta)\, d\eta \Big)\, d\xi \\
&= M_o + V_o\, x - \int_0^x \int_0^\xi q(\eta)\, d\eta\, d\xi
\end{aligned}
$$

where $M_o = M(0)$ is the value of the moment at $x = 0$. Observe that we need to be careful with the dummy variables of integration since the limits of integration are variable. The double integral in the last term can be integrated by parts as we did in Eq. 3.12 for the axial bar problem (see Sect. 3.3). To wit,

$$
\int_0^x \int_0^\xi q(\eta)\, d\eta\, d\xi = \int_0^x (x - \xi) q(\xi)\, d\xi
$$

Therefore, the equation for the bending moment is

$$
M(x) = M_o + V_o\, x - \int_0^x (x - \xi) q(\xi)\, d\xi \tag{10.3}
$$

The equations for moment and shear, while derived from integration of the differential equations of equilibrium, make sense in the context of the free body diagram shown

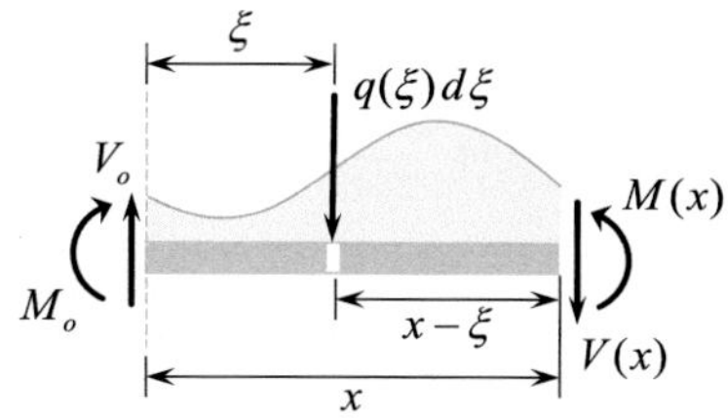

Fig. 10.1 *Moment and shear equations*. Free body diagram of a beam with distributed loading $q(\xi)$ with a cut at x

in Fig. 10.1. The shear and moment at $x=0$ are V_o and M_o, respectively. Taking a cut at x exposes the shear $V(x)$ and moment $M(x)$. Summing forces yields Eq. 10.2. Moment equilibrium with moments taken at the cut yields Eq. 10.3. The moment of the force $q(\xi)d\xi$ on the typical particle at ξ is $\big(x-\xi\big)q(\xi)\,d\xi$ and acts in a counterclockwise direction. The total moment caused by the load is then the integral over the particles from 0 to x, which yields Eq. 10.3.

We can follow the same process for the rotation equation. Substituting $M(x)$ from Eq. 10.3 into the rotation equation, Eq. 10.1(c), and simplifying terms, we get

$$\theta(x)=\theta_o+\frac{M_o\,x}{EI}+\frac{V_o\,x^2}{2EI}-\frac{1}{2EI}\int_0^x\big(x-\xi\big)^2q(\xi)\,d\xi \tag{10.4}$$

where $\theta_o=\theta(0)$ is the value of the rotation at $x=0$. The integration by parts for the last term is a little more complicated in this case but works out in a similar way. Finally, for the transverse displacement, we can substitute $\theta(x)$ from Eq. 10.4 into Eq. 10.1(d) and integrate to get

$$w(x)=w_o-\theta_o\,x-\frac{M_o\,x^2}{2EI}-\frac{V_o\,x^3}{6EI}+\frac{1}{6EI}\int_0^x\big(x-\xi\big)^3q(\xi)\,d\xi \tag{10.5}$$

where $w_o=w(0)$ is the value of the transverse displacement at $x=0$. One can verify that Eqs. 10.2 through 10.5 are correct by differentiating each one to show that it satisfies Eq. 10.1. The last term can be differentiated using the Leibniz rule for differentiating an integral with variable limits:

$$\frac{d}{dx}\int_0^x g(x,\xi)\,d\xi=g(x,x)+\int_0^x\frac{\partial}{\partial x}\Big(g(x,\xi)\Big)\,d\xi \tag{10.6}$$

which is particularly simple in this case because the $g(x,x)$ terms all go to zero. The following example illustrates the application of this formula to the loading terms that appear in Eqs. 10.4 and 10.5.

Example 10.1
Show that

$$\frac{d}{dx}\int_0^x \tfrac{1}{6}\big(x-\xi\big)^3 q(\xi)\,d\xi = \int_0^x \tfrac{1}{2}\big(x-\xi\big)^2 q(\xi)\,d\xi$$

and that

$$\frac{d}{dx}\int_0^x \tfrac{1}{2}\big(x-\xi\big)^2 q(\xi)\,d\xi = \int_0^x \big(x-\xi\big) q(\xi)\,d\xi$$

using the Leibniz rule in Eq. 10.6.

Solution. For the first case, identify the function $g(x,\xi)=\frac{1}{6}(x-\xi)^3 q(\xi)$. Now, evaluate

$$g(x,x) = \tfrac{1}{6}\big(x-x\big)^3 q(x) = 0$$

and

$$\frac{\partial}{\partial x}\Big(g(x,\xi)\Big) = \frac{\partial}{\partial x}\Big(\tfrac{1}{6}\big(x-\xi\big)^3 q(\xi)\Big) = \tfrac{1}{2}\big(x-\xi\big)^2 q(\xi)$$

Notice that ξ acts as a constant in the partial derivative. For the second case, identify the function $g(x,\xi)=\frac{1}{2}(x-\xi)^2 q(\xi)$. Now, evaluate

$$g(x,x) = \tfrac{1}{2}\big(x-x\big)^2 q(x) = 0$$

and

$$\frac{\partial}{\partial x}\Big(g(x,\xi)\Big) = \frac{\partial}{\partial x}\Big(\tfrac{1}{2}\big(x-\xi\big)^2 q(\xi)\Big) = \big(x-\xi\big) q(\xi)$$

Finally, substitute these results into Eq. 10.6 to complete the demonstration.

Equations 10.2 through 10.5 are often called *the method of initial parameters* because the solution is expressed in terms of the values $\{V_o, M_o, \theta_o, w_o\}$, which represent the state variables at the left end of the beam, i.e., the initial point $x=0$. This method has the vibe of an initial-value problem, wherein the state is known at the initial point and the differential equations determine the evolution from there. The equations of rigid body dynamics are an example of an initial-value problem. The position and velocity at time zero are known and the equations describe the evolution from there. However, the differential equations governing the beam are *not* an initial-value problem, they are a boundary-value problem. Each problem has specified *boundary conditions*, as laid out in Table 8.1. What that means is that only two of the four state variables are known from the boundary conditions at $x=0$. The other two conditions are at the end $x=L$.

In our code, we will use the generalized trapezoidal rule to integrate the equations as if they were an initial-value problem in order to generate the plots of the state variables as a function of x. To get into a position to do that, we will first use Eqs. 10.2 through 10.5 to find all the values of the state variables at both ends of the beam. These equations hold for any value of x, and thus must hold for the value $x = L$. Therefore, we can write

$$\begin{aligned}
V_L &= V_o - \int_0^L q(\xi)\, d\xi \\
M_L &= M_o + V_o L - \int_0^L (L-\xi) q(\xi)\, d\xi \\
\theta_L &= \theta_o + \frac{M_o L}{EI} + \frac{V_o L^2}{2EI} - \frac{1}{2EI}\int_0^L (L-\xi)^2 q(\xi)\, d\xi \\
w_L &= w_o - \theta_o L - \frac{M_o L^2}{2EI} - \frac{V_o L^3}{6EI} + \frac{1}{6EI}\int_0^L (L-\xi)^3 q(\xi)\, d\xi
\end{aligned} \tag{10.7}$$

where $V_L = V(L)$, $M_L = M(L)$, $\theta_L = \theta(L)$, and $w_L = w(L)$ are the values of the state variables at the right end, i.e., $x = L$. Let us designate the four definite integrals that appear in these equations as

$$\begin{aligned}
\mathcal{I}_0 &= \int_0^L q(\xi)\, d\xi & \mathcal{I}_2 &= \frac{1}{2EI}\int_0^L (L-\xi)^2 q(\xi)\, d\xi \\
\mathcal{I}_1 &= \int_0^L (L-\xi)\, q(\xi)\, d\xi & \mathcal{I}_3 &= \frac{1}{6EI}\int_0^L (L-\xi)^3 q(\xi)\, d\xi
\end{aligned} \tag{10.8}$$

These definite integrals result in numerical values that depend upon the specific form of the loading function $q(x)$, as the following example shows.

Example 10.2
Compute $\mathcal{I}_3$ for a linearly decreasing load that goes from intensity q_o at $\xi = 0$ to zero at $\xi = L$.

Solution. The loading function with linearly decreasing magnitude has the expression:

$$q(\xi) = q_o\left(1 - \frac{\xi}{L}\right)$$

Let us first use a change of variable $\zeta = \xi/L$. Note that $d\xi = L\, d\zeta$. The load function can be expressed as $q(\zeta) = q_o(1-\zeta)$. Also note that the expression in the integral $(L-\xi)$ can be written as $L(1-\zeta)$. The integral can now be evaluated as

$$\begin{aligned}\mathcal{I}_3 &= \frac{1}{6EI}\int_0^L (L-\xi)^3 q(\xi)\,d\xi \\ &= \frac{1}{6EI}\int_0^1 L^3(1-\zeta)^3 q_o(1-\zeta)\,L\,d\zeta \\ &= \frac{q_oL^4}{6EI}\int_0^1 (1-\zeta)^4\,d\zeta \\ &= \frac{q_oL^4}{6EI}\Big[-\tfrac{1}{5}(1-\zeta)^5\Big]_0^1 = \tfrac{1}{30}\frac{q_oL^4}{EI}\end{aligned}$$

The change of variable strategy is useful for *any* function $q(\xi)$. The benefit is that it results in an integral from 0 to 1 and all terms with units collect out front. One would expect the leading term to be the same for $\mathcal{I}_3$ in each case (e.g., q_oL^4/EI) and only the numerical coefficient resulting from the integral to change for different load distributions. Hence, one can get a clear idea of the basic character of these integrals. The computations for $\mathcal{I}_0$, $\mathcal{I}_1$, and $\mathcal{I}_2$ follow a similar pattern, giving

$$\mathcal{I}_0 = \tfrac{1}{2}q_oL, \qquad \mathcal{I}_1 = \tfrac{1}{3}q_oL^2, \qquad \mathcal{I}_2 = \tfrac{1}{8}\frac{q_oL^3}{EI}$$

for this load distribution. The details are left to the reader.

Using the notation defined in Eqs. 10.8 and 10.7 can be written in the following form:

$$\begin{aligned} V_o - V_L &= \mathcal{I}_0 \\ M_o + V_o\,L - M_L &= \mathcal{I}_1 \\ \theta_o + \frac{M_o\,L}{EI} + \frac{V_o\,L^2}{2EI} - \theta_L &= \mathcal{I}_2 \\ w_o - \theta_o\,L - \frac{M_o\,L^2}{2EI} - \frac{V_o\,L^3}{6EI} - w_L &= -\mathcal{I}_3 \end{aligned} \tag{10.9}$$

These equations constitute a system of four equations relating the eight unknown state variables at the two ends of the beam (four at each end). In any specific case, we will know four of the state variables, two at each end, from the boundary conditions. Thus, we will be able to solve these equations in most cases (the free-free case will present some problems).

Equations 10.2 through 10.5 can be used to find the complete solution to a beam problem with a specific loading function, as the following example shows. But before we get to the algorithm, let us demonstrate how these equations can be used to solve a beam problem by hand.

Example 10.3
Find the deflection $w(x)$ for a simply supported beam of length L and bending modulus EI, subjected to a uniformly distributed load of magnitude q_o over its entire length, as shown in Fig. 10.2. Find the displacement at midspan. Use the method of initial parameters.

Solution. The transverse displacement $w(x)$ is given by Eq. 10.5. This equation involves w_o, θ_o, M_o, and V_o, two of which are known from boundary conditions and the other two can be computed from Eqs. 10.9. For Eq. 10.9, we must compute $\mathcal{I}_0$, $\mathcal{I}_1$, $\mathcal{I}_2$, and $\mathcal{I}_3$ for the specific load form of this problem. For $q(\xi) = q_o$, we can compute

$$\mathcal{I}_0 = q_o L, \qquad \mathcal{I}_1 = \tfrac{1}{2} q_o L^2, \qquad \mathcal{I}_2 = \tfrac{1}{6}\frac{q_o L^3}{EI}, \qquad \mathcal{I}_3 = \tfrac{1}{24}\frac{q_o L^4}{EI}$$

The details are left to the reader. The boundary conditions for the simply supported beam are

$$w_o = 0, \qquad M_o = 0, \qquad w_L = 0, \qquad M_L = 0$$

from Eq. 10.8. Now, the value of V_o can be determined from Eq. 10.9(b). Noting that $M_o = 0$ and $M_L = 0$, we get

$$V_o L = \mathcal{I}_1 \qquad \rightarrow \qquad V_o = \tfrac{1}{2} q_o L$$

Similarly, we can find the value θ_o from Eq. 10.9(d). Noting that $w_o = 0$, $M_o = 0$, and $w_L = 0$, and using the value of V_o just found, we get

$$-\theta_o L - \frac{V_o L^3}{6EI} = -\mathcal{I}_3 \qquad \rightarrow \qquad \theta_o = -\frac{q_o L^3}{24EI}$$

Substituting these values into Eq. 10.5, noting that

$$\frac{1}{6EI}\int_0^x (x-\xi)^3 q_o \, d\xi = \frac{q_o}{6EI}\Big[-\tfrac{1}{4}(x-\xi)^4\Big]_0^x = \frac{q_o x^4}{24EI}$$

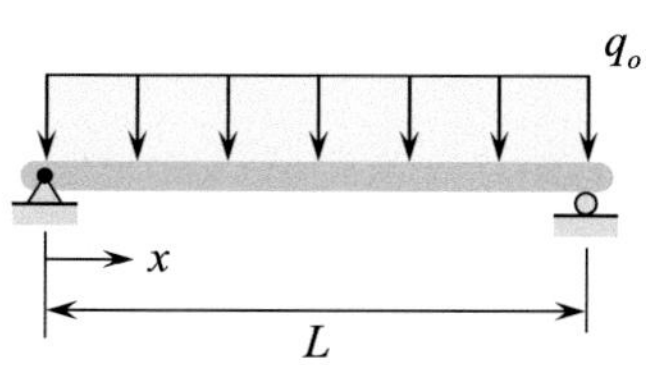

Fig. 10.2 *Example.* Simply supported beam of length L and bending modulus EI, subjected to a uniform distributed load of magnitude q_o over its length

we get the final result

$$w(x) = \frac{q_o}{24EI}(x^4 - 2x^3L + xL^3)$$

We can find the displacement at midspan by substituting $x = \frac{1}{2}L$ into the equation for deflection. Doing so gives

$$w(\tfrac{1}{2}L) = \frac{5\,q_o L^4}{384EI}$$

This example illustrates a general approach to solving beam problems using the method of initial parameters. Each different load form requires the computation of $\mathcal{I}_0$ through $\mathcal{I}_3$. Each different set of boundary conditions requires a slightly different manipulation of Eq. 10.9 to determine the remaining unknown end states. Our aim in this chapter is to automate that task. But before we complete the formulation, we will first take a look at a classical result that is intimately related to the developments so far.

10.2 The Moment-Area Equations

There is a classical method for computing deflections and rotations called the *moment-area method* that is a slight deviation from the method of initial parameters. Let us assume that we have obtained the moment function $M(x)$ by some means, perhaps by integrating the equations of equilibrium as we did in the previous section, or from a statics calculation on a finite free body diagram with a cut at x. Either way, we will consider the moment function to be known. Using the same strategy as before, we can integrate the rotation equation $\theta' = M/EI$ as follows:

$$\theta(x) = \theta_o + \int_0^x \frac{M(\xi)}{EI}\,d\xi$$

The integral can be interpreted as the area under the M/EI diagram, which is the reason that *area* is part of the name of the method. Next, we can integrate the displacement-rotation relationship $w' = -\theta$ to get

$$
\begin{aligned}
w(x) &= w_o - \theta_o\, x - \int_0^x \int_0^\xi \frac{M(\eta)}{EI}\, d\eta\, d\xi \\
&= w_o - \theta_o\, x - \int_0^x (x-\xi)\frac{M(\xi)}{EI}\, d\xi
\end{aligned}
$$

where the double integral has been reduced to a single integral though integration by parts, in the same way we did in Eq. 3.12 for the axial bar problem (see Sect. 3.3). This equation is the reason that *moment* is part of the name of the method, but that needs a little explanation. If we interpreted $\hat{q}(\xi) = M(\xi)/EI$ as a fictitious distributed loading on the beam (and *not* the actual loading that gave rise to the bending moment $M(\xi)$ in the first place), then the integral term is the moment caused by the fictitious loading relative to the cut at x with moment arm $(x - \xi)$. This interpretation is illustrated in the sketch on the left in Fig. 10.3, where the fictitious shears are the rotations θ and the fictitious moments are the deflections w. So, the method is not called moment-area because the moment function is involved in the equation. It is called moment-area because we can reinterpret the integral as a moment equilibrium calculation with a fictitious set of loads, shears, and moments. The method essentially converts the kinematic problem into an analogous static equilibrium problem.

This method is historically important, and it is often still one of the methods used for computing deflections of beams in courses on structural analysis. The method is most often written as a computation of relative rotation and deflection between points $x = a$ and $x = b$. Replacing x with b and making the lower limit of the integral a rather than zero, we can write the moment-area equations:

$$
\begin{aligned}
\theta_b &= \theta_a + \int_a^b \frac{M(\xi)}{EI}\, d\xi \\
w_b &= w_a - (b-a)\theta_a - \int_a^b (b-\xi)\frac{M(\xi)}{EI}\, d\xi
\end{aligned}
\tag{10.10}
$$

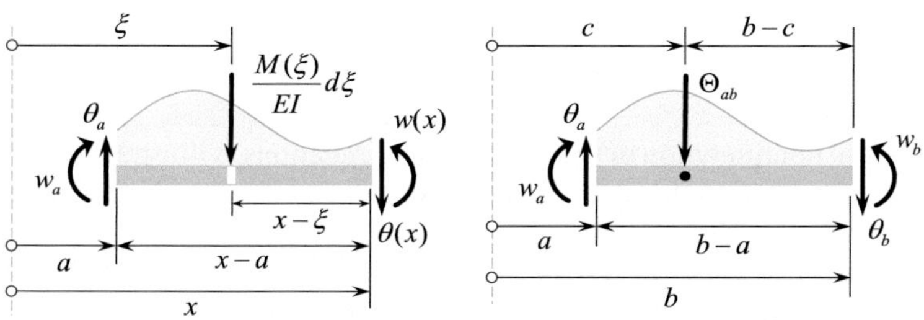

Fig. 10.3 *Moment-area equations*. We can reinterpret the kinematic calculation as an analogous statics problem if we view the fictitious loading as M/EI, the fictitious shears as θ_a and θ_b, and the fictitious moments as w_a and w_b. The left sketch is the fictitious distributed loading, with cut at x. The right sketch has the fictitious loading resultant, i.e., the *area*, lumped at the centroid located at c, with cut at $x = b$

The method is often further refined by finding the centroid of the M/EI diagram. In particular, we can seek a value c such that

$$\int_a^b (b-\xi)\frac{M(\xi)}{EI}\,d\xi = (b-c)\int_a^b \frac{M(\xi)}{EI}\,d\xi$$

The value c is the location of the centroid of the M/EI diagram, as shown in the sketch on the right side of Fig. 10.3. Now let us rename the area under the M/EI diagram and the moment of that area relative to point b as

$$\Theta_{ab} = \int_a^b \frac{M(\xi)}{EI}\,d\xi, \qquad \Delta_{ab} = (b-c)\int_a^b \frac{M(\xi)}{EI}\,d\xi = (b-c)\Theta_{ab}$$

where Θ_{ab} is the area and Δ_{ab} is the moment of the area about the right end. With this notation, we can write an equation to determine the location of the centroid of the distribution, i.e., c, as

$$(b-c)\Theta_{ab} = \int_a^b (b-\xi)\frac{M(\xi)}{EI}\,d\xi$$

which can be simplified to solve for c as

$$c = \frac{1}{\Theta_{ab}}\int_a^b \xi\,\frac{M(\xi)}{EI}\,d\xi$$

The computations involved in finding the centroid c are basically the same complexity as the computations one would do to directly solve for deflections using Eq. 10.10. However, for moment distributions where the area and location of centroid are known (e.g., constant or linear functions), the lumped version of the moment-area equations has significant advantages.

We can write the final form of the moment-area equations as

$$\theta_b = \theta_a + \Theta_{ab}, \qquad w_b = w_a - (b-a)\theta_a - \Delta_{ab} \tag{10.11}$$

The geometric interpretation of the moment-area equations is illustrated in Fig. 10.4. The area Θ_{ab} represents the angle between the tangent lines to the deflection curve at $x=a$ and $x=b$. The moment Δ_{ab} is often called the *tangential deviation* and represents the deflection between the tangents at a and b that is not explained by the rotation at a. In other words, Δ_{ab} is the deflection caused by the curvature of the beam alone. Since $\Delta_{ab}=(b-c)\Theta_{ab}$, the black dot at the root of the angle Θ_{ab} on the sketch must be located at the centroid c of the M/EI diagram. The use of the moment-area method is illustrated in the following example.

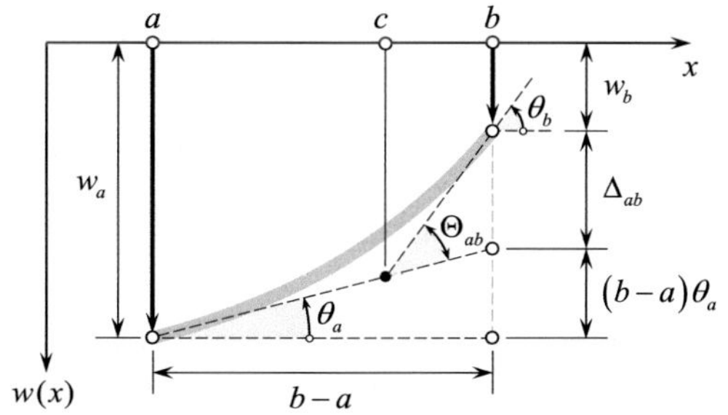

Fig. 10.4 *Interpretation of moment-area equations*. The geometric interpretation of the area Θ_{ab} and moment Δ_{ab} and their role in computing deflections and rotations

Example 10.4
Consider the cantilever beam of length L and bending modulus EI subjected to a downward point load of magnitude P at the end. Compute the rotation and deflection at the tip using the moment-area equations.

Solution. The beam geometry and loading are shown in Fig. 10.5 along with the M/EI diagram. The area of the M/EI diagram is the area of a triangle. Thus,

$$\Theta_{ab} = \tfrac{1}{2}(L)\Big(-\frac{PL}{EI}\Big) = -\frac{PL^2}{2EI}$$

The centroid of the area is at $2L/3$ from the right end. Noting that $\theta_o = 0$ and $w_o = 0$ at the fixed end, the tip rotation and deflection can then be computed as

$$\theta_L = -\frac{PL^2}{2EI}, \qquad w_L = -\big(\tfrac{2}{3}L\big)\Big(-\frac{PL^2}{2EI}\Big) = \frac{PL^3}{3EI}$$

where the algebraic signs indicate that the rotation is clockwise and the deflection is downward, as expected.

The moment-area method is well suited to the task of computing the deflection and rotation at specific points like the ends of the beam. Again, this has been an

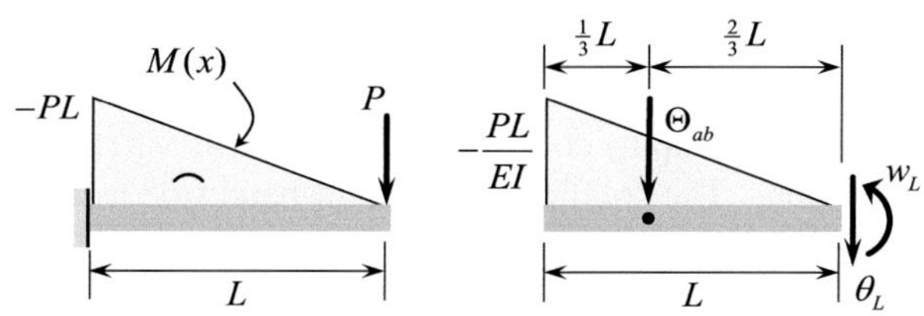

Fig. 10.5 *Example*. Beam of length L and bending modulus EI subjected to a point load at the tip. The sketch on the left shows the moment diagram and loading. The sketch on the right shows the M/EI diagram with area Θ_{ab} lumped at the centroid of the triangular distribution

aside on a historically important hand calculation method. Let us now return to the original goal of automating the computations.

10.3 Automating the Solution Process

Our aim is to write a code that will automate the computation and plotting of the V, M, θ, and w diagrams for any boundary conditions and for any applied loads. We can write Eqs. 10.9 in matrix form as

$$\begin{bmatrix} 1 & 0 & 0 & 0 & -1 & 0 & 0 & 0 \\ a & 1 & 0 & 0 & 0 & -1 & 0 & 0 \\ c & b & 1 & 0 & 0 & 0 & -1 & 0 \\ -d & -c & -a & 1 & 0 & 0 & 0 & -1 \end{bmatrix} \begin{Bmatrix} V_o \\ M_o \\ \theta_o \\ w_o \\ V_L \\ M_L \\ \theta_L \\ w_L \end{Bmatrix} = \begin{Bmatrix} \mathcal{I}_0 \\ \mathcal{I}_1 \\ \mathcal{I}_2 \\ -\mathcal{I}_3 \end{Bmatrix} \tag{10.12}$$

where

$$a = L, \qquad b = \frac{L}{EI}, \quad c = \frac{L^2}{2EI}, \qquad d = \frac{L^3}{6EI} \tag{10.13}$$

Let us assume that the end-state variables that are known are equal to zero (e.g., zero displacement and zero rotation at one end means that it is a fixed end). By the rules of matrix multiplication, the zero values multiply their respective coefficients in the matrix and the result is just zero. Hence, we can find a reduced system by removing the columns of the matrix associated with the four known values and removing the known values from the list of unknown state variables. For a fixed-simple beam, we have $w_o = 0$, $\theta_o = 0$, $w_L = 0$, and $M_L = 0$. So, Eq. 10.12 reduces to

$$\begin{bmatrix} 1 & 0 & -1 & 0 \\ a & 1 & 0 & 0 \\ c & b & 0 & -1 \\ -d & -c & 0 & 0 \end{bmatrix} \begin{Bmatrix} V_o \\ M_o \\ V_L \\ \theta_L \end{Bmatrix} = \begin{Bmatrix} \mathcal{I}_0 \\ \mathcal{I}_1 \\ \mathcal{I}_2 \\ -\mathcal{I}_3 \end{Bmatrix} \tag{10.14}$$

which can be solved to give the four *unknown* state variables at the ends of the beam. Now, all end-state variables are known. In particular, the values at $x = 0$ are all known.

To get set up to code these equations in MATLAB, let us make some notational definitions. The coefficient and right-hand side matrices are

$$\mathbf{B} = \begin{bmatrix} 1 & 0 & 0 & 0 & -1 & 0 & 0 & 0 \\ a & 1 & 0 & 0 & 0 & -1 & 0 & 0 \\ c & b & 1 & 0 & 0 & 0 & -1 & 0 \\ -d & -c & -a & 1 & 0 & 0 & 0 & -1 \end{bmatrix}, \qquad \mathbf{z} = \begin{Bmatrix} \mathcal{I}_0 \\ \mathcal{I}_1 \\ \mathcal{I}_2 \\ -\mathcal{I}_3 \end{Bmatrix} \tag{10.15}$$

where $a, b, c,$ and d are defined in Eq. 10.13. When we apply the boundary conditions, the reduced system has the form $\mathbf{Cs} = \mathbf{z}$, where $\mathbf{C}$ is a 4×4 matrix that can be computed directly from $\mathbf{B}$ and $\mathbf{s}$ is a 4×1 array containing the unknown state variables. We can accomplish the task of reducing $\mathbf{B}$ to $\mathbf{C}$ by defining an index array `Var` with eight entries (one for each of the eight state variables and, therefore, one for each column of $\mathbf{B}$). An entry of 1 means the associated variable is unknown and, therefore, we need to keep that column of $\mathbf{B}$. An entry of 0 means the associated variable is known to be zero and that column of $\mathbf{B}$ should be deleted to create $\mathbf{C}$. For the example with fixed-simple ends, the array would be

$$\texttt{Var} = \begin{bmatrix} 1 & 1 & 0 & 0 & 1 & 0 & 1 & 0 \end{bmatrix}$$

Now, with one line of MATLAB code, we can compute $\mathbf{C}$ from $\mathbf{B}$ as

```
C = B(:,Var==1)
```

In MATLAB, that line means take all rows (the colon in the first slot of the arguments of `B`) and take any column that is associated with a value of 1 in the array `Var` (the second slot of the arguments of `B`). In this case, it picks out the first, second, fifth, and seventh columns of $\mathbf{B}$. In our code, we will need to articulate the 16 cases of the `Var` matrix in accord with the 16 possible boundary conditions.

Now that we have determined all the state variables at the ends of the beam, most importantly all the values at $x = 0$, we can integrate the differential equations numerically as if they were an initial-value problem. We will use the generalized trapezoidal rule to carry out the integration. This approach is attractive because if we address the calculation this way, then all we need to do is specify the loading function $q(x)$ to generate the solutions.

The basic framework for integration by generalized trapezoidal rule is to convert the actual integral into a numerical approximation. Specifically,

$$\int_{x_n}^{x_{n+1}} g(x)\,dx \approx \Delta x\big[\beta\, g(x_n) + \big(1-\beta\big)\, g(x_{n+1})\big]$$

where $\Delta x = x_{n+1} - x_n$ and β is the numerical analysis parameter. This algorithm allows us to create a sequence $\{y_0, y_1, y_2, \ldots, y_N\}$ that approximately solves the differential equation $y' = g(x)$ at N equally spaced points $\{x_0, x_1, x_2, \ldots, x_N\}$ from the known starting value y_0 at x_0, with the recursion

$$y_{n+1} = y_n + \Delta x\big[\beta\, g(x_n) + \big(1-\beta\big)\, g(x_{n+1})\big]$$

as outlined in Appendix A. The best value of β is a half. We can, for example, write the integral version of the shear equation (Eq. 10.2) as

$$V_{n+1} = V_n - \int_{x_n}^{x_{n+1}} q(\xi)\, d\xi$$

where $V_n = V(x_n)$ and $V_{n+1} = V(x_{n+1})$. The other equations have similar expressions. Specifically,

$$\begin{aligned} M_{n+1} &= M_n + \int_{x_n}^{x_{n+1}} V(\xi)\, d\xi \\ \theta_{n+1} &= \theta_n + \int_{x_n}^{x_{n+1}} \frac{M(\xi)}{EI}\, d\xi \\ w_{n+1} &= w_n - \int_{x_n}^{x_{n+1}} \theta(\xi)\, d\xi \end{aligned}$$

The integrals in each of these equations can be approximated by the generalized trapezoidal rule. Hence, the beam equations can be integrated from left to right with the recursion

$$\begin{aligned} V_{n+1} &= V_n - \Delta x\big[\beta\, q_n + (1-\beta)q_{n+1}\big] \\ M_{n+1} &= M_n + \Delta x\big[\beta\, V_n + (1-\beta)\, V_{n+1}\big] \\ \theta_{n+1} &= \theta_n + \frac{\Delta x}{EI}\big[\beta\, M_n + (1-\beta)\, M_{n+1}\big] \\ w_{n+1} &= w_n - \Delta x\big[\beta\, \theta_n + (1-\beta)\theta_{n+1}\big] \end{aligned} \tag{10.16}$$

We can get this recursion started because we know $\{V_o, M_o, \theta_o, w_o\}$ from the analysis of the end states, and we can always compute the value of the load, including the initial value and any subsequent value $q_n = q(x_n)$, because $q(x)$ is a known function. We can use this computation to generate the plots of the state variables over the length. If we carry this integral from 0 to L, then we will get state variables at the far end, which we can compare with the previously computed end-state values as a check on accuracy of the numerical integration.

10.4 Code Implementation for Distributed Loads

In this section, we implement the computational strategy in a MATLAB code and provide some commentary on the coding choices. Most of the variable names correspond closely with the mathematical symbols used in the equations presented in the preceding sections. The main code for computing the shear, moment, rotation, and transverse displacement under a distributed load with any boundary conditions is Code 10.1, listed below.

Code 10.1 MATLAB code `BeamCode1` to compute the shear, moment, rotation, and transverse deflection of a beam with any end conditions and any applied loading

```
   clear; clc;

%. Physical parameteters
   L = 10;                              % Beam length
   EI = 100;                            % Beam bending stiffness
   qo = 3;                              % Load magnitude
   dLoadType = 4;                       % For DistrLoadFcn function

%. BC Code: =1 fixed, =2 free, =3 simple, =4 slide
   BCL = 1;                             % Left end
   BCR = 3;                             % Right end

%  BCs:     V  M  T  W :    =0 value is 0, =1 value is unknown
   BCs = [ 1, 1, 0, 0 ;                 % fixed
           0, 0, 1, 1 ;                 % free
           1, 0, 1, 0 ;                 % simple
           0, 1, 0, 1 ];                % slide

%. Establish known and unknown variables
   Var = [BCs(BCL,:), BCs(BCR,:)];

%. Establish load name
   [~,Load] = DistrLoadFcn(0,qo,dLoadType);

%. Numerical integration parameters
   nSimp = 21;                          % Number of Simpson points
   [wt,st,nSimp] = NumInt(3,nSimp);     % Simpson weights and stations

%. Compute I0, I1, I2, I3 by Simpson's rule
   z = [0; 0; 0; 0];
   for i=1:nSimp
     x = st(i);
     [q,~] = DistrLoadFcn(x,qo,dLoadType);
     Xi = (1-x)*L;
     z = z + q*L*[ 1; Xi; Xi^2/(2*EI); -Xi^3/(6*EI)]*wt(i);
   end

%. Construct Matrix B, reduce to C
   a = L; b = L/EI; c = L^2/(2*EI); d = L^3/(6*EI);

   B = [ 1,  0,  0,  0, -1,  0,  0,  0 ;...
         a,  1,  0,  0,  0, -1,  0,  0 ;...
         c,  b,  1,  0,  0,  0, -1,  0 ;...
        -d, -c, -a,  1,  0,  0,  0, -1 ];

   C = B(:,Var==1);

%. Compute missing end states and store in F
   s = C\z;
   F = zeros(8,1);
   F(Var==1) = s;

%. Compute the V, M, Theta, and W diagrams
   nGTR = 51;                                     % Number of GTR points
   beta = 0.5;                                    % GTR parameter
```

```
%. Initialize values for GTR
   Store = zeros(nGTR,6);
   dx = L/(nGTR-1);
   x = 0;                                          % Initial x value
   [qold,~] = DistrLoadFcn(x/L,qo,dLoadType);      % Initial load
   Vold = F(1);                                    % Initial shear
   Mold = F(2);                                    % Initial moment
   Told = F(3);                                    % Initial rotation
   Wold = F(4);                                    % Initial displacement

%. Integrate by GTR
   for j=1:nGTR

     Store(j,:) = [x,Wold,Told,Mold,Vold,qold];
     x = x + dx;
     [qnew,~] = DistrLoadFcn(x/L,qo,dLoadType);
     Vnew = Vold - dx*(beta*qold + (1-beta)*qnew);
     Mnew = Mold + dx*(beta*Vold + (1-beta)*Vnew);
     Tnew = Told + dx*(beta*Mold + (1-beta)*Mnew)/EI;
     Wnew = Wold - dx*(beta*Told + (1-beta)*Tnew);

     Vold = Vnew;
     Mold = Mnew;
     Told = Tnew;
     Wold = Wnew;
     qold = qnew;

   end

%. Output results
   BeamEchoInput(L,EI,qo,BCL,BCR,Load,nSimp,nGTR,beta);
   BeamPrintResults(F,Store,z);
   BeamCreatePlots(Store,L,BCL,BCR,Load);
```

The code starts by specifying the key physical parameters: length `L`, bending stiffness `EI`, magnitude of the distributed load `qo`, and `dLoadType` indicating which load type will be used in the function `DistrLoadFcn`. The boundary conditions are specified through the boundary codes `BCL` for the left end and `BCR` for the right end, wherein the values 1, 2, 3, or 4 indicate fixed, free, simple, or slide supports, respectively (see Fig. 8.2 for definitions of those support types).

The `Var` array establishes which variables are known and which are unknown. It is constructed from the array `BCs`, whose rows are associated with the fixed, free, simple, and slide conditions. The array `Var` is built by picking out the appropriate rows of `BCs`, using the values of `BCL` and `BCR`. The loading is defined and computed through the function `DistrLoadFcn` (Code 10.2).

The first computational task is to evaluate $\mathcal{I}_0$, $\mathcal{I}_1$, $\mathcal{I}_2$, and $\mathcal{I}_3$ from Eq. 10.8. These integrals are computed numerically using a composite *Simpson rule*. The weights `wt` and stations `st` are established by calling the function `NumInt`, which is Code A.2 in Appendix A.4. The number of integration points (`nSimp`) is set by the user. The number of points should be sufficient to ensure accurate integration. Integration is accomplished by looping over the integration points and accumulating the contributions to each sum.

The second task is to compute the unknown end states by first creating the matrix **C** and then solving the equations **Cs** = **z**, as outlined in the previous section. Once the values of the unknown end states have been computed, they are stored back into the array F in their proper locations, obeying the order $\{V_o, M_o, \theta_o, w_o, V_L, M_L, \theta_L, w_L\}$. We store these values in proper order because we need to identify the initial values at $x = 0$ to start the generalized trapezoidal integration. In the array F, those will always be the first four values.

The third task is to compute the V, M, θ, and w diagrams. We accomplish this goal by numerically integrating the governing equations using the generalized trapezoidal rule (GTR), starting at the left end and integrating to the right end. The initial values are established from F along with the initial value of the load qold, which can be obtained from DistrLoadFcn by specifying $x = 0$. The GTR algorithm only needs the previous value to compute the next one, so the nomenclature uses old for the values at location x_n and new for values at location x_{n+1}. The new values are computed from the old values, in accord with Eqs. 10.16. The update puts the new values into the old locations to get ready for the next integration step. The results of each step are recorded in the array Store for later plotting.

Finally, the results are plotted. The function BeamCreatePlots generates the diagrams of V, M, θ, and w. This function is not included in the book, but what it does should be evident through the plots presented in the examples.

The function DistrLoadFcn is given in Code 10.2. The function is invoked first to establish the name and type of loading, stored in the "struct" Load, which is used later in the function BeamEchoInput. Subsequently, the function is used to evaluate the loading given the location x.

Code 10.2 MATLAB code DistrLoadFcn to specify and compute the distributed loading for the beam code

```
function [q,Load] = DistrLoadFcn(x,qo,Type)
%  Evaluate q(x) for load type specified by Type.

%. Load function names for interactive menu
   Names = {'Constant'; 'Linear ramp up';...
            'Linear ramp down'; 'Half sinusoid';...
            'Unit cosine patch'; 'Quarter sine load';...
            'Self-equilibrating load'};

%. Set load type name and number
   Load.Type = Type;
   Load.Name = Names{Type};

%. Evaluate load
   switch Load.Type

     case 1 % Constant load
       q = 1;

     case 2 % Linear ramp up
       q = x;

     case 3 % Linear ramp down
       q = 1 - x;
```

```
        case 4 % Half-sine load
          q = sin(pi*x);

        case 5 % Unit area cosine patch
          a = 0.4; b = 0.6; h = b - a; q = 0;
          if (x>a) && (x<b)
            q = (1-cos(2*pi*(x-a)/h))/(2*h);
          end

        case 6 % Quarter Sinusoidal load
          q = sin(pi*x/2);

        case 7 % self-equilibrating load
          q = cos(2*pi*x);

      end % switch

%. Scale load
   q = q*qo;

end
```

The load type is determined by `Type`, and the value of the function, given the input `x`, which is assumed to be a fraction of the length L, is computed from the associated function in the `switch` construct. Additional load forms can be easily added by adding more cases, both to the array `Names` and as an additional `case` in the `switch` statement. Observe that most of the functions are defined to have unit maximum values, which then get multiplied by the load magnitude `qo` to complete the calculation. The one exception is the cosine patch load (case 5) that is formulated to have unit area, so that the magnitude decreases as the spread of the load increases, always giving the same total load.

Example 10.5
Apply the code to find the moment, shear, and deflected shape of a fixed-simple beam of length $L = 10$ and bending stiffness $EI = 100$ subjected to a half-sine loading of magnitude $q_o = 3$.

Solution. The inputs for Code 10.1 are `L=10`, `EI=100`, `qo=3`, and `LoadType=4` (to get the half-sine loading). The boundary codes are `BCL=1` (fixed) and `BCR=3` (simple). The results are shown in Fig. 10.6. The displacement plot shows $-w(x)$ because we take $\mathbf{e}_3$ downward, which defines $w(x)$ in our sign convention as being positive downward. In MATLAB for an (x, y) plot, the ordinate is assumed positive upward. The actual values of w computed in this problem are positive. Thus, to show downward deflection, we plot $-w$.

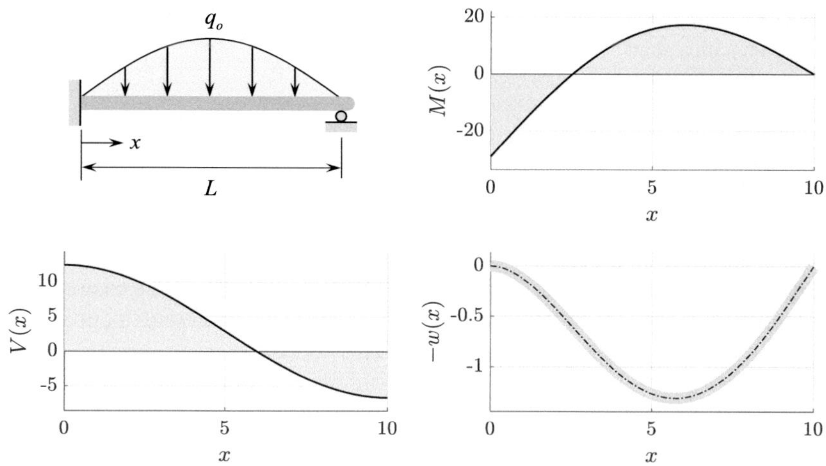

Fig. 10.6 *Example.* A fixed-simple beam of length $L = 10$ and bending modulus $EI = 100$ subjected to a half-sine load of magnitude $q_o = 3$. The sketch on the upper left shows the support conditions and loading. The plots show $V(x)$, $M(x)$, and $w(x)$

10.5 Point Loads

While point loads do not exist in nature, they provide a useful idealization that works well in the context of beam theory. A point load is the limit of a distributed load as the region over which it is distributed approaches infinitesimal size. In this section, we extend the code from the previous one to allow the incorporation of point loads. To specify a point load, we must give its magnitude F_i and its location x_i, as shown in Fig. 10.7. There can be multiple point loads, so $i = 1, 2, \ldots, n$.

If we take a free body diagram of the beam with a cut at x to the left of x_i, and keep the left piece, then the load F_i is not included. Therefore, it does not appear in the equation of equilibrium and, hence, does not appear in the function $V(x)$ from $0 \leq x < x_i$. On the other hand, if we take a free body diagram of the beam with the cut at x to the right of x_i, then the load F_i *is* included and it *does* appear in the function $V(x)$ from $x_i \leq x \leq L$. The shear jumps at x_i by the amount F_i. We need to account for this feature in our code.

Fig. 10.7 *Point loads.* A beam with a point load F_i located at a distance x_i from the left end

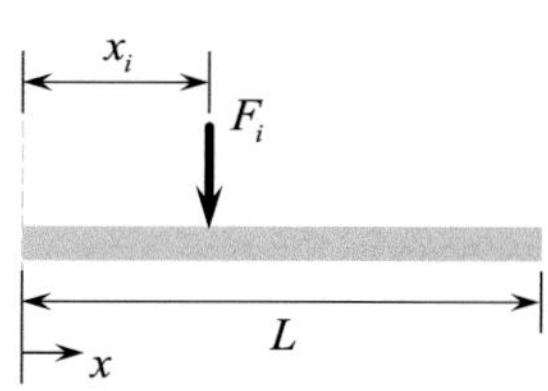

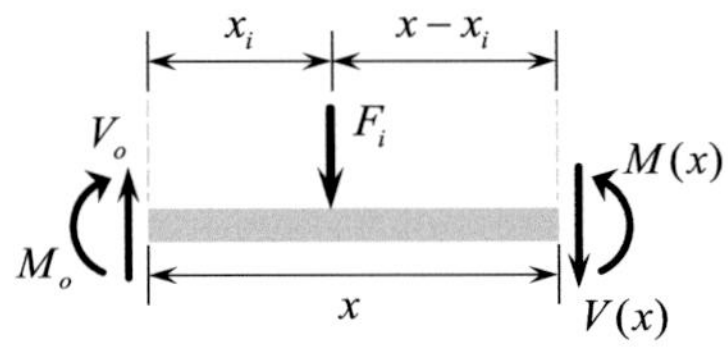

Fig. 10.8 *Effect of a point load.* A beam with a point load F_i located a distance x_i from the left end with a cut at x to the right of the load, exposing $V(x)$ and $M(x)$ (there is no axial force)

To see the effect of the point load, let us retrace the steps we took in integrating the governing equations for the distributed load. We will consider the case of a single point load, as shown in Fig. 10.8, which is a free body diagram with a cut at x to the right of the load F_i. The internal shear force is

$$V(x) = V_o - F_i \tag{10.17}$$

where, again, V_o is the shear at the left end of the beam. Compare this equation with Eq. 10.2 for the distributed load $q(x)$. Summing moments at the cut gives

$$M(x) = M_o + V_o x - F_i\big(x - x_i\big) \tag{10.18}$$

where M_o is the moment at the left end of the beam. Compare this equation with Eq. 10.3.

Now, we can integrate the rotation equation $\theta' = M/EI$. First, evaluate the rotation at x_i by integrating from zero to x_i. For this region, the moment is $M(x) = M_o + V_o x$. Thus, the rotation at x_i is

$$\theta(x_i) = \theta_o + \frac{1}{EI}\int_0^{x_i} \big(M_o + V_o\,\xi\big)\,d\xi = \theta_o + \frac{M_o x_i}{EI} + \frac{V_o x_i^2}{2EI}$$

We can compute the rotation for $x \geq x_i$ as

$$\begin{aligned}
\theta(x) &= \theta(x_i) + \frac{1}{EI}\int_{x_i}^{x} \big(M_o + V_o\,\xi - F_i(\xi - x_i)\big)d\xi \\
&= \theta(x_i) + \frac{1}{EI}\Big[M_o\,\xi + V_o\big(\tfrac{1}{2}\xi^2\big) - F_i\big(\tfrac{1}{2}\xi^2 - x_i\,\xi\big)\Big]_{x_i}^{x} \\
&= \theta(x_i) + \frac{M_o\big(x - x_i\big)}{EI} + \frac{V_o\big(x^2 - x_i^2\big)}{2EI} - \frac{F_i}{2EI}\big(x^2 - 2x_i x + x_i^2\big) \\
&= \theta_o + \frac{M_o x}{EI} + \frac{V_o x^2}{2EI} - \frac{F_i}{2EI}\big(x - x_i\big)^2
\end{aligned}$$

where, in the last step, we substituted the value of $\theta(x_i)$ we had computed previously. Thus, the final expression for the rotation at x is

$$\theta(x) = \theta_o + \frac{M_o x}{EI} + \frac{V_o x^2}{2EI} - \frac{F_i}{2EI}\big(x - x_i\big)^2 \tag{10.19}$$

for $x > x_i$. This equation can be verified by differentiating and multiplying by EI to recover Eq. 10.18.

We can follow a similar process to integrate the equation $w' = -\theta$ to get

$$w(x) = w_o - \theta_o x - \frac{M_o x^2}{2EI} - \frac{V_o x^3}{6EI} + \frac{F_i}{6EI}\big(x - x_i\big)^3 \tag{10.20}$$

for $x > x_i$, where w_o and θ_o are the transverse displacement and rotation, respectively, at $x = 0$. The details of the derivation are left to the reader.

Equations 10.17 through 10.20 can be evaluated at $x = L$ to give

$$\begin{aligned}
V_L &= V_o - F_i \\
M_L &= M_o + V_o\, L - F_i\big(L - x_i\big) \\
\theta_L &= \theta_o + \frac{M_o\, L}{EI} + \frac{V_o\, L^2}{2EI} - \frac{F_i}{2EI}\big(L - x_i\big)^2 \\
w_L &= w_o - \theta_o\, L - \frac{M_o\, L^2}{2EI} - \frac{V_o\, L^3}{6EI} + \frac{F_i}{6EI}\big(L - x_i\big)^3
\end{aligned} \tag{10.21}$$

The difference between these equations and Eq. 10.7 is that the distributed load terms have been replaced by the terms involving F_i. What that implies is that the only modification needed to solve the end-state problem is in the computation of $\mathcal{I}_0$ through $\mathcal{I}_3$. It is straightforward to generalize these expressions for point loads to multiple loads because all the loads are to the left of the right end of the beam. Hence, we can write

$$\begin{aligned}
\mathcal{I}_0 &= \sum_{i=1}^{n} F_i & \mathcal{I}_2 &= \frac{1}{2EI}\sum_{i=1}^{n} F_i\big(L - x_i\big)^2 \\
\mathcal{I}_1 &= \sum_{i=1}^{n} F_i\big(L - x_i\big) & \mathcal{I}_3 &= \frac{1}{6EI}\sum_{i=1}^{n} F_i\big(L - x_i\big)^3
\end{aligned} \tag{10.22}$$

These expressions can be compared with Eq. 10.7. Because integration is a linear operation, the effects of distributed and point loads can simply be added together.

Code Implementation for Point Loads To implement a code that includes point loads, we first define a function `PointLoadFcn` that is analogous to the function `DistrLoadFcn`. Code 10.3 defines the loads and their locations for various load cases. The point-load information is stored in the "struct" `Load`. Additional cases can be added.

Code 10.3 MATLAB code `PointLoadFcn` to specify applied point loads for the beam code

```
function [Load] = PointLoadFcn(L,Po,Ptype,Load)
%  Establish locations and magnitudes of point loads

%. Load function names for interactive menu
   Names = {'None'; 'Couple'; 'Equally spaced loads';...
            'End load'; 'Load at midspan'};

%. Select load type by menu
   Load.Type = Ptype;
   Load.LoadName = Names{Ptype};

   switch Load.Type

     case 1 % None
       P = 0;                               % Value of point load
       Loc = 0;                             % Location of point load

     case 2 % Couple
       P = [-1, 1];                         % Value of point load
       Loc = [0.40, 0.60];                  % Location of point load

     case 3 % Equally spaced loads
       n = 4;                               % Number of loads
       P = ones(n,1);                       % Value of loads
       Loc = (1:n)/(n+1);                   % Location of loads

     case 4 % end load (must have free end at x=L)
       P = 1;                               % Value of point load
       Loc = 1;                             % Location of point load

     case 5 % Load at midspan
       P = 1;                               % Value of loads
       Loc = 0.5;                           % Location of loads

   end % switch

%. Scale and store in struct PLoad
   Load.nLoads = length(P);                 % Number of point loads
   Load.P = Po*P;                           % Load values
   Load.Loc = Loc*L;                        % Load locations

end
```

The cases can be set up for certain configurations of point-load scenarios. Each case specifies the load values `P` and their locations `Loc`. At the end of the function, the loads are scaled by `Po`, which is an input to the function, and the information is stored in the "struct" `Load` for use in the main program. The reason `Load` is passed into the function is that the information for the distributed load from the function `DistrLoadFcn` already exists. The function `PointLoadFcn` simply adds to that information.

Let us now modify Code 10.1 to incorporate both distributed and point loads. There are three main adjustments needed: (1) the definition of the point loads, as already mentioned; (2) adding the contributions of point loads to $\mathcal{I}_0$ through $\mathcal{I}_3$; and (3) changing the approach to integration over the length of the beam to generate the plots. Code 10.4 is the modified code.

Code 10.4 MATLAB code `BeamCode2` to compute the shear, moment, rotation, and transverse deflection of a beam with any end conditions and any applied loading, including both distributed loads and point loads

```
   clear; clc;

%. Physical parameteters
   L = 10;                                   % Beam length
   EI = 2000;                                % Beam bending stiffness
   dLoadType = 1;                            % Distributed load type
   pLoadType = 5;                            % Point load type
   qo = 0;                                   % Distributed load multiplier
   Po = 5;                                   % Point load multiplier

%. BC Code: =1 fixed, =2 free, =3 simple, =4 slide
   BCL = 1;                                  % Left end
   BCR = 1;                                  % Right end

%  BCs:     V  M  T  W :     =0 value is 0, =1 value is unknown
   BCs = [ 1, 1, 0, 0 ;                      % fixed
           0, 0, 1, 1 ;                      % free
           1, 0, 1, 0 ;                      % simple
           0, 1, 0, 1 ];                     % slide

%. Establish known and unknown variables
   Var = [BCs(BCL,:), BCs(BCR,:)];

%. Establish loads
   [~,Load] = DistrLoadFcn(0,qo,dLoadType);
   [Load] = PointLoadFcn(L,Po,pLoadType,Load);

%. Numerical integration parameters
   nSimp = 21;                               % Number of Simpson points
   [wt,st,nSimp] = NumInt(3,nSimp);          % Simpson weights and stations

%. Compute I0, I1, I2, I3 by Simpson integration
   z = [0; 0; 0; 0];
   for i=1:nSimp
     xi = st(i);
     [q,~] = DistrLoadFcn(xi,qo,dLoadType);
     Lmx = (1-xi)*L;
     z = z + q*L*[ 1; Lmx; Lmx^2/(2*EI); -Lmx^3/(6*EI)]*wt(i);
   end

%. Add contributions of point loads
   for j=1:Load.nLoads
     Pj = Load.P(j);
     Xj = L - Load.Loc(j);
     z = z + Pj*[1; Xj; Xj^2/(2*EI); -Xj^3/(6*EI)];
   end

%. Construct Matrix B, reduce to C
   a = L; b = L/EI; c = L^2/(2*EI); d = L^3/(6*EI);

   B = [ 1,  0,  0,  0, -1,  0,  0,  0 ;...
         a,  1,  0,  0,  0, -1,  0,  0 ;...
         c,  b,  1,  0,  0,  0, -1,  0 ;...
        -d, -c, -a,  1,  0,  0,  0, -1 ];

   C = B(:,Var==1);
   CheckC(C);

%. Compute missing end states and store in F
   s = C\z;
   F = zeros(8,1);
   F(Var==1) = s;
```

```
%. Compute the V, M, Theta, and W diagrams
   nGTR = 51;                                     % Number of GTR points
   beta = 0.5;                                    % GTR parameter
   eps = 1.e-8;                                   % Very small number

%. Initialize values for GTR
   Store = zeros(nGTR*Load.nLoads,6);
   x = 0;                                         % Initial x value
   n = 0;                                         % Counter for storage
   Vold = F(1);                                   % Initial shear
   Mold = F(2);                                   % Initial moment
   Told = F(3);                                   % Initial rotation
   Wold = F(4);                                   % Initial displacement

%. Set up segments between point loads for GTR integration
   EndPts = [0,Load.Loc,L];
   for nSeg=1:Load.nLoads+1
     Ln = EndPts(nSeg+1) - EndPts(nSeg);
     dx = Ln/(nGTR-1);
     [qold,~] = DistrLoadFcn(x/L,qo,dLoadType);

%... Integrate over current segment
     for j=1:nGTR
       n = n + 1;
       Store(n,:) = [x,Wold,Told,Mold,Vold,qold];
       x = x + dx;
       [qnew,~] = DistrLoadFcn(x/L,qo,dLoadType);
       Vnew = Vold - dx*(beta*qold + (1-beta)*qnew);
       Mnew = Mold + dx*(beta*Vold + (1-beta)*Vnew);
       Tnew = Told + dx*(beta*Mold + (1-beta)*Mnew)/EI;
       Wnew = Wold - dx*(beta*Told + (1-beta)*Tnew);

%..... Update state variables, unless it is the last point in the
     segment
       if j<nGTR
         Vold = Vnew;
         Mold = Mnew;
         Told = Tnew;
         Wold = Wnew;
         qold = qnew;
       end

     end

%... Adjust shear by amount of point load, unless it is the last segment
     if (nSeg<=Load.nLoads)
       x = x - dx + eps;
       Vold = Vold - Load.P(nSeg);
     end
   end

%. Output results
   BeamEchoInput(L,EI,qo,BCL,BCR,Load,nSimp,nGTR,beta);
   BeamPrintResults(F,Store,z);
   BeamCreatePlots(Store,L,BCL,BCR,Load);
```

The first observation is that the code is structured exactly the same way as the previous code. There are only a couple of additions to the physical inputs: the load magnitude `Po` multiplier for the point loads and the value of `pLoadType` which indicates what load type for point loads will be used in the function `PointLoadFcn`. The contributions of the point loads to `z` (i.e., $\mathcal{I}_0$ through $\mathcal{I}_3$) are added in a loop over the number of point loads using Eq. 10.22.

The other major change is the computation to get the V, M, θ, and w diagrams. An efficient way to deal with the discontinuities in the shear and moment plots is to break the beam up into segments with the locations of the point loads defining the ends of the segments. The array `EndPts` is constructed from 0, the point-load locations, and L. Inside the loop over segments, the loop over GTR points is the same as the previous code, with a couple of exceptions. First, we define `n` as a counter that increments across all the segments to define the row of the array `Store` to record the results. Second, we avoid the update for the last point in the segment so that the `old` values align with the location x_i. Finally, at the end of each GTR loop, we adjust x back one notch (since it was updated to one Δx past the endpoint of the segment). We add a small amount `eps` for plotting purposes and update the shear by the amount of the point load.

Example 10.6
Apply Code 10.4 to analyze a fixed-fixed beam of length $L = 10$, bending modulus $EI = 2000$, and subjected to a point load of magnitude $P = 8$ at midspan.

Solution. The physical inputs for Code 10.4 are

```
L=10,    EI=2000,    Po=8,    pLoadType=5
```

To have no distributed load, we set `qo=0` (it does not matter what the distributed load type is, then). The boundary codes are

`BCL=1` (fixed),   `BCR=1` (fixed)

The results are shown in Fig. 10.9. As discussed in the previous example, the displacement plot shows $-w(x)$.

The maximum moment and deflection for this case can be found analytically to be

$$M_{max} = \frac{PL}{8} = \frac{(8)(10)}{8} = 10$$
$$w_{max} = \frac{PL^3}{192EI} = \frac{(8)(10)^3}{192(2000)} = 0.021$$

These values can be observed on the plots (the maximum moment occurs at both ends and in the middle), thereby verifying the code. The units are not specified, but all quantities must have consistent units. For example, EI has units of force times length squared. The units of force and length must be the same as those used for P and L, respectively.

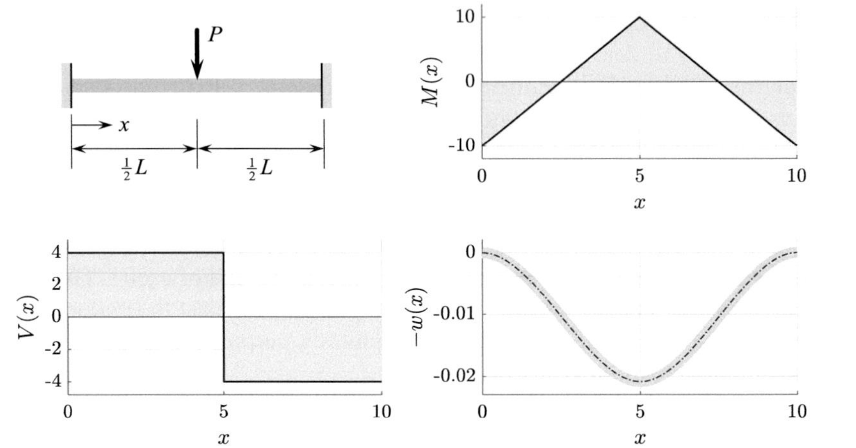

Fig. 10.9 *Example.* A fixed-fixed beam of length $L = 10$ and constant bending modulus $EI = 2000$ subjected to a point load of magnitude $P = 8$. The sketch on the upper left shows the support conditions and loading. The plots show $V(x)$, $M(x)$, and $w(x)$

10.6 Multispan Beam

Let us now extend our approach from the previous section to the *multispan beam* with N spans. The case with $N = 4$ is illustrated in Fig. 10.10. The length, flexural stiffness, and applied loading of each beam can be different, and we can have any boundary conditions at the two ends. The example shown in the figure has simple supports at both ends. The beams are continuous over the intermediate roller supports. There are no axial loads, so axial deformations are zero. The aim is to find the internal state, i.e., $V_n(x)$, $M_n(x)$, $\theta_n(x)$, and $w_n(x)$, for each span n, caused by the applied loads.

The most important observation for the multispan beam is that each beam segment obeys the relationship among the end states given by Eq. 10.7, with the positive sign convention indicated in Fig. 10.11. Thus, the end states also satisfy Eq. 10.12, which

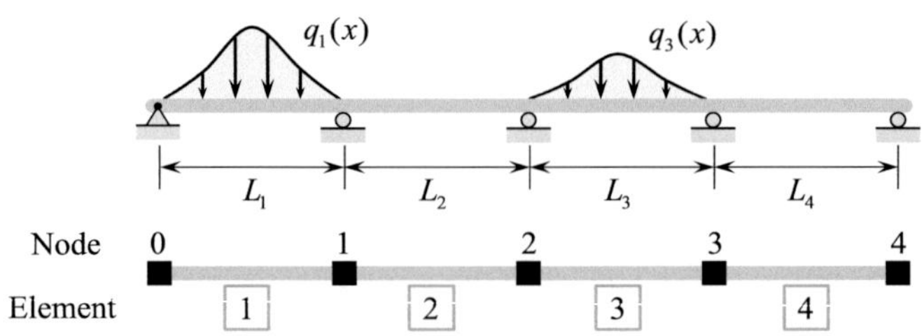

Fig. 10.10 *Multispan beam.* Four-span beam with different lengths, cross-sectional properties, and loading conditions, with the numbering convention for nodes and elements

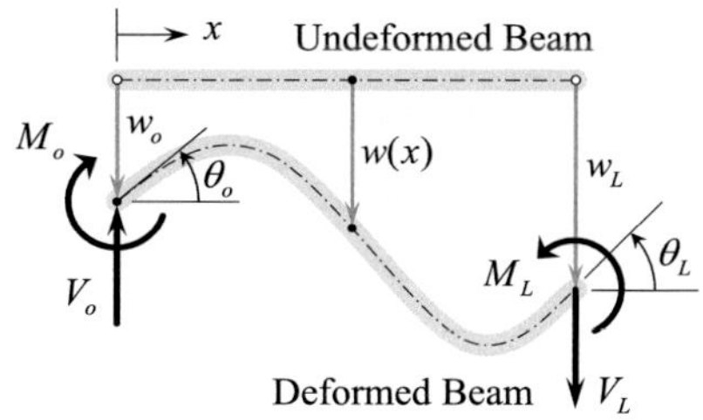

Fig. 10.11 *Beam end states.* Each beam in the multispan system has the end states shown. The state $\mathcal{S}_n(x)$ is reckoned relative to the local coordinate x, measured from the left end of the nth span

is simply the matrix version of Eq. 10.7. To set up the notation for the multispan beam, let

$$\mathbf{A}_n = \begin{bmatrix} 1 & 0 & 0 & 0 \\ a_n & 1 & 0 & 0 \\ c_n & b_n & 1 & 0 \\ -d_n & -c_n & -a_n & 1 \end{bmatrix}, \qquad \mathbf{z}_n = \begin{Bmatrix} \mathcal{I}_0^n \\ \mathcal{I}_1^n \\ \mathcal{I}_2^n \\ -\mathcal{I}_3^n \end{Bmatrix} \tag{10.23}$$

where $a_n = L_n$, $b_n = L_n/EI_n$, $c_n = L_n^2/2EI_n$, and $d_n = L_n^3/6EI_n$. The expressions for $\mathcal{I}_0^n$, $\mathcal{I}_1^n$, $\mathcal{I}_2^n$, and $\mathcal{I}_3^n$ are given by Eq. 10.8, each corresponding with the loading on that span. With this notation, Eq. 10.12 for the nth span can be written as

$$\mathbf{A}_n\mathcal{S}_n(0) - \mathcal{S}_n(L) = \mathbf{z}_n \tag{10.24}$$

where the state variables in span n have been designated as

$$\mathcal{S}_n(x) = \begin{Bmatrix} V_n(x) \\ M_n(x) \\ \theta_n(x) \\ w_n(x) \end{Bmatrix} \tag{10.25}$$

and x is measured from the left end of span n. Since we will only be using the end states at this stage, we will reuse the symbol x for the local axial coordinate for every span to simplify the notation.

The individual beam end states must also satisfy continuity equations at the interior nodes where adjacent elements meet. Let us define

$$\mathbf{Q}_o = \mathcal{S}_1(0), \qquad \mathbf{Q}_n = \mathcal{S}_n(L) \tag{10.26}$$

for $n = 1, \ldots, N$. What that means is that we will keep track of the state at the right end of span n with $\mathbf{Q}_n$. In addition, we will keep track of the state at the left end of the *first* span with $\mathbf{Q}_o$. We can then relate the state at the end of span n and the beginning of span $n+1$, for $n = 1, \ldots, N-1$, as

$$\mathcal{S}_{n+1}(0) = \mathcal{S}_n(L) - R_n\mathbf{p} = \mathbf{Q}_n - R_n\mathbf{p} \tag{10.27}$$

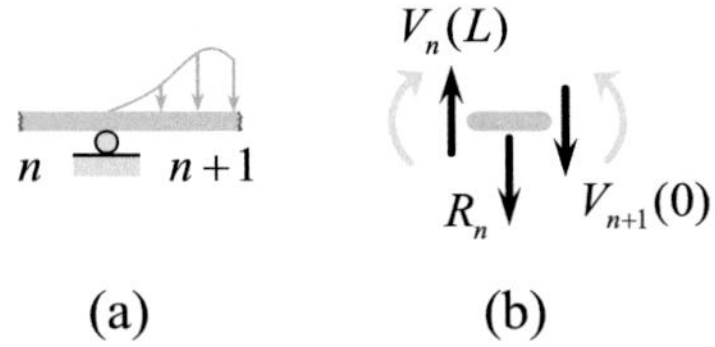

Fig. 10.12 *Shear discontinuity*. **a** The roller support between span n and span $n+1$. **b** Free body diagram showing shears and the reaction force R_n

where $\mathbf{p} = [\,1, 0, 0, 0\,]^T$ and R_n is the reaction force at node n associated with the roller support. Equation 10.27 represents the *continuity conditions* of the state variables (along with the discontinuity relationship for the shear) at the nodes. Figure 10.12 shows how the discontinuity condition for shear is determined. From the free body diagram in Fig. 10.12b, we can write the equilibrium equation $V_{n+1}(0) = V_n(L) - R_n$, which is the first component of Eq. 10.27. This equation also establishes that the reaction force R_n is positive downward.

Substituting the continuity conditions into Eq. 10.24 gives, for the four-span beam,

$$\begin{aligned} \mathbf{A}_1\mathbf{Q}_o - \mathbf{Q}_1 &= \mathbf{z}_1 \\ \mathbf{A}_2\big(\mathbf{Q}_1 - R_1\mathbf{p}\big) - \mathbf{Q}_2 &= \mathbf{z}_2 \\ \mathbf{A}_3\big(\mathbf{Q}_2 - R_2\mathbf{p}\big) - \mathbf{Q}_3 &= \mathbf{z}_3 \\ \mathbf{A}_4\big(\mathbf{Q}_3 - R_3\mathbf{p}\big) - \mathbf{Q}_4 &= \mathbf{z}_4 \end{aligned}$$

Note that we have one matrix equation for each element. The subscripts on the element quantities refer to the element number. The subscript on the nodal quantities (the reaction forces) refers to the node number. The unknown quantities are the state variables $\mathbf{Q}_n$ and the support reactions R_n. Two of the four state variables from $\mathbf{Q}_o$ and two from $\mathbf{Q}_N$, the end of the last span, are known from the boundary conditions at the ends of the multispan structure. We can write all these equations in a single matrix equation in block form (for the case with four spans) as

$$\begin{bmatrix} \mathbf{A}_1 & -\mathbf{I} & \mathbf{0} & \mathbf{0} & \mathbf{0} & \mathbf{0} & \mathbf{0} & \mathbf{0} \\ \mathbf{0} & \mathbf{A}_2 & -\mathbf{I} & \mathbf{0} & \mathbf{0} & -\mathbf{r}_2 & \mathbf{0} & \mathbf{0} \\ \mathbf{0} & \mathbf{0} & \mathbf{A}_3 & -\mathbf{I} & \mathbf{0} & \mathbf{0} & -\mathbf{r}_3 & \mathbf{0} \\ \mathbf{0} & \mathbf{0} & \mathbf{0} & \mathbf{A}_4 & -\mathbf{I} & \mathbf{0} & \mathbf{0} & -\mathbf{r}_4 \end{bmatrix} \begin{Bmatrix} \mathbf{Q}_o \\ \mathbf{Q}_1 \\ \mathbf{Q}_2 \\ \mathbf{Q}_3 \\ \mathbf{Q}_4 \\ R_1 \\ R_2 \\ R_3 \end{Bmatrix} = \begin{Bmatrix} \mathbf{z}_1 \\ \mathbf{z}_2 \\ \mathbf{z}_3 \\ \mathbf{z}_4 \end{Bmatrix}$$

where $\mathbf{r}_n = \mathbf{A}_n\mathbf{p} = [\,1, a_n, c_n, -d_n\,]^T$ is just the first column of $\mathbf{A}_n$. All the submatrices in the first five columns are 4×4 and the submatrices in the last three columns are all 4×1. As such, the coefficient matrix is 16×23. In each row, the matrix $[\,\mathbf{A}_n, -\mathbf{I}\,]$ is what we call $\mathbf{B}_n$ in the code for the single span. In the multispan case, these matrices overlap. For the multispan beam code, we will call the whole coefficient matrix $\mathbf{B}$ and the right-hand side $\mathbf{Z}$, which comprises the element matrices $\mathbf{z}_1$, $\mathbf{z}_2$, $\mathbf{z}_3$, and $\mathbf{z}_4$.

To reduce the equation to solvable form, we must recognize the values in the state array that are known to be zero. The boundary conditions work the same way as they did for the single beam code. For the multispan code, we will need to augment the `Var` array with interior support conditions. In the case of an interior roller support, we know that the transverse displacement w_n (the deflection at the right end of span n) is zero. The value of w_n is stored in the fourth slot of $\mathbf{Q}_n$. If that value is equal to zero, then it is sufficient to eliminate the corresponding column of $\mathbf{B}$, as we have done for the boundary conditions. Therefore, we can take care of the condition of zero transverse displacement at the interior supports by adding the four columns `[1,1,1,0]` to the `Var` array for each support point, which indicates that the shear, moment, and rotation are unknown, and the displacement is known to be zero. We must have an entry in the `Var` array associated with all the unknowns in the equation. Thus, we also need to add `[1,1,1]` at the end of `Var` to indicate that we want to keep all the columns associated with the unknown reaction forces R_n. The number of ones added must be equal to the number of support reactions at the rollers, which is one fewer than the number of spans.

Absent roller support There is a direct correspondence between the zeros we place in `Var` for $w_n = 0$ and the inclusion of a reaction force associated with that support location. The reaction force is what prevents displacement. We can model the absence of a roller, i.e., a node connecting two elements that does not have a roller support. One reason to include such a case is that it opens up the possibility of having adjoining elements with different EI values. Another reason might be that it allows the modeling of additional load cases, like a constant distributed load over a small segment of the beam (what we call a *patch loading*), which tends to cause difficulties with the numerical integration scheme when we model that loading on a single span with the `DistrLoadFcn` function.

To account for the possibility of an absent roller support, we can select an intermediate support condition of `[1,1,1,1]` (rather than `[1,1,1,0]`, as mentioned for the case with a roller support) to reflect that the transverse displacement is unknown (along with the other three state variables, of course). Then we must include a strategy that eliminates the corresponding reaction force by using, say, `[1,0,1]` if the middle of three roller supports is absent (rather than `[1,1,1]`, as mentioned for the case with a roller support) to reflect that the reaction force R_2 is known to be zero.

Internal hinge One additional internal condition that we can incorporate in our code is the *internal hinge*. The internal hinge is a device that forces the internal moment to be zero at that point. There is an associated concentrated rotation of the hinge that makes the rotation on the left and right sides of the hinge unequal, as illustrated in Fig. 10.13. In essence, the rotation field $\theta(x)$ is discontinuous at an internal hinge, similar to how the shear $V(x)$ is discontinuous at a roller support due to the reaction force.

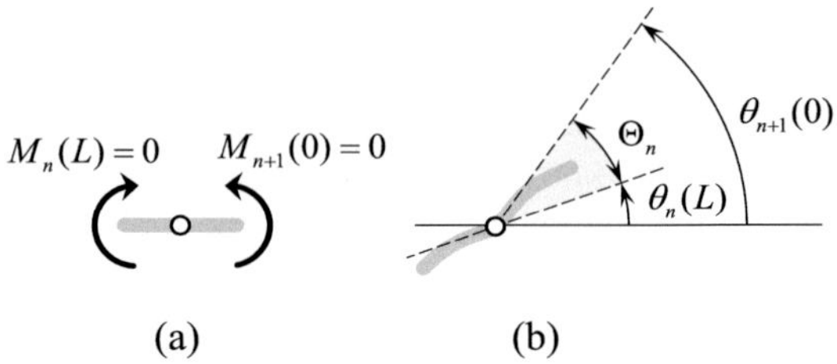

Fig. 10.13 *Internal hinge.* An internal hinge is a device that allows relative rotation between the ends of the beams connected to it but transmits shear force. **a** The internal moment at a hinge is zero. **b** The concentrated rotation of the hinge is $\Theta_n = \theta_{n+1}(0) - \theta_n(L)$

We can enforce the zero-moment conditions by changing our interior support conditions to `[1,0,1,1]`. The zero in the second slot indicates that the moment at that point is equal to zero (and, therefore, the associated column of the **B** array will be deleted). We also need to add the computation of the discrete rotation Θ_n of the hinge. To accomplish this goal, we can add variables similar to the reaction forces associated with roller supports. Let us adjust the continuity conditions of Eq. 10.27 to be

$$\begin{aligned} \mathcal{S}_{n+1}(0) &= \mathcal{S}_n(L) - R_n\mathbf{p} + \Theta_n\mathbf{q} \\ &= \mathbf{Q}_n - R_n\mathbf{p} + \Theta_n\mathbf{q} \end{aligned} \tag{10.28}$$

where Θ_n is the hinge rotation at node n and $\mathbf{q} = [\,0, 0, 1, 0\,]^T$. The adjustment maintains the possibility of a roller at node n and adds the possibility of an internal hinge at that same point. The last term enforces the condition

$$\theta_{n+1}(0) = \theta_n(L) + \Theta_n$$

Now, we can write the equations for all spans as

$$\begin{aligned} \mathbf{A}_1\mathbf{Q}_o - \mathbf{Q}_1 &= \mathbf{z}_1 \\ \mathbf{A}_2\big(\mathbf{Q}_1 - R_1\mathbf{p} + \Theta_1\mathbf{q}\big) - \mathbf{Q}_2 &= \mathbf{z}_2 \\ \mathbf{A}_3\big(\mathbf{Q}_2 - R_2\mathbf{p} + \Theta_2\mathbf{q}\big) - \mathbf{Q}_3 &= \mathbf{z}_3 \\ &\vdots \\ \mathbf{A}_n\big(\mathbf{Q}_{n-1} - R_{n-1}\mathbf{p} + \Theta_{n-1}\mathbf{q}\big) - \mathbf{Q}_n &= \mathbf{z}_n \end{aligned}$$

which can be put into matrix form (e.g., for a three-span beam) as

$$\begin{bmatrix} \mathbf{A}_1 & -\mathbf{I} & \mathbf{0} & \mathbf{0} & \mathbf{0} & \mathbf{0} & \mathbf{0} & \mathbf{0} \\ \mathbf{0} & \mathbf{A}_2 & -\mathbf{I} & \mathbf{0} & -\mathbf{r}_2 & \mathbf{0} & \mathbf{t}_2 & \mathbf{0} \\ \mathbf{0} & \mathbf{0} & \mathbf{A}_3 & -\mathbf{I} & \mathbf{0} & -\mathbf{r}_3 & \mathbf{0} & \mathbf{t}_3 \end{bmatrix} \begin{Bmatrix} \mathbf{Q}_o \\ \mathbf{Q}_1 \\ \mathbf{Q}_2 \\ \mathbf{Q}_3 \\ R_1 \\ R_2 \\ \Theta_1 \\ \Theta_2 \end{Bmatrix} = \begin{Bmatrix} \mathbf{z}_1 \\ \mathbf{z}_2 \\ \mathbf{z}_3 \end{Bmatrix}$$

where $\mathbf{r}_n = \mathbf{A}_n \mathbf{p}$ as before and $\mathbf{t}_n = \mathbf{A}_n \mathbf{q} = [\,0, 0, 1, -a_n\,]^T$ is just the third column of $\mathbf{A}_n$.

Code 10.5 implements the multispan beam for an arbitrary number of spans. Each span can have a distributed load from the function `DistrLoadFcn` (i.e., from Code 10.2).

Code 10.5 MATLAB code `MultispanBeam` to compute the shear, moment, rotation, and transverse deflection of a multispan beam with any end conditions, hinge or roller support continuity conditions, and any applied distributed loading

```
   clear; clc;

%. Physical parameteters
   nSpans = 4;                              % Number of spans
   L = [10,10,10,10];                       % Beam lengths
   EI = [200,200,200,200];                  % Beam bending stiffnesses
   qo = [2,0,2,0];                          % Load magnitudes
   dLoadType = [1,1,1,1];                   % Distributed load types

%. BC Code: =1 Fixed, =2 Free, =3 Simple, =4 Slide
   BCL = 3;                                 % BC at Left end
   BCR = 3;                                 % BC at Right end

%. Continuity code: =1 no support, =2 roller, =3 hinge, =4 roller+hinge
   Cont = [2,2,2];

%. BCs:      V  M  T  W                     Boundary conditions
   BCs = [ 1, 1, 0, 0 ;                     % Fixed
           0, 0, 1, 1 ;                     % Free
           1, 0, 1, 0 ;                     % Simple
           0, 1, 0, 1 ];                    % Slide

%. CCs:      V  M  T  W                     Continuity conditions
   CCs = [ 1, 1, 1, 1 ;                     % No support (R=0)
           1, 1, 1, 0 ;                     % Roller support (w=0)
           1, 0, 1, 1 ;                     % Internal hinge (M=0)
           1, 0, 1, 0 ];                    % Hinge + Roller (M=0, w=0)

%. Form BC array
   IntSupport = zeros(1,4*(nSpans-1));      % Interior Support conditions
   Reaction = zeros(1,nSpans-1);            % Reaction at node
   Rotation = zeros(1,nSpans-1);            % Relative rotation at node
   for i=1:nSpans-1
     ii = 4*(i-1) + [1,2,3,4];
     IntSupport(ii) = CCs(Cont(i),:);
     Reaction(i) = not(CCs(Cont(i),4));
     Rotation(i) = not(CCs(Cont(i),2));
   end
   Var = [BCs(BCL,:), IntSupport, BCs(BCR,:), Reaction, Rotation];

%. Initialize arrays for equations
   nVars = length(Var);                     % Number of variables
   B = zeros(4*nSpans,nVars);               % Initialize coefficient matrix
   Z = zeros(4*nSpans,1);                   % Initialize right hand side

%. Numerical integration parameters
   nSimp = 21;                              % Number of Simpson points
   [wt,st,nSimp] = NumInt(3,nSimp);         % Simpson weights and stations

%. Loop over spans to assemble arrays
   for n=1:nSpans

%... Compute I0, I1, I2, I3 by Simpson's rule
```

```
      z = zeros(4,1);
      for i=1:nSimp
        x = st(i);
        [q,~] = DistrLoadFcn(x,qo(n),dLoadType(n));
        Xi = (1-x)*L(n);
        z = z + q*L(n)*[1; Xi; Xi^2/(2*EI(n)); -Xi^3/(6*EI(n))]*wt(i);
      end

%... Construct span matrices Bn, rn, and tn
      a = L(n);
      b = L(n)/EI(n);
      c = L(n)^2/(2*EI(n));
      d = L(n)^3/(6*EI(n));

      Bn = [ 1,  0,  0,  0, -1,  0,  0,  0 ;
             a,  1,  0,  0,  0, -1,  0,  0 ;
             c,  b,  1,  0,  0,  0, -1,  0 ;
            -d, -c, -a,  1,  0,  0,  0, -1 ];

      rn = [ 1;  a;  c; -d ];              % For reactions
      tn = [ 0;  0;  1; -a ];              % For rotations

%... Establish indices for assembly into the global B and Z matrices
      ii = 4*(n-1) + [1,2,3,4];            % Row indices
      jj = 4*(n-1) + [1,2,3,4,5,6,7,8];    % Column indices
      nColr = 4*(nSpans+1) + n - 1;        % Column for rn
      nColt = nColr + nSpans - 1;          % Column for tn

%... Place Bn, rn, tn, and z in appropriate slots in B and Z
      B(ii,jj) = Bn;
      if (n > 1)
        B(ii,nColr) = -rn;
        B(ii,nColt) =  tn;
      end
      Z(ii) = z;

    end

%. Trim columns of B to get C
    C = B(:,Var==1);

%. Compute missing response values s and place in order in F
    s = C\Z;
    F = zeros(nVars,1);
    F(Var==1) = s;

%. Extract span initial values, adjust for reactions and hinge rotations
    So = zeros(4,nSpans);
    for n=1:nSpans
      ii = 4*(n-1) + [1,2,3,4];
      nColr = 4*(nSpans+1) + n - 1;
      nColt = nColr + (nSpans-1);
      So(:,n) = F(ii);
      if n>1
        So(1,n) = So(1,n) - F(nColr);      % Adjust shear for reaction
        So(3,n) = So(3,n) + F(nColt);      % Adjust rotation for hinge
      end
    end

%. Compute the V, M, T, and W diagrams
    nGTR = 51;                             % Number of GTR points
    beta = 0.5;                            % GTR parameter
    Store = zeros(nGTR*nSpans,6);          % Storage for state variables
    nn = 0;                                % Row index for Store
    xx = 0;                                % x for left end of span
    eps = 1.e-7;                           % A small number
```

```
%. Loop over spans
   for n=1:nSpans
     Vold = So(1,n);                          % Initial shear, span n
     Mold = So(2,n);                          % Initial moment, span n
     Told = So(3,n);                          % Initial rotation, span n
     Wold = So(4,n);                          % Initial displacement, span n
     dx = L(n)/(nGTR-1);                      % Step size
     x = 0;
     [qold,~] = DistrLoadFcn(x,qo(n),dLoadType(n));

%... Integrate by GTR
     for j=1:nGTR

       nn = nn + 1;
       Store(nn,:) = [xx+x,-Wold,Told,Mold,Vold,qold];
       x = x + dx;
       [qnew,~] = DistrLoadFcn(x,qo(n),dLoadType(n));
       Vnew = Vold - dx*(beta*qold + (1-beta)*qnew);
       Mnew = Mold + dx*(beta*Vold + (1-beta)*Vnew);
       Tnew = Told + dx*(beta*Mold + (1-beta)*Mnew)/EI(n);
       Wnew = Wold - dx*(beta*Told + (1-beta)*Tnew);

       Vold = Vnew;
       Mold = Mnew;
       Told = Tnew;
       Wold = Wnew;
       qold = qnew;

     end
     xx = xx + L(n) + eps;

   end

%. Output results
   EchoInput(nSpans,L,EI,qo,dLoadType,BCL,BCR,Cont)
   CreatePlots(Store,Reaction,Rotation,L,BCL,BCR);
```

The code starts with a declaration of the number of spans, `nSpans`, followed by the physical inputs. The lengths, stiffnesses, load magnitudes, and loading types are now specified in arrays of length equal to the number of spans. Each span can have its own values of these input variables.

Boundary conditions are handled in the same way as the single beam code of the previous sections, including the specification of the possible boundary conditions in the array `BCs`, with the appropriate row selected using the values of `BCL` and `BCR`. The array `Cont`, which is included to handle the continuity conditions, must have length `nSpans-1` because that is the number of interior nodes. An additional array called `CCs` is included to represent the continuity conditions for various cases. For each span, the code looks at the entry in `Cont` and then selects that row of the `CCs` array to give the appropriate continuity condition to place in the array `IntSupport`. Those values are then included in the `Var` array between the end conditions.

The `CCs` array is also scanned for the presence of roller or internal hinge conditions. If, for example, there is a zero in the fourth column of `CCs` for that span, then there is a roller support there and we must include a reaction force to go along with the restraint of displacement. Otherwise, the reaction at that node is zero. Similarly, if there is a zero in the second column of `CCs` for that span, that means there is an internal hinge there, and we must include a discrete rotation to go along with the zero-moment condition. These two goals are accomplished simply by logical negation (i.e., with the `not` function that gives `not`(0) = 1 and `not`(1) = 0) of the values in array `CCs` and placing those values in either the arrays `Reaction` or `Rotation`. In short, if there is a 1 in the fourth column of `CCs`, then there must be a zero in the associated slot in `Reaction` and vice versa. The same applies for the rotations.

The number of entries in the `Var` array is equal to the number of state variables, which is four times the number of spans plus one, plus twice the number of interior nodes (to make room for the reactions and rotations). The order reflects the ordering of the state variables, reactions, and relative rotations at the nodes. Again, the `Var` array contains ones associated with the variables that are not known and zeros for the variables that are known to be zero. Thus, the sum of the values in `Var` is equal to the number of unknowns in the system after the known zero values have been eliminated. This value must be equal to the number of equations, which is four times the number of spans.

The formation of the arrays `B` and `Z` is done through a loop over the spans. The calculations within each span are the same as those done for the single beam, with the Simpson integration to compute the values of $\mathcal{I}_0^n$ through $\mathcal{I}_3^n$ (stored in the array `z`). The indices `ii` and `jj` are created to establish which rows and columns of `B` the span contribution `Bn` should go in. Also, the columns at the end of `B` come from putting the arrays `rn` and `tn` in the appropriate slots, which start in the column immediately after the `4*(nSpans+1)` columns associated with the $\mathbf{Q}$ variables. Once `B` is formed, the coefficient matrix `C` is created by eliminating the columns of `B` associated with zero values in the `Var` array. The resulting system of equations is then solved for the problem unknowns, and those unknowns are placed in their appropriate locations in the array `F`. Finally, the $\mathbf{Q}_n$ values are extracted from `F`, adjusted by the reaction to give $\mathcal{S}_n(0)$, which are the initial values needed in each span for the generalized trapezoidal rule (GTR) integration to get the plots.

The organization of the code to do the GTR integration is similar to how we handled the point loads in Code 10.4, with the integration done independently over each span. In this case, we know the initial values for each span because they are stored in the array `So`. In order to store the actual x values in the array `Store`, which is used for plotting, we define the variable `xx`, which is the value of x at the left end of the current span, measured from the far left end of the multispan beam. After completing the GTR loop, `xx` is incremented by the length of the current span to get ready for the next trip through the loop. A small value `eps` is added to resolve a glitch in the MATLAB `area` function, which shades under a plot, just moving the start of the next span to the right of the last point of the previous span so that they do not share exactly the same x value.

Finally, the input values are printed to the command window with the function `EchoInput`, and the results are expressed as plots of the state variables across all spans through a call to the function `CreatePlots`. Neither of these functions are included here. The following example illustrates what the plotting function produces.

Example 10.7
Apply the multispan beam code to analyze a four-span beam that is simply supported at both ends. The individual span lengths are all $L = 10$ and the individual bending moduli are all $EI = 200$. The first and third spans are loaded with a uniformly distributed load of magnitude $q_o = 2$. The second and fourth spans are not loaded. Plot $q(x)$, $V(x)$, $M(x)$, and $w(x)$.

Solution. The physical inputs for Code 10.5 are established as

```
nSpans = 4;
L = [10, 10, 10, 10];
EI = [200, 200, 200, 200];
qo = [2, 0, 2, 0];
LoadType = [2, 2, 2, 2];
```

Load type 2 is the constant load. The boundary codes are `BCL=3` (simple) and `BCR=3` (simple). The continuity conditions are `Cont=[2,2,2]`, which corresponds to all intermediate nodes being supported on rollers. The results are shown in Fig. 10.14. As discussed in the previous example, the displacement plot shows $-w(x)$ because the positive direction for displacement is downward (i.e., in the $\mathbf{e}_3$ direction) and positive in a MATLAB plot is up. Thus, $-w$ looks like the deflected shape of the beam.

With this simple loading, we can make some observations about the shear and moment diagrams. In the spans with nonzero loading, the shear diagram is linear and the moment diagram is quadratic. In the spans with no load, the shear is constant and the moment is linear. The moment is zero at the two ends of the multispan beam but not at the supports. The shear jumps by an amount equal to the reaction force at the interior support points. The reactions computed by the code are `-11.4286` for the outer two roller supports and `-9.2857` for the middle roller support, an interestingly symmetric response for a nonsymmetric loading.

The loaded spans deflect downward under the loads, but the unloaded spans deflect upward. The loaded spans tend to rotate clockwise at the left end and counterclockwise at the right end. The unloaded spans accommodate that tendency by deflecting upward. The zero deflection at the rollers must be met, so the rotation in the unloaded spans must reverse to get to the zero deflection at the right end.

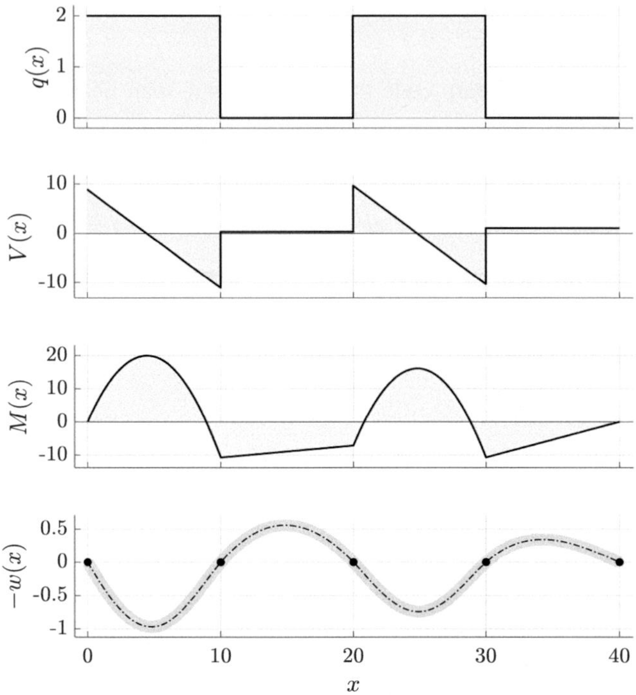

Fig. 10.14 *Example.* A four-span beam simply supported at both ends. The span lengths are all $L = 10$ and bending moduli are all $EI = 200$. The first and third spans are loaded with a uniformly distributed load of magnitude $q_o = 2$. The second and fourth spans are not loaded. The plots show $q(x)$, $V(x)$, $M(x)$, and $-w(x)$

10.7 Summary

This chapter formulates the equations needed to write codes to automate the calculations for single and multispan beams subjected to any type of loading with any boundary conditions and several important intermediate support conditions. Automation of the calculations associated with the theories of solid mechanics is a primary focus of the subject *structural analysis*. The codes we have developed here have two main purposes. First, they follow closely the steps we use to do similar calculations by hand. As such, the codes can verify hand calculations (and vice versa). Second, the codes provide an environment to study the behavior of beams without getting tangled in the details of calculation. With these tools, the student can begin to develop an understanding of structural behavior through observations of response and to deepen understanding of the underlying theory.

10.8 Problems

10.1 *Computation.* Use Code 10.4 to investigate the behavior of a beam with different boundary conditions and different loading conditions. Explore how changing the boundary conditions alters the distribution of internal forces. Consider different load profile shapes, including point loads, with the same amount of total load, and with the same boundary conditions. How much difference does the profile of the load make to the maximum deflection, moment, and shear if the total load is the same?

10.2 *Computation.* Use Code 10.5 to investigate how loads are shared among the beams that make up a multispan system. Consider a seven-span beam with a uniform load only in the middle span. How quickly do the moments, shears, and displacements dissipate with distance from the loaded span? Compare a case with all spans loaded to a case where every other span is loaded with loads of the same magnitude. Which case has the largest moments and deflections?

10.3 *Computation.* Write a multispan beam code that extends Code 10.5 to include point loads in any span. This code is an amalgamation of Code 10.5, which includes only distributed loads, and Code 10.4, which is a single beam with both distributed loads and point loads. Use the function `PointLoadFcn` in Code 10.3 to generate the point loads.

10.4 *Computation.* Convert Code 10.5 to be a function that takes in the physical properties `EI`, `L`, `qo`, `Type`, and the boundary condition codes and returns the maximum moment, shear, rotation, and deflection in each span. Create a driver code that establishes the input values and calls the function. Verify that the code works by passing input values for a case with a known solution to the function.

10.5 *Computation.* Write a code to support the design of a multispan beam with multiple load cases. The code should do the following:

- Input the design variables. Use the dimensions of the cross section as the design variables. For example, the design variables for an I-beam would be d, b_f, t_f, and t_w for the beam depth, flange width, flange thickness, and web thickness.
- Compute the properties of the cross section A, I, and Q so that the volume of material, deflections, and stresses can be computed for each trial design.
- Establish allowable values for deflection and the normal and shear stresses. The allowable deflection can be specified as a fraction of the span length, e.g., $w_{allow} < L/800$. The allowable stresses can be specified as a fraction of the yield value, e.g., $\sigma_{allow} < 0.67\sigma_y$ and $\tau_{allow} < 0.67\tau_y$, where σ_y is the uniaxial yield stress of the material and $\tau_y = \sigma_y/\sqrt{3}$ is the shear yield stress. Include Young's modulus and the quantities required to specify the allowable values among the physical input values.

- Establish a means of specifying the load cases to be included in the analysis. The code should loop over all those load cases and extract the maximum values of the response quantities that are limited by the deflection and stress design criteria. Use the function from Problem 10.4 to perform the analysis for each load case.
- Compare the maximum deflection and stress against the design criterion. Compute design ratios as the actual value divided by the allowable value. The goal of design is to get those ratios as close to one as possible.
- To simplify the specification of load cases, assume that each span can have a constant uniform load and each span can be either loaded or not, as illustrated in the figure. The challenge for enumerating the load cases reduces to specifying all the possible load combinations with this restriction.

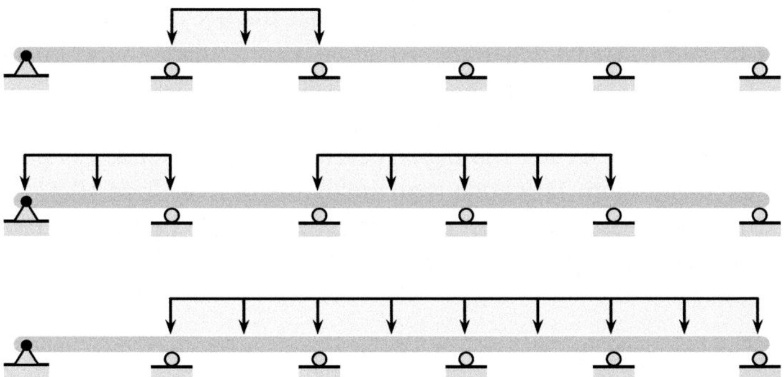

One approach is to use the binary representation of the load case number for a given number of spans. The following code creates an array with each row being the binary representation of the row number. Each row can then be used to specify a load case.

```
function [LoadCase] = BinCase(nSpans)
%  Create LoadCase array from binary digits of the load case number

%. Compute number of load cases, initialize LoadCase
   nLoadCases = 2^nSpans-1;
   LoadCase = zeros(nLoadCases,nSpans);

%. Create LoadCase array
   for i=1:nLoadCases
     n = i;
     for j=1:nSpans
        m = 2^(nSpans-j);
        a = floor(n/m);
        n = n - a*m;
        LoadCase(i,j) = a;
     end
   end

end
```

The task of design involves an iterative process wherein you select candidate beam cross-sectional properties, analyze the response, and judge the performance against the design criteria. If the performance does not meet the design criteria, then you select a better design and repeat the process. The best design will be the one that uses the least material (i.e., has the lowest weight) and meets the design criteria.

Use the code for a specific design scenario. Specify the number of spans, the span lengths, material properties, boundary conditions, and load magnitudes for the span loads. Consider a case where all spans have the same cross section and then a case where each span can have its own cross section. How much difference is there in the total material used in the two cases?

Chapter 11
Torsion

The twisting of a bar about its own axis without bending or extension is called pure torsion. Like the axial bar and beam problems, the theory of torsion is built on a kinematic hypothesis that results in a relationship between rotation and strain. This kinematic relationship combines with equilibrium and the constitutive behavior of the material to form a complete mathematical model of torsion. In this chapter, we formulate the theory of pure torsion and show how to solve the resulting equations. We start with bars with circular cross sections, for which the theory is exact. In the latter part of the chapter, we formulate approximate theories for a few important cases of bars that have non-circular cross sections.

There are some similarities between the axial bar, beam, and torsion problems, but there are also some significant differences. Figure 11.1 illustrates the three cases to put them into context. For the axial bar, the applied load $p(x)$ is oriented along the axis of the bar. As a result, the bar displaces only in the longitudinal direction. For beams, loads $q(x)$ are applied transverse to the beam axis. As a result, the beam bends with associated displacement transverse to the longitudinal axis of the beam and rotation of the cross sections in the plane of bending. For a bar in torsion, the applied torque $t(x)$ is oriented along the axis of the bar, causing the bar to rotate or twist about its own axis. The area of the cross section is the primary geometric feature of the axial bar. For the beam, the first and second moments of the area relative to the centroid are also important. For torsion, we will find that the polar moment of inertia is the relevant property of the cross section. Young's modulus E is the only elastic material property that comes into play for the axial bar and the beam. The shear modulus G is important for torsion.

The axial bar is a second-order differential equation with two boundary conditions, one at each end. The torsion problem is also a second-order differential equation. The methods of solving axial bar and torsion problems have many similarities, and the reader will benefit from comparing those solution processes side by side. In fact, many of the examples in the two chapters are parallel in structure.

K. D. Hjelmstad, *Mechanics of Deformable Solids: An Introduction*,
https://doi.org/10.1007/978-3-031-89873-0_11

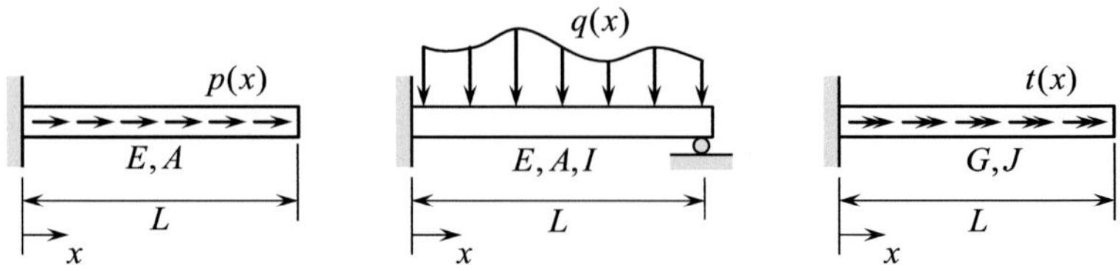

Fig. 11.1 *Context.* The axial bar (left) is subjected to purely axial forces and responds axially. The beam (middle) is subjected to transverse forces and responds by bending. The torsion bar (right) is subjected to applied torques and responds by twisting about its own axis

11.1 What is Torsion?

Torsion is an action associated with twisting. When we use the term *twist* in the context of the deformation of a bar in torsion, we mean that the bar rotates about its own axis without bending or changing length. To twist a body, we can apply equal and opposite torques (twisting moments) at the ends—the kind of action you use to get the lid off of a jar, wring out a rag, or unscrew your custom-made pool cue so you can put it back into its handsome carrying case. Figure 11.2 illustrates twisting of a bar of square cross section. It is easier to visualize twisting for a square bar than for a circular bar because you can see what happens to the corners and sides.

Figure 11.2 shows twisting caused by torques applied at the ends of the bar. In our formulations, we will consider the more general case of applied torques that are distributed over the length of the bar. The figure shows a prismatic bar. Through examples we will also consider cross sections that vary along the length, including both a gradual variation and a step variation where the cross section abruptly changes at a certain point. The focus of the first part of the chapter is on bars with circular cross sections, for which the theory derived is exact. We consider non-circular cross sections in the latter part of the chapter.

Fig. 11.2 *Torsion.* Torsion is a problem that involves twisting of a bar about its own axis due to applied torques

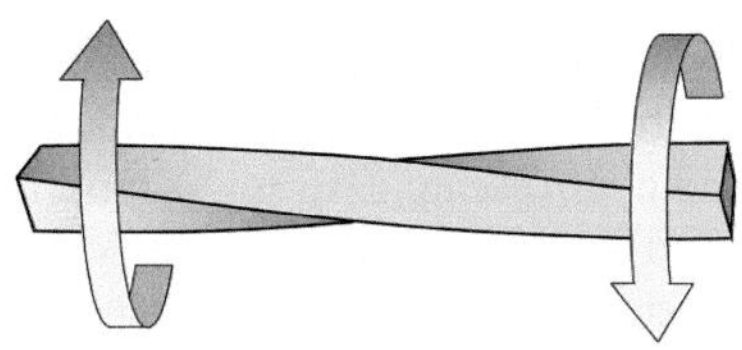

11.2 Torsion of Bars with Circular Cross Sections

In this section, we develop the equations that govern the torsion of circular bars. Consider the circular bar subjected to end torques shown in Fig. 11.3. The term *torque* is used for moments that cause rotation about the axis of the bar, distinguishing them from *bending moments* in beams. Let us designate the angle of rotation about the x axis as $\varphi(x)$. This angle of twist is the primary motion variable for torsion. Twisting of a circular bar does not change the orientation of the circular lines that represent the cross sections. A line initially oriented parallel to the axis in the undeformed configuration (e.g., the dark gray line) distorts due to the relative rotation of adjacent cross sections, as shown in the figure. These deformations give rise to shear strains that induce shear stresses in the bar. Because the cross sections do not warp out of their planes, each element in the bar is in a state of pure shear. For pure torsion, no normal stresses are present.

11.2.1 Kinematic Hypothesis

The kinematic hypothesis for a circular bar in torsion is that each cross section rotates rigidly about the axis of the bar but does not change shape or warp out of its own plane. This hypothesis is exact for circular cross sections, including hollow tubes, but not for any other shape of cross section. Cross sections of other shapes tend to warp out of their own planes, as we will see later. Hence, the analysis of this section applies rigorously only to bars with circular cross sections. It does, however, contain the basic ingredients of all torsion problems.

11.2.2 Strain-Rotation Relationship

To understand the nature of twisting, consider the segment between stations x and $x + \Delta x$ of a circular bar at radius r, shown in Fig. 11.4. This radius is not the exterior radius of the bar, but rather some general r in the domain of the bar. Let us suppose

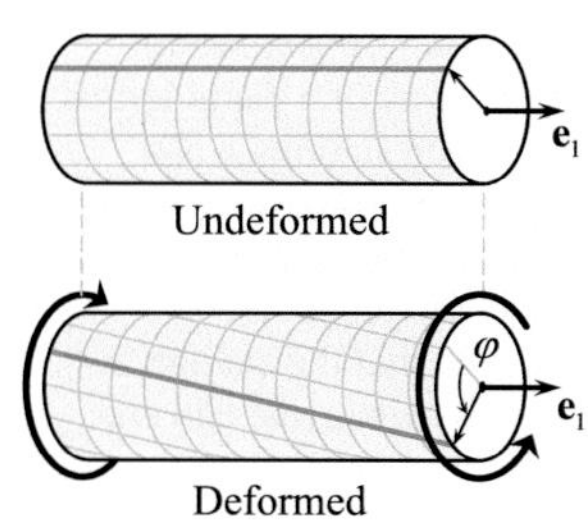

Fig. 11.3 *Angle of twist.* Twist of a bar about its own axis in torsion is designated as $\varphi(x)$, which is the primary measure of motion in a torsion problem

that we have etched a line parallel to the axis and radial lines on the ends in the undeformed bar (the black lines on the sketch). When the bar is twisted, the right end rotates relative to the left end by an amount $\Delta\varphi = \varphi(x+\Delta x) - \varphi(x)$. The gray lines are where the black lines go due to twisting.

The angle between the black line and the gray line along the longitudinal axis in Fig. 11.4a is called the *shear angle* and is denoted as γ. The angle between the black line and the gray line in the cross section in Fig. 11.4b is $\Delta\varphi$. There is a very simple connection between the angles. Note that the distance s is the arc length on the circle between the ends of the black line and the gray line in both views. This distance can be computed two ways: (1) angle γ times length Δx on the lateral side, which is a good approximation for small angles γ, or (2) angle $\Delta\varphi$ times radial distance r. They must give the same result. Therefore,

$$s = \gamma \Delta x = r \Delta\varphi$$

Divide by Δx and take the limit as Δx goes to zero to get

$$\gamma(x) = \lim_{\Delta x \to 0} \left(r \frac{\Delta\varphi}{\Delta x} \right) = r \lim_{\Delta x \to 0} \frac{\varphi(x+\Delta x) - \varphi(x)}{\Delta x}$$

Noting the definition of the derivative of a function, this equation reduces to

$$\gamma(x) = r \frac{d\varphi}{dx} \tag{11.1}$$

This equation is the mathematical realization of the kinematic hypothesis. It holds for any radius r in the cross section. Hence, the shear strain varies linearly from zero at the center to a maximum at the exterior surface of the bar.

In general, we can expect the rotation of the cross section, $\varphi(x)$, to vary along the length of the bar due to nonuniform torque or variable radius of the cross section. By taking the limit, we find that the shear strain is proportional to the first derivative of the angle of rotation. This relationship is called the *strain-rotation relationship*. Note that the angle of twist $\varphi(x)$ has units of radians, dx has units of length, and r has units of length. Therefore, the shear strain has units of radians (i.e., it is an angle measure).

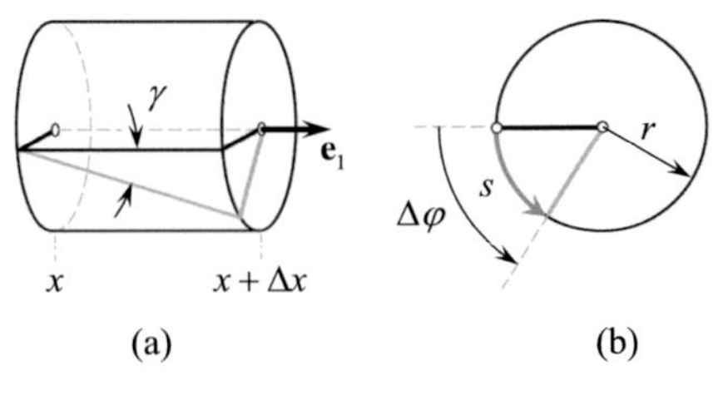

Fig. 11.4 *Kinematics of twist*. Twist of a bar at radius r between stations x and $x+\Delta x$. **a** Longitudinal view showing shearing angle γ. **b** Cross-sectional view showing that the change in angle between one end and the other is $\Delta\varphi$

11.2.3 Resultant Torque and Equilibrium

Consider a thin circular portion of the cross section at radius r with thickness dr, shown Fig. 11.5. Because each element (the gray squares) is in a state of pure shear, the only stresses that develop are shear stresses τ. If we make a cut through a cross section, each deformed square element contributes to shear stress at the cut because one of the four edges of the deformed element gets exposed by the cut. The magnitude of the shear stress τ is constant at a given radius due to the symmetry of the circular geometry. The direction of each traction τ on the cross section is tangent to the circle. Each radius r contributes similar shears, but the stress varies with r. Hence, we will view $\tau(r)$ as a function of the radial distance from the center.

The tractions acting on a cross section exposed by cutting a free body diagram contribute to a resultant torque T acting on the cross section, as shown in Fig. 11.6. In the case of pure torsion, the tractions on an elemental patch of the cross section are $\tau(r)\mathbf{m}\,dA$, where $\tau(r)$ is the shear stress, $dA = r\,dr\,d\theta$ is the elemental area in polar coordinates, and $\mathbf{m}$ is a unit vector tangent to the circle at r. The moment arm is $r\mathbf{n}$, where $\mathbf{n}$ is a unit vector pointing in the radial direction. The net moment acting on the cross section is the sum of the moments caused by the elemental forces, taken at the center of the circle. The sum is an integral over the area of the cross section. Hence, the net torque is

$$\begin{aligned}\mathbf{T} &= \int_A r\mathbf{n} \times \tau(r)\mathbf{m}\,dA \\ &= \int_A r\,\tau(r)\big(\mathbf{n} \times \mathbf{m}\big)\,dA \\ &= \int_A r\,\tau(r)\,dA\,\mathbf{e}_1\end{aligned}$$

since $\mathbf{n} \times \mathbf{m} = \mathbf{e}_1$. The magnitude of the net torque resultant is, therefore,

$$T = \int_A r\,\tau(r)\,dA \tag{11.2}$$

The net torque is the *stress resultant* for torsion.

The shear stresses acting on the cross section do not cause a net shear force because each traction has an equal and opposite twin diametrically opposed to it. Therefore, for pure torsion problems, there is no net force acting on the cross section (i.e., the

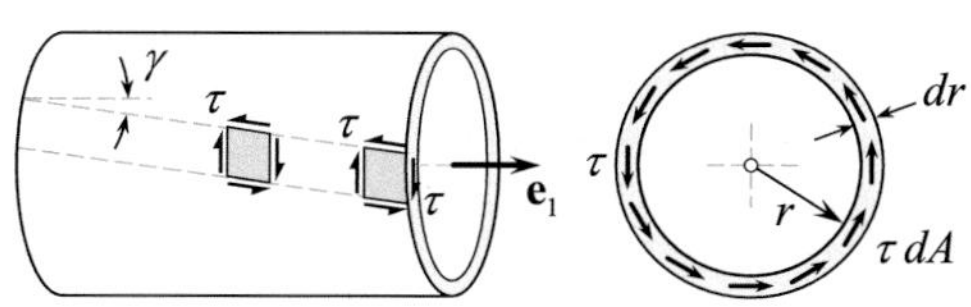

Fig. 11.5 *Shear acting on a cross section.* The shear strain causes shear stress on each element. One of the four shear stresses τ contributes to the cut at the cross section

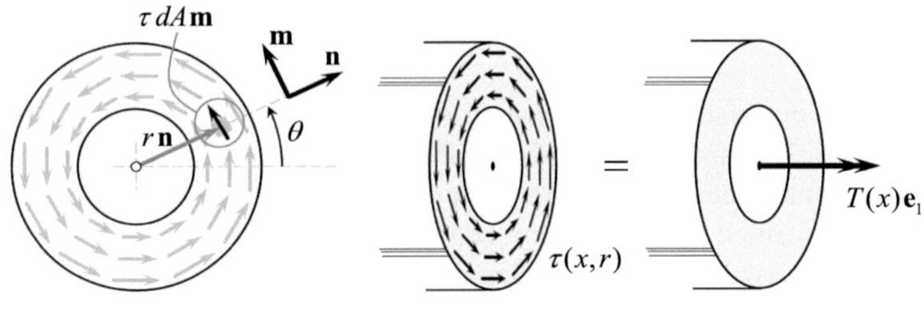

Fig. 11.6 *Resultant torque.* The net torque stress resultant is the sum over the cross section of the moments caused by the individual shear stresses relative to the axis

forces N and V in axial bars and beams) and therefore, force equilibrium plays no role in the theory.

Torque is a vector quantity that results from the cross product of moment arm and force vectors. The direction of the torque vector is $\mathbf{e}_1$, as the preceding derivation showed. There are two main ways of drawing a torque on a diagram, as shown in Fig. 11.7. The circular arrow representation gives a sense of the direction of the tractions that contribute to the torque. This representation is not a vector because a vector must be a straight line. The more rigorous notation is the double-headed arrow that points in the vector direction of the torque. We use a double-headed arrow to distinguish moment vectors from force vectors on a free body diagram because, while you can add force vectors together and you can add moment vectors together, you cannot add force vectors to moment vectors. As we saw frequently in beam problems, you *can* add moment vectors to moments of force vectors (i.e., terms involving $\mathbf{r} \times \mathbf{F}$). In this chapter we confine our attention to pure torsion problems, and in this context either depiction can be used. In three-dimensional beam problems, torque combines with bending moments about two axes. Thus, in that context it is better to represent moment as a vector with a double-headed arrow.

To establish *equilibrium* equations for the torsion bar, consider Fig. 11.8, which shows a piece of the bar of length Δx isolated as a free body diagram. The torque acting on the left end of the piece is $-T(x)\mathbf{e}_1$ and the torque acting on the right end is $T(x+\Delta x)\mathbf{e}_1$. The distributed applied torques are represented as $t(\xi)\Delta x\mathbf{e}_1$ where, by the mean value theorem of calculus, ξ is some value between x and $x + \Delta x$.

Note that the torques on the free body diagram are all shown in accord with the positive sign convention. The applied torques are positive if the double-headed arrow points in the $\mathbf{e}_1$ direction. The internal resultant torques T are defined as positive with

Fig. 11.7 *Depicting torque.* Torque can be represented by a circular arrow or by a double-headed arrow

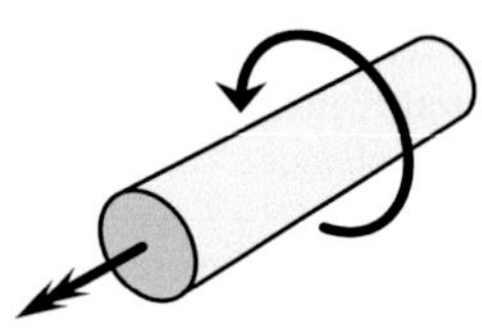

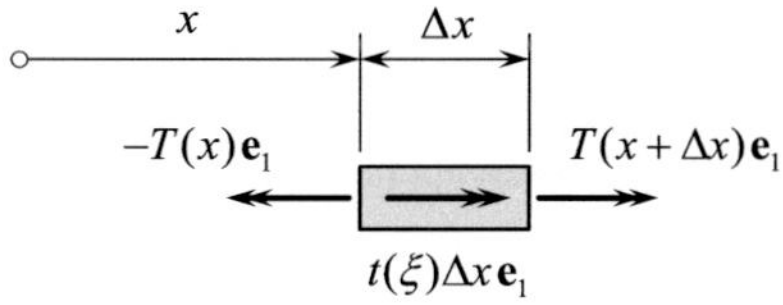

Fig. 11.8 *Equilibrium.* Free body diagram of a small segment of bar between x and $x + \Delta x$, with torques at the cut and the applied torque $t(x)$ along the length

the sign convention shown in Fig. 11.9. If $T(x)$ has a positive value, then the torque is represented as $T(x)\mathbf{e}_1$ on the face with outward normal $\mathbf{e}_1$ and as $-T(x)\mathbf{e}_1$ on the face with outward normal $-\mathbf{e}_1$. If the $T(x)$ has a negative value, then the vector directions are reversed (with corresponding reversal of the circular arc arrow version of the torque).

For static equilibrium, the moments on the free body diagram must sum to zero. To wit,

$$T(x + \Delta x)\mathbf{e}_1 - T(x)\mathbf{e}_1 + t(\xi)\Delta x\,\mathbf{e}_1 = \mathbf{0}$$

Taking the dot product of this equation with $\mathbf{e}_1$, dividing through by Δx, and taking the limit as $\Delta x \to 0$ gives

$$\lim_{\Delta x \to 0} \frac{T(x + \Delta x) - T(x)}{\Delta x} + t(\xi) = 0$$

In the limit, the first term becomes the derivative of $T(x)$, the value of ξ in the second term approaches x, and we get the equilibrium equation

$$\frac{dT}{dx} + t(x) = 0 \tag{11.3}$$

The value of x is arbitrary, so this equation holds for all values of x. We call this a *local* equilibrium equation because it holds for a small free body diagram in the local neighborhood of x. The equilibrium equation is a differential equation. It says that the rate of change of the internal torque $T(x)$ plus the applied distributed torque $t(x)$ is equal to zero. Hence, we have an equation that relates the applied torques to the internal stress resultants. If we can find a function $T(x)$ that satisfies this equation and also satisfies any torque boundary conditions at the ends of the bar, then *any* free body diagram cut from the bar will satisfy equilibrium.

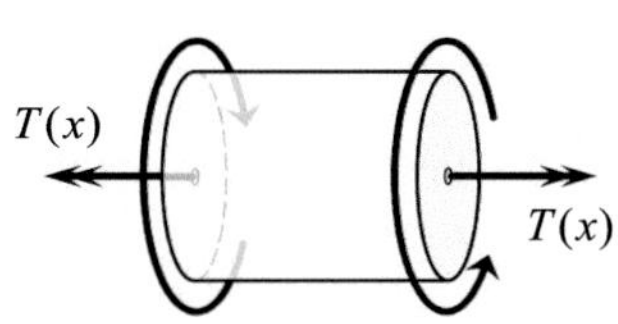

Fig. 11.9 *Torque sign convention*. Definition of the positive sense of torque $T(x)$. A positive torque is $T(x)\mathbf{e}_1$ on the face with normal $\mathbf{e}_1$ and $-T(x)\mathbf{e}_1$ on the face with normal $-\mathbf{e}_1$

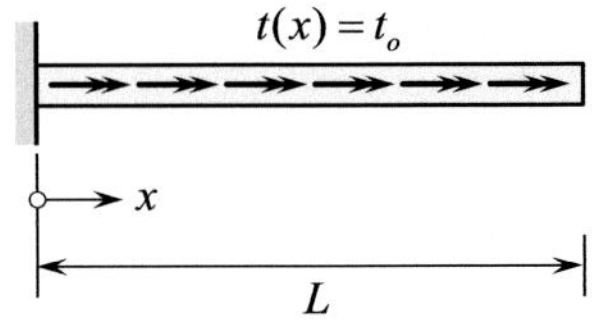

Fig. 11.10 *Example.* Torsion bar of length L subjected to a constant uniform external applied torque $t(x) = t_o$

Example 11.1
Consider the bar of length L subjected to a constant, uniformly distributed external applied torque $t(x) = t_o$, shown in Fig. 11.10. The bar is fixed at the left end and free at the right end. Compute the internal torque $T(x)$ by directly integrating the equation of equilibrium, Eq. 11.3.

Solution. Since the applied torque is a known function, we can integrate Eq. 11.3 directly as follows:

$$T'(x) = -t_o$$
$$T(x) = -t_o x + C$$

where C is the constant of integration. We can evaluate the constant of integration using the condition that the internal torque vanishes at $x = L$, as it is a free end. Using this condition gives

$$T(L) = -t_o L + C = 0 \quad \rightarrow \quad C = t_o L$$

Substituting this value back into the general equation gives the final expression for the internal torque as

$$T(x) = t_o(L - x)$$

We can evaluate $T(0) = t_o L$. This value is equal to the magnitude of the reaction torque at the fixed support at the left end.

We can also solve equilibrium problems by taking a finite free body diagram of the bar (as opposed to the infinitesimal one used to derive Eq. 11.3) with a cut at x to expose the internal torque $T(x)$, as the following example shows.

Example 11.2
Consider the same problem as the previous example, depicted in Fig. 11.10. Find the internal torque by taking a free body diagram with a cut at x.

Solution. Figure 11.11 shows two free body diagrams of the bar. The one on top is a free body diagram of the entire bar, showing the reaction torque $\mathbf{T}_A$ at

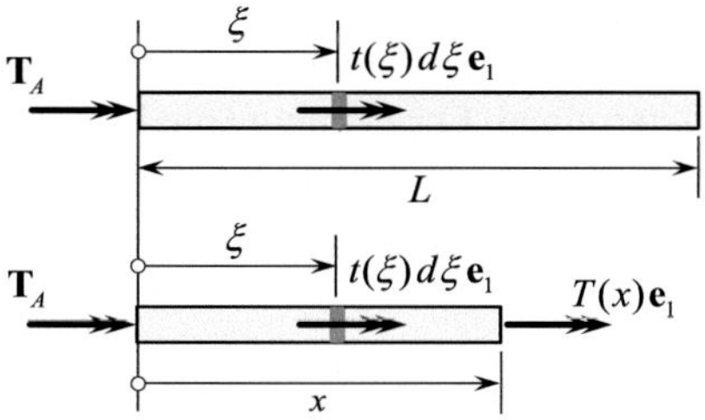

Fig. 11.11 *Example.* Torsion bar subjected to a constant uniform external torque $t(x) = t_o$. Free body diagram of the whole bar (top) and free body diagram with cut at x (bottom), keeping the left piece and discarding the right piece

the left end of the bar and the applied torque associated with the typical particle $t(\xi)d\xi\mathbf{e}_1$, which is located a distance ξ from the left end. We can use this free body diagram to find the reaction torque. Establishing moment equilibrium of the entire bar, noting that $t(\xi) = t_o$, gives

$$\mathbf{T}_A + \int_0^L t_o\, d\xi\, \mathbf{e}_1 = \mathbf{0}$$

$$\mathbf{T}_A + t_o L\, \mathbf{e}_1 = \mathbf{0}$$

We can solve for the reaction torque as $\mathbf{T}_A = -t_o L\mathbf{e}_1$, which tells us that the magnitude is $t_o L$ and the vector direction is $-\mathbf{e}_1$. That makes sense because if all the *applied* torques are in the $\mathbf{e}_1$ direction, then the reaction torque must be in the $-\mathbf{e}_1$ direction to equilibrate them. Of course, the reaction torque points along the axis of the bar because the applied torques are all in that direction.

Establishing equilibrium on the second free body diagram gives

$$\mathbf{T}_A + \int_0^x t_o\, d\xi\, \mathbf{e}_1 + T(x)\mathbf{e}_1 = \mathbf{0}$$

$$-t_o L\mathbf{e}_1 + t_o x\, \mathbf{e}_1 + T(x)\mathbf{e}_1 = \mathbf{0}$$

where we have substituted the value of the reaction torque just found. Dotting the equation with $\mathbf{e}_1$ and solving for the internal torque gives

$$T(x) = t_o\big(L - x\big)$$

which is identical to what we got in the previous example. Observe that in the second free body diagram, the range of the integration is from 0 to x, not 0 to L, as it was in the first free body diagram. When we take the cut at x and remove the right piece, we take the applied torque that was on that piece away, too. The effect of the applied torques that were removed is implicit in the internal torque $T(x)$ at the cut because it represents the action of the piece removed on the piece that remains.

These two examples show the two basic ways of dealing with equilibrium. The first example solved the differential equation governing $T(x)$ while the second example determined $T(x)$ from a free body diagram, integrating the applied torque over the region of the free body diagram. The second example reminds us that we do not see the internal torque on a free body diagram until we take a cut to expose it.

11.2.4 Boundary Conditions

Boundary conditions are a statement about the rotation or internal torque at the two ends of the bar. For the torsion problem there are exactly two boundary conditions, one at each end. The nature of the boundary conditions is generally conveyed through a symbol on the sketch, as shown in Fig. 11.12. A bar end is either *fixed* against rotation or *free* to rotate. The fixed end is shown with a shaded block and the free end with the absence of any symbol. It is possible for a concentrated, but known, torque to be applied at a free end.

At a fixed end, we know the motion at that point is constrained to be zero. That means $\varphi(\bullet) = 0$, where $\bullet$ can be either 0 or L. At a free end, we know something about the internal torque $T(x)$. The actual condition for the internal torque can be derived through an equilibrium argument, as shown in the following example.

Example 11.3
What are the boundary conditions associated with a torsion problem with a bar of length L fixed at the left end and subjected to a concentrated torque T_o applied at the right end?

Solution. The fixed end implies that $\varphi(0) = 0$. An end that is not fixed is *free*. The right end is free to rotate but has a concentrated applied torque there. If we establish equilibrium on a very small free body diagram of the bar end (to expose the internal torque), as shown in Fig. 11.13, we find that $T(L) = T_o$.

One of the boundary conditions is a *kinematic* condition, a constraint on the rotation $\varphi(x)$. The other is a *kinetic* condition, i.e., it involves the internal torque and is established through equilibrium.

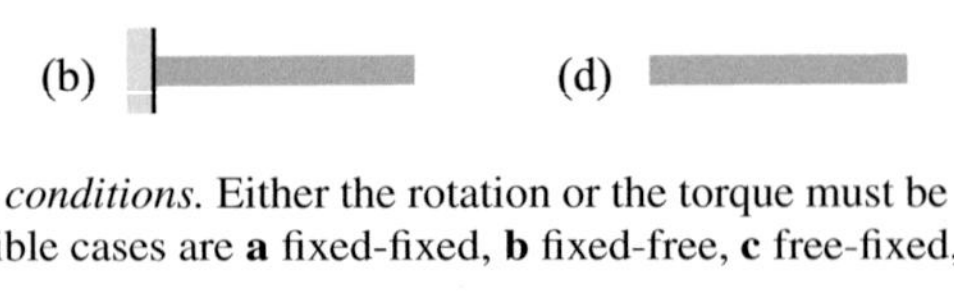

Fig. 11.12 *Boundary conditions.* Either the rotation or the torque must be specified at each end of the bar. The four possible cases are **a** fixed-fixed, **b** fixed-free, **c** free-fixed, and **d** free-free

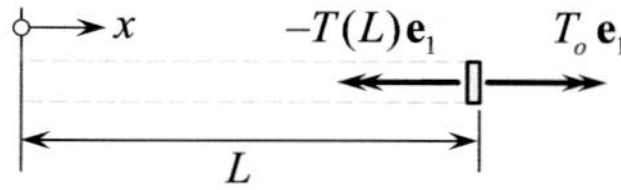

Fig. 11.13 *Example.* Free body diagram of a small piece at the right end of the bar for a problem with a concentrated applied torque at the end

11.2.5 Constitutive Model for Torsion

Take a close look at the grid of squares etched on the surface of the circular bar in Fig. 11.3. In the undeformed configuration, the grid squares have sides defined by the lines of latitude (the circular edges of the cross sections) and lines of longitude (the horizontal lines parallel to the bar axis). When the bar twists, the latitude lines remain in their plane, but the horizontal lines rotate by an amount γ (see Fig. 11.4). The length of the bar gets a little shorter because the deformed lines of longitude remain the same length. Thus, the grid squares deform into parallelograms with angle γ, as shown in Fig. 11.14a. From Chap. 4, we know that $\gamma = 2\varepsilon_{xy}$ is the shear strain. The lengths of the edges do not change (i.e., there is no axial strain), but the angles between the edges do change because the edges of the parallelogram are no longer perpendicular. The shear strain and stress that happens due to twisting of a circular bar is called *pure shear*.

The stress associated with the pure shear deformation is τ, shown in Fig. 11.14b. The stress and strain are related through Hooke's law as

$$\tau = G\gamma \tag{11.4}$$

where G is the *shear modulus*. The shear modulus is the slope of the stress-strain curve in the linear region of behavior, as illustrated in Fig. 11.14c. For most materials, there will be some level of stress and strain for which the linear constitutive hypothesis is valid. Eventually, the relationship between stress and strain becomes nonlinear, due to some mechanism like yielding or cracking of the material. In this chapter, we will confine our attention to problems for which the assumption of linear response holds. The consequences of yielding are explored briefly in Chap. 12.

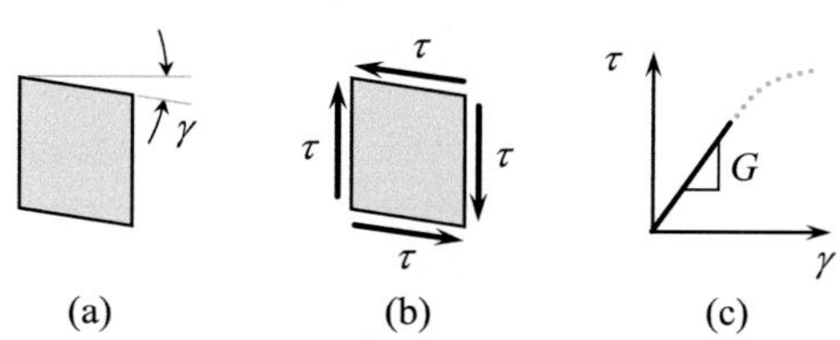

Fig. 11.14 *Stress and strain.* **a** The rotation of the bar causes a pure state of shear strain γ. **b** The stress associated with pure shear is τ. **c** Hooke's law relates shear stress to shear strain

Recall from Chap. 6 that for Hooke's law, the shear modulus for a linear isotropic material can be expressed in terms of Young's modulus, E, and Poisson's ratio, ν, as

$$G = \frac{E}{2(1+\nu)} \tag{11.5}$$

For isotropic materials, the practical range of values for Poisson's ratio is $0 < \nu < 0.5$. Therefore, the shear modulus ranges between one-half and one-third of Young's modulus for most common materials.

11.2.6 Torque-Rotation Relationship

We can put equilibrium, kinematics, and constitution together to form a complete theory of the torsion bar. These three ingredients can be put together to form a direct relationship between torque and rotation, as the following derivation shows:

$$\begin{aligned} T(x) &= \int_A r\,\tau\,dA \\ &= \int_A r\,\big(G\gamma\big)\,dA \\ &= \int_A rG\big(r\varphi'(x)\big)\,dA \\ &= \int_A Gr^2\,dA\;\varphi'(x) \end{aligned}$$

This derivation starts with the definition of the stress resultant $T(x)$. In the next step, we substitute the constitutive equation $\tau = G\gamma$. We then substitute the strain-rotation relationship $\gamma = r\varphi'(x)$, where the prime indicates derivative with respect to x. Finally, we pull $\varphi'(x)$ out of the integral over the area because it does not depend on the variables of integration over the cross section. Let us define the *effective torsional stiffness* of the cross section as

$$\overline{GJ} = \int_A Gr^2\,dA \tag{11.6}$$

which recognizes that G might vary over the cross section. Now, the relationship between the internal torque and the rotation can be written as

$$T(x) = \overline{GJ}\;\varphi'(x) \tag{11.7}$$

which is a first-order differential equation governing the rotation function $\varphi(x)$. We call this equation the *torque-rotation* relationship. It tells us that the internal torque is proportional to the rate of change of the angle of twist, and that the constant of proportionality is the effective torsional stiffness.

Often, the elastic modulus is constant over the cross section and can be pulled out of the integral so that $\overline{GJ} = GJ$, i.e., the elastic shear modulus G multiplied by the polar moment of the area

$$J = \int_A r^2 \, dA \tag{11.8}$$

Thus, the key geometric property of the cross section for torsion is the *polar moment of the area*, often called the polar moment of inertia. It appears in the equation governing the angle of twist of the bar, Eq. 11.7, and it will also appear in the torsion stress formula, which will be derived in the next section. The polar moment of area is defined in Eq. 11.8 as the integral of the radial distance r squared over the area. As such, it is a measure both of how much material is in the cross section and where that material is located. With the r^2 weighting of the element dA, the polar moment of inertia strongly favors material distant from the center of the cross section over material close to the center. That explains why a hollow tube section is more efficient at resisting torque than the same amount of material invested in a solid circular section. Since $r^2 = y^2 + z^2$, the polar moment of inertia is equal to the sum of the moments of inertia about the y and z axes, i.e., $J = I_{yy} + I_{zz}$ (see Eq. 9.18).

Example 11.4
Find the polar moment of area for the hollow circular cross section with inside radius R_i and outside radius R_o, shown in Fig. 11.15.

Solution. The polar moment of area is defined in Eq. 11.8. It is most convenient to carry out the integration using polar coordinates (r, θ). The element of area in polar coordinates is $dA = r\,dr\,d\theta$. Thus, we can calculate

$$\begin{aligned} J &= \int_0^{2\pi}\int_{R_i}^{R_o} r^2\, r\, dr\, d\theta \\ &= \int_0^{2\pi}\int_{R_i}^{R_o} r^3\, dr\, d\theta \\ &= \Big[\,\theta\,\Big]_0^{2\pi}\Big[\tfrac{1}{4}r^4\Big]_{R_i}^{R_o} \end{aligned}$$

Substituting the limits gives the result

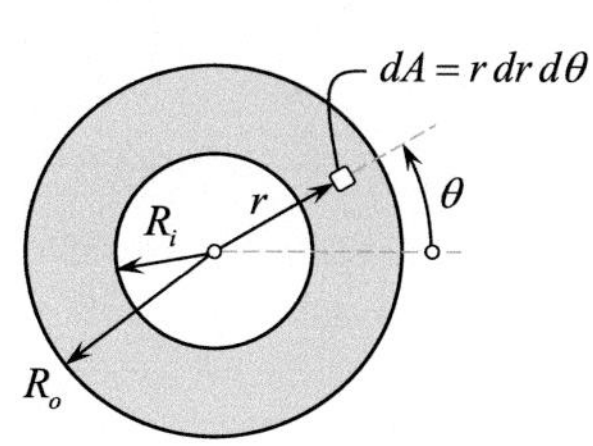

Fig. 11.15 *Example.* Circular tube cross section with inside radius R_i and outside radius R_o. Use polar coordinates (r, θ) to compute J

$$J = \tfrac{1}{2}\pi\left(R_o^4 - R_i^4\right) \tag{11.9}$$

If the cross section is solid with radius R, then using $R_o = R$ and $R_i = 0$ gives the result

$$J = \tfrac{1}{2}\pi R^4$$

We can put a finer point on the efficiency of a hollow tube compared with a bar with a solid circular cross section by comparing a solid bar to a hollow bar with the same area. If the area is the same, then the amount of material in the bar is the same (volume is length multiplied by area). The calculation of the two polar moments of inertia can then be compared, as shown in the following example.

Example 11.5
Find the ratio of the polar moment of area of a circular tube of inside radius R_i and outside radius R_o to a solid circular cross section with the same area.

Solution. The polar moment of area for the hollow, J_h, and solid, J_s, sections are, respectively,

$$J_h = \tfrac{1}{2}\pi\left(R_o^4 - R_i^4\right), \qquad J_s = \tfrac{1}{2}\pi R^4$$

where R is the radius of the solid section. The areas of the two sections are, respectively,

$$A_h = \pi\left(R_o^2 - R_i^2\right), \qquad A_s = \pi R^2$$

Setting the two areas equal to each other (to give the same amount of material), we have

$$\pi\left(R_o^2 - R_i^2\right) = \pi R^2 \qquad \to \qquad R^2 = R_o^2 - R_i^2$$

Using this value, we can write polar moment of area of the solid section as

$$J_s = \tfrac{1}{2}\pi\left(R_o^2 - R_i^2\right)^2$$

Note that $R_o^4 - R_i^4 = (R_o^2 + R_i^2)(R_o^2 - R_i^2)$, and compute the ratio of the two polar moments of area to get

$$\frac{J_h}{J_s} = \frac{\tfrac{1}{2}\pi\left(R_o^4 - R_i^4\right)}{\tfrac{1}{2}\pi\left(R_o^2 - R_i^2\right)^2} = \frac{(R_o^2 + R_i^2)(R_o^2 - R_i^2)}{(R_o^2 - R_i^2)(R_o^2 - R_i^2)} = \frac{R_o^2 + R_i^2}{R_o^2 - R_i^2}$$

Obviously, when $R_i \to 0$, the ratio approaches one. For small values of the wall thickness $t = R_o - R_i$, the ratio is approximately

$$\frac{J_h}{J_s} = \frac{(\bar{R}+\frac{1}{2}t)^2 + (\bar{R}-\frac{1}{2}t)^2}{(\bar{R}+\frac{1}{2}t)^2 - (\bar{R}-\frac{1}{2}t)^2} \approx \frac{\bar{R}}{2t}$$

where $\bar{R} = (R_o + R_i)/2$ is the radius of the centerline of the tube. This result shows that a hollow tube can be much more efficient than a solid section. However, if t gets too small, then the wall of the tube will buckle. So, there is a practical limit to how thin the wall of the tube can be.

11.2.7 Torsion Stress Formula

Not only is the kinematic hypothesis the key to the kinematics of motion, it also provides a route to finding the stress distribution associated with the internal torque. Recall that the internal torque is the *resultant* of the shear stresses acting on a cross section (see Fig. 11.6). In other words, $T(x)$ is the aggregate effect of the field $\tau(x, r)$. In general, one cannot determine a distribution from its net effect (e.g., you cannot determine individual exam scores from the class average). However, the kinematic hypothesis enables the determination of the stress distribution.

Start the derivation with the constitutive relationship

$$\tau(x, r) = G(r)\gamma(r, x)$$

where we have acknowledged that the shear modulus might vary over the cross section. Substitute $\gamma = r\,\varphi'$ and then use Eq. 11.7 in place of φ' to get

$$\tau(x, r) = \frac{G(r)T(x)r}{\overline{GJ}} \tag{11.10}$$

If G is constant over the cross section, then $\overline{GJ} = GJ$ and the G in the numerator cancels the one in the denominator, leaving

$$\tau(x, r) = \frac{T(x)r}{J} \tag{11.11}$$

In other words, the shear stress is the net torque times r divided by the polar moment of inertia of the cross section. The relationships in Eqs. 11.10 and 11.11 are called the *torsion stress formulas*, the latter being a specialized version of the former. The shear stress in the cross section increases linearly with the radial distance from the center, as shown in Fig. 11.16. The most important implication of this distribution is

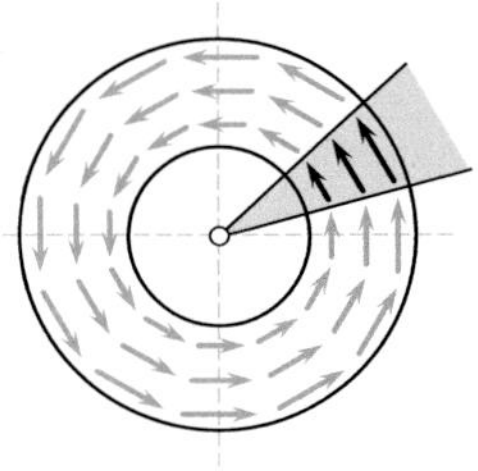

Fig. 11.16 *Stress distribution.* For a cross section with constant G, the shear stress increases linearly with the radial distance r

that the largest shear stress occurs at the largest radius, which is the exterior surface of the bar if G is constant. For a solid bar, the stress at the center is zero.

It is important to realize that we can compute the stress from the net torque only because of the kinematic hypothesis. The derivation above is the proof. If the shear modulus is not constant, then the stresses at a cross section are affected, as the following example shows.

Example 11.6

Consider the circular cross section with variable modulus shown in Fig. 11.17. The outside layer has modulus $G_2 = G_o$ which is smaller than the modulus of the core $G_1 = 2G_o$, where G_o is a reference value. The inside radius is $R_i = 0.6R_o$. Find the effective modulus $\overline{GJ}$ and the shear stress for torque T.

Solution. The effective torsional stiffness can be determined from Eq. 11.6 as

$$\begin{aligned}
\overline{GJ} &= \int_A G(r)\, r^2\, dA \\
&= \int_0^{2\pi}\int_0^{0.6R_o} (2G_o) r^3\, dr\, d\theta + \int_0^{2\pi}\int_{0.6R_o}^{R_o} (G_o) r^3\, dr\, d\theta \\
&= 2\pi(2G_o)\left(\tfrac{1}{4}(0.6R_o)^4\right) + 2\pi(G_o)\left(\tfrac{1}{4}\left[R_o^4 - (0.6R_o)^4\right]\right) \\
&= 1.774\, G_o R_o^4
\end{aligned}$$

To deal with the variable modulus, we broke the integral into two regions, each of which has constant, albeit different, values of the modulus. The shear stress distribution is

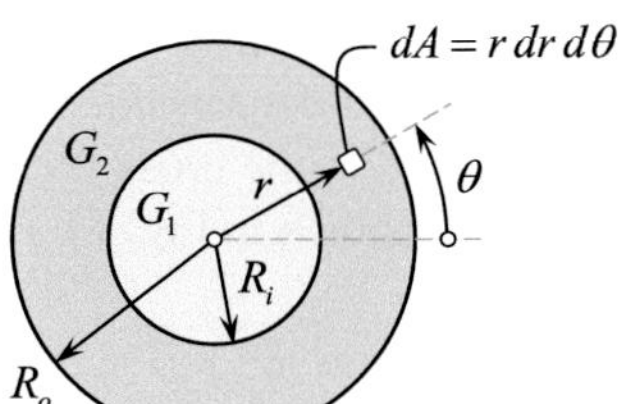

Fig. 11.17 *Example.* Circular cross section with variable shear modulus. The center part has modulus G_1 and the outer part has modulus G_2

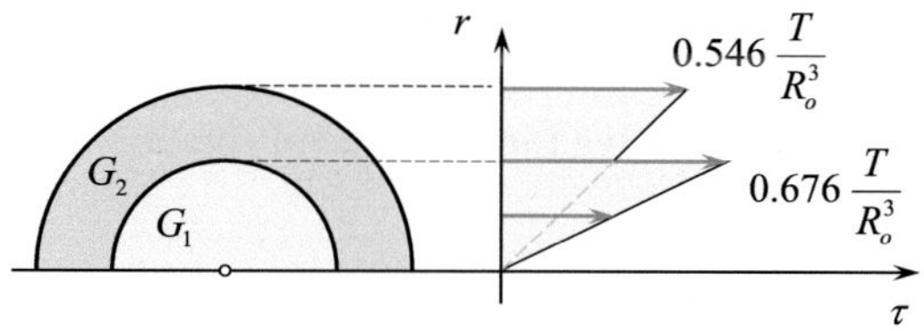

Fig. 11.18 *Example.* Stress distribution on a cross section with variable modulus with the maximum shear stresses shown

$$\tau(r) = \frac{G(r)\,T\,r}{\overline{GJ}} = \frac{T}{1.774\,G_o R_o^4}\begin{cases} 2G_o r & 0 \le r \le 0.6R_o \\ G_o r & 0.6R_o \le r \le R_o \end{cases}$$

Observe that the shear stress depends upon r. It is not obvious where the maximum value occurs. There are two candidates: (1) the outside of the inner material, and (2) the outside of the outer material. For the outside of the core, we have

$$\tau(0.6R_o) = \frac{T}{1.774\,G_o R_o^4}\big((2G_o)(0.6R_o)\big) = 0.676\frac{T}{R_o^3}$$

The shear stress at the outside of the outer material is

$$\tau(R_o) = \frac{T}{1.774\,G_o R_o^4}\big(G_o R_o\big) = 0.564\frac{T}{R_o^3}$$

Therefore, the greatest stress happens at the outer surface of the core. The shear stress distribution is shown in Fig. 11.18. Notice that the reference value for the shear modulus does not appear in the stress. Only the *relative* stiffness of the parts matters for the stress.

The stress in the region with larger elastic modulus has a greater slope than the stress in the region with lower elastic modulus. This result is a direct consequence of the kinematic hypothesis. The strain is proportional to r, and because the stress is modulus times strain, the resulting stresses are proportional to their respective moduli.

11.3 Solving Torsion Problems

In this section, we will explore the solution of torsion of circular bars by walking through some specific examples. These problems illustrate a range of important issues, including problems with static indeterminacy and torsion of nonprismatic bars.

We can find $T(x)$ either by taking a finite-sized free body diagram with a cut at x, which exposes $T(x)$, or we can solve Eq. 11.3. Once $T(x)$ is known, we can use it in Eq. 11.7 and solve for the angle of rotation by integration, as the following example illustrates.

Example 11.7
Consider the fixed-free bar of length L and torsional stiffness GJ, shown in Fig. 11.10. Find the rotation function $\varphi(x)$ and the rotation at the end of the bar.

Solution. The internal torque was determined in previous examples to be

$$T(x) = t_o\big(L - x\big)$$

We can determine the rotation function by solving Eq. 11.7. This differential equation is separable. Hence, we can integrate both sides to get

$$\varphi'(x) = \frac{t_o\big(L - x\big)}{GJ}$$
$$\varphi(x) = \frac{t_o}{GJ}\big(Lx - \tfrac{1}{2}x^2\big) + C$$

The constant of integration C can be determined from the boundary condition $\varphi(0) = 0$. To wit,

$$\varphi(0) = \frac{t_o}{GJ}\big(L(0) - \tfrac{1}{2}(0)^2\big) + C = 0$$

We can solve this equation to get $C = 0$. Substituting this value into the general expression gives the final result for the rotation:

$$\varphi(x) = \frac{t_o}{GJ}\big(Lx - \tfrac{1}{2}x^2\big)$$

The rotation at the end of the bar can be obtained by substituting $x = L$. To wit,

$$\varphi(L) = \frac{t_o}{GJ}\big(L(L) - \tfrac{1}{2}(L)^2\big) = \frac{1}{2}\frac{t_o L^2}{GJ}$$

It is always a good idea to check the units. First, t_o has units of moment per length (force times length per length, which is just force). The modulus has units force per area or force per length squared. The polar moment of area has units of length to the fourth power. Hence, the effective modulus GJ has units of force times length squared. Thus, the right side has units of radians (with units of length per length), which is appropriate for the rotation angle $\varphi(x)$.

This example illustrates the basic framework for solving torsion problems. The problem was statically determinate, so the only unknown in the expression for the rotation was the constant of integration. We used the single boundary condition involving rotation to find that constant. The following sections apply the same basic strategy to statically indeterminate problems and nonprismatic bars.

11.3.1 Static Indeterminacy

In the previous example, the internal torque $T(x)$ was completely determined from the equations of static equilibrium. That function was then used in Eq. 11.7 to solve for the rotation $\varphi(x)$. The fixed-fixed boundary condition leads to a statically indeterminate system (see Fig. 11.12a). In that case, it is not possible to uniquely determine the reaction torques from equations of static equilibrium alone. Hence, the problem is *statically indeterminate*. There are two unknown reactions and only one equation of moment equilibrium for the free body diagram of the entire bar. Each time we take a cut to expose the internal torque $T(x)$, the internal torque is unknown, and we get one new equation. Hence, there is no possibility of finding a free body diagram that will allow us to find both reaction torques.

To solve a statically indeterminate problem, we need to use the equations of kinematics along with the boundary conditions on rotation. The number of boundary conditions on rotation is two for the fixed-fixed torsion bar. The additional condition will allow us to find the unknown reaction torque in addition to the constant of integration, as the following example demonstrates.

Example 11.8
Consider the fixed-fixed bar of length L subjected to a linearly increasing applied torque of magnitude t_o, shown in Fig. 11.19. The effective modulus of the bar is GJ. Determine the reaction torques, the internal torque $T(x)$, and the rotation $\varphi(x)$.

Solution. We need a mathematical expression for the loading function that represents what the picture of the loading indicates. It is a linear function, i.e., $t(x) = ax + b$. We need to determine the values of a and b for this particular case. We know that $t(0) = 0$ and $t(L) = t_o$. Substituting these two points into the equation gives two equations in the unknowns a and b. Solving these equations and substituting back into the original expression gives

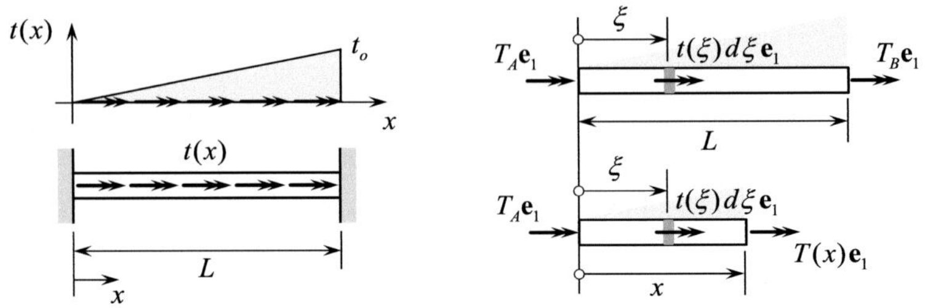

Fig. 11.19 *Example.* Fixed-fixed bar of length L with linearly distributed applied torque of magnitude t_o at the right end. The effective modulus of the bar is GJ. The sketch on the left is the problem definition (showing the applied torque distribution). The sketches on the right are free body diagrams

$$t(x) = t_o \frac{x}{L}$$

If we need $t(\xi)$, we will just substitute ξ for x in this equation.

We can establish equilibrium of the whole bar using the free body diagram of the entire system (the top right diagram in Fig. 11.19). To wit,

$$T_A \mathbf{e}_1 + \int_0^L t_o \frac{\xi}{L} \mathbf{e}_1 \, d\xi + T_B \mathbf{e}_1 = \mathbf{0}$$

$$T_A + \tfrac{1}{2} t_o L + T_B = 0$$

In the second step, we have dotted the equation with $\mathbf{e}_1$ and evaluated the integral. This equation relates the two unknown reaction torques T_A and T_B, but this equation is not adequate to uniquely determine both. We need $T(x)$ for the torque-rotation equation, which we will integrate to get the rotation. We can establish equilibrium using the second free body diagram as follows:

$$T_A \mathbf{e}_1 + \int_0^x t_o \frac{\xi}{L} \mathbf{e}_1 \, d\xi + T(x) \mathbf{e}_1 = \mathbf{0}$$

$$T_A + \tfrac{1}{2} t_o \frac{x^2}{L} + T(x) = 0$$

With this equation, we can solve for the internal torque $T(x)$ as

$$T(x) = -T_A - \tfrac{1}{2} t_o \frac{x^2}{L}$$

This equation shows that we can determine $T(x)$ in terms of the reaction torque T_A. Having an unknown reaction torque will not be a problem for integrating the differential equation governing rotation because, while T_A is unknown, it is not a function of x. Thus, we can compute

$$\varphi'(x) = \frac{1}{GJ}\Big(-T_A - \tfrac{1}{2}t_o\frac{x^2}{L}\Big)$$
$$\varphi(x) = \frac{1}{GJ}\Big(-T_A x - \tfrac{1}{6}t_o\frac{x^3}{L} + C\Big)$$

This equation has two unknowns: the constant of integration C and the reaction torque T_A. We have two boundary conditions: $\varphi(0)=0$ and $\varphi(L)=0$. The first gives

$$\varphi(0) = \frac{1}{GJ}\Big(-T_A(0) - \tfrac{1}{6}t_o\frac{(0)^3}{L} + C\Big) = 0$$

We can solve this equation to get $C=0$. The second condition gives

$$\varphi(L) = \frac{1}{GJ}\Big(-T_A(L) - \tfrac{1}{6}t_o\frac{(L)^3}{L}\Big) = 0$$

where we have noted that $C=0$. Solve this equation for the reaction torque to get

$$T_A = -\tfrac{1}{6}t_o L$$

Now, with the reaction torque known, we can complete the solution by substituting it back to find T_B, $T(x)$, and $\varphi(x)$. To wit,

$$T_B = -\tfrac{1}{3}t_o L$$
$$T(x) = t_o\Big(\tfrac{1}{6}L - \tfrac{1}{2}\frac{x^2}{L}\Big)$$
$$\varphi(x) = \frac{t_o}{GJ}\Big(\tfrac{1}{6}Lx - \tfrac{1}{6}\frac{x^3}{L}\Big)$$

Once you have the solution, it is a good idea to differentiate the functions found to verify that $\varphi'(x)=T(x)/GJ$ and $T'(x)=-t(x)$. Also, notice that both reactions are negative, which means that they both point in the $-\mathbf{e}_1$ direction. That makes sense because the applied torques are in the positive $\mathbf{e}_1$ direction and the reactions must resist them. The total applied torque, which is $\frac{1}{2}t_o L$, is equal to the sum of the two reaction torques.

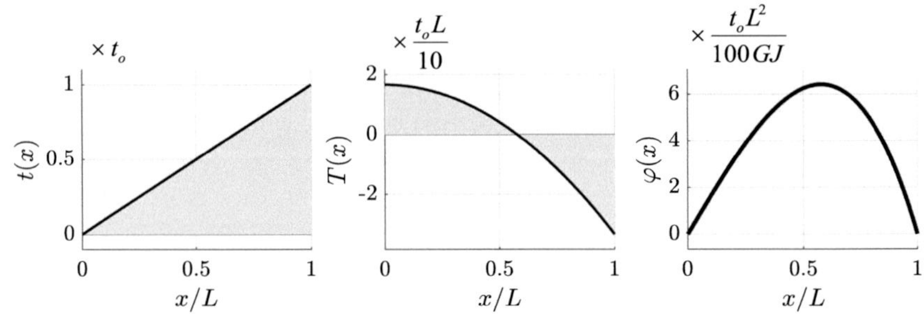

Fig. 11.20 *Example.* Results of the analysis of the fixed-fixed bar shown in Fig. 11.19. Note the scaling of the axes

The results are shown in Fig. 11.20. The slope of the $T(x)$ diagram is equal to the negative of the value of the $t(x)$ diagram for every value of x. Since $t(x)$ is positive over the entire length, the slope of the torque diagram must be negative at every point. The slope starts at zero at $x=0$ because $t(0)=0$ and gets increasingly negative with the largest slope at the right end. The slope of $\varphi(x)$ is proportional to the value of $T(x)$. Hence, the point where $T=0$ corresponds to the maximum value of $\varphi(x)$. The location of this maximum can be found by setting the expression for T equal to zero and solving for x. In this case, we get $x=L/\sqrt{3}=0.577L$. Finally, observe that the values of the internal torque at the ends, i.e., $T(0)$ and $T(L)$, are equal to the reaction forces.

This example demonstrates that there is a balance between the unknowns (reactions and constants of integration) and the boundary conditions available to determine them. This fact suggests that there is really no difference between a statically determinate and statically indeterminate problem beyond the nature of the boundary conditions. The fixed-fixed problem has two boundary conditions on *rotation* while the fixed-free problem has one boundary condition on torque and one on rotation. The information known from boundary conditions always matches the number of unknowns in the problem.

11.3.2 Nonprismatic Bars

A bar that has the same cross section for its entire length is called *prismatic*. If the cross section varies along the length of the bar, then it is called *nonprismatic*. There is no change to the preceding derivation of the governing equations except to note that the effective modulus GJ is now a function of x. There are two basic kinds of problems that arise for nonprismatic bars: (1) abrupt changes in the effective modulus, and (2) continuous variation of the effective modulus. We will show what is at stake through two examples.

A bar can have an abrupt change in cross section if, for example, bars with different cross sections or different materials are connected together to form the entire bar. The following example shows how to solve a problem with discontinuity in the cross section. This approach requires the definition of *piecewise functions*. Because discontinuities in the internal torque also require the introduction of piecewise functions, we will consider an example that has both forms of discontinuity.

Example 11.9
Consider the fixed-fixed bar of length $3L$ subjected to a concentrated torque T_o at the one-third point with an abrupt change in the torsional stiffness from GJ_o to $2GJ_o$ at the two-thirds point, shown in Fig. 11.21. Find the internal torque, rotation, and shear stress at radius R.

Solution. Because there is a point torque, the internal torque $T(x)$ is discontinuous, and because the cross section has an abrupt change, $\varphi'(x)$ is discontinuous. To deal with these discontinuities, we will divide the bar into three segments and define the internal torque and rotation with piecewise functions as follows:

$$T(x) = \begin{cases} T_1(x) & 0 \le x \le L \\ T_2(x) & L \le x \le 2L \\ T_3(x) & 2L \le x \le 3L \end{cases}, \qquad \varphi(x) = \begin{cases} \varphi_1(x) & 0 \le x \le L \\ \varphi_2(x) & L \le x \le 2L \\ \varphi_3(x) & 2L \le x \le 3L \end{cases}$$

Figure 11.21 shows the free body diagrams needed to solve the problem. The one on top is of the entire bar and is useful for relating the two reaction torques to each other. The second free body diagram has a cut to the left of the point torque, thereby defining the component function $T_1(x)$. The bottom free body diagram has a cut to the right of the point torque, thereby defining the component functions $T_2(x) = T_3(x)$. The presence of the discontinuity in the cross section does not affect the equations of equilibrium. Establishing equilibrium of the

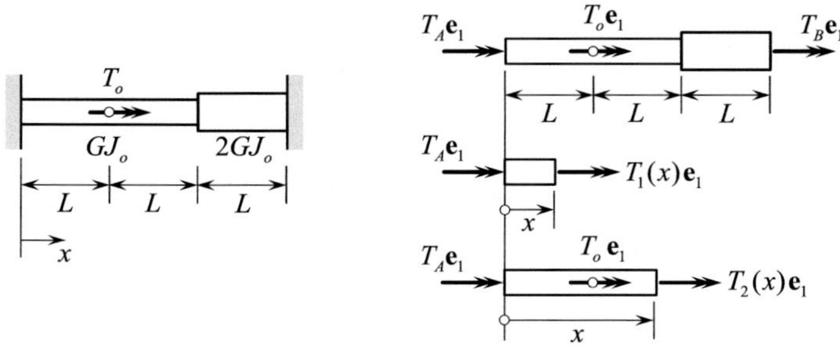

Fig. 11.21 *Example.* Fixed-fixed bar of length $3L$ with point torque T_o and abrupt change in cross section. Problem definition is on the left. Free body diagrams are on the right

entire bar results in the equation

$$T_A + T_o + T_B = 0$$

which we can view as an equation to find T_B once we have determined T_A. This problem is statically indeterminate, so we will carry T_A as an unknown until we can find it using boundary conditions on the rotations. The equilibrium equations that result from the other two free body diagrams determine the internal torque to be

$$T(x) = \begin{cases} -T_A & 0 \le x \le L \\ -T_A - T_o & L \le x \le 2L \\ -T_A - T_o & 2L \le x \le 3L \end{cases}$$

Again, the internal torque is known in terms of the reaction T_A, which is unknown. The torque-rotation relationship must be integrated for each segment. To wit,

$$\varphi_1' = \frac{-T_A}{GJ_o} \qquad \varphi_2' = \frac{-(T_A + T_o)}{GJ_o} \qquad \varphi_3' = \frac{-(T_A + T_o)}{2GJ_o}$$

$$\varphi_1 = \frac{-T_A x}{GJ_o} + C_1 \quad \varphi_2 = \frac{-(T_A + T_o)x}{GJ_o} + C_2 \quad \varphi_3 = \frac{-(T_A + T_o)x}{2GJ_o} + C_3$$

Each segment has a different constant of integration. Thus, there are four unknown quantities: the three constants of integration plus the reaction torque T_A. We have two boundary conditions and two *continuity conditions* associated with the piecewise function $\varphi(x)$. While the derivative of $\varphi(x)$ is discontinuous at the location of the abrupt change in cross section, the rotation itself is continuous. Therefore, we have the following conditions:

$$\varphi_1(0) = 0, \qquad \varphi_1(L) = \varphi_2(L), \qquad \varphi_2(2L) = \varphi_3(2L), \qquad \varphi_3(3L) = 0$$

The first and last equations are the boundary conditions, and the two in the middle are continuity conditions. We do not know the value of the rotation at $x = L$ or $x = 2L$, but we do know that whatever it is, the two functions that meet there must give the same value at that point. Implementing the first boundary condition $\varphi_1(0) = 0$ gives

$$\frac{-T_A(0)}{GJ_o} + C_1 = 0, \qquad \rightarrow \qquad C_1 = 0$$

The second condition, $\varphi_1(L) = \varphi_2(L)$, noting that $C_1 = 0$, gives

$$\frac{-T_A(L)}{GJ_o} = \frac{-(T_A + T_o)(L)}{GJ_o} + C_2, \qquad \rightarrow \qquad C_2 = \frac{T_o L}{GJ_o}$$

The third condition, $\varphi_2(2L) = \varphi_3(2L)$, gives

$$\frac{-(T_A + T_o)(2L)}{GJ_o} + \frac{T_o L}{GJ_o} = \frac{-(T_A + T_o)(2L)}{2GJ_o} + C_3$$

where we have used the value of C_2 already found. We can solve this equation to give

$$C_3 = -\frac{T_A L}{GJ_o}$$

Finally, substituting this constant, the fourth condition $\varphi_3(3L) = 0$ gives

$$\frac{-(T_A + T_o)(3L)}{2GJ_o} - \frac{T_A L}{GJ_o} = 0, \qquad \rightarrow \qquad T_A = -\tfrac{3}{5} T_o$$

Now that we have all the constants and the reaction torque, we can go back and complete the expressions for the reaction, the internal torque, and the rotation. To wit, $T_B = -0.4T_o$ and the internal torque is

$$T(x) = \begin{cases} 0.6T_o & 0 \le x \le L \\ -0.4T_o & L \le x \le 2L \\ -0.4T_o & 2L \le x \le 3L \end{cases}$$

and the rotation is

$$\varphi(x) = \frac{T_o}{GJ_o} \begin{cases} 0.6x & 0 \le x \le L \\ L - 0.4x & L \le x \le 2L \\ 0.6L - 0.2x & 2L \le x \le 3L \end{cases}$$

The results are shown graphically in Fig. 11.22. The rotation is continuous, but the derivative of rotation (i.e., the strain) is discontinuous. The rotation is linear within each segment. Since the third segment is twice as stiff as the second one, the shear stress at radius R in the third segment is smaller than in the second segment, even though the internal torque is the same in both.

In reality, the stress and strain fields are complicated in the region near the abrupt change in cross section and depend upon the precise nature of the geometry. In practice, a design that has a smooth transition will better avoid stress concentrations.

The following example shows how the response of the torsion bar is affected by an effective stiffness $GJ(x)$ that varies continuously along the length of the bar.

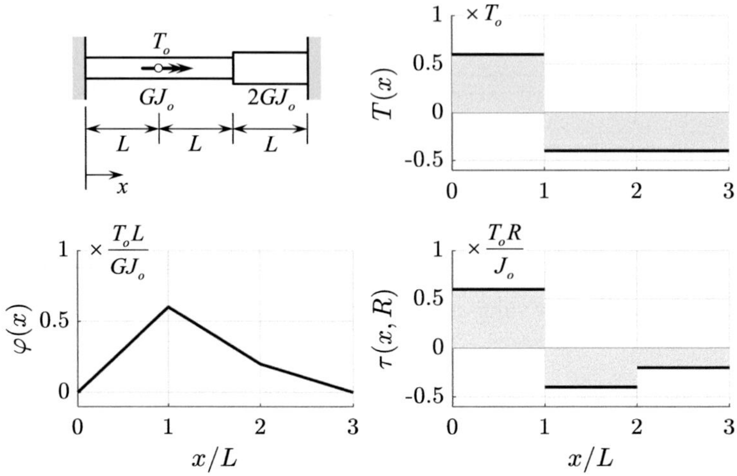

Fig. 11.22 *Example.* Fixed-fixed bar of length $3L$ with point torque T_o and abrupt change in cross section. Plots of internal torque $T(x)$, the shear stress $\tau(x, R)$, and rotation $\varphi(x)$

Example 11.10
Let us reconsider the loading and boundary conditions of the example shown in Fig. 11.19, but now for a nonprismatic bar with a solid circular cross section with radius varying according to the equation $R(x) = R_o\big(1 + x/L\big)$. In other words, the radius increases linearly from a value R_o at the left end to $2R_o$ at the right end. Find the reaction torques, internal torque, and rotation.

Solution. If the radius varies linearly, then the polar moment of area varies quartically as

$$J = \tfrac{1}{2}\pi\big(R(x)\big)^4 = \tfrac{1}{2}\pi R_o^4\big(1 + x/L\big)^4 = J_o\big(1 + x/L\big)^4$$

where $J_o = \frac{1}{2}\pi R_o^4$ is the polar moment of area at the left end of the bar. The free body diagrams and the equations of equilibrium do not change. Hence, internal torque is

$$T(x) = -T_A - \tfrac{1}{2}t_o\frac{x^2}{L}$$

as before. Now, we need to solve for the rotation from Eq. 11.7 which, in this case, takes the form

$$\varphi(x) = \int \frac{-T_A - \frac{1}{2}t_o x^2/L}{GJ_o\big(1 + x/L\big)^4}\,dx + C$$

where the integral is an indefinite integral and C is the constant of integration. It will be convenient to do a change of variable to help with the integral. Let

$$\zeta = 1 + x/L, \qquad d\zeta = dx/L$$

Note that $x = 0$ corresponds with $\zeta = 1$ and $x = L$ corresponds with $\zeta = 2$. Also, we can solve for $x = L(\zeta - 1)$. Substituting the change of variable gives

$$\begin{aligned}
\varphi(\zeta) &= -\int \frac{T_A + \frac{1}{2}t_o L(\zeta - 1)^2}{GJ_o\zeta^4} L\, d\zeta + C \\
&= -\frac{T_A L}{GJ_o}\int \frac{d\zeta}{\zeta^4} - \frac{1}{2}\frac{t_o L^2}{GJ_o}\int \left(\frac{1}{\zeta^2} - \frac{2}{\zeta^3} + \frac{1}{\zeta^4}\right) d\zeta + C \\
&= \frac{T_A L}{GJ_o}\frac{1}{3\zeta^3} + \frac{1}{2}\frac{t_o L^2}{GJ_o}\left(\frac{1}{\zeta} - \frac{1}{\zeta^2} + \frac{1}{3\zeta^3}\right) + C
\end{aligned}$$

Before we transform back to the original coordinate x, we can implement the boundary conditions to solve for C and T_A. At the left end of the bar, $\zeta = 1$ ($x = 0$), and we have

$$\varphi(1) = \frac{T_A L}{3GJ_o} + \frac{t_o L^2}{6GJ_o} + C = 0 \qquad \rightarrow \qquad C = -\frac{T_A L}{3GJ_o} - \frac{t_o L^2}{6GJ_o}$$

At the right end of the bar, $\zeta = 2$ ($x = L$), so we can compute

$$\varphi(2) = \frac{T_A L}{GJ_o}\frac{1}{24} + \frac{1}{2}\frac{t_o L^2}{GJ_o}\left(\frac{1}{2} - \frac{1}{4} + \frac{1}{24}\right) - \frac{T_A L}{3GJ_o} - \frac{t_o L^2}{6GJ_o} = 0$$

where we have substituted the value of C found previously. Solving this equation for the reaction torque gives

$$T_A = -\tfrac{1}{14}t_o L = -0.0714\, t_o L$$

Thus, $T_B = -\frac{3}{7}\, t_o L$. Substituting T_A and C, we get the final expression for rotation and torque in terms of ζ. Specifically,

$$\begin{aligned}
\varphi(\zeta) &= \frac{t_o L^2}{GJ_o}\left(\frac{1}{7\zeta^3} - \frac{1}{2\zeta^2} + \frac{1}{2\zeta} - \frac{1}{7}\right) \\
T(\zeta) &= t_o L\left(\tfrac{1}{14} - \tfrac{1}{2}(\zeta - 1)^2\right)
\end{aligned}$$

We can transform back to the original variable $x = L(\zeta - 1)$ to get the final expressions for the internal torque and rotation. Let $\xi = x/L$. Then,

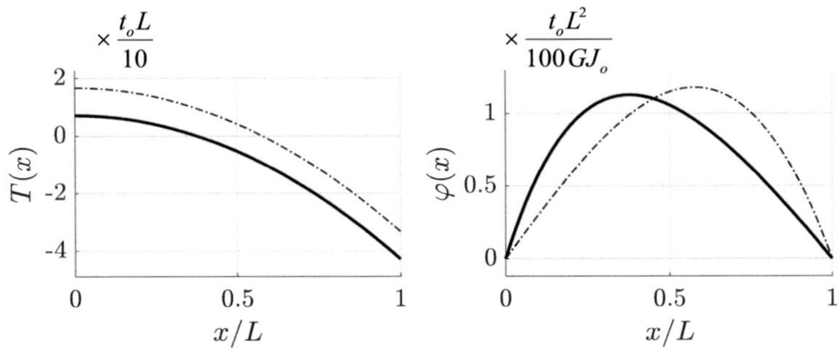

Fig. 11.23 *Example.* Results of the analysis of the fixed-fixed bar shown in Fig. 11.19. The solid line is the nonprismatic bar, and the dotted line is the prismatic bar with $GJ = 5.444GJ_o$

$$\varphi(\xi) = \tfrac{1}{14}\frac{t_o L^2}{GJ_o}\left[\frac{\xi + \xi^2 - 2\xi^3}{(1+\xi)^3}\right]$$
$$T(\xi) = \tfrac{1}{14} t_o L\big(1 - 7\xi^2\big)$$

We can differentiate these functions to prove that they are compatible with the differential equations they come from.

Figure 11.23 compares the torque and rotation functions for the prismatic and nonprismatic cases. The solid line is the nonprismatic case and the dotted line is the prismatic case, with $GJ = 5.444GJ_o$. The reason for selecting that stiffness for the prismatic bar is to make the comparison between two bars with the same total amount of material in them. The variation of the cross section does not affect the differential equation for $T(x)$ and, therefore, does not change the *shape* of the curve. However, as we have seen, the distribution of material *does* affect the reaction torques. That is why the $T(x)$ plots are shifted by a constant amount.

The rotation $\varphi(x)$ is affected by the distribution of the cross section. The thinner part of the bar is easier to rotate than the thicker part because it has less resistance. Thus, the nonprismatic $\varphi(x)$ curve is shifted left from the prismatic one. The peak of the nonprismatic case is slightly lower than the prismatic case.

This example serves to illustrate the additional complexity that arises in the nonprismatic bar due to the variable polar moment of inertia. All aspects of the solution, except for finding the integrals of the functions, were similar to the example with the prismatic bar.

11.4 Torsion of Bars with Non-circular Cross Sections

Torsion of circular bars is the simplest case of torsion, and it is indicative of torsion problems in general. The circular cross section is the most common one for bars used to transmit torque, e.g., as the axle transmitting torque to a wheel. However, not all bar cross sections are circular, and it is reasonable to ask if the theory derived in the first part of this chapter works for those cases. The short answer to that question is no, but there are some fairly simple modifications to the theory that can be used in some important cases. In this section, we outline the modifications needed for cross sections that can be classified as non-circular solid sections, thin-walled open sections (e.g., the I-beam), and thin-walled closed sections (e.g., rectangular tubes). In most cases, these non-circular bars are not designed to transmit torque as their primary mechanical objective but must resist torque because it is always part of our three-dimensional world. You would not make a truck axle out of an I-beam, but an I-beam in a building structure might twist due to eccentric loads.

11.4.1 Non-circular Solid Sections

The problem with non-circular sections is that their cross sections warp out of their initial planes when the bar is twisted, as illustrated in Fig. 11.24. Hence, the kinematic hypothesis that plane sections remain plane does not hold for these cross sections. The exact solution to the problem of prismatic bars subjected to constant torque was formulated by Saint-Venant.[1] He showed that there is no change of shape of the cross-sectional plane viewed head on (e.g., a square would remain a square). The cross section still rotates about the bar axis, as it does for the circular cross section. However, the cross sections warp out of their original plane. The warping function $u(y, z)$ that characterizes the deviation from the plane of the cross section in the $\mathbf{e}_1$ direction is governed by a partial differential equation.

The main issue with torsion is that the shear stresses in the cross section cannot have a component normal to the boundary of the cross section because the sides

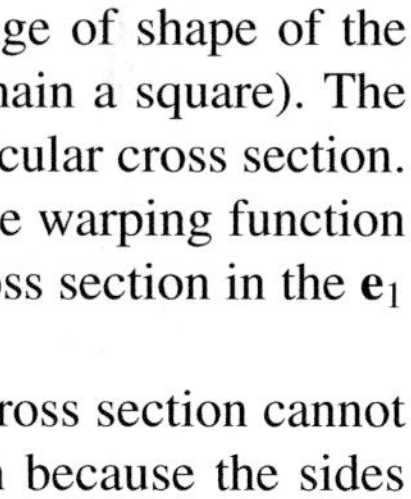

Fig. 11.24 *Warping of cross section.* Twisting of a non-circular bar causes the cross section to warp out of its original plane

[1] Adhémar Jean Claude Barré de Saint-Venant (1797–1866) was a French mathematician who made many contributions to the theory of deformable solids.

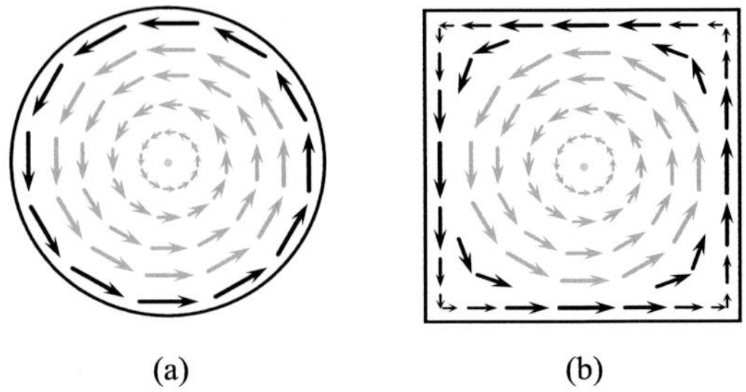

Fig. 11.25 *Shear stresses in cross section.* **a** Shear stress in a circular cross section. **b** Qualitative shear stresses in a square cross section

of the bar are free of traction. For the circular cross section, Fig. 11.25a, the shear stresses follow a circular contour, and thus satisfy this condition. The shear stresses also increase with distance from the center of twist. For a square cross section in Fig. 11.25b, the shear stresses can increase with distance from the center of twist in the center of the cross section, but near the corners, the shear stresses must decrease to zero to meet the condition of the traction-free boundary. Thus, those portions of the cross section are less effective in carrying shear and contribute less to the net torque than a point in a circular cross section at the same distance from the center of twist. The gradients of the shear stresses in these regions are what causes the cross section to warp. The decrease in effectiveness at the corners makes the actual torsional stiffness smaller than the polar moment of inertia, which is the predicted stiffness from the kinematic hypothesis of plane sections remain plane.

Saint-Venant did suggest an approximate expression for the torsional stiffness $G\hat{J}$, with *torsion constant*

$$\hat{J} = \frac{1}{4\pi^2}\frac{A^4}{J} \tag{11.12}$$

where A is the area and J is the polar moment of inertia of the cross section. For a circular cross section, the area is $A = \pi R^2$ and the polar moment of inertia is $J = \frac{1}{2}\pi R^4$. Thus, the torsion constant from Eq. 11.12 is

$$\hat{J} = \frac{1}{4\pi^2}\frac{\left(\pi R^2\right)^4}{\left(\frac{1}{2}\pi R^4\right)} = \frac{1}{2}\pi R^4$$

which is exactly the same value obtained earlier in the chapter. Hence, the approximation is exact for the circular cross section. For a rectangular cross section of height h and width b, the area is $A = bh$ and the polar moment of inertia is $J = \frac{1}{12}(b^3h + bh^3)$. Thus, Eq. 11.12 gives

$$\hat{J} = \frac{1}{4\pi^2}\frac{(bh)^4}{\frac{1}{12}\left(b^3h + bh^3\right)} = \frac{3}{\pi^2}\frac{b^3h^3}{\left(b^2 + h^2\right)}$$

For a square cross section of dimension a, this formula reduces to

$$\hat{J} = \frac{3}{\pi^2}\frac{a^3a^3}{(a^2+a^2)} = \frac{3a^4}{2\pi^2} = 0.152\,a^4$$

which can be compared to the exact value of $0.141a^4$ from elasticity theory. The polar moment of inertia for the square cross section is $J = 0.166\,a^4$. For solid non-circular cross sections, the polar moment of inertia is at least in the neighborhood of the correct value if the ratio of the largest and smallest dimensions of the cross section is near one. As that ratio gets farther from one, the approximation deteriorates.

11.4.2 Open Thin-Walled Cross Sections

A *thin-walled* cross section is any cross section composed of relatively thin elements. There are two basic categories of thin-walled sections: open and closed. An *open thin-walled section* does not enclose any area (and, therefore, would not make a good fence to keep your sheep in). Some examples of open thin-walled sections are shown in Fig. 11.26. The gray region is the material of the cross section. At every location in the cross section, one can discern a thin direction where there is a distance t from one edge to the other. The following discussion focuses on these open sections.

To start the discussion of open thin-walled sections, let us first consider the pure twist of a solid rectangular cross section of height h and width t, shown in Fig. 11.27. This problem has an exact elasticity solution, and the solution can be used to build an approximate solution to the general open thin-walled torsion problem. If h is much greater than t, then the rectangle is representative of a part of a thin-walled cross section. For the thin rectangle, the shear stresses vary linearly through the thickness, except at the ends. Because the moment arms of the shear stresses are very small, the thin rectangle is very flexible in torsion.

The exact elasticity solution for the rectangle includes the relationships for net torque and maximum shear stress τ_{max}. The torque is related to the rate of twist as

$$T = G\hat{J}\,\varphi' \qquad \hat{J} = \tfrac{1}{3}ht^3(1-\beta_J) \tag{11.13}$$

where T is the torque, G is the shear modulus of elasticity, φ' is the rate of twist, and $\hat{J}$ is the torsion constant. The maximum shear stress is

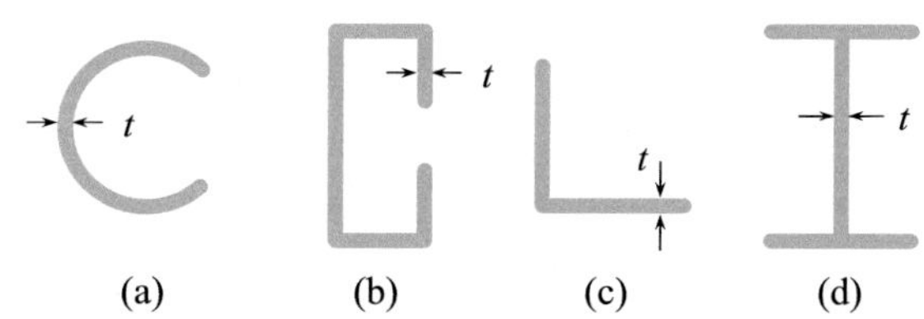

Fig. 11.26 *Open thin-walled cross sections.* Typical open thin-walled cross sections. **a** Open circular tube. **b** Channel section. **c** Angle section. **d** I-beam

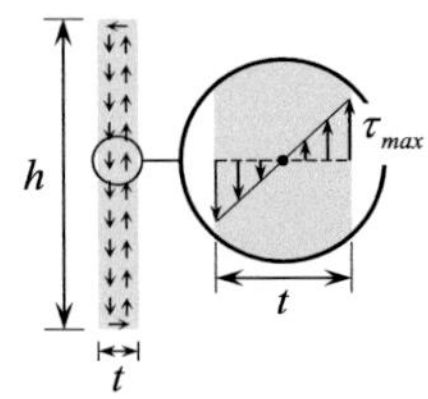

Fig. 11.27 *Shear Stresses.* Shear stresses in a thin rectangular cross section due to torsion

$$\tau_{\max} = Gt\big(1-\beta_\tau\big)\,\varphi' = \frac{Tt}{\hat{J}}\big(1-\beta_\tau\big) \tag{11.14}$$

The coefficients β_J and β_τ that appear in these equations depend on h/t and are given by the expressions

$$\beta_J = \frac{96}{\pi^4}\sum_{n=1,3,5\ldots}\frac{\tanh(n\zeta)}{n^5\,\zeta}, \qquad \beta_\tau = \frac{8}{\pi^2}\sum_{n=1,3,5\ldots}\frac{1}{n^2\cosh(n\zeta)}$$

where $\zeta = \pi h/2t$. While these are infinite series, they converge quickly, needing only a few terms to give accurate results. The variation of the coefficients β_J and β_τ with the ratio h/t are shown in Fig. 11.28. The open circle on the left is $h/t = 1$, i.e., the solid square cross section. The values of β_J and β_τ decrease quickly as h/t increases, and the curves in the figure are asymptotic to one, with β_τ approaching zero faster than β_J.

It is evident that for a thin rectangular cross section, the approximations for the torsion constant and shear stress,

$$\hat{J} = \tfrac{1}{3}ht^3 \qquad \tau_{\max} = \frac{Tt}{\hat{J}} \tag{11.15}$$

respectively, are quite good. We can use these results to generalize to any thin-walled open cross section composed of multiple rectangular segments. An approximate formula for the torsion constant $\hat{J}$ for an open thin-walled section is

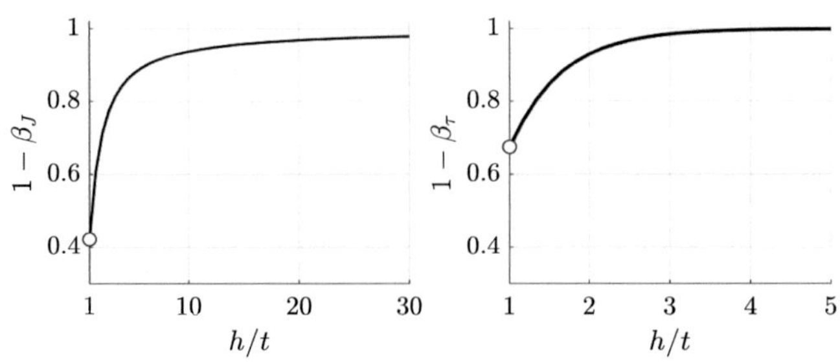

Fig. 11.28 *Rectangular Cross Section.* Coefficients for the exact solution to the torsion of a solid rectangular cross section

$$\hat{J} = \sum_{i=1}^{N} \tfrac{1}{3} h_i t_i^3 \tag{11.16}$$

where h_i and t_i are the length and thickness of the ith piece of the cross section and N is the number of rectangular pieces. If a piece is curved, then h_i is the arc length of its middle line. An approximation for the maximum shear stress in an open thin-walled cross section is

$$\tau_{\max}^{i} = \frac{T t_i}{\hat{J}} \tag{11.17}$$

where T is the torque and $\hat{J}$ is given by Eq. 11.16. Thus, it is evident that the largest shear stress happens in the thickest piece.

Example 11.11
Find the torsion constant $\hat{J}$ of an I-beam of height h, width $b_f = h/2$, flange thickness $t_f = h/10$, and web thickness of $t_w = h/15$, shown in Fig. 11.33. Consider the height to be the distance between the middle lines of the flange. Compare the result with the polar moment of inertia (Fig. 11.29).

Solution. The torsional stiffness $\hat{J}$ is given by Eq. 11.16. Thus,

$$\hat{J} = \tfrac{1}{3} h t_w^3 + (2)\tfrac{1}{3} b_f t_f^3 = \tfrac{1}{3} h \left(\tfrac{1}{15} h\right)^3 + \tfrac{2}{3}\left(\tfrac{1}{2} h\right)\left(\tfrac{1}{10} h\right)^3 = 0.000321\, h^4$$

Contrast this value with the polar moment of inertia for the I-beam. Neglecting terms involving t_w^3 and t_f^3, the polar moment of inertia is approximately

$$\begin{aligned} J &= \tfrac{1}{12} h^3 t_w + \tfrac{1}{2} h^2 b_f t_f + \tfrac{1}{6} b_f^3 t_f \\ &= \tfrac{1}{12} h^3 \left(\tfrac{1}{15} h\right) + \tfrac{1}{2} h^2 \left(\tfrac{1}{2} h\right)\left(\tfrac{1}{10} h\right) + \tfrac{1}{6}\left(\tfrac{1}{2} h\right)^3 \left(\tfrac{1}{10} h\right) \\ &= 0.0326\, h^4 \end{aligned}$$

The ratio $J/\hat{J} \approx 100$ shows how much more flexible open thin-walled cross sections are in torsion than solid sections and how much deviation there must be from the hypothesis that plane sections remain plane.

The results in this section apply to problems with constant torsion where warping is free to develop. If there is a mechanism to restrain warping (e.g., a fixed end), then normal stresses σ_{xx} will develop and the shear stresses can be significantly altered from the pure torsion case. Warping restraint increases the stiffness of a bar subjected to torsion, but the effects vary along the axis of the bar and, hence, cannot be summarized as a single factor to adjust the torsion constant $\hat{J}$. The shear stresses are

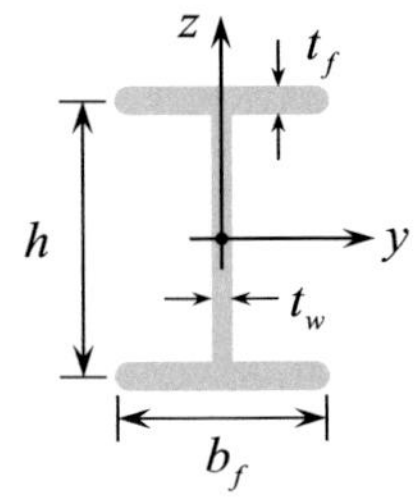

Fig. 11.29 *I-Beam example.* Thin-walled I-beam of height h, width b_f, flange thickness t_f, and web thickness t_w

also significantly altered from those shown in Fig. 11.27. The analysis of thin-walled bars with warping restraint is beyond the scope of this book.

11.4.3 Closed Thin-Walled Cross Sections

A *closed thin-walled cross section* encloses a region that is not part of the cross section. It has an inside and an outside. If the closed section were a fence, you could put your sheep on the inside, and they could not get out. You could use it as a pipe to convey water. Some examples are shown in Fig. 11.30. Again, the gray region is the material of the cross section and at every location in the cross section, one can discern a thin direction where there is a distance t from one edge to the other. The thickness is not necessarily constant.

Consider the general closed thin-walled cross section shown in Fig. 11.31a. Distance along the middle line of the contour is measured with s, which starts at 0 and ends at b. Therefore, the total length of the middle line is b. The unit vector field $\mathbf{n}(s)$ is perpendicular to the contour and $\mathbf{m}(s)$ is along the contour, perpendicular to $\mathbf{n}(s)$. The x axis of the bar is oriented in the $\mathbf{e}_1$ direction. The thickness of the cross section, $t(s)$, can vary along the contour. The shear stresses σ_{xs} on the cross section must be oriented in the direction of the contour because there are no shear tractions on the outer or inner surfaces of the wall of the cross section. The normal stresses σ_{xx} are zero. The shear stresses develop in a manner similar to the circular bar shown in Fig. 11.5. We assume that the thickness of the bar is small compared with the size of the cross section. Therefore, the variation of the shear stress across the thickness is nearly constant.

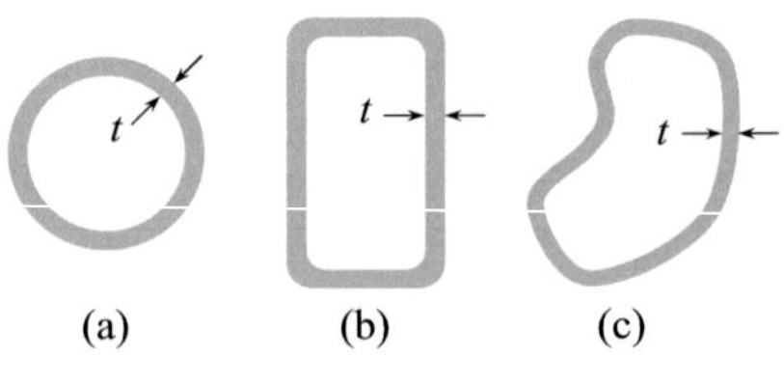

Fig. 11.30 *Closed thin-walled cross sections.* Typical thin-walled cross sections. **a** Hollow circular tube. **b** Hollow rectangular tube. **c** General closed tube

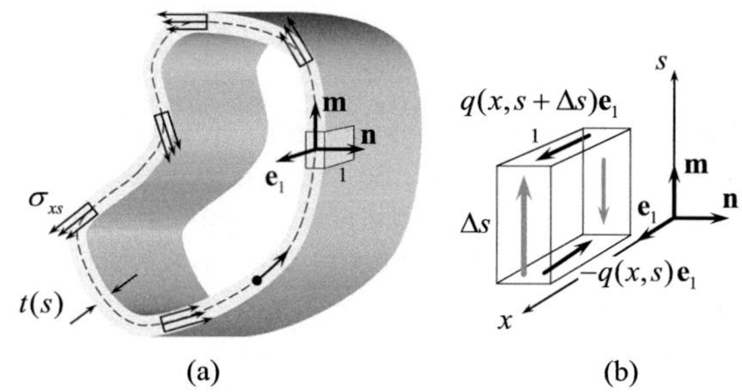

Fig. 11.31 *Closed thin-walled cross section.* **a** Illustration of closed thin-walled cross section. **b** Piece cut from the closed thin-walled cross section. Free body diagram, showing the forces on the sides

Consider the free body diagram shown in Fig. 11.31b, which is a piece of unit length in the x direction, cut from the wall of the cross section between s and $s + \Delta s$. There are no normal stresses, so the only forces acting on the piece are due to shear. The two shear tractions in the $\mathbf{e}_1$ direction are labeled (the two vertical forces are not). Define the *shear flow* to be the shear stress multiplied by the thickness, i.e.,

$$q(x, s) = \sigma_{xs}(x, s)t(s)$$

where σ_{xs} is the average shear stress acting on the piece and t is the thickness. Thus, the total force acting on the top face is $q(x, s + \Delta s)\mathbf{e}_1$ (recall that the length of the piece is one), and the force acting on the bottom face is $-q(x, s)\mathbf{e}_1$. The forces must be in equilibrium. Summing the forces, dotting with $\mathbf{e}_1$ to eliminate the two vertical forces, dividing through by Δs, and taking the limit as $\Delta s \to 0$ gives

$$\lim_{\Delta s\to 0} \frac{q(x, s + \Delta s) - q(x, s)}{\Delta s} = \frac{\partial q}{\partial s} = 0$$

which proves that q is constant around the contour of the cross section (i.e., it does not depend upon s). Hence, we can write $q(x)$ rather than $q(x, s)$ for the shear flow. This observation is the key to the analysis of closed thin-walled bars.

The bar twists about $\mathcal{O}$, which we will take to be the axis of the bar, as shown in Fig. 11.32a. The force acting on the typical particle at $\mathbf{r}$ is $q\mathbf{m}\,ds$. We can compute the net torque $\mathbf{T}$ acting on the cross section. The magnitude of the torque is $T = \mathbf{T} \cdot \mathbf{e}_1$. Thus,

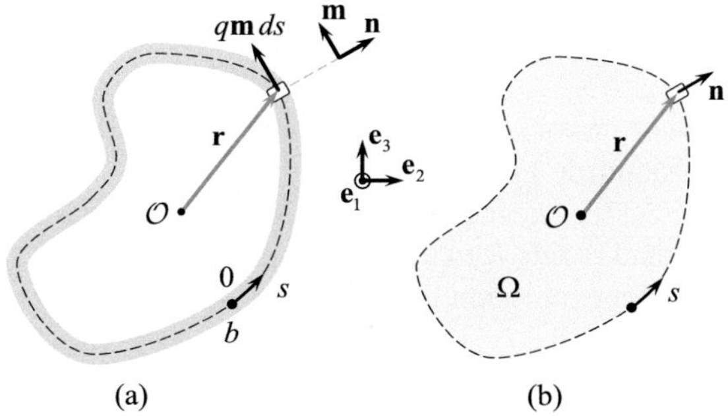

Fig. 11.32 *Closed thin-walled cross section.* **a** The cross section rotates about the axis $\mathcal{O}$ without distorting its shape. The force acting on the typical particle at $\mathbf{r}$ is $q\,\mathbf{m}\,ds$. **b** The enclosed area of the region defined by the middle line is Ω

$$T = \mathbf{e}_1 \cdot \oint \mathbf{r} \times q\mathbf{m}\, ds$$

where the integral symbol with a circle stands for a line integral around the entire contour of the middle line. The shear flow q is constant and can be taken outside of the integral. The triple scalar product obeys the cyclic property and, therefore we can write $\mathbf{e}_1 \cdot (\mathbf{r} \times \mathbf{m}) = \mathbf{r} \cdot (\mathbf{m} \times \mathbf{e}_1)$. Note, also, that $\mathbf{m} \times \mathbf{e}_1 = \mathbf{n}$. Therefore, the expression for the net torque is

$$T = q \oint \mathbf{r} \cdot \mathbf{n}\, ds \tag{11.18}$$

We can apply the *divergence theorem*, Eq. 9.22, to compute the integral. Let $\mathbf{x}$ be a vector to any point in the region enclosed by the middle line and note that $\mathbf{x} = \mathbf{r}$ for points on the boundary. The divergence of $\mathbf{x}$ is $\text{div}(\mathbf{x}) = 2$ (see the calculation leading to Eq. 9.24). Therefore, we can compute

$$\oint \mathbf{r} \cdot \mathbf{n}\, ds = \int_{\Omega} \text{div}(\mathbf{x})\, dA = 2\Omega \tag{11.19}$$

where Ω is the area enclosed by the middle line of the contour of the cross section, shown in Fig. 11.32b. Note that this area is *not* the same as the area of the cross section A, the shaded area in Fig. 11.32a, which can be computed as

$$A = \oint t(s)\, ds$$

The final expression for the net torque is

$$T = 2\Omega q \tag{11.20}$$

which is often referred to as *Bredt's formula*. This equation can be used to determine the shear flow from the net torque and, consequently, it gives a formula for the shear stress

$$\sigma_{xs} = \frac{T}{2\Omega t} \tag{11.21}$$

Evidently, the maximum shear stress in a closed thin-walled section occurs at the place with the smallest thickness t.

Example 11.12
Find the shear stress due to torque T in a circular cross section with thickness t and radius R to the middle line using Eq. 11.21. Compare the result with the exact expression in Eq. 11.11.

Solution. The area enclosed by the middle line of the circular cross section is $\Omega = \pi R^2$. Therefore, from Eq. 11.21, we get

$$\sigma_{xs} = \frac{T}{2\pi R^2 t}$$

The exact solution, noting that $J = \frac{1}{2}\pi(R_o^4 - R_i^4)$ for the shear stress at radius $r = R$, is

$$\tau = \frac{TR}{J} = \frac{TR}{\frac{1}{2}\pi(R_o^4 - R_i^4)}$$

The outside and inside radii are $R_o = R + t/2$ and $R_i = R - t/2$, respectively, so that

$$\begin{aligned} R_o^4 - R_i^4 &= \left(R_o^2 + R_i^2\right)\left(R_o^2 - R_i^2\right) \\ &= \left(2R^2 + \tfrac{1}{2}t^2\right)\left(2Rt\right) \approx 4R^3 t \end{aligned}$$

Substituting this result into the shear stress formula gives

$$\tau = \frac{TR}{\frac{1}{2}\pi(4R^3 t)} = \frac{T}{2\pi R^2 t}$$

which is the same as the value obtained from Eq. 11.11. The main difference is that we neglected the t^2 term in the calculation of J. This example confirms that the equation for the thin-walled section corresponds with the exact solution for the circular cross section.

We can develop a relationship between the torque and rate of rotation by recognizing that the cross section rotates about point $\mathcal{O}$ as a rigid body but that there is a warping displacement $u(x, s)$ out of the plane of the cross section. The component of displacement at location $\mathbf{r}$ in the direction $\mathbf{m}$ (i.e., in the contour) due to rotation is the cross product between the rotation vector $\varphi(x)\mathbf{e}_1$ and the radial vector $\mathbf{r}$ dotted with $\mathbf{m}$. Therefore,

$$v = \left(\varphi(x)\mathbf{e}_1 \times \mathbf{r}\right) \cdot \mathbf{m}$$

The triple scalar product can be rearranged as $(\mathbf{e}_1 \times \mathbf{r}) \cdot \mathbf{m} = (\mathbf{m} \times \mathbf{e}_1) \cdot \mathbf{r} = \mathbf{n} \cdot \mathbf{r}$. Thus,

$$v(x, s) = \varphi(x)\, \mathbf{r}(s) \cdot \mathbf{n}(s)$$

The distance $\mathbf{r} \cdot \mathbf{n}$ is the component of the distance from $\mathcal{O}$ to the typical particle that is perpendicular to $\mathbf{m}$. The warping function is $u(x, s)$, which is the component of displacement in the $\mathbf{e}_1$ direction. Now, we can compute the shear strain at the typical particle (see Eq. 4.58) as

$$2\varepsilon_{xs} = \frac{\partial u}{\partial s} + \frac{\partial v}{\partial x} = \frac{\partial u}{\partial s} + \varphi'(x)\,\mathbf{r}\cdot\mathbf{n} \tag{11.22}$$

where we have noted that **r** and **n** do not depend upon x. If we integrate $\partial u/\partial s$ around the entire contour, we must get zero because

$$\int_0^b \frac{\partial u}{\partial s}\,ds = u(x,b) - u(x,0) = 0$$

since 0 and b are the same point in space. From Hooke's law and Eq. 11.21, we know that $2\varepsilon_{xs} = \sigma_{xs}/G = T/2\Omega G t$. Substituting this expression into Eq. 11.22 and integrating over the entire contour gives

$$\oint \frac{T}{2\Omega G t}\,ds = \oint \varphi'(x)\,\mathbf{r}\cdot\mathbf{n}\,ds \tag{11.23}$$

Finally, noting Eq. 11.19, we can write

$$\frac{T}{2\Omega G}\oint \frac{ds}{t(s)} = 2\Omega\,\varphi'(x) \tag{11.24}$$

which can be written in the form

$$T = G\hat{J}\,\varphi'(x) \tag{11.25}$$

if we define the torsion constant as

$$\hat{J} = \frac{4\Omega^2}{S} \tag{11.26}$$

where Ω is the area enclosed by the middle line of the cross section and S is defined as

$$S = \oint \frac{ds}{t(s)} \tag{11.27}$$

where $t(s)$ is the thickness of the cross section and s measures distance along the perimeter. The integral is a line integral around the cross section. If the thickness is constant, then S is simply the length of the contour divided by the thickness, i.e., b/t.

Example 11.13
Find the torsional constant $\hat{J}$ of a rectangular tube of height h, width $w = h/2$, and thickness t, shown in Fig. 11.33. Consider the height and width to be the dimensions of the middle line. Compare the result with the polar moment of inertia.

Fig. 11.33 *Rectangular tube example.* Thin-walled rectangular tube of height h, width w, and uniform thickness t

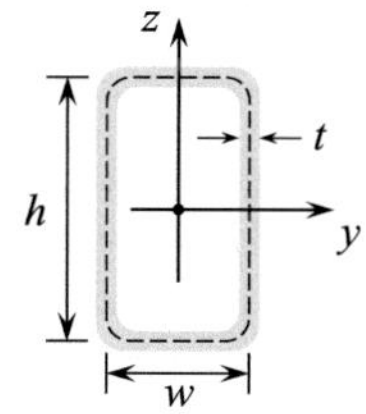

Solution. The perimeter length of the middle line is $b = 2h + 2w = 3h$. The area enclosed by the middle line is $\Omega = wh = 0.5h^2$. Thus, the torsional stiffness is $G\hat{J}$ with

$$\hat{J} = \frac{4\Omega^2}{S} = \frac{4(0.5h^2)^2}{3h/t} = \tfrac{1}{3}h^3 t$$

This value can be compared to the polar moment of inertia of the cross section relative to the centroid, which is at the center. Assuming that the thickness is small compared with the other dimensions, we can compute

$$\begin{aligned} J &= \int_A r^2\, dA = \oint (y^2 + z^2)\, t\, ds \\ &= 2\int_{-h/2}^{h/2} \left(\tfrac{1}{4}w^2 + z^2\right) t\, dz + 2\int_{-w/2}^{w/2} \left(y^2 + \tfrac{1}{4}h^2\right) t\, dy \\ &= 2t\left[\tfrac{1}{4}w^2 z + \tfrac{1}{3}z^3\right]_{-h/2}^{h/2} + 2t\left[\tfrac{1}{3}y^3 + \tfrac{1}{4}h^2 y\right]_{-w/2}^{w/2} \end{aligned}$$

We have noted that the y value for the two vertical pieces in the first integral is $w/2$ and the z value for the two horizontal pieces in the second integral is $h/2$. Substituting the limits of integration gives the final result

$$J = \tfrac{1}{2}t\left(w^2 h + h^2 w\right) + \tfrac{1}{6}t\left(h^3 + w^3\right)$$

Using $w = \frac{1}{2}h$ for the present example, the polar moment of inertia relative to the centroid is

$$J = \tfrac{1}{2}t\left(\left(\tfrac{1}{2}h\right)^2 h + h^2\left(\tfrac{1}{2}h\right)\right) + \tfrac{1}{12}t\left(h^3 + \left(\tfrac{1}{2}h\right)^3\right) = \tfrac{9}{16}h^3 t$$

The ratio $J/\hat{J} = 1.69$ shows how important the warping effect is to torsion of thin-walled closed cross sections.

11.5 Summary

This chapter focused on the deformation and stress in bars subjected to pure torsion. The kinematic hypothesis that plane sections remain plane after deformation was used to derive the equations governing bars with circular cross sections. The resulting equations were analogous to the axial bar problem, with torque replacing the axial force and rotation about the axis replacing displacement along the axis.

The second part of the chapter considered the problem of torsion of bars with non-circular cross sections. We found that the plane-sections kinematic hypothesis does not lead to accurate results, but the assumption that the cross section rotates without distorting its shape and warps out of its own plane does. We developed approximations for the important special cases of open and closed thin-walled cross sections, which are common in structural applications. While the treatment of torsion in this chapter is far from complete, it is evident that applications using thin-walled open sections (e.g., I-beams in a framed structure) should minimize torsion where possible. On the other hand, if transmitting torque is the main purpose (e.g., in a transmission shaft), then closed thin-walled sections or solid sections are more appropriate and circular sections are optimal.

11.6 Problems

11.1 A tapered bar $\mathcal{AB}$ of solid circular cross section is twisted by end torques of magnitude T_o. The diameter of the bar varies linearly from d_A at the left end to d_B at the right end. Derive the expression for the angle of twist relative to its value at the left end of the tapered bar. For a ratio $d_B/d_A = 1.5$, what is the total relative twist between the ends of the tapered bar compared with prismatic bars of diameters d_A and d_B made of the same material?

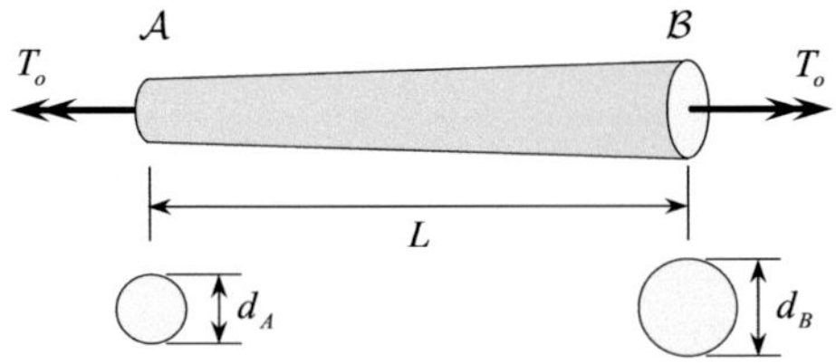

11.2 A solid circular bar $\mathcal{ABCD}$ is fixed at both ends and is acted upon by two torques T_o and $2T_o$ at $\mathcal{B}$ and $\mathcal{C}$, respectively. Determine the support reactions at $\mathcal{A}$ and $\mathcal{D}$ and compute the maximum shear stress in each segment. Find the angle of twist at points $\mathcal{B}$ and $\mathcal{C}$. Find where the maximum value of the angle of twist occurs and what the value is. Plot the distribution of the internal torque and angle of twist along the axis.

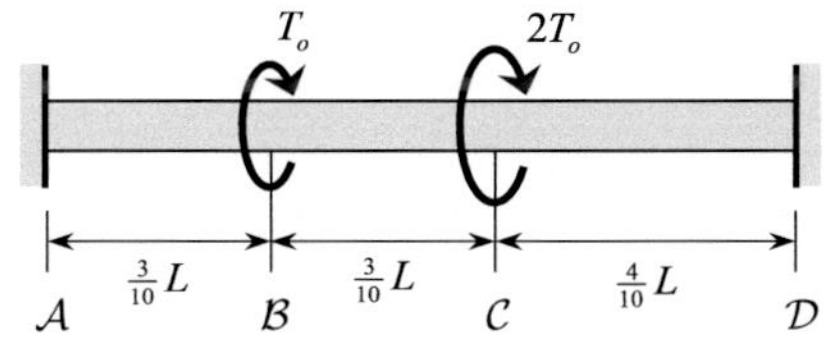

11.3 A prismatic bar $\mathcal{AB}$ of length L is loaded by distributed torque $t(x)$ that varies linearly from zero at end $\mathcal{A}$ to a maximum value t_o at end $\mathcal{B}$. The bar is fixed at both ends and made of two materials. The inner material has shear modulus $2G_o$ and radius R, and the outer material has shear modulus G_o and radius $1.5R$, where G_o and R are reference values. Find the reaction torques at $\mathcal{A}$ and $\mathcal{B}$ and the angle of twist φ as a function of x. Express results in terms of t_o, L, G_o, and R. Plot the torque and angle of twist along the bar. Find the location and value of the maximum shear stress in the bar.

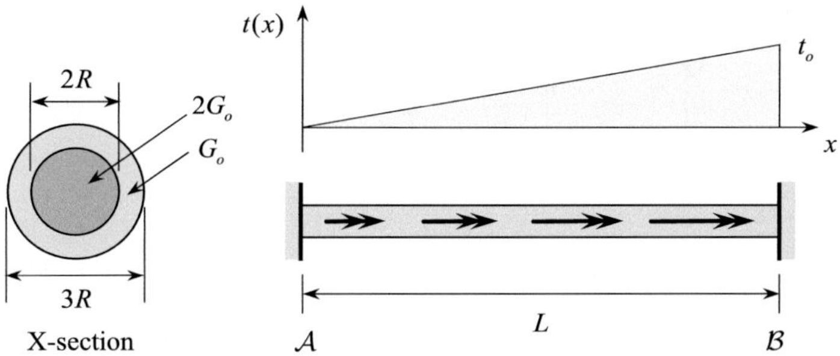

11.4 A circular bar $\mathcal{ABCD}$ of length L has three segments, and the middle segment is loaded by a constant distributed torque of magnitude t_o. The bar is fixed at both ends. The two end segments have a diameter of d and the middle segment has a diameter of $2d$. All three segments have shear modulus G. Find the reacting torques at $\mathcal{A}$ and $\mathcal{D}$. Find the internal torque and angle of twist along the bar. Find the maximum shear stress in the bar. Express results in terms of t_o, L, G, and d.

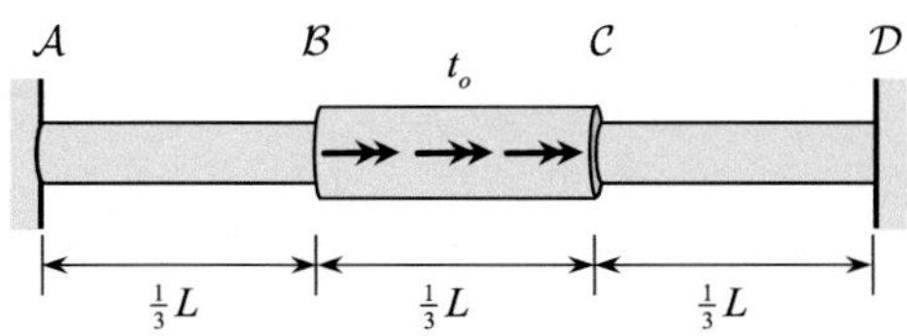

11.5 A rod of length L is to be twisted by torques T_o at the ends until the difference in the angle of rotation between the ends of the rod is φ_o. If the allowable shear strain in the rod is γ_a, find the minimum allowable diameter d of the rod. Find the shear stress associated with the allowable shear strain. Evaluate the problem for the specific values of $L = 18$ in, $\varphi_o = 3.0°$, $\gamma_a = 0.0006$ rad, and $G = 5800$ ksi.

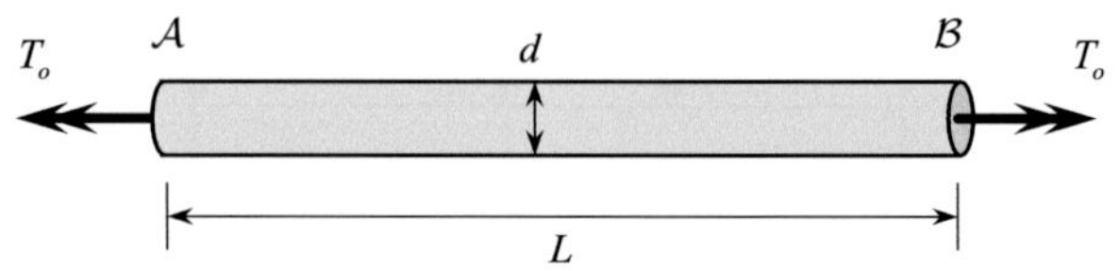

11.6 A stepped shaft $\mathcal{ABCDE}$ consists of solid and hollow circular segments. There are concentrated torques applied at $\mathcal{A}$, $\mathcal{B}$, and $\mathcal{C}$, of magnitudes T_o, $2T_o$, and $4T_o$, respectively, in the directions shown. The shaft is fixed at $\mathcal{E}$. Segment $\mathcal{AC}$ has diameter d, $\mathcal{CE}$ has diameter $2d$, and there is a bore in $\mathcal{DE}$ of diameter d. All segments have shear modulus G. Determine the internal torques in each segment along with the maximum shear stress and strains. Plot the shear stress and angle of twist along the axis of the shaft. Express the results in terms of T_o, L, d, and G.

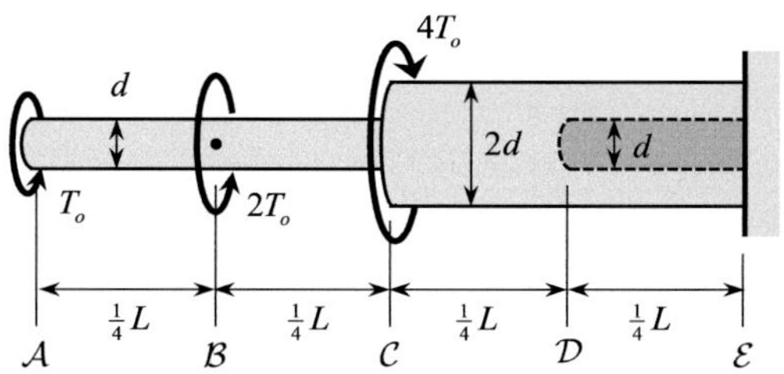

11.7 A bar $\mathcal{AB}$ of length L with solid circular cross section is fixed against rotation at the ends and loaded by a distributed torque $t(x)$ that varies linearly from zero at end $\mathcal{A}$ to a maximum value t_o at end $\mathcal{B}$. Find the reaction torques at the fixed ends. Determine the expression for the angle of twist and find the maximum angle of twist. Plot the response fields.

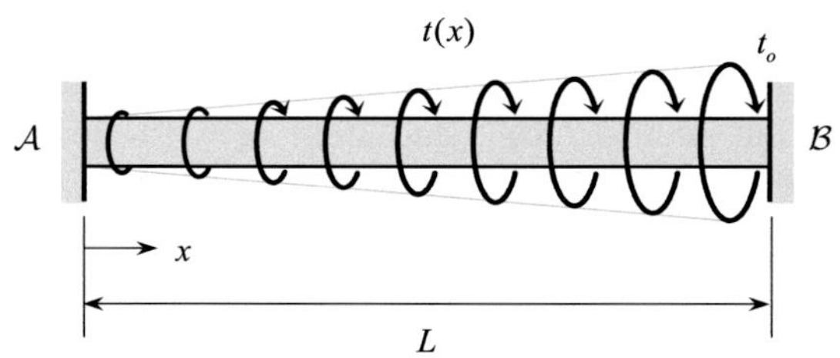

11.8 A fixed-fixed prismatic bar of length L is loaded by a constant distributed torque t_o along its entire length. At $\mathcal{B}$ a concentrated torque of magnitude T_o is also applied. The bar is made of two materials. The inner material has shear modulus G_1 and diameter d_1, and the outer material has shear modulus G_2 and diameter d_2. If $G_1 = 2G_2$ and $d_1 = 0.75d_2$, find the maximum shear stress just to the left of point $\mathcal{B}$. Solve for the reacting torques at $\mathcal{A}$ and $\mathcal{C}$ and find the angle of twist as a function of x. Express results in terms of T_o, t_o, L, G_1, and d_1.

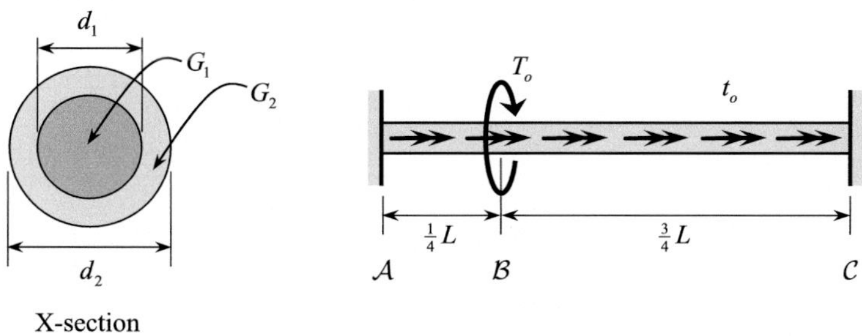

11.9 A stepped shaft $\mathcal{ABC}$ is made with a bar of diameter d and shear modulus G_1 that extends the full length L. Over the second half of the shaft, the inner bar is bonded to a casing of diameter $2d$ with shear modulus G_2. The materials used have $G_1 = 5G_o$ and $G_2 = G_o$, where G_o is a reference value. The composite shaft is fixed against rotation at both ends and has a torque of T_o applied at $\mathcal{B}$ in the direction shown. Compute and plot the internal torque $T(x)$ and the rotation $\varphi(x)$ along the length of the shaft. Find the maximum shear stress. Express the results in terms of T_o, L, d, and G_o.

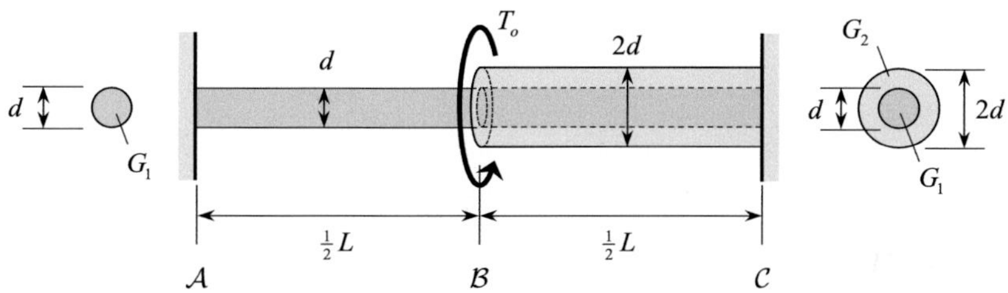

11.10 A prismatic bar $\mathcal{AB}$ of length L is fixed against rotation at both ends and is loaded by a sinusoidally distributed torque,

$$t(x) = t_o \sin(\pi x/L)$$

along its length. The bar is composed of two materials, the inner part with diameter d_1 and shear modulus G_1 and outer part with diameter d_2 and shear modulus G_2. If $d_1 = 0.75d_2$ and $G_1 = 5G_2$, find the maximum shear stress in the bar. Solve for the reacting torques at $\mathcal{A}$ and $\mathcal{B}$ and find the torque and angle of twist as a function of x. Express results in terms of t_o, L, G_2, and d_2.

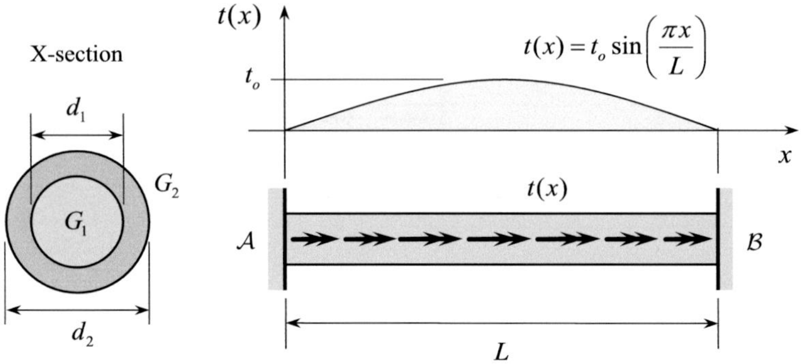

11.11 A prismatic bar $\mathcal{AB}$ of length L is fixed against rotation at both ends and is loaded by a sinusoidally distributed torque along its length, as shown. The bar is composed of two materials. The inner part has diameter $d_1 = h$ and shear modulus $G_1 = 3G_o$ and the outer part has diameter $d_2 = 2h$ and shear modulus $G_2 = G_o$, where h and G_o are given constant values. Solve for the reacting torques at $\mathcal{A}$ and $\mathcal{B}$ and find the torque and angle of twist as a function of x. Find the maximum shear stress in the bar. Express results in terms of t_o, L, G_o, and h.

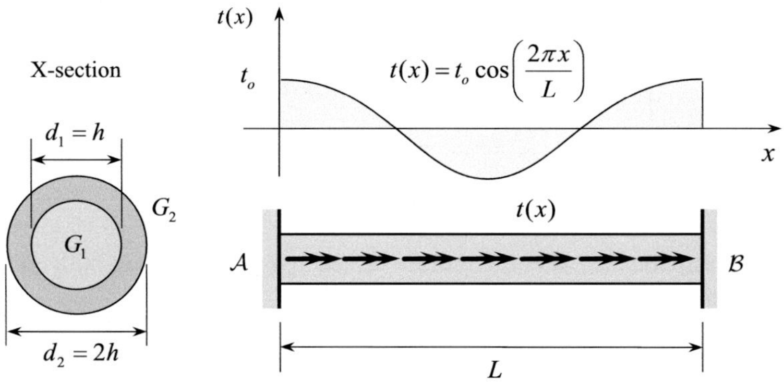

11.12 The bar $\mathcal{ABC}$ of length $2L$ is fixed at both ends and consists of two segments with the cross sections shown. Both segments have shear modulus G. The two parts are welded together at point $\mathcal{B}$ where a point torque T_o is applied. Determine the internal torques in each bar along with the maximum shear stresses. Plot the torques and angle of twist along the axial direction of the bar. Express results in terms of T_o, L, G, and a.

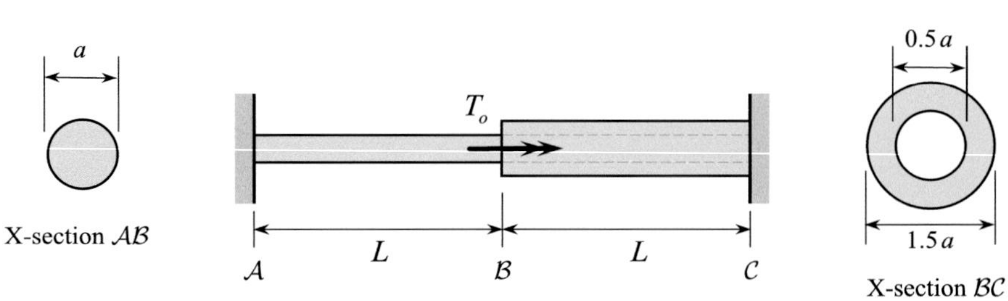

11.13 The bar $\mathcal{ABCDE}$ of length $4L$ is fixed at both ends and consists of two segments with the cross sections shown. The bar has radii and shear moduli as shown in the sketch, with $G_1 = 369G_o$ and $G_2 = 256G_o$, where G_o is a reference value. The two parts are continuous at point $\mathcal{C}$. Point torques of magnitudes T_o, $2T_o$, and T_o are applied at $\mathcal{B}$, $\mathcal{C}$, and $\mathcal{D}$, respectively. Determine the internal torque and angle of twist in the bar. Find the maximum shear stress. Plot the torques and angle of twist along the bar. Express results in terms of T_o, L, G_o, and a.

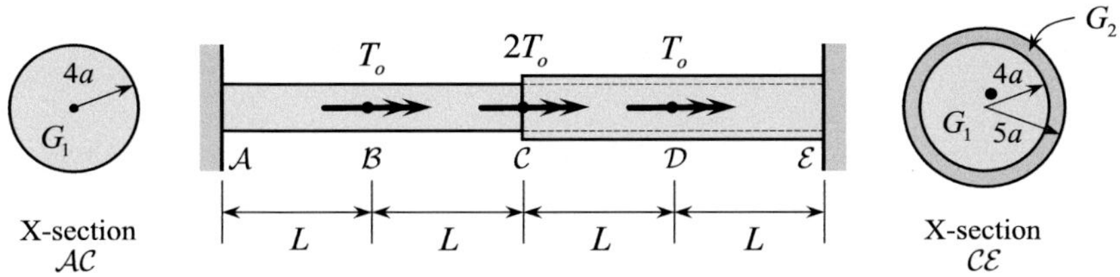

Chapter 12
Strength and Stability

The central theme of this book is the mechanics of *elastic* deformable solids, built on the three-legged stool of equilibrium, kinematics, and constitutive behavior of materials. What we have found is that with these three elements, along with boundary conditions, we can form a complete mathematical model of how solid bodies behave. To solve practical problems, we have linearized the equations. When deformations are assumed to be small, we can justify writing equilibrium equations in the reference configuration and the nonlinear components of the strain-displacement relationships can be neglected without introducing significant modeling error. We have also used linear constitutive models. The linear equations allow us to see how forces flow through solid bodies and what deformations accrue in the process. However, the equations do not allow us to explore the *strength* of the solid body or to understand the consequences of linearizing the equations. This chapter is a very brief introduction to the strength and stability of solids, addressing the capacity of systems, not just their linear response.

There are two basic features that limit the capacity of a solid body. The first is the material itself. Real materials cannot sustain ever-increasing levels of stress and strain. The etiology of material failure is complex, including phenomena from material damage like yielding, which is common in metals, to crushing and cracking, which is common for cementitious materials like concrete. At some point, the material transitions to a different mode of behavior, causing the relationship between stress and strain to depart from linearity. Analysis of response with nonlinear material models is much more complex than for linear models. Also, there are *many* models that have been developed to imitate different phenomena in different materials. In this chapter we will take a look at one of those models, elastoplasticity, to see the impact of material nonlinearity on the theories of the previous chapters.

The second feature that limits the capacity of solid bodies is *instability*. In our linear models, we write the equations of equilibrium in the reference configuration. However, equilibrium must hold in the deformed configuration (ask anyone who lifts weights). As a system deforms, the moment arms of some of the loads can increase to

K. D. Hjelmstad, *Mechanics of Deformable Solids: An Introduction*,
https://doi.org/10.1007/978-3-031-89873-0_12

the point of instability. This phenomenon is central to buckling of beams subjected to compressive axial loads. Buckling can occur even for materials in their elastic range of behavior. We will show how buckling instability occurs in beams based upon the equations we have derived previously.

Of course, buckling and material failure can combine to create additional failure states. This chapter is only an invitation to these subjects. The aim is to bring them up, not to exhaustively explore them. Thus, this chapter is an invitation to further study of the mechanics of deformable solids.

12.1 Material Strength

The simplest model of nonlinear constitutive behavior is *elastoplasticity*, illustrated in Fig. 12.1. In uniaxial tension, Hooke's law reduces to $\sigma = E\varepsilon$. Stress is proportional to strain, and Young's modulus E is the constant of proportionality. The idealized elastoplastic material reaches a level of stress σ_o upon which inelastic straining commences without any increase in stress. In a real material, this phenomenon might be caused by slipping of the crystal lattice structure at the microscopic level. On the macro scale, we observe the response shown in Fig. 12.1. If the load causing the stress is removed, the elastoplastic material will unload along a line parallel to the original one with slope E. Thus, when it reaches zero stress, a residual strain ε_p, called the *plastic strain*, remains. Unlike elastic materials, inelastic materials do not return to their original shape once loads are removed.

Uniaxial tension is a very special case of stress and strain. What happens if the body is in a multiaxial state of stress? For the elastic response, Hooke's law, which we express as

$$\mathbf{S} = \lambda\,\mathrm{tr}(\mathbf{E})\mathbf{I} + 2\mu\,\mathbf{E}$$

governs the response (see Sect. 6.1). Recall that $\mathbf{S}$ is the stress tensor and $\mathbf{E}$ is the strain tensor, with λ and μ being elastic material constants. This equation governs the evolution of the six independent components of stress as a function of the six independent components of strain. Will the material yield when σ_{xx} reaches the value σ_o? Will the other components influence when yielding will happen? One way to characterize yielding in a multiaxial state is to postulate that when a certain combination of the stress tensor components reaches a fixed value, then yielding will

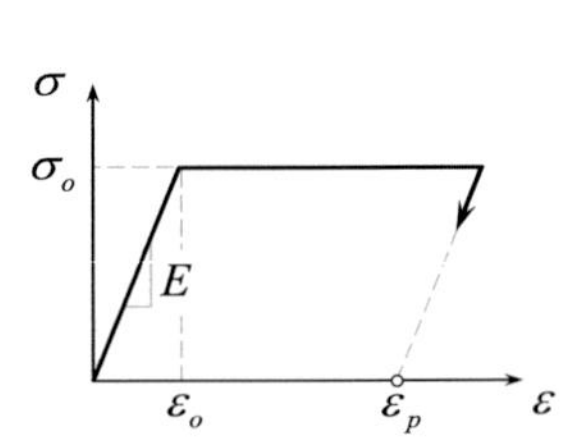

Fig. 12.1 *Elastoplasticity.* Stress-strain curve for a uniaxial tension test showing a perfectly elastoplastic response. The yield stress is σ_o and the yield strain is ε_o. Young's modulus is E

commence. The function that describes this combination is called the *yield function*, $\varphi(\mathbf{S})$. The material will yield when the stress state $\mathbf{S}$ reaches the *yield surface*, i.e., the condition

$$\varphi(\mathbf{S}) = k^2 \tag{12.1}$$

where the constant k, a material property, is the single measure of the level at which the material will yield. All stress states that satisfy $\varphi(\mathbf{S}) < k^2$ are elastic. It is reasonable to expect that k is related to σ_o from Fig. 12.1, because among the infinite possible multiaxial states, uniaxial stress is one of them. Hence, the yield criterion should work for that case, too.

The specific nature of the yield function depends upon the material. For isotropic elastoplastic materials, one would expect the yield surface to also be isotropic. The orientation of the coordinate system that defines the stress components (e.g., σ_{xx} and σ_{xy}) should not matter. We can assure that by having the yield function depend only on the *invariants* of the stress tensor (see Sect. 1.2.3). For example, the *von Mises*[1] yield function, which has been shown to represent metal plasticity well, is

$$\varphi(\mathbf{S}) = \tfrac{1}{2}\mathcal{I}_2(\mathbf{S}) - \tfrac{1}{6}\mathcal{I}_1^2(\mathbf{S}) \tag{12.2}$$

where $\mathcal{I}_1(\mathbf{S}) = \text{tr}(\mathbf{S})$ and $\mathcal{I}_2(\mathbf{S}) = \text{tr}(\mathbf{S}^2)$ are the first two primary invariants of the stress tensor (see Eq. 1.40). This function can also be written as

$$\varphi(\mathbf{S}) = \tfrac{1}{2}\,\mathcal{I}_2(\mathbf{S}')$$

where

$$\mathbf{S}' = \mathbf{S} - \tfrac{1}{3}\mathcal{I}_1(\mathbf{S})\mathbf{I}$$

is the so-called *deviatoric stress*, which is defined as it is because $\text{tr}(\mathbf{S}') = 0$. The deviatoric stress, therefore, primarily concerns shear stresses. The yield function can also be expressed in terms of the principal stresses σ_1, σ_2, and σ_3 by noting that

$$\mathcal{I}_2(\mathbf{S}') = \tfrac{1}{3}\big[\,(\sigma_1 - \sigma_2)^2 + (\sigma_2 - \sigma_3)^2 + (\sigma_3 - \sigma_1)^2\,\big]$$

The von Mises yield function treats tension and compression the same. There are other yield functions that are useful to define the capacity for materials that behave differently in tension and compression. For example, the *Mohr-Coulomb* function is often used for granular materials, and the *Drucker-Prager* function is often used for concrete.

We can develop a better understanding of the meaning of the yield level parameter k in Eq. 12.1 by looking at a couple of special stress states. First, consider a state of pure shear. The stress tensor components and the components of $\mathbf{S}^2$ are, respectively,

---

[1] Richard von Mises (1883–1953) was an Austrian scientist and mathematician who made important contributions to solid mechanics.

$$\mathbf{S} = \begin{bmatrix} 0 & \tau & 0 \\ \tau & 0 & 0 \\ 0 & 0 & 0 \end{bmatrix}, \qquad \mathbf{S}^2 = \begin{bmatrix} \tau^2 & 0 & 0 \\ 0 & \tau^2 & 0 \\ 0 & 0 & 0 \end{bmatrix}$$

Thus, $\mathcal{I}_1(\mathbf{S}) = \text{tr}(\mathbf{S}) = 0$ and $\mathcal{I}_2(\mathbf{S}) = \text{tr}(\mathbf{S}^2) = 2\tau^2$. Using these values in Eq. 12.2, and calling τ_o the value of the shear stress when the material yields in pure shear, we find that the material will yield when

$$\varphi(\mathbf{S}) = \tfrac{1}{2}\left(2\tau_o^2\right) - \tfrac{1}{6}\left(0\right)^2 = \tau_o^2 = k^2$$

From this special case, we can see that $k = \tau_o$, i.e., it is the yield level for a state of pure shear.

Now, consider the case of uniaxial stress. The stress tensor components and the components of $\mathbf{S}^2$ are, respectively,

$$\mathbf{S} = \begin{bmatrix} \sigma & 0 & 0 \\ 0 & 0 & 0 \\ 0 & 0 & 0 \end{bmatrix}, \qquad \mathbf{S}^2 = \begin{bmatrix} \sigma^2 & 0 & 0 \\ 0 & 0 & 0 \\ 0 & 0 & 0 \end{bmatrix}$$

Thus, $\mathcal{I}_1(\mathbf{S}) = \text{tr}(\mathbf{S}) = \sigma$ and $\mathcal{I}_2(\mathbf{S}) = \text{tr}(\mathbf{S}^2) = \sigma^2$. Using these values in Eq. 12.2, and calling σ_o the value of the normal stress when the material yields in uniaxial tension (shown in Fig. 12.1), we find that the material will yield when

$$\varphi(\mathbf{S}) = \tfrac{1}{2}\left(\sigma_o^2\right) - \tfrac{1}{6}\left(\sigma_o\right)^2 = \tfrac{1}{3}\sigma_o^2 = k^2$$

Thus, k also satisfies $k = \sigma_o/\sqrt{3}$ in uniaxial tension. It is the same k in both cases, so $\sigma_o = \sqrt{3}\,\tau_o$. In other words, the uniaxial yield stress is $\sqrt{3}$ times the value of the shear yield stress. The value of the material parameter k could be determined from either a uniaxial tension test or a pure shear test (which could be executed, for example, on a thin tube in a torsion testing machine).

Hydrostatic pressure presents an interesting and informative case for the von Mises yield function. The stress tensor for a state of hydrostatic pressure is $\mathbf{S} = p\,\mathbf{I}$, where p is the pressure. The square of the stress tensor is $\mathbf{S}^2 = p^2\mathbf{I}$. The first and second primary invariants are $\mathcal{I}_1(\mathbf{S}) = 3p$ and $\mathcal{I}_2(\mathbf{S}) = 3p^2$. Substituting these into Eq. 12.2 gives $\varphi(\mathbf{S}) = 0$, *independent* of the pressure p. What this result means is that a material whose yielding behavior is governed by the von Mises yield function cannot yield in a state of hydrostatic pressure, which is the only stress state that has no shear stress on any plane (in three dimensions, the Mohr circle is just a point on the σ axis).

Upon yielding, the elastoplastic material strains without increasing the value of $\varphi(\mathbf{S})$. A complete model of inelasticity must also include a rule that describes how the plastic strains evolve in appropriate proportion. For example, in uniaxial tension you would not expect shear strains to develop, and in pure shear you would not expect normal strains to develop. One such rule that accomplishes this goal is the *normal flow rule*, which says that the inelastic strains will be proportional to the normal to the yield surface. Because the yield surface concerns stress and not strain, this rule is also called an *associated* flow rule. It associates strain with stress.

12.2 Effects of Inelasticity

The response shown in Fig. 12.1 applies to stress at a point. In applications like torsion and beam bending, the stresses vary throughout the body. The point with the highest stress will yield first. Because that point cannot increase its stress, the internal stresses redistribute upon further deformation. We can make some analytical headway to illustrate the impact of progressive yielding in situations involving states of stress in pure torsion and beam bending. We will explore those two cases through specific examples.

12.2.1 Torsion

As a first example of the consequences of yielding, consider the problem of torsion of a circular bar with shear modulus G and yield shear stress of τ_o, shown in Fig. 12.2. At some level of deformation, characterized by the rate of twist φ', the particles within the radius $r < r_o$ are still elastic while the particles with $r \geq r_o$ have yielded and, therefore, are at the fixed shear stress τ_o. We can express the torque as a function of the rate of twist to see how progressive yielding affects the response.

Recall that for the elastic response in torsion of a solid circular bar of radius R made of one material, the equation governing the shear strain versus rate of twist (Eq. 11.1), shear stress versus strain ($\tau = G\gamma$), and torque versus rate of twist (Eq. 11.7) are, respectively,

$$\gamma = r\varphi', \qquad \tau = Gr\varphi', \qquad T = GJ\varphi' \tag{12.3}$$

where γ is the shear strain, r is the radial distance from the center, φ' is the rate of twist about the axis of the bar, G is the shear modulus, T is the net torque, and $J = \pi R^4/2$ is the polar moment of inertia of the cross section. The first equation

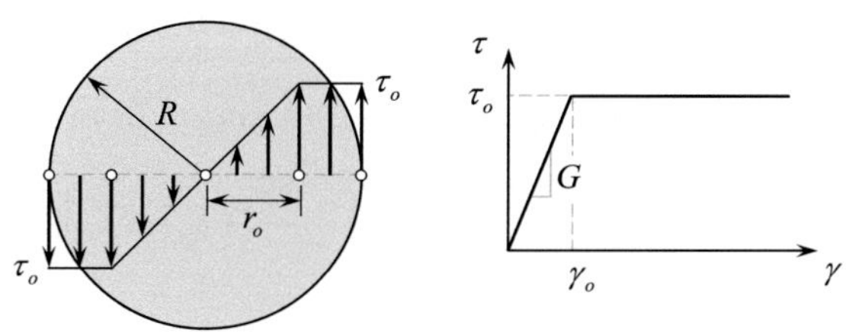

Fig. 12.2 *Elastoplastic torsion.* Circular cross section showing the shear stress along the axis. For $r < r_o$ the response is elastic. For $r \geq r_o$ the shear stress is constant at τ_o

holds independent of the state of stress, the second equation holds only for stresses and strains in the elastic region, and the third equation holds only if the entire cross section is elastic.

The point of *first yield* is the instant in the loading process where the most highly stressed fiber in the cross section just reaches the yield stress τ_o. Because every fiber at a given radius has the same stress, all the fibers on the outer surface of the bar reach yield at the same time. We can determine the rate of twist at first yield, φ'_o, by observing that the fiber with $r = R$ has a stress just equal to $\tau = \tau_o$. Thus, from Eq. 12.3(b), we can write

$$\tau_o = GR\varphi'_o \quad \rightarrow \quad \varphi'_o = \frac{\tau_o}{GR} \tag{12.4}$$

Now, consider the state shown in Fig. 12.2. The yielding has progressed to the radius r_o. From Eq. 12.3(b) we can compute the rate of twist in terms of r_o or we can compute r_o in terms of the rate of twist. To wit,

$$\varphi' = \frac{\tau_o}{Gr_o} \quad \rightarrow \quad r_o = \frac{\tau_o}{G\varphi'} \tag{12.5}$$

We can compute the net torque from its definition (Eq. 11.2), as follows:

$$\begin{aligned} T &= \int_A \tau\, r\, dA \\ &= \int_0^{2\pi}\int_0^{r_o} \tau\, r^2\, dr\, d\theta + \int_0^{2\pi}\int_{r_o}^{R} \tau_o\, r^2\, dr\, d\theta \\ &= 2\pi \int_0^{r_o} (Gr\varphi')\, r^2\, dr + 2\pi \int_{r_o}^{R} \tau_o\, r^2\, dr \end{aligned}$$

The second line separates the elastic part from the plastic part, and the third line substitutes $\tau = Gr\varphi'$ in the first term because those fibers are still elastic. Integrating and substituting the limits of integration gives

$$T = \tfrac{1}{2}\pi G r_o^4 \varphi' + \tfrac{2}{3}\pi\tau_o\left(R^3 - r_o^3\right) \tag{12.6}$$

Let us denote the torque at first yield as T_o and compute it from Eq. 12.6 by noting that $r_o = R$ and the rate of twist is φ'_o at this state. Also, from Eq. 12.4, we have $G\varphi'_o = \tau_o/R$. Thus,

$$T_o = \tfrac{1}{2}\pi G R^4 \varphi'_o = \tfrac{1}{2}\pi\tau_o R^3 \tag{12.7}$$

We can simplify Eq. 12.6 by substituting r_o from Eq. 12.5 and factoring out T_o. Doing this gives

$$T = T_o\left[\tfrac{4}{3} - \tfrac{1}{3}\left(\varphi'_o/\varphi'\right)^3\right], \qquad \varphi' > \varphi'_o \tag{12.8}$$

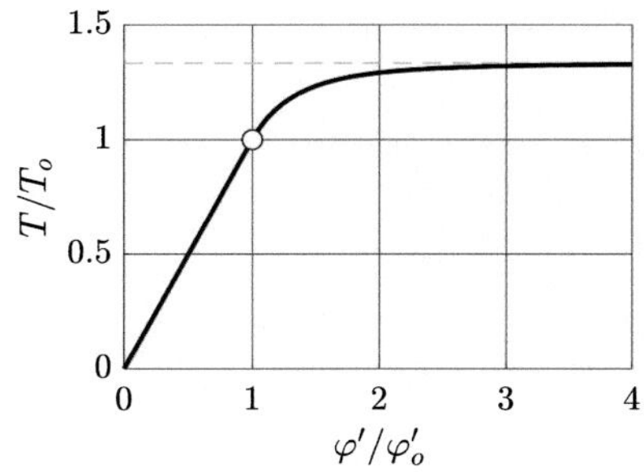

Fig. 12.3 *Elastoplastic torsion.* Response curve for elastoplastic torsion of a solid circular cross section. The yield torque is T_o and the rate of twist at first yield is φ'_o

When the bar is entirely elastic, the torque is $T = GJ\varphi'$. We can write the polar moment of inertia as $J = T_o R/\tau_o$, and we know that $\varphi'_o = \tau_o/GR$. Thus, we can write $GJ = T_o/\phi'_o$. Using this expression, Eq. 12.3(c) can be written as

$$T = T_o(\varphi'/\varphi'_o), \qquad \varphi' \le \varphi'_o \tag{12.9}$$

These results are plotted in Fig. 12.3. The abscissa is the rate of twist normalized by the rate of twist at first yield, and the ordinate is the torque normalized by the torque at first yield. The point of first yield is shown with the open circle. The response is nonlinear after yielding, and the torque increases with rate of twist. However, the slope of the response curve flattens more and more as the rate of twist increases, asymptotically approaching a value $T_u = 1.33T_o$, which is the *ultimate strength* of the bar. In other words, the bar can carry 33% additional torque beyond the point of first yield. While yielding is a change in the mode of behavior, it is *not* the ultimate torsional strength of the bar, T_u is. The bar gets additional strength due to the kinematics of deformation. The bar will fail when the shear strain at the outside fiber reaches the breaking strain of the material. The torque approaches the ultimate torque asymptotically, so the bar will fail at a torque less than T_u. However, Fig. 12.3 shows that by a value of $\varphi' \approx 3\varphi'_o$, the bar has very nearly reached its full capacity.

12.2.2 Beam Bending

As a second example of the consequences of yielding, consider the problem of pure bending (i.e., there is no shear or axial force) of a rectangular bar of depth h and width b, shown in Fig. 12.4. The material has Young's modulus E and yield stress σ_o, with elastoplastic response shown in Fig. 12.1. At some level of deformation, characterized by the curvature κ,[2] the particles with $|z| < z_o$ are still elastic, while the particles with $|z| \ge z_o$ have yielded and, therefore, are at the fixed normal stress

[2] In the chapter on beams, we referred to the curvature as κ_o to correspond with ε_o as a strain resultant. Here we will refer to curvature simply as κ and use κ_o to indicate the curvature at first yielding. This revised notation will parallel the notation in the previous section.

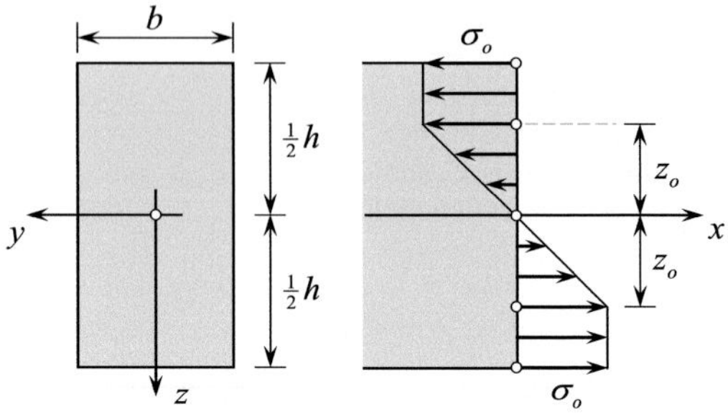

Fig. 12.4 *Elastoplastic beam bending.* Rectangular cross section showing the normal stress at a cut. For $|z| < z_o$ the response is elastic. For $|z| \geq z_o$ the normal stress is constant at σ_o (compressive on the top and tensile on the bottom)

σ_o (or $-\sigma_o$ for negative values of z). We can calculate the response curve with bending moment as a function of curvature to see how progressive yielding affects the response.

Recall that for the elastic response of a beam in planar bending, the equations governing strain and curvature (Eq. 7.13), the normal stress formula in absence of net axial strains (Eq. 7.24), and the bending moment-curvature relationship (Eq. 7.23) for a prismatic beam made of one material are, respectively,

$$\varepsilon = z\kappa, \qquad \sigma = E\kappa z, \qquad M = EI\kappa \tag{12.10}$$

where E is Young's modulus, κ is the curvature, and $I = bh^3/12$ is the moment of inertia of the cross section relative to the centroid. The first equation holds independent of the state of stress, the second equation holds only for stresses and strains in the elastic region, and the third equation holds only if the entire cross section is elastic.

We can determine the curvature at first yield, κ_o, by observing that the fiber with $z = h/2$ has a stress just equal to $\sigma = \sigma_o$. Thus, from Eq. 12.10(b), we can write

$$\sigma_o = \tfrac{1}{2}E\kappa_o h \qquad \rightarrow \qquad \kappa_o = \frac{2\sigma_o}{Eh} \tag{12.11}$$

Now, consider the state shown in Fig. 12.4. The yielding has progressed to the depth z_o. From Eq. 12.10(b), we can compute the curvature κ in terms of z_o, or we can compute z_o in terms of the curvature. To wit,

$$\kappa = \frac{\sigma_o}{Ez_o} \qquad \rightarrow \qquad z_o = \frac{\sigma_o}{E\kappa} \tag{12.12}$$

We can compute the net bending moment from its definition (Eq. 7.14), as follows:

$$\begin{aligned} M &= \int_A \sigma\, z\, dA \\ &= 2\int_0^{z_o} \sigma\, z\, b\, dz + 2\int_{z_o}^{h/2} \sigma_o\, z\, b\, dz \\ &= 2\int_0^{z_o} \big(E\kappa z\big)\, z\, b\, dz + 2\int_{z_o}^{h/2} \sigma_o\, z\, b\, dz \end{aligned}$$

The factor of two in the second step recognizes the antisymmetry in the integrals. In the third step, we have substituted the elastic stress relationship $\sigma = Ez\kappa$ from Eq. 12.10(b) in the first term because those fibers are still elastic. Carrying out the integrals and substituting the limits of integration gives

$$M = \tfrac{2}{3}Ebz_o^3\kappa + \sigma_o b\left(\tfrac{1}{4}h^2 - z_o^2\right) \tag{12.13}$$

which relates the net bending moment M to the curvature κ. This equation can be simplified. Let us denote the bending moment at first yield as M_o, which we can compute from Eq. 12.13 by noting that $z_o = h/2$ and the curvature is κ_o at this state. From Eq. 12.11, we have $E\kappa_o = 2\sigma_o/h$. Thus,

$$M_o = \tfrac{2}{3}Eb\left(\tfrac{1}{2}h\right)^3\kappa_o = \tfrac{1}{6}\sigma_o bh^2 \tag{12.14}$$

Finally, substituting z_o from Eq. 12.12 and factoring out M_o allows us to write Eq. 12.13 in the form

$$M = M_o\left[\tfrac{3}{2} - \tfrac{1}{2}\left(\kappa_o/\kappa\right)^2\right], \qquad \kappa > \kappa_o \tag{12.15}$$

When the bar is entirely elastic, the torque is $M = EI\kappa$. We can write this equation in a different form by noting that $I = M_o h/2\sigma_o$ and $\kappa_o = 2\sigma_o/Eh$, which can be rearranged to give $EI = M_o/\kappa_o$. Using these expressions, the elastic moment versus curvature equation can be written as

$$M = M_o\left(\kappa/\kappa_o\right), \qquad \kappa \leq \kappa_o \tag{12.16}$$

This equation is exactly the same as Eq. 12.10(c). Equation 12.10 makes it easier to see that the bending stiffness is *EI*, whereas in Eq. 12.16, one can see more clearly that the moment M reaches the yield moment M_o when the curvature κ reaches the yield curvature κ_o.

These results are plotted in Fig. 12.5. The abscissa is the curvature normalized by the curvature at first yield and the ordinate is the bending moment normalized by the bending moment at first yield. The point of first yield is shown with the

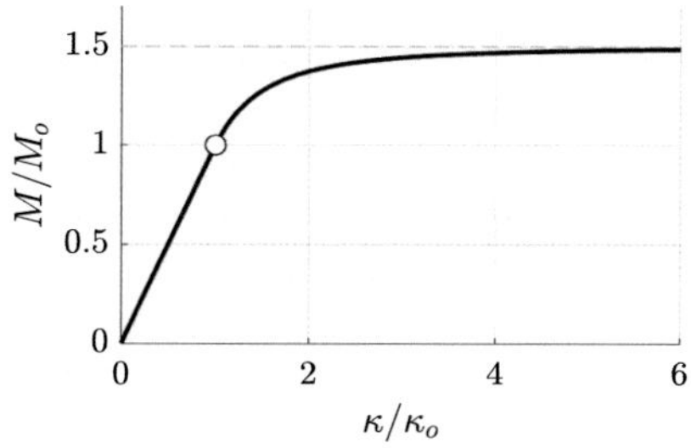

Fig. 12.5 *Elastoplastic beam bending.* Response curve for elastoplastic bending of a solid rectangular cross section. The yield moment is M_o and the curvature at first yield is κ_o

open circle. The response is nonlinear after yielding, and the moment increases with curvature. However, the slope of the response curve flattens more and more as the curvature increases. The response curve is asymptotic to a value $M_u = 1.5M_o$, which is the *ultimate strength* of the beam in pure bending. In other words, the beam can carry 50% additional moment beyond the point of first yield. However, it requires significant deformation to achieve that value. By a curvature of $\kappa \approx 6\kappa_o$, the moment is near its ultimate value. While yielding is a change in the mode of behavior, it does *not* reflect failure of the beam. The beam gets additional strength due to the kinematics of deformation. The beam will fail when the normal strain at the outside fiber reaches the breaking strain of the material.

For beam bending problems that include shear, the shear stresses at the cross section would reduce the amount of normal stress through the yield function in Eq. 12.1. For beams in common applications, the influence of the shear stresses is small, in part because the shear stresses are small compared with the bending stresses and in part because the shear stresses are maximum at the centroid, while the normal stresses are maximum at the extreme fibers. For short beams with high shear, the influence can be important. When axial forces are present, the reduction in moment capacity can be considerable.

12.3 Stability of Static Equilibrium

Throughout the book, equilibrium has played a central role as one of the three legs of the mechanics stool. All the theories we have considered concern systems in static equilibrium. As we learned in Chap. 2, static equilibrium is simply a state where all the forces balance and the system, therefore, has no tendency to accelerate. With no acceleration, the system does not develop velocity and, therefore, it does not move from the static configuration. It turns out that some static equilibrium configurations are stable and some are not. Static stability is best understood through the lens of dynamics.

Not all dynamical systems possess equilibrium configurations. For example, a particle in projectile motion will never find an equilibrium configuration. A system must have *restoring forces* in order to have configurations that are in equilibrium. Gravity provides a restoring force for a pendulum. No matter which side the pendulum swings, gravity acts to recenter it. The static equilibrium position is hanging straight down. Springs provide restoring forces because they pull back when stretched and push back when compressed, resisting motion in both directions. Elastic deformable solids have restoring forces because the material acts to return to its undeformed configuration when stressed. When restoring forces are present, the system tends to oscillate around the equilibrium position if disturbed by a small amount, and the motions eventually die out due to some mechanism like damping. In fact, the most basic *definition of stability* concerns the dynamic response to a small perturbation of the system: If a system is perturbed by a small amount and the response is correspondingly small (small input gives small output), then the system is stable. If the response to a small disturbance is large (small input gives large output), then the system is unstable.

It is possible for a system to have an *unstable* equilibrium configuration. For example, imagine that you have perfect cubic blocks to stack on a perfectly level surface. There will be a height of the stack that, no matter how carefully you place the next block, the tower will collapse. All the weights of the forces align with each other and with the reaction force at the base. The forces and moments sum to zero. And yet the slightest disturbance will cause the system to fail, often dramatically. This is an example of a system that satisfies the equations of equilibrium but is not stable.

In this section we will examine cases for which *conservation of energy* can be used to unravel the mystery of whether or not a system in equilibrium is stable. We will illustrate the concepts using a discrete system with rigid elements and elastic springs. Then, we will connect the ideas to deformable solids by noting that elastically deformed solids store potential energy like the spring in the simple discrete system does. The end result will be a method to find the static buckling loads for beams subjected to large axial forces.

12.3.1 Stability of a Simple System

To warm up to the ideas of static stability theory, consider the idealized system shown in Fig. 12.6a. The bar of length L is rigid and has negligible mass. A mass m with negligible rotational inertia is attached to the top of the bar. The mass has weight mg, where g is the acceleration of gravity. The bar is pinned at the lower end and restrained by an elastic rotational spring of stiffness k. The moment induced in the spring by rotation of the bar is $M = k\theta$, where θ is the rotation of the bar from vertical, as shown in Fig. 12.6b. The angle θ is the only parameter required to characterize the motion of the system. The spring is the only deformable element, providing a restoring force that tries to push the system back to vertical, no matter which direction it moves. The weight of the mass acts to increase θ (if $\theta \neq 0$), using the moment arm $L \sin\theta$.

Based on the free body diagram in Fig. 12.6c, balance of moments about the base shows that the system is in static equilibrium if

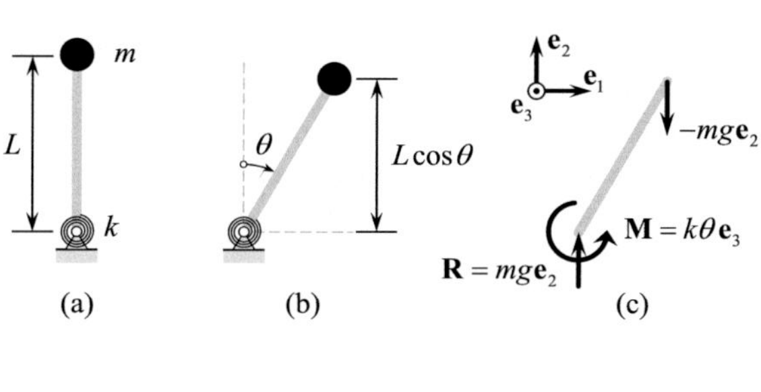

Fig. 12.6 *Simple system.* **a** Rigid bar of negligible mass, length L, and with mass m on top. The bar is pinned at the bottom and restrained by an elastic rotational spring of stiffness k. **b** System rotated by θ from initial position. **c** Free body diagram in displaced configuration

$$k\theta = mgL\sin\theta \tag{12.17}$$

Balance of forces proves that the reaction at the pin is $\mathbf{R} = mg\mathbf{e}_2$, independent of θ. Observe that the straight configuration, i.e., $\theta = 0$, satisfies Eq. 12.17 no matter what the values of the physical parameters (L, m, g, k) are. Equilibrium in the bent configuration $(\theta \neq 0)$ depends on the relative magnitudes of the system's physical parameters. The moment provided by the spring can balance the eccentric force of the mass.

One can show that the potential energy, U, and the kinetic energy, T, of the system in Fig. 12.6 are, respectively,

$$U = \tfrac{1}{2}k\theta^2 + mgL\cos\theta, \qquad T = \tfrac{1}{2}mL^2\dot{\theta}^2 \tag{12.18}$$

The first term in the potential energy is the energy stored in the spring. The second term is the potential energy of the mass m due to gravity (i.e., mgh, where $h = L\cos\theta$ is the height above datum). For a dynamic context, we will consider $\theta(t)$ to vary with time t. Therefore, both the potential and kinetic energies are functions of time, implicitly through the angle $\theta(t)$. If energy is conserved, then the total energy $\mathcal{E} = U + T$ is constant. Therefore, the rate of change of energy must be zero. Computing the time derivative of the total energy gives

$$\begin{aligned} \frac{d\mathcal{E}}{dt} &= k\theta\dot{\theta} - mgL\sin\theta\dot{\theta} + mL^2\dot{\theta}\ddot{\theta} \\ &= \big[k\theta - mgL\sin\theta + mL^2\ddot{\theta}\big]\dot{\theta} = 0 \end{aligned} \tag{12.19}$$

where a dot above a symbol denotes derivative with respect to time. The response is static if $\dot{\theta} = 0$ and $\ddot{\theta} = 0$. If those rates are not zero, then energy is conserved only if

$$k\theta - mgL\sin\theta + mL^2\ddot{\theta} = 0 \tag{12.20}$$

which is the dynamic equation of motion of the system. If θ satisfies the equation of static equilibrium (Eq. 12.17), then the dynamic equation of motion reduces to $\ddot{\theta} = 0$ because the first two terms add to zero. Conversely, if the acceleration is zero, then the dynamic equation of motion reduces to the equation of static equilibrium. Viewed that way, we can see static equilibrium configurations lurking in the dynamic equation of motion.

We can use conservation of energy to gain insight on the stability of static equilibrium. If energy is conserved, then any *increase* in potential energy must be balanced with an equal *decrease* in kinetic energy, and vice versa. If a ball rolls uphill, it gains potential energy, and therefore must lose kinetic energy. It slows down and eventually comes to a stop before rolling back down the hill.

The Taylor series expansion of the potential energy around the configuration θ_o can be written as

$$U(\theta(t)) = U(\theta_o) + t\,\dot{U}(\theta_o) + \tfrac{1}{2}t^2\,\ddot{U}(\theta_o) + \cdots \tag{12.21}$$

where the notation $U(\theta_o)$ means the potential energy at the state θ_o, which we also take to correspond with time $t = 0$. The time derivatives of the potential energy are, using the chain rule for differentiation on Eq. 12.18(a),

$$\begin{aligned} \dot{U}(\theta(t)) &= \big[\,k\theta - mgL\sin\theta\,\big]\dot{\theta} \\ \ddot{U}(\theta(t)) &= \big[\,k\ - mgL\cos\theta\,\big]\dot{\theta}^2 + \big[\,k\theta - mgL\sin\theta\,\big]\ddot{\theta} \end{aligned} \tag{12.22}$$

If θ_o is an equilibrium configuration, then these derivatives are

$$\begin{aligned} \dot{U}(\theta_o) &= \big[\,k\theta_o - mgL\sin\theta_o\,\big]\dot{\theta}_o = 0 \\ \ddot{U}(\theta_o) &= \big[\,k\ - mgL\cos\theta_o\,\big]\dot{\theta}_o^2 \end{aligned} \tag{12.23}$$

where $\dot{\theta}_o$ is the angular velocity at the state θ_o and represents the perturbation to the system that will test the stability of equilibrium. We will give the system a small initial angular velocity $\dot{\theta}_o$ to see if the subsequent response is small. Notice that the first derivative of potential energy vanishes, as does the second term in the second derivative because θ_o satisfies the equilibrium equation. Now, the Taylor series expansion of the potential energy reduces to

$$U(\theta(t)) = U(\theta_o) + \tfrac{1}{2}t^2\,\ddot{U}(\theta_o) + \cdots \tag{12.24}$$

The value $U(\theta_o)$ is the amount of potential energy stored in the system at equilibrium. In the bent position, for example, the spring possesses potential energy and the mass is elevated above the base. Therefore, the system possesses potential energy. The next term in this expansion determines if the potential energy will increase or decrease from $U(\theta_o)$ when the system is perturbed. The terms associated with the $+\cdots$ will be negligible if we restrict our attention to small values of t because they all involve higher powers of t. Since t^2 is always positive, the algebraic sign of the second derivative, $\ddot{U}(\theta_o)$, determines the outcome. Since $\dot{\theta}^2$ is also always positive in Eq. 12.23(b), the term in square brackets, i.e.,

$$A = k - mgL\cos\theta_o \tag{12.25}$$

determines the sign of the second derivative of the potential energy. If it is positive, then the potential energy increases and the kinetic energy decreases. An initial disturbance, $\dot{\theta}_o$, puts the system in motion, thereby endowing it with kinetic energy, and pushes the system out of equilibrium. As the system gains potential energy, it loses kinetic energy and, as a result, slows down and eventually stops, reverses direction, and starts to move back toward the equilibrium position. That is *stable* behavior. On the other hand, if A is negative, then the potential energy decreases and the kinetic energy increases, i.e., the velocity of the system increases, and it moves faster and displaces more. The more it displaces, the faster it goes. That is *unstable* behavior.

Let us examine the stability of the equilibrium configuration of the system in Fig. 12.6. Consider first the straight configuration $\theta_o = 0$. For this state, $A = k - mgL$, which is positive as long as $k > mgL$, and it is negative if $k < mgL$. The condition where $k = mgL$ is called the *critical state*. If you think of the problem through the lens of how much mass can the system carry for given k, L, and g values, then

$$m_{cr} = \frac{k}{gL}$$

is the critical mass. That is the value that divides stable response from unstable response for the straight configuration. The system can have mass m greater than the critical value and still be in equilibrium. However, the equilibrium is unstable, and the slightest perturbation will start a motion that will move the system away from that equilibrium configuration. For values of m less than the critical mass, the system will simply oscillate around the equilibrium configuration when perturbed. If there is an energy dissipation mechanism like damping, the motion will eventually settle back into the static equilibrium configuration. Both k and L affect the critical value. A stiffer spring will increase the value, as will a shorter length.

Viewed as a design problem, if you want to safely carry a mass m at a given height L, then you need to select a spring k that makes m_{cr} greater than m by some amount. The value $m_{cr}/m > 1$ is called the *factor of safety*. The size of the factor of safety depends on the level of uncertainty of the knowledge of any of the parameters that affect the problem. Unexpected things happen during the design life of an engineered system. In this simple example, the spring might not turn out to have the specified stiffness or the mass might increase over time due to changes in usage. The factor of safety is there to keep the system away from states with unwanted behavior like instability.

Now, consider the bent configuration, where equilibrium requires that θ_o satisfy the equilibrium equation $k\theta_o = mgL \sin\theta_o$. Thus, in this case,

$$\begin{aligned} A &= k - mgL\cos\theta_o \\ &= \frac{mgL\sin\theta_o}{\theta_o} - mgL\cos\theta_o \\ &= mgL\left(\frac{\sin\theta_o}{\theta_o} - \cos\theta_o\right) \end{aligned} \tag{12.26}$$

Figure 12.7 shows a plot of the values of A, normalized by mgL for a large range of θ_o values. The function is positive for all θ_o between -4.5 and 4.5. Any negative values are well beyond the range of practical interest. Of course, mgL is positive. Therefore, the bent configuration is always stable.

The results of this entire analysis are summarized in Fig. 12.8, which plots the mass m that can be equilibrated for a given θ, assuming that k and L are fixed. The mass is normalized by the critical mass $m_{cr} = k/gL$. The solid black lines are stable, and the dotted gray lines are unstable equilibrium configurations. Only pairings of

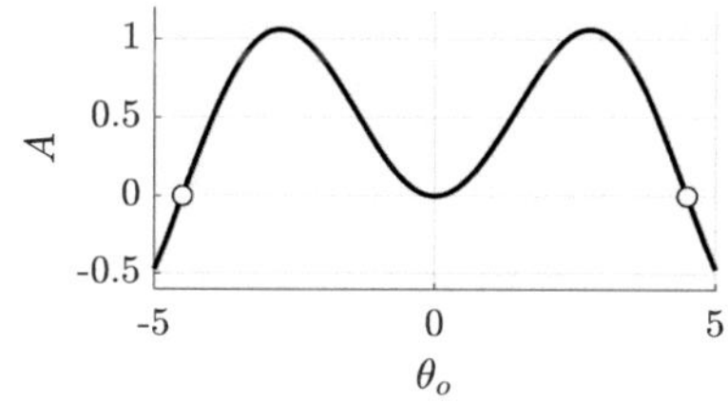

Fig. 12.7 *Second derivative function.* Plot of function A for the system in Fig. 12.6. The ordinates are normalized by mgL

m and θ that lie *on* those lines represent equilibria. Any (m, θ) pairs not on the lines would have unbalanced forces and, therefore, would cause acceleration. For example, if an external agent pulled the system of mass $m/m_{cr} = 1.1$ to an angle $\theta = 1.5$ (the black dot on Fig. 12.8), the moment in the spring would overwhelm the downward force of the weight, causing the bar to accelerate back toward the nearest equilibrium configuration. Figure 12.8 is called a *bifurcation diagram* because it shows that there are multiple solutions to the equilibrium of the system and they branch at the critical value, shown as an open circle in the figure. The branching, or bifurcation, happens at the critical mass. If we think of the mass as providing the load, then that value is often called the *buckling load*.

This example shows that there is a lot going on, even for a very simple system. When we establish equilibrium in the deformed configuration, we open up the possibility that there is more than one solution to the problem for a given set of parameters (e.g., there are three equilibrium configurations for any mass greater than the critical mass in the example). Using the *energy criterion* for stability, we can discern the stability of all equilibrium configurations. The most important aspect of what we found was that the straight configuration was only stable up to the critical value. The force of the weight causes the bar to be in compression. Compression often leads to instability.

In this example, the bent configurations happened to be stable. However, the amount of rotation required to carry even small amounts of load above the critical mass is considerable. Rotations that large would likely break the spring. For this reason, the critical load is usually viewed as the practical capacity of the system. In essence, it is a *strength* limit. The system has a maximum load carrying capacity, even if the material (i.e., the spring) remains elastic. The mechanics of instability are

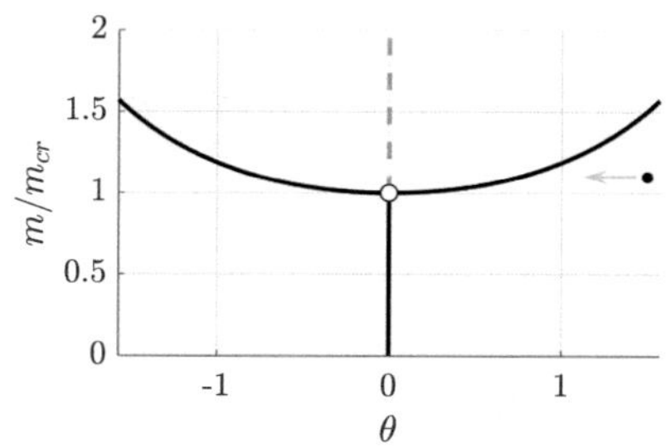

Fig. 12.8 *Bifurcation diagram.* Static equilibrium configurations for the rigid bar, mass, and spring system for fixed values of k, L, and g. The mass is m and the critical weight is $m_{cr} = k/gL$. Solid black lines are stable. Dotted gray lines are unstable

quite different from those associated with inelasticity, but both lead to limits on the load carrying capacity of the system.

12.3.2 Discrete Systems with Multiple Variables

The results of the simple example in the previous section can be generalized. Notice first that the energy criterion for static stability does not require knowledge of the kinetic energy of the system, only the potential energy. As long as energy is conserved, the argument concerning stability is the same. There is also no need for a time derivative to compute the relevant part of the second derivative of the potential energy because

$$A = \frac{d^2U}{d\theta^2} \tag{12.27}$$

We can generalize the stability test to a system that has more than one motion parameter. Let $\boldsymbol{\theta} = \{\theta_1, \ldots, \theta_N\}$ be the list of variables that completely describe the motion of the system, and N is the number of those motion variables. One can show that an $N \times N$ matrix $\mathbf{A}(\boldsymbol{\theta})$, with components

$$A_{ij}(\boldsymbol{\theta}) = \frac{\partial^2 U(\boldsymbol{\theta})}{\partial\theta_i\, \partial\theta_j} \tag{12.28}$$

takes the place of the scalar A as the second derivative for testing stability of equilibrium. The main difference between the single and multiple degree-of-freedom systems is that the former system can move in only one way, while the latter system can move in many different ways, i.e., different combinations of the θ_i values. If there is an unstable configuration, nature will find it. To be stable, then, we must show that the potential energy increases no matter what configuration the system moves into as a result of the perturbation. Mathematically, an equilibrium configuration $\boldsymbol{\theta}_o$ of the system is stable if

$$\bar{\boldsymbol{\theta}}^T \mathbf{A}(\boldsymbol{\theta}_o)\bar{\boldsymbol{\theta}} > 0 \qquad \text{for all } \bar{\boldsymbol{\theta}} \tag{12.29}$$

where the perturbation $\bar{\boldsymbol{\theta}}$ plays the role of $\dot{\theta}$ from the previous case. Equation 12.29 is satisfied if the smallest eigenvalue of $\mathbf{A}(\boldsymbol{\theta}_o)$ is positive. If even one eigenvalue is negative, then the system is unstable, because then there will exist a $\bar{\boldsymbol{\theta}}$ that makes $\bar{\boldsymbol{\theta}}^T\mathbf{A}(\boldsymbol{\theta}_o)\bar{\boldsymbol{\theta}}$ less than zero. Thus, to determine the stability of an equilibrium configuration, we solve the eigenvalue problem

$$\mathbf{A}(\boldsymbol{\theta}_o)\boldsymbol{\varphi} = \mu\boldsymbol{\varphi}$$

The solution to the eigenvalue problem yields N pairs $(\mu_n, \boldsymbol{\varphi}_n)$. If the smallest μ_n is greater than zero, then the system is stable. The critical load can be found by setting $\mu_1 = 0$ and solving for the associated load parameter. The eigenvector $\boldsymbol{\varphi}_1$ is the bent shape, often called a *buckling mode shape*, that the system must adopt to follow the bifurcated path analogous to the one shown in Fig. 12.8 for the simple system.

The eigenvalues and eigenvectors for $n > 1$ are not of great interest physically because the system will buckle when it hits the first critical value. However, the eigenvectors are orthogonal (see Appendix A.5) and, therefore, helpful in proving the stability condition. The eigenvectors span N-dimensional space. Therefore, we can represent the perturbed state $\bar{\boldsymbol{\theta}}$ as

$$\bar{\boldsymbol{\theta}} = \sum_{n=1}^{N} a_n \boldsymbol{\varphi}_n \tag{12.30}$$

Now, the shape of $\bar{\boldsymbol{\theta}}$ is determined by the coefficients a_n. For simplicity, assume that the eigenvectors have been normalized so that $\boldsymbol{\varphi}_n^T \boldsymbol{\varphi}_n = 1$. Substituting the expression for $\bar{\boldsymbol{\theta}}$ from Eq. 12.30 into Eq. 12.29, noting the eigenvalue property $\mathbf{A}(\boldsymbol{\theta}_o)\boldsymbol{\varphi}_n = \mu_n \boldsymbol{\varphi}_n$, and using orthogonality, we can write

$$\bar{\boldsymbol{\theta}}^T \mathbf{A}(\boldsymbol{\theta}_o)\bar{\boldsymbol{\theta}} = \sum_{n=1}^{N} \mu_n a_n^2 \tag{12.31}$$

All the a_n values are squared and, hence, positive. We could select a shape $\bar{\boldsymbol{\theta}}$ that has $a_1 \neq 0$ and all other $a_n = 0$. If μ_1 is less than zero, then the sum is less than zero, and the system is unstable. This proof is also informative in the sense that $\boldsymbol{\varphi}_1$ is the eigenvector associated with μ_1 and, therefore, $\bar{\boldsymbol{\theta}} = \boldsymbol{\varphi}_1$ is the shape that the system moves into when it buckles at the critical load.

While the equations of equilibrium can always be determined by Newtonian balance of forces and moments, we can also state equilibrium as the configuration with stationary potential energy. Thus, equilibrium holds if

$$\frac{\partial U(\boldsymbol{\theta})}{\partial \theta_i} = 0 \qquad i = 1, \ldots, N \tag{12.32}$$

where N is the number of motion variables. There are N equations governing static equilibrium, corresponding to the N motion variables. Since reaction forces are not included in the potential energy, they do not appear in these equations.

Example 12.1
Consider the system with two rigid bars of length L, hinged together and pin supported at the base, shown in Fig. 12.9. The system is subjected to a downward load of magnitude P at the top and is restrained with rotational springs of stiffness k that inhibit the relative rotation of the bars at the hinges. The springs are unstretched in the straight configuration. Find the buckling load of the system. Neglect the self-weight of the bars.

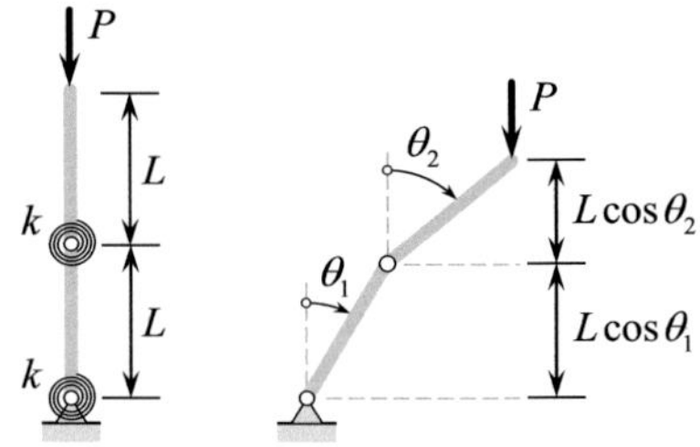

Fig. 12.9 *Example*. Two rigid bars of length L, hinged together and restrained by springs of stiffness k, and subjected to a load P at the top. **a** Straight configuration. **b** Bent configuration, showing motion variables θ_1 and θ_2

Solution. The potential energy of the two springs and the load P is

$$U(\theta_1, \theta_2) = \tfrac{1}{2}k\,\theta_1^2 + \tfrac{1}{2}k(\theta_2 - \theta_1)^2 + PL(\cos\theta_1 + \cos\theta_2)$$

where we have reflected that the relative rotation of the bottom spring is θ_1 and the relative rotation of the middle spring is $\theta_2 - \theta_1$. The potential energy of the load is P multiplied by the height above the base. The equations of equilibrium can be computed from the potential energy, according to Eq. 12.32, as

$$\frac{\partial U}{\partial \theta_1} = 2k\theta_1 - k\theta_2 - PL\sin\theta_1 = 0$$

$$\frac{\partial U}{\partial \theta_2} = -k\theta_1 + k\theta_2 - PL\sin\theta_2 = 0$$

The straight configuration $\theta_1 = \theta_2 = 0$ is a solution to this system of equations. Therefore, the straight configuration is an equilibrium configuration. The second derivative matrix $\mathbf{A}$ is obtained from Eq. 12.28 as

$$\mathbf{A} = \begin{bmatrix} 2k - PL\cos\theta_1 & -k \\ -k & k - PL\cos\theta_2 \end{bmatrix}$$

Let us test the stability of the straight configuration by finding the smallest eigenvalue of $\mathbf{A}$. For $\theta_1 = \theta_2 = 0$, we can write

$$\mathbf{A} - \mu\mathbf{I} = \begin{bmatrix} 2k - PL - \mu & -k \\ -k & k - PL - \mu \end{bmatrix}$$

since $\cos(0) = 1$. The eigenvalues can be found from $\det(\mathbf{A} - \mu\mathbf{I}) = 0$. To simplify the algebra, let $\lambda = PL + \mu$. Now,

$$\det\left(\mathbf{A}-\mu\,\mathbf{I}\right)=\det\begin{bmatrix}2k-\lambda & -k\\ -k & k-\lambda\end{bmatrix}=(2k-\lambda)(k-\lambda)-k^2=0$$

Thus, the characteristic equation governing the values of λ (and therefore μ) that satisfy the eigenvalue problem is

$$\lambda^2-3k\lambda+k^2=0$$

This equation can be solved by the quadratic formula. The smaller of the two solutions is

$$\lambda_1=\tfrac{1}{2}(3-\sqrt{5})k=0.3820\,k$$

Translating back to the original parameters, that means

$$\lambda_1=PL+\mu_1=0.3820\,k \qquad \rightarrow \qquad \mu_1=0.3820\,k-PL$$

The eigenvalue μ_1 of $\mathbf{A}$ will be positive if $0.3820\,k>PL$. Thus, the critical load of the system (i.e., when $\mu_1=0$) is

$$P_{cr}=0.3820\,\frac{k}{L}$$

which is the buckling load of the system, defined as the load for which equilibrium transitions from stable to unstable in the straight configuration.

12.3.3 Planar Beam Buckling

One of the most important observations about stability of equilibrium that we can glean from the discrete problems of the previous section is that writing the equations of equilibrium and the potential energy in the *deformed* configuration is crucial to getting the information needed to assess stability. Chapters 8 and 10, where we developed methods to solve beam bending problems, focused on the *linearized* theory. As mentioned there (see Fig. 8.1), the linearized equations of equilibrium are the same as writing the equations in the undeformed configuration. Consequently, those equations will *not* yield any information about the stability of equilibrium. For beams subjected only to transverse loads, such an assumption is reasonable. However, there are situations for which it is not. In particular, beams with large compressive axial loads can exhibit instability. In this section, we consider the buckling of beams.

To prepare for the analysis of beams subjected to axial loads, we must return to the nonlinear equations of beam theory from Chap. 7. Recall the kinematic relationships of Eqs. 7.11 and 7.10 are

$$w' = -\big(1 + \varepsilon_o\big)\sin\theta \qquad \kappa_o = \theta' \qquad u' = \big(1 + \varepsilon_o\big)\cos\theta - 1 \tag{12.33}$$

where ε_o and κ_o are the axial strain and curvature,[3] respectively, u is the displacement along the axis of the beam, w is the displacement transverse to the axis of the beam, and θ is the rotation of the cross section. The constitutive equation for moment from Eq. 7.23 is

$$M = EI\kappa_o \tag{12.34}$$

Finally, the equilibrium equations of Eqs. 7.18 and 7.21 are

$$Q' + q = 0, \qquad M' + Hw' - Q\big(1+u'\big) + m = 0 \tag{12.35}$$

where we have included only the transverse part of the force equilibrium equations. This system of equations is nonlinear.

Some of the nonlinearities have an important influence on beam buckling and some do not, so we will selectively retain nonlinear terms to form what is often called a *second-order theory*. For simplicity, we will assume that $\varepsilon_o \approx 0$ and that θ is small, but not so small that we can make a linear approximation. If we expand the sine and cosine up to quadratic terms, we get

$$\sin\theta \approx \theta, \qquad \cos\theta \approx 1 - \tfrac{1}{2}\theta^2 \tag{12.36}$$

With these approximations, Eq. 12.33 reduce to

$$\theta = -w' \qquad \kappa_o = -w'' \qquad u' = -\tfrac{1}{2}(w')^2 \tag{12.37}$$

The first two equations are nothing more than the Bernoulli-Euler assumptions of linear beam theory. The third equation tells us that a beam has a shorter straight line distance from end to end due to the fact that it is bent. We have neglected this effect in the linear theory, but it will be important to finding the potential energy of the loads in beam buckling problems. We will continue to use the linear constitutive equation for moment, i.e., $M = -EIw''$, where we have substituted the approximate expression for the curvature. For moment equilibrium, we will neglect u' relative to one, and we will consider only cases with $m = 0$. With these approximations and restrictions, the moment equilibrium equation reduces to

$$M' + Hw' - Q = 0 \tag{12.38}$$

To advance the discussion, let us consider the problem shown in Fig. 12.10. The beam has length L and bending stiffness EI. It is pinned at the bottom and supported by a roller at the top. The load on the system is P, which aligns with the original axis of the beam. Figure 12.10b shows the beam in a bent configuration. The length of the

[3] Note that we are returning to the notation where κ_o is the curvature, not the curvature at yield, as in the previous section on inelasticity.

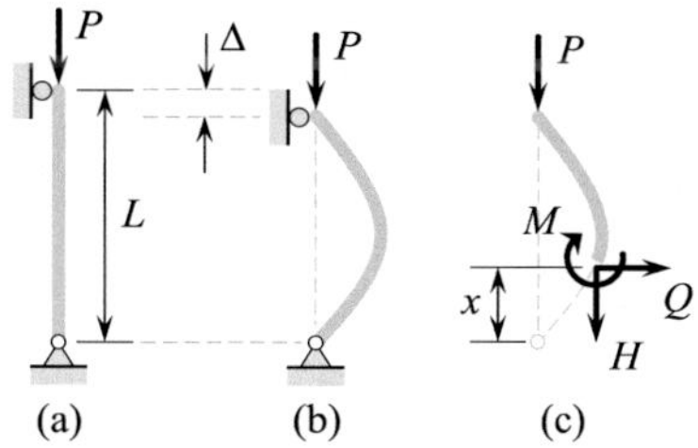

Fig. 12.10 *Beam buckling.* A beam of length L and bending stiffness EI, pinned at the base, supported by a roller at the top, and subjected to a load P at the top. **a** Straight configuration. **b** Bent configuration. **c** Free body diagram

beam does not change because $\varepsilon_o = 0$, but the point of load displaces by an amount Δ due to the bending of the beam. The reaction force at the pin is P directed along the original axis of the beam in both the straight and bent configurations. Thus, the applied force and reaction are in equilibrium, even when the beam is bent. Summing moments at the pin proves that the reaction force at the roller is zero. Figure 12.10c shows a free body diagram of the system with a cut at x, exposing the internal stress resultants M, H, and Q.

Equilibrium of forces shows that $Q = 0$ and $H = -P$. Therefore, the moment equilibrium equation, Eq. 12.38, can be written as

$$M' - Pw' = 0 \tag{12.39}$$

Differentiating Eq. 12.39 once and substituting $M = -EIw''$ gives

$$EIw'''' + Pw'' = 0 \tag{12.40}$$

The reason for differentiating the equation is to prepare for other cases where Q is not equal to zero and to accommodate boundary conditions other than the ones shown in Fig. 12.10. It should be evident that if we summed moments at the cut in the free body diagram, noting that the transverse displacement at that point is w, we would obtain $M - Pw = 0$, which is consistent with Eqs. 12.39 and 12.40, the latter being the most general form of the equation.

Equation 12.40 is a fourth-order ordinary differential equation with constant coefficients. The solution to such an equation is in the form $w(x) = Ce^{ax}$, where the constants C and a are to be determined. To find a, we differentiate the candidate solution and substitute it back into the differential equation. To wit,

$$\begin{aligned} EI\left(a^4Ce^{ax}\right) + P\left(a^2Ce^{ax}\right) &= 0 \\ \left(EIa^4 + Pa^2\right)Ce^{ax} &= 0 \end{aligned}$$

If C is zero, then there is no deflection. The exponential is never zero. Therefore, the only way the candidate solution can satisfy the equation is if

$$a^2\left(EIa^2 + P\right) = 0$$

which has solutions $a=0$ and $a=\pm i\lambda$, where i is the imaginary unit and λ is a positive value defined as

$$\lambda=\sqrt{\frac{P}{EI}}$$

There are four solutions to the differential equation, so the general solution is the linear combination of those solutions. Since the zero root is repeated, we can write the general solution as

$$w(x)=C_1+C_2x+C_3e^{i\lambda x}+C_4e^{-i\lambda x}$$

Using Euler's formula, $e^{i\lambda x}=\cos\lambda x+i\sin\lambda x$, to convert the complex exponentials to sines and cosines, we can write this equation in the more convenient form

$$w(x)=D_1+D_2x+D_3\cos\lambda x+D_4\sin\lambda x \tag{12.41}$$

This solution satisfies Eq. 12.40 for any boundary conditions. To find a solution in a particular case, we must implement the boundary conditions. In the case shown in Fig. 12.10, the boundary conditions are

$$w(0)=0,\qquad w(L)=0,\qquad M(0)=0,\qquad M(L)=0$$

Using the constitutive equation, the last two can be recast as $w''(0)=0$ and $w''(L)=0$. The second derivative of the displacement is

$$w''(x)=-\lambda^2D_3\cos\lambda x-\lambda^2D_4\sin\lambda x$$

Now, the boundary conditions can be written as

$$\begin{aligned}w(0)&=0=D_1+D_3\\w(L)&=0=D_1+D_2L+D_3\cos\lambda L+D_4\sin\lambda L\\w''(0)&=0=-\lambda^2D_3\\w''(L)&=0=-\lambda^2D_3\cos\lambda L-\lambda^2D_4\sin\lambda L\end{aligned}$$

The third equation gives $D_3=0$. Using this value in the first equation gives $D_1=0$. Substituting the fourth equation into the second gives $D_2=0$. That leaves, from either the second or fourth equation, $D_4\sin\lambda L=0$. If D_4 is zero, then there is no deflection. Thus, to have a bent configuration, we must have

$$\sin\lambda L=0 \tag{12.42}$$

This equation is called the *characteristic equation*, and it provides a way to determine the specific values of λ that satisfy the boundary conditions. The solutions to the characteristic equation are $\lambda L=\pi, 2\pi, 3\pi, \ldots$, which can be summarized as

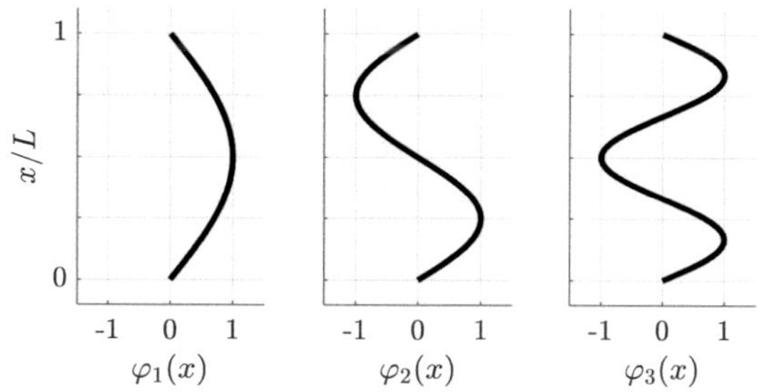

Fig. 12.11 *Buckling mode shapes.* The first three eigenfunctions, or mode shapes, for a simply supported beam of length L and constant bending stiffness *EI*

$$\lambda_n = \frac{n\pi}{L} \tag{12.43}$$

for $n = 1, 2, 3, \ldots$. There are an infinite number of these values. Substituting the original definition of $\lambda_n^2 = P_n/EI$, in which we have added the index n to reflect the fact that we found multiple λ values, gives $P_n = n^2\pi^2 EI/L^2$. These are discrete values of the axial force P for which a bent configuration is possible. We call these values the *critical loads*. The first of these loads is the smallest. For $n = 1$, we call

$$P_{cr} = \frac{\pi^2 EI}{L^2} \tag{12.44}$$

the *buckling load*. Since this solution was first found by Euler, it is also known as the *Euler load*.

Recall from Eq. 12.41 that $w(x) = D_4 \sin \lambda x$, since all the other coefficients were found to be zero. The coefficient D_4 cannot be determined because we used up all the boundary condition equations finding D_1, D_2, D_3, and λ. We found multiple values of λ. For each λ_n, there is as associated function

$$\varphi_n(x) = \sin \lambda_n x \tag{12.45}$$

that is called an *eigenfunction*. Each one is a version of $w(x)$ without the D_4 coefficient. Thus, each one is a bent shape of the beam that satisfies all the boundary conditions. One can think of each eigenvector as a mode of behavior that is possible for the system, and the eigenvectors are often called *mode shapes*. The first three eigenfunctions are shown in Fig. 12.11. In many ways, eigenfunctions are like the eigenvectors we encountered in the discrete matrix setting, except that they are continuous functions, not vectors. Many of the properties of eigenvectors carry over to eigenfunctions. For example, the eigenfunctions are orthogonal.

The proof of orthogonality is based upon the original differential equation, Eq. 12.40. Each of the eigenfunctions $\varphi_n(x)$ satisfies this equation and the boundary conditions. Dividing the equation through by *EI* and noting that $P/EI = \lambda^2$, we can write

$$\varphi_n'''' + \lambda_n^2 \varphi_n'' = 0 \tag{12.46}$$

Multiply this equation by $\varphi_m(x)$ and integrate over the domain of the bar to get

$$\int_0^L [\varphi_n'''' + \lambda_n^2 \varphi_n'']\varphi_m \, dx = 0 \tag{12.47}$$

Since the integrand is zero, the integral is zero. Integrating the first term by parts twice and the second term once, we get

$$\int_0^L [\varphi_n''\varphi_m'' - \lambda_n^2 \varphi_n'\varphi_m'] \, dx + [(\varphi_n''' + \lambda_n^2 \varphi_n')\varphi_m]_0^L - [\varphi_n''\varphi_m']_0^L = 0$$

Because the eigenfunctions satisfy all the boundary conditions, the boundary terms that result from integration by parts all vanish. It turns out that this is true of *all* boundary conditions, not just the ones for the simple-simple beam. Thus, the equation reduces to

$$\int_0^L [\varphi_n''\varphi_m'' - \lambda_n^2 \varphi_n'\varphi_m'] \, dx = 0 \tag{12.48}$$

We can repeat the process, reversing the roles of n and m to get

$$\int_0^L [\varphi_m''\varphi_n'' - \lambda_m^2 \varphi_m'\varphi_n'] \, dx = 0$$

Finally, subtract the first equation from the second, noting that the terms with second derivatives cancel out, to get

$$(\lambda_n^2 - \lambda_m^2) \int_0^L \varphi_n'\varphi_m' \, dx = 0 \tag{12.49}$$

If $n \neq m$, then $\lambda_n \neq \lambda_m$. Thus, we have the orthogonality relationship

$$\int_0^L \varphi_n' \, \varphi_m' \, dx = 0 \qquad m \neq n \tag{12.50}$$

If the first derivatives are orthogonal, then so are the second derivatives. To see it, return to Eq. 12.48 and implement the orthogonality of the first derivatives. To wit,

$$\int_0^L \varphi_n'' \, \varphi_m'' \, dx = 0 \qquad m \neq n \tag{12.51}$$

Also, setting $m = n$ in Eq. 12.48 gives

$$\int_0^L (\varphi_n'')^2 \, dx = \lambda_n^2 \int_0^L (\varphi_n')^2 \, dx \tag{12.52}$$

The eigenvalues and eigenfunctions for $n > 1$ can be very useful mathematically for solving problems and making proofs, as we will see in the sequel.

Discrete eigenvectors span N-dimensional space. Similarly, the eigenfunctions span the space of functions of x defined in the range $x \in [0, L]$ that satisfy the boundary conditions. One consequence is that we can represent functions in the space as infinite series using the eigenfunctions. To wit,

$$w(x) = \sum_{n=1}^{\infty} a_n \, \varphi_n(x) \tag{12.53}$$

We will use this expansion to prove that the beam is stable in the straight configuration only up to the first critical load.

Stability of Equilibrium The straight configuration, $w(x) = 0$, satisfies equilibrium for any load level P. However, the straight configuration is not stable for all values of P. In fact, we can show that for $P > P_{cr}$, the beam is unstable. If the beam is loaded from zero, then the beam will buckle into the bent configuration when it reaches the critical value P_{cr}.

The stability argument for discrete systems was based upon a potential energy function. One can show that the potential energy stored in a bent elastic beam is

$$U_{beam} = \int_0^L \tfrac{1}{2} EI (w'')^2 \, dx \tag{12.54}$$

We also need an expression for the potential energy of the load. In Fig. 12.10, the drop in the load due to bending is Δ. Hence, the potential energy of the load relative to the base is $P(L - \Delta)$, basically *mgh* for the load P. The reference point for potential energy is arbitrary. If we take it to be at the top of the straight beam, then the potential energy is $-P\Delta$. To determine what Δ is, note that Eq. 12.37(c) can be integrated to give

$$\int_0^L u' \, dx = u(L) - u(0) = \int_0^L -\tfrac{1}{2}(w')^2 \, dx \tag{12.55}$$

The displacement at the base is $u(0) = 0$, therefore $\Delta = -u(L)$. Hence, we can express the potential energy associated with P as

$$U_{load} = -P \int_0^L \tfrac{1}{2}(w')^2 \, dx \tag{12.56}$$

Adding the two potential energies together, we get the total potential energy of the system

$$U = \int_0^L \left(\tfrac{1}{2} EI (w'')^2 - \tfrac{1}{2} P (w')^2 \right) dx \tag{12.57}$$

The next problem we face is how to take derivatives of the potential energy since it is not an ordinary function. In fact, it is a function of a function, which we call a *functional*. A functional takes as input a function $w(x)$ and produces a single number as output. In this case, the function $w(x)$ represents a particular configuration of the system. That configuration has a potential energy U, which is just a single scalar value. A different configuration would be associated with a different potential energy.

We need a way to determine the rate of change of potential energy as we move from an equilibrium configuration to some perturbed configuration so that we can tell if the potential energy goes up or down when the system is perturbed. We can take the directional derivative of a functional as follows:

$$\mathcal{D}U(w)\cdot\bar{w} = \frac{d}{d\epsilon}\Big[U(w+\epsilon\,\bar{w})\Big]_{\epsilon=0} \tag{12.58}$$

The directional derivative gives the rate of change of U at the configuration w, just as it starts to move in the direction of the perturbation $\bar{w}$. If the rate of change of potential energy is zero no matter what direction you perturb it, then it must be an equilibrium configuration. The beauty of the directional derivative is that it is an ordinary derivative with respect to ϵ, and the variable x associated with the functions w and $\bar{w}$ is treated as a constant during that differentiation. For the beam potential energy function, we can compute

$$\begin{aligned}
\mathcal{D}U(w)\cdot\bar{w} &= \frac{d}{d\epsilon}\int_0^L \Big(\tfrac{1}{2}EI(w''+\epsilon\bar{w}'')^2 - \tfrac{1}{2}P(w'+\epsilon\bar{w}')^2\Big)\,dx\,\Big|_{\epsilon=0} \\
&= \int_0^L \Big(EI(w''+\epsilon\bar{w}'')\bar{w}'' - P(w'+\epsilon\bar{w}')\bar{w}'\Big)\,dx\,\Big|_{\epsilon=0} \\
&= \int_0^L \Big(EIw''\bar{w}'' - Pw'\bar{w}'\Big)\,dx
\end{aligned} \tag{12.59}$$

To execute this calculation, we first augment w to $w+\epsilon\bar{w}$ everywhere it appears. In the first line we have noted that $(w+\epsilon\bar{w})' = w' + \epsilon\bar{w}'$ because the prime indicates a derivative with respect to x and ϵ does not depend on x. The same goes for the second derivative term. The second line takes the derivative inside the integral, which is legal because the limits of integration do not depend on ϵ. Finally, any remaining ϵ are set to zero. Let us call this functional $G(w,\bar{w}) = \mathcal{D}U(w)\cdot\bar{w}$. Thus,

$$G(w,\bar{w}) = \int_0^L \big(EIw''\bar{w}'' - Pw'\bar{w}'\big)\,dx \tag{12.60}$$

In physical terms, $G(w,\bar{w})$ measures the rate of change of energy at the configuration described by $w(x)$ as it just begins to move in the direction of $\bar{w}(x)$. If that rate of change is zero no matter which direction you go, then the potential energy must be stationary, i.e., at either a maximum or minimum. Potential energy is stationary only at equilibrium configurations. To show that setting the first derivative of the potential energy equal to zero is equivalent to the governing equations, integrate the first term in $G(w,\bar{w})$ by parts twice and the second term once to get

$$\begin{aligned}
G(w,\bar{w}) = \int_0^L &\big(EIw'''' + Pw''\big)\bar{w}\,dx \\
&+ EIw''\bar{w}'\Big|_0^L - \big(EIw''' + Pw'\big)\bar{w}\Big|_0^L
\end{aligned}$$

If w and $\bar{w}$ satisfy the boundary conditions, then the last two terms vanish. Actually, $\bar{w}$ only needs to satisfy boundary conditions on displacement and rotation because only $\bar{w}$ and $\bar{w}'$ show up in the boundary terms from integration by parts. These boundary conditions are often called *essential* boundary conditions, perhaps because they are essential to making those boundary terms vanish. Boundary conditions on moment and shear are called *natural* boundary conditions, as they are satisfied naturally. Therefore, we can also write the equilibrium condition $G(w, \bar{w}) = 0$ as

$$G(w, \bar{w}) = \int_0^L \big(EIw'''' + Pw''\big)\bar{w}\, dx = 0$$

If this quantity must vanish no matter what $\bar{w}$ is, then the only possibility is that the term in parentheses must itself be zero. In other words,

$$EIw'''' + Pw'' = 0$$

which is identical to Eq. 12.40. This step is called the *fundamental theorem of the calculus of variations*. The argument is simple. If $\bar{w}$ can be anything, then it can be the same as the term in parentheses. The result is an integral of some quantity squared. A function squared is positive everywhere. Therefore, if the integral is equal to zero, then the only possibility is that the term in parentheses is zero.

There are a couple of implications of the previous analysis. The first is that the definition of the potential energy of the bent beam, U_{beam}, in Eq. 12.54 is appropriate. There are other ways to derive that expression for the potential energy, but we will not pursue those here. The second is that the directional derivative is the appropriate way to characterize the rate of change needed for the potential energy argument. It is also evident that the statement $G(w, \bar{w}) = 0$ for all $\bar{w}$ is an alternative way of stating the governing differential equation.

Finally, we can compute the second derivative of the potential energy as the second directional derivative of the potential energy. To wit,

$$A(w, \bar{w}) = \frac{d^2}{d\epsilon^2}\Big[\, U(w + \epsilon\bar{w})\,\Big]_{\epsilon=0} \tag{12.61}$$

For our functional, this definition gives

$$A(w, \bar{w}) = \int_0^L \big[\, EI(\bar{w}'')^2 - P(\bar{w}')^2 \,\big]\, dx \tag{12.62}$$

The energy criterion for stability, then, follows the same pattern as the discrete system. If $A(w, \bar{w}) > 0$ for all $\bar{w}$, then the system is stable. Otherwise, it is unstable. The easiest way to put a fine point on this stability criterion is to expand $\bar{w}$ using the eigenfunctions as

$$\bar{w}(x) = \sum_{n=1}^{\infty} a_n\, \varphi_n(x)$$

The first and second derivatives are

$$\bar{w}'(x) = \sum_{n=1}^{\infty} a_n \varphi_n'(x), \qquad \bar{w}''(x) = \sum_{n=1}^{\infty} a_n \varphi_n''(x)$$

Because the eigenfunctions are orthogonal,

$$\begin{aligned}\int_0^L (\bar{w}')^2 \, dx &= \int_0^L \Big(\sum_{n=1}^{\infty} a_n \varphi_n' \sum_{m=1}^{\infty} a_m \varphi_m' \Big) \, dx \\ &= \sum_{n=1}^{\infty} \sum_{m=1}^{\infty} a_n a_m \int_0^L \varphi_n' \varphi_m' \, dx \\ &= \sum_{n=1}^{\infty} a_n^2 \int_0^L (\varphi_n')^2 \, dx\end{aligned}$$

A similar relationship holds true for $\bar{w}''$. Substitute these relationships into Eq. 12.62 and use Eq. 12.52 to eliminate the φ'' terms. The result is

$$A(w, \bar{w}) = \sum_{n=1}^{\infty} a_n^2 \big(\lambda_n^2 EI - P \big) \int_0^L (\varphi_n')^2 \, dx \tag{12.63}$$

Each integral in the sum is positive because it is the integral of the square of a function. If we select a $\bar{w}$ that has $a_1 \neq 0$ with all other $a_n = 0$, then the second derivative functional will be greater than zero if $\lambda_1^2 EI - P > 0$. For the simply supported beam of Fig. 12.10, $\lambda_1^2 = \pi^2/L^2$. Therefore, the beam is stable in the straight configuration as long as

$$P < \frac{\pi^2 EI}{L^2} = P_{cr} \tag{12.64}$$

i.e., when the load is below the first critical load of the system. The system is unstable for any load values $P > P_{cr}$.

This analysis shows the complications that arise in dealing with continuous systems, but it also demonstrates the similarity between the discrete case and the continuous case. Because we were selective in which terms we retained in our theory, the second derivative of the potential energy does not contain w. Therefore, we cannot determine if the bent configuration is stable or not. A more detailed analysis keeping all the terms would show that the post-buckling response of an elastic beam is stable, similar to the simple system of Fig. 12.6. We will consider some numerical strategies in the next section, after the following example.

Example 12.2
Compute the buckling load of a beam of length L and stiffness EI subjected to compressive force P acting at the top. The beam is fixed at both ends (but able to move along the axis of the beam), as shown in Fig. 12.12.

Solution. The general solution and its first derivative, which is needed to implement the boundary conditions, are (from Eq. 12.41)

$$\begin{aligned} w(x) &= D_1 + D_2 x + D_3 \cos \lambda x + D_4 \sin \lambda x \\ w'(x) &= D_2 - \lambda D_3 \sin \lambda x + \lambda D_4 \cos \lambda x \end{aligned} \tag{12.65}$$

For the fixed-fixed case, the boundary conditions are

$$w(0) = 0, \qquad w(L) = 0, \qquad w'(0) = 0, \qquad w'(L) = 0$$

Substituting the boundary conditions into the general solution gives

$$\begin{aligned} w(0) &= 0 = D_1 + D_3 \\ w(L) &= 0 = D_1 + D_2 L + D_3 \cos \lambda L + D_4 \sin \lambda L \\ w'(0) &= 0 = D_2 + \lambda D_4 \\ w'(L) &= 0 = D_2 - \lambda D_3 \sin \lambda L + \lambda D_4 \cos \lambda L \end{aligned}$$

First, let $\mu = \lambda L$ to simplify the notation. From the first and third equations, we get $D_1 = -D_3$ and $D_2 L = -\mu D_4$. Substituting these results into the second and fourth equations, respectively, gives the system of equations

$$\begin{bmatrix} \cos \mu - 1 & \sin \mu - \mu \\ -\sin \mu & \cos \mu - 1 \end{bmatrix} \begin{Bmatrix} D_3 \\ D_4 \end{Bmatrix} = \begin{Bmatrix} 0 \\ 0 \end{Bmatrix} \tag{12.66}$$

These equations have a nonzero solution only if the determinant of the coefficient matrix is zero. Thus, to have a bent configuration, μ must satisfy the *characteristic equation*

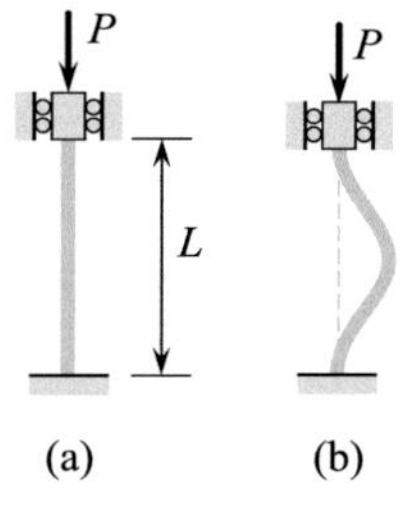

Fig. 12.12 *Example.* A fixed-fixed beam with axial load P applied at the top. **a** Straight configuration. **b** Bent configuration

$$g(\mu) = (\cos\mu - 1)^2 + \sin\mu(\sin\mu - \mu)$$
$$= 2(1 - \cos\mu) - \mu\sin\mu = 0$$

The following code solves this equation by Newton's method. The results for the first seven roots are shown in Fig. 12.13. The graph shows the roots of the function with open circles. The table gives the numerical values of those roots.

```
%. Set tolerance and starting value for Newton's method
   tol = 1.e-8;
   StartValues = [2,3,4,5,6,7,8]*pi;
   nCases = length(StartValues);

%. Newton iteration to solve
   Ev = zeros(nCases,1);
   for i=1:nCases
     err = 1; mu = StartValues(i);
     while err>tol
       g = 2 - 2*cos(mu) - mu*sin(mu);
       A =  sin(mu) - mu*cos(mu);
       mu = mu - g/A;
       err = norm(g);
     end
     Ev(i) = mu;
   end
   FxFxOut(Ev,StartValues)
```

The function `FxFxOut` prints the output to the command window and plots the function $g(\mu)$ along with the computed roots to make sure none have been missed (see Fig. 12.13). The smallest value gives the buckling load. Recalling that the load is $P = \lambda^2 EI$ and $\mu = \lambda L$, we get

$$P_{cr} = \frac{\mu_1^2 EI}{L^2} = \frac{4\pi^2 EI}{L^2} = 39.4784\frac{EI}{L^2}$$

which is four times the value of the buckling load for the simple-simple beam.

To find the eigenfunctions, observe that we can write an equation for the bent shape as

$$w(\zeta) = D_3\big(\cos\mu\zeta - 1\big) + D_4\big(\sin\mu\zeta - \mu\zeta\big)$$

where $\zeta = x/L$. This equation is Eq. 12.65(a), reflecting that $D_1 = -D_3$ and $D_2 L = -\mu D_4$. We need to be careful in eliminating one of the two remaining constants in favor of the other to get the eigenfunction. For the odd values of n, both $\sin\mu$ and $\cos\mu - 1$ are zero. Therefore, we can pick $D_3 = -1$ and, from the first of Eq. 12.66, D_4 must be zero. In these cases, the eigenfunctions are

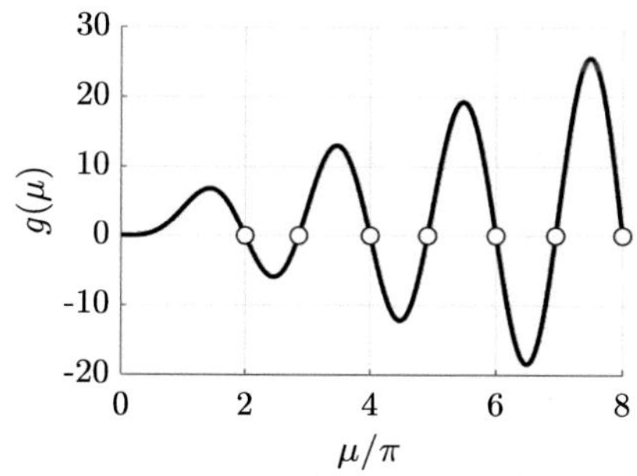

n	μ_n/π	μ_n^2
1	2.0000	39.4784
2	2.8606	80.7629
3	4.0000	157.9137
4	4.9180	238.7181
5	6.0000	355.3058
6	6.9418	475.5995
7	8.0000	631.6547

Fig. 12.13 *Example*. Plot of $g(\mu)$ versus μ/π for the characteristic equation. The table shows the values of the first seven roots, which are the open circles on the graph

Fig. 12.14 *Eigenfunctions*. The first three eigenfunctions for buckling of a fixed-fixed beam with axial load P

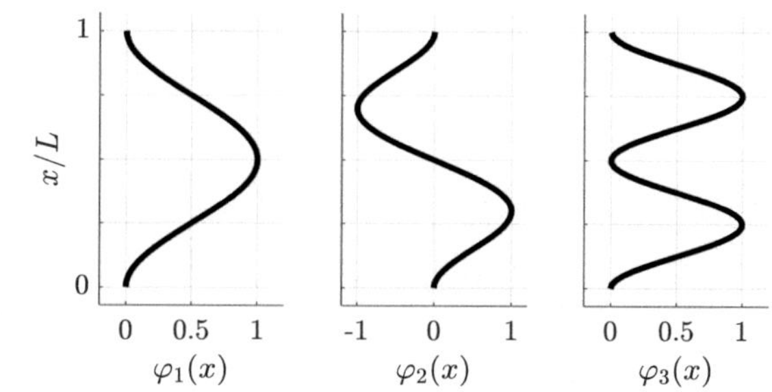

$$\varphi_n(\zeta) = 1 - \cos \mu_n \zeta$$

for $n = 1, 3, 5, \ldots$. For the even values of n, we can pick $D_3 = \cos \mu - 1$ and $D_4 = \sin \mu$. In these cases, the eigenfunctions have the form

$$\varphi_n(\zeta) = \big(\cos \mu_n - 1\big)\big(\cos \mu_n \zeta - 1\big) + \sin \mu_n \big(\sin \mu_n \zeta - \mu_n \zeta\big)$$

for $n = 2, 4, 6, \ldots$. The first three eigenfunctions for this problem, normalized to have maximum value of one, are shown in Fig. 12.14.

This example exhibits a more complicated characteristic equation and illustrates how to find the eigenfunctions associated with the eigenvalues.

12.3.4 The Rayleigh-Ritz Method

Perhaps inspired by Eq. 12.52, Lord Rayleigh[4] defined what we now call the *Rayleigh quotient* to be

$$\mathcal{R}(w) = \frac{\int_0^L (w'')^2\,dx}{\int_0^L (w')^2\,dx} \tag{12.67}$$

and observed that $\mathcal{R}(w) \geq \lambda_1^2$ for all w that satisfy the essential boundary conditions. In fact, $\mathcal{R}(\varphi_1) = \lambda_1^2$, which shows that the minimum is achieved when w is the first eigenfunction. Thus, over all functions w for which this quotient could be evaluated, the one that gives the smallest value produces a value $\mathcal{R}$ that is closest to the smallest eigenvalue of $A(w, \bar{w})$. We can use the Rayleigh quotient to search for the buckling load by trying different w functions and using the smallest one as the best estimate of the first critical load. The following example illustrates this point.

Example 12.3
Compute the Rayleigh quotient for the simply supported beam of length L for the two functions

$$w_1(\zeta) = \zeta - \zeta^2, \qquad w_2(\zeta) = \zeta - 2\zeta^3 + \zeta^4$$

where $\zeta = x/L$. Compare the results with the exact value of π^2/L^2.

Solution. First, compute the derivatives

$$\begin{aligned} w_1'(\zeta) &= (1-2\zeta)/L & \quad w_2'(\zeta) &= (1-6\zeta^2+4\zeta^3)/L \\ w_1''(\zeta) &= -2/L^2 & \quad w_2''(\zeta) &= (-12\zeta+12\zeta^2)/L^2 \end{aligned}$$

Now, compute the integrals needed, noting that $dx = L\,d\zeta$. For the first function, we have

$$\begin{aligned} \int_0^L (w_1'')^2\,dx &= \frac{1}{L^3}\int_0^1 (-2)^2\,d\zeta = \frac{4}{L^3} \\ \int_0^L (w_1')^2\,dx &= \frac{1}{L}\int_0^1 (1-2\zeta)^2\,d\zeta = \frac{1}{3L} \end{aligned}$$

Computing the Rayleigh quotient, we get $\mathcal{R}(w_1) = 12/L^2$. For the second function, we have

[4] John William Strutt, 3rd Baron Rayleigh (1842–1919) was a British mathematician and physicist who made numerous contributions to mechanics.

$$\int_0^L (w_2'')^2\,dx = \frac{1}{L^3}\int_0^1 (-12\zeta + 12\zeta^2)^2\,d\zeta = \frac{24}{5L^3}$$

$$\int_0^L (w_2')^2\,dx = \frac{1}{L}\int_0^1 (1 - 6\zeta^2 + 4\zeta^3)^2\,d\zeta = \frac{17}{35L}$$

Computing the Rayleigh quotient, we get $\mathcal{R}(w_2) = 168/17L^2 = 9.8824/L^2$.

These values can be compared with the exact value of $\pi^2/L^2 = 9.8696/L^2$. Note that the first function satisfies the boundary conditions on displacement at the ends, but not moment. The second function satisfies the displacement and moment boundary conditions at the ends. Both functions obey the Rayleigh inequality. The second function gives a very good estimate of the critical load.

Walther Ritz[5] improved on Rayleigh's method by considering the possibility that the test functions could be linear combinations of known functions, i.e.,

$$w(x) = \sum_{n=1}^{N} a_n h_n(x) = \mathbf{a}^T\mathbf{h}(x) \tag{12.68}$$

where a_n is an unknown (but constant) coefficient and $h_n(x)$ is a specified function, called a *Ritz function*. The coefficients and Ritz functions in matrix notation are

$$\mathbf{a} = \begin{Bmatrix} a_1 \\ a_2 \\ \vdots \\ a_N \end{Bmatrix}, \qquad \mathbf{h}(x) = \begin{Bmatrix} h_1(x) \\ h_2(x) \\ \vdots \\ h_N(x) \end{Bmatrix}$$

The beauty of the Ritz approximation is that it is easy to differentiate and integrate. The derivatives of w with respect to x are

$$w'(x) = \mathbf{a}^T\mathbf{h}'(x), \qquad w''(x) = \mathbf{a}^T\mathbf{h}''(x) \tag{12.69}$$

Now, let us apply the Ritz approximation to the potential energy to get

$$\begin{aligned} U &= \int_0^L \big[\tfrac{1}{2}EI(w'')^2 - \tfrac{1}{2}P(w')^2\big]dx \\ &= \int_0^L \big[\tfrac{1}{2}EI(\mathbf{a}^T\mathbf{h}'')(\mathbf{a}^T\mathbf{h}'') - \tfrac{1}{2}P(\mathbf{a}^T\mathbf{h}')(\mathbf{a}^T\mathbf{h}')\big]dx \\ &= \tfrac{1}{2}\mathbf{a}^T\Big[\int_0^L EI\,\mathbf{h}''\mathbf{h}''^T\,dx - \int_0^L P\,\mathbf{h}'\mathbf{h}'^T\,dx\Big]\mathbf{a} \end{aligned} \tag{12.70}$$

[5] Walther Heinrich Wilhelm Ritz (1878–1909) was a Swiss physicist.

In the second line we substituted the Ritz approximation. In the third line we noted that $\mathbf{a}^T\mathbf{h} = \mathbf{h}^T\mathbf{a}$, and we pulled the $\mathbf{a}$ terms outside the integral because they are constant. The integrals can be evaluated because the Ritz functions are known. Let us define the matrices

$$\mathbf{K} = \int_0^L EI\,\mathbf{h}''\mathbf{h}''^T\,dx, \qquad \mathbf{G} = \int_0^L \mathbf{h}'\mathbf{h}'^T\,dx \tag{12.71}$$

These matrices are $N \times N$ in size because the number of elements of $\mathbf{h}$ is N. With these definitions, we can write the potential energy as

$$U(\mathbf{a}) = \tfrac{1}{2}\mathbf{a}^T\mathbf{A}\,\mathbf{a} \tag{12.72}$$

where the matrix $\mathbf{A}$ is defined as

$$\mathbf{A} = \mathbf{K} - P\mathbf{G} \tag{12.73}$$

The most important observation is that the potential energy is now an *ordinary* function of the parameters $\mathbf{a}$. The Ritz process is called *discretization* because it makes a continuous problem into a discrete one. Now, we can take the first derivative and set it equal to zero to find the equilibrium configuration. To wit,

$$\frac{\partial U(\mathbf{a})}{\partial \mathbf{a}} = \big[\mathbf{K} - P\mathbf{G}\big]\mathbf{a} = \mathbf{0} \tag{12.74}$$

which is a matrix eigenvalue problem. The solution $\mathbf{a} = \mathbf{0}$ is the straight configuration, which is in equilibrium no matter what the value of P is. For bent configurations to exist, we must find solutions $\mathbf{a} \neq \mathbf{0}$ by solving the eigenvalue problem. Note also that if we take the function $\bar{w} = \bar{\mathbf{a}}^T\mathbf{h}$ in Eq. 12.62, then the second derivative function is

$$A(w, \bar{w}) = \bar{\mathbf{a}}^T\mathbf{A}\,\bar{\mathbf{a}} \tag{12.75}$$

Now, the test for stability is identical to what we saw for the discrete systems. The scalar $A(w, \bar{w})$ is positive for all possible $\bar{w}$ if and only if the smallest eigenvalue of the matrix $\mathbf{A}$ is greater than zero. The eigenvalues of $\mathbf{A}$ can be found by solving the eigenvalue problem $\mathbf{A}\boldsymbol{\varphi} = \mu\boldsymbol{\varphi}$. To find the smallest value of P for which the system is stable, set $\mu_1 = 0$. Doing so yields

$$\mathbf{A}\boldsymbol{\varphi} = \big[\mathbf{K} - P\mathbf{G}\big]\boldsymbol{\varphi} = \mathbf{0} \tag{12.76}$$

which is exactly the same eigenvalue problem as Eq. 12.74. Thus, the mode shape $\boldsymbol{\varphi}_i$ can be interpreted as the shape the beam must be in to buckle at load level P_i. As usual, the P_1 and $\boldsymbol{\varphi}_1$ are the most important. The following example illustrates how to compute the estimate of the buckling load using the Ritz method.

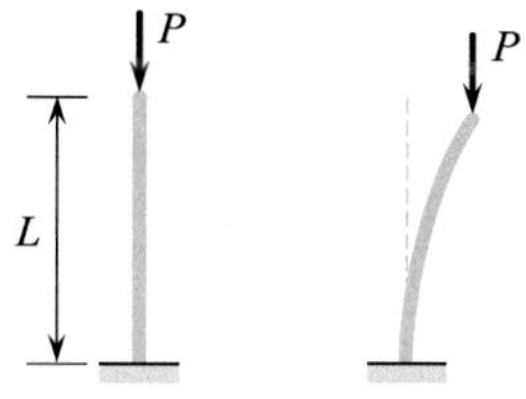

Fig. 12.15 *Example.* A cantilever column with fixed support at the base and an axial load P applied at the top. **a** Straight configuration. **b** Bent configuration

Example 12.4

Compute the buckling load of a cantilever beam of length L and bending stiffness EI subjected to axial force P at the top, as shown in Fig. 12.15. Use the Ritz method with a cubic approximation

$$w(x) = a_1\zeta^2 + a_2\zeta^3$$

where $\zeta = x/L$.

Solution. The Ritz functions are $h_1 = \zeta^2$ and $h_2 = \zeta^3$. Observe that they satisfy the essential boundary conditions $w(0) = 0$ and $w'(0) = 0$. Specifically, $h_i(0) = 0$ and $h_i'(0) = 0$. We can compute the following derivatives of the Ritz functions:

$$\mathbf{h}' = \begin{Bmatrix} 2\zeta \\ 3\zeta^2 \end{Bmatrix} \frac{1}{L}, \qquad \mathbf{h}'' = \begin{Bmatrix} 2 \\ 6\zeta \end{Bmatrix} \frac{1}{L^2}$$

The $\mathbf{K}$ matrix can be computed as

$$\mathbf{K} = \frac{EI}{L^3} \int_0^1 \begin{bmatrix} 4 & 12\zeta \\ 12\zeta & 36\zeta^2 \end{bmatrix} d\zeta = \frac{EI}{L^3} \begin{bmatrix} 4 & 6 \\ 6 & 12 \end{bmatrix}$$

where the change of variable $\zeta = x/L$ changes the limits of integration to 0 and 1 and makes $dx = L\,d\zeta$. The $\mathbf{G}$ matrix can be computed as

$$\mathbf{G} = \frac{1}{L} \int_0^1 \begin{bmatrix} 4\zeta^2 & 6\zeta^3 \\ 6\zeta^3 & 9\zeta^4 \end{bmatrix} d\zeta = \frac{1}{60L} \begin{bmatrix} 80 & 90 \\ 90 & 108 \end{bmatrix}$$

We can find the eigenvalues by computing the roots of $\det[\mathbf{K} - P\mathbf{G}] = 0$. Thus,

$$\begin{aligned} \det[\mathbf{K} - P\mathbf{G}] &= \det \begin{bmatrix} 4 - 80\beta & 6 - 90\beta \\ 6 - 90\beta & 12 - 108\beta \end{bmatrix} \\ &= (4 - 80\beta)(12 - 108\beta) - (6 - 90\beta)^2 \\ &= 540\beta^2 - 312\beta + 12 = 0 \end{aligned}$$

To simplify the algebra, we multiplied the eigenvalue problem through by L^3/EI and defined $\beta = PL^2/60EI$ before taking the determinant. The smallest root can be computed with the quadratic formula and is

$$\beta = \frac{13 - 2\sqrt{31}}{45} = 0.0414327$$

Now, $P = 60EI\beta/L^2$. Thus, the buckling load is

$$P_{cr} = 2.4860\frac{EI}{L^2}$$

which compares to the exact solution of $P_{cr} = \pi^2 EI/4L^2 = 2.4674EI/L^2$. The two-term polynomial Ritz approximation is within 1% of the exact result.

The estimate of the critical load for the cantilever column problem can be made more accurate by taking more terms. Define the nth Ritz function and its derivatives as

$$h_n(\zeta) = \zeta^{n+1}, \qquad h'_n = (n+1)\zeta^n\frac{1}{L}, \qquad h''_n = n(n+1)\zeta^{n-1}\frac{1}{L^2} \tag{12.77}$$

where, again, $\zeta = x/L$ and $dx = Ld\zeta$. Now, we can compute the components of $\mathbf{K}$ and $\mathbf{G}$ exactly as

$$K_{nm} = EI\int_0^1 h''_n h''_m \, L\, d\zeta = \frac{nm(n+1)(m+1)}{n+m-1}\frac{EI}{L^3}$$

and

$$G_{nm} = \int_0^1 h'_n h'_m \, L\, d\zeta = \frac{(n+1)(m+1)}{n+m+1}\frac{1}{L}$$

The computation of the critical load is done by the following code:

Code 12.1 MATLAB code `CantileverRitz` for computing the buckling load of a cantilever column

```
%. Number of Ritz functions
   nRitz = 4;

%. Form K and G matrices
   K = zeros(nRitz,nRitz);
   G = zeros(nRitz,nRitz);
   for n=1:nRitz
     for m=1:nRitz
       K(n,m) = n*m*(n+1)*(m+1)/(n+m-1);
       G(n,m) = (n+1)*(m+1)/(n+m+1);
     end
   end
```

```
%. Compute critical load
   Pcr = eig(K,G);
   fprintf(' Cantilever column critical load = %8.4f EI/L^2\n',Pcr(1))
```

Of course, the code replicates the value computed in this example with two terms (i.e., nRitz=2). With three terms we get $P_{cr} = 2.4677\, EI/L^2$, and with four terms we get $P_{cr} = 2.4674\, EI/L^2$, which matches the exact value to five significant digits.

Generalizing the Ritz Method The biggest challenge in implementing the Ritz method is selecting Ritz functions that satisfy the essential boundary conditions. The cantilever column in the previous example was the simplest case, wherein we could use the monomial functions ζ^n, where $\zeta = x/L$, to give an approximation

$$w(\zeta) = a_0 + a_1\zeta + a_2\zeta^2 + \cdots + a_n\zeta^n$$

and then simply set $a_0 = a_1 = 0$ to satisfy the conditions $w(0) = 0$ and $w'(0) = 0$. Other boundary conditions can be dealt with simply if we build the Ritz functions on the four cubic Hermitian functions

$$\begin{aligned} h_1(\zeta) &= 1 - 3\zeta^2 + 2\zeta^3 \\ h_2(\zeta) &= \zeta - 2\zeta^2 + \zeta^3 \\ h_3(\zeta) &= 3\zeta^2 - 2\zeta^3 \\ h_4(\zeta) &= -\zeta^2 + \zeta^3 \end{aligned} \tag{12.78}$$

Each of these functions has either its value or first derivative at either $\zeta = 0$ or $\zeta = 1$ equal to one while the other three values and/or first derivatives are equal to zero at the ends. For example, $h_1(0) = 1$ while $h_1(1) = 0$, $h_1'(0) = 0$, and $h_1'(1) = 0$. Now, if we write the Ritz approximation as

$$w(\zeta) = a_1 h_1(\zeta) + a_2 h_2(\zeta) + a_3 h_3(\zeta) + a_4 h_4(\zeta)$$

then setting any of the a_n values to zero will restrain either displacement or rotation at one of the ends of the beam, depending on the associated Ritz function. The order is as follows: a_1 is the displacement at $x = 0$, a_2/L is the rotation at $x = 0$, a_3 is the displacement at $x = L$, and a_4/L is the rotation at $x = L$. Now, we can enforce essential boundary conditions simply by setting the appropriate a_n value to zero, as we had done for the cantilever beam. To improve the approximation beyond cubic functions, we can add what we call *bubble functions*. The most basic bubble function is

$$b(\zeta) = \zeta^2(1 - \zeta)^2 \tag{12.79}$$

This function has its value and first derivative equal to zero at both ends. Therefore, it does not affect the displacement or rotation at the ends of the beam. It is a quartic function, so it adds an additional polynomial degree to the approximation beyond the cubic Hermitian functions. Now, we can add additional Ritz functions as

$$h_{5+n}(\zeta) = c_n b(\zeta)\zeta^n, \qquad n = 0, \ldots, N-1 \tag{12.80}$$

where N is the number of bubble functions and the coefficient

$$c_n = \frac{(4+n)^{4+n}}{4(2+n)^{2+n}} \tag{12.81}$$

is chosen so that the maximum value of the nth Ritz function is one. All the additional functions have zero value and zero first derivative at the ends. Code 12.2 computes the Ritz functions along with the first, second, and third derivatives at a point x between 0 and 1.

Code 12.2 MATLAB code `RitzFcns` for computing Ritz functions based on the cubic Hermitian functions with added bubble functions

```
function h = RitzFcns(x,nBubbles)
%   Compute the cubic Hermitian functions extended by bubbles
%
%          x = location to evaluate functions x = [0,1]
%   nBubbles = Number bubble functions
%
%     h(i,:) =   [h, h', h'', h'''] of the ith Ritz function

%. Initialize functions and derivatives
   nFcns = nBubbles + 4;                        % Number of Ritz functions
   h = zeros(nFcns,4);                          % Third derivative

%. Cubic Hermitian functions, order: w(0), w'(0), w(1), w'(1)
   A = [ 1,      0,     -3,      2 ;...
         0,      1,     -2,      1 ;...
         0,      0,      3,     -2 ;...
         0,      0,     -1,      1 ];

   H = [ 1,      0,      0,      0 ;...
         x,      1,      0,      0 ;...
       x^2,    2*x,      2,      0 ;...
       x^3, 3*x^2,     6*x,      6 ];

%. First four Ritz functions
   h(1:4,:) = A*H;

%. Compute bubble functions only if nBubbles > 0
   if nBubbles>0

%... Compute the bubble functions and derivatives
     for i=5:nFcns
       n = i-5;
       nm1 = n-1;
       np1 = n+1; np2 = n+2; np3 = n+3; np4 = n+4;

%..... Constant to make magnitude of each function equal to one
       cn = (np4^np4)/(4*np2^np2);

%..... h(i,j) j=1,2,3,4 is function, 1st, 2nd, and 3rd derivatives
       h(i,1) = cn*x^np2*(1 - 2*x + x^2);
       h(i,2) = cn*x^np1*(np2 - 2*np3*x + np4*x^2);
       h(i,3) = cn*x^n*(np1*np2 - 2*np2*np3*x + np3*np4*x^2);
```

```
        h(i,4) = cn*x^n*(-2*np1*np2*np3 + np2*np3*np4*x);
        if n>0; h(i,4) = h(i,4) + cn*x^nm1*n*np1*np2; end
      end
    end

end
```

With the Ritz functions defined, we can implement a general solution to the beam buckling problem. Code 12.3 does this computation.

Code 12.3 MATLAB code `RitzBuckling` for computing the buckling load for a beam with any boundary conditions

```
%. Ritz/FE discretization
   nBubbles = 4;                              % Number of bubble functions
   nRitz = 4 + nBubbles;                      % Number of Ritz functions

%. Establish BCs: =1 fixed, =2 simple, =3 slide =4 free
   BC0 = 1;                                   % BC at x=0
   BCL = 2;                                   % BC at x=L

%. BCs:    W   Th  : =0 Value known to be zero, =1 Value is unknown
   BC = [ 0,   0 ;                            % Fixed
          0,   1 ;                            % Simple
          1,   0 ;                            % Slide
          1,   1 ];                           % Free

%. Create array to trim K and G matrices to reflect BCs
   BCindex = [BC(BC0,:), BC(BCL,:), ones(1,nBubbles)];

%. Compute the K and G matrices
   [K,G] = AssembleEqns(BCindex,nBubbles);

%. Compute critical loads and mode shapes
   Pcr = eig(K,G);

%. Print results
   BCtext = {'Fixed';'Simple';'Slide';'Free'};
   fprintf(' Buckling Analysis by Ritz Method\n\n')
   fprintf('   Base boundary condition :  %s\n'   ,BCtext{BC0})
   fprintf('   Tip boundary condition  :  %s\n\n',BCtext{BCL})
   fprintf('   Critical Load, Pcr = %12.5f EI/L^2\n',Pcr(1))
```

The number of bubble functions to include in the approximation is `nBubbles`. If this value is zero, then the approximation is based only on the four cubic Hermitian functions. It is not possible to do an approximation of order lower than cubic with this code. The implementation of the boundary conditions is based on setting the appropriate coefficients of the Ritz expansion equal to zero. Each row of the array `BC` has either a one or zero associated with displacement (`W`) or rotation (`Th`). If the value is one, then that motion is unknown and, hence, must be retained in the list of variables. If the value is zero, then we will reflect that by eliminating the associated row and column in the **K** and **G** matrices. All the coefficients associated with the bubble functions are unknown, as they are not affected by the boundary conditions. The array `BCindex` tells us which coefficients to keep and which to eliminate to enforce the restraint implied by the boundary conditions.

Code 12.4 computes the matrices **K** and **G**. The eigenvalue problem is solved using the MATLAB function `eig` in the form of the generalized eigenvalue problem. Finally, the first critical load `Pcr` is extracted. The constant factors EI/L^3 (for **K**) and $1/L$ (for **G**) are not included in the calculation, which means that the code actually computes $\lambda = PL^2/EI$. The output reports the numerical value, indicating the critical load is that number multiplied by EI/L^2. Thus, the result depends only on the boundary conditions with the effects of the physical parameters being a multiplier.

Code 12.4 MATLAB code `AssembleEqns` for computing the matrices **K** and **G** for the buckling eigenvalue problem

```
function [K,G] = AssembleEqns(BCindex,nBubbles)
%. Form equations by numerical integration

%. Initialize K and G matrices
   nRitz = 4 + nBubbles;
   K = zeros(nRitz,nRitz);
   G = zeros(nRitz,nRitz);

%. Get numerical integration parameters
   [wt,st,nQpts] = NumInt(1,nRitz);

%. Numerically integrate to get K and G
   for i=1:nQpts
     h = RitzFcns(st(i),nBubbles);
     K = K + wt(i)*(h(:,3)*h(:,3)');
     G = G + wt(i)*(h(:,2)*h(:,2)');
   end

%. Trim K and G to reflect boundary conditions
   K = K(BCindex==1,BCindex==1);
   G = G(BCindex==1,BCindex==1);

end
```

This function computes the integrals associated with **K** and **G** numerically. The weights and stations for the numerical quadrature are obtained from the function `NumInt`, which is Code A.2 in Appendix A. For each quadrature point, the integrand is formed for the value of x at the integration station `st`, and the contribution of that integration point, weighted by `wt`, is added to the arrays `K` and `G`. The one in the first slot of the call to `NumInt` selects Gauss-Legendre integration. If the number of Gauss-Legendre quadrature points is set equal to `nRitz`, then the integrals will be computed exactly.

Example 12.5

Compute the buckling load of a beam of length $L = 10$ and bending stiffness $EI = 100$ subjected to axial force P at the top, shown in Fig. 12.16. The beam is fixed at the base and supported by a roller at the tip. Use Code 12.3 to execute the calculation.

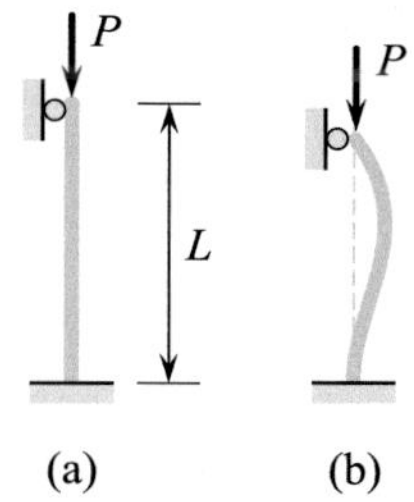

Fig. 12.16 *Example.* A beam with fixed support at the base and roller support at the top with an axial load P applied at the top. **a** Straight configuration. **b** Bent configuration

Solution. The boundary conditions must be set with `BC0=1` and `BCL=2` to give fixed and simple conditions, respectively. We do not know in advance how many Ritz functions are needed to get an accurate result. Therefore, we will start with the basic Hermitian functions and add bubble functions until we get a steady result. The results are shown in Table 12.1.

The solution with $n = 4$ is poor, but the accuracy increases quickly with the addition of more functions. With $n = 10$, the solution is accurate to seven significant digits.

The classical approach of solving the differential equation can be carried out for any boundary conditions to compute the buckling load of a prismatic beam. The proof of stability is the same for all boundary conditions. In fact, the buckling loads for the simple-simple, fixed-free (cantilever), fixed-fixed, and fixed-simple boundary conditions are well known and tabulated. Why bother with numerical methods, then? The classical approach to the solution of differential equations is much more difficult for problems with nonprismatic beams, where $EI(x)$ is a function of x, and beams with variable axial load, e.g., the buckling of a cantilever beam under its own weight. The Ritz approach extends gracefully to those more complex cases. In fact, the function `AssembleEqns` (Code 12.4) can be modified to do these two cases. The Ritz method can also be readily adapted to complex structures via the

Table 12.1 *Example.* Predicted buckling load P_{cr} for fixed-simple beam with number of Ritz functions n used in the approximation. The values in the table should be multiplied by EI/L^2

n	P_{cr}
4	30.00000
5	20.91868
6	20.28579
7	20.19432
8	20.19093
9	20.19073
10	20.19073

finite element method, which is the foundation of most structural analysis codes capable of estimating buckling loads of complex structures.

12.4 Epilogue

Accounting for the *stiffness* of a deformable body allows us to determine how forces flow through a solid body. It opens the door to analyzing statically indeterminate systems. That was the main focus of the book. The aim of this final chapter was to reveal a little about what is at stake in determining the capacity or *strength* of a deformable body. The strength of a system can be governed by material failure, instability, or both.

Estimating the strength of a system increases the complexity of the task of modeling deformable solids. Many different materials can be modeled with Hooke's law in their linear range of behavior, but few share the same failure modes. Constitutive models have been developed to determine the strength of systems made of different materials. Estimating strength also significantly increases the level of mathematical difficulty. Setting up the calculations to deal with nonlinearities in the kinematics and kinetics pushes the solution methods into the realm of numerical analysis. The Ritz method in this chapter gives a glimpse of the power of such methods and illustrates how numerical approaches change the way we organize the calculations.

Strength and stability are essential to understand in order to use the mechanics of deformable solids to design real systems. This chapter, then, represents the end of this book but serves as a prologue and invitation to further study of these important ideas.

Appendix A
Math Supplement

A.1 Newton's Method

Newton developed an iterative method for solving nonlinear algebraic equations. Consider a single equation with a single unknown, written in the form

$$g(x) = 0 \tag{A.1}$$

An example function, $g(x) = x^3 - 3x^2 - x + 3$, is shown in Fig. A.1. The solutions to Eq. A.1 are the open circles where the graph crosses the x axis. Unlike linear equations, there may be multiple solutions to the nonlinear equations. The graph shows how many solutions there are and where they are located. Newton's method is an algorithm to find those solutions.

We know from calculus that any function can be expanded in a Taylor series around some value x_o as

$$g(x) = g(x_o) + (x - x_o)\, g'(x_o) + \tfrac{1}{2}(x - x_o)^2 g''(x_o) + \cdots$$

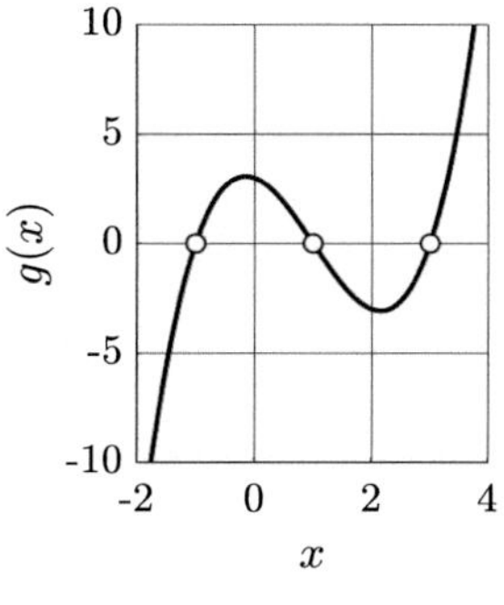

Fig. A.1 *Example nonlinear function.* The cubic function $g(x)$ has three roots, shown by the open circles. Hence, the solution to $g(x) = 0$ is not unique

K. D. Hjelmstad, *Mechanics of Deformable Solids: An Introduction*,
https://doi.org/10.1007/978-3-031-89873-0

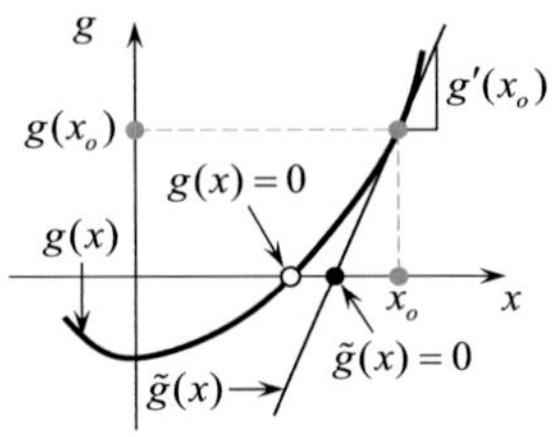

Fig. A.2 *Newton's method.* We approximate the original function $g(x)$ with a linear function that matches the value and tangent of the original function at the point x_o. We then solve the linear equation to get an estimate of the solution

where $g'(x_o)$ and $g''(x_o)$ are the first and second derivatives of $g(x)$ with respect to x, evaluated at the point x_o. The result is a polynomial. We can create a *linear* function by taking the first two terms of the Taylor series expansion of the original nonlinear function. Let us call that linear function

$$\tilde{g}(x) = g(x_o) + (x - x_o)\, g'(x_o)$$

You can get an idea of the relationship between the original function $g(x)$ and the associated linear function $\tilde{g}(x)$ from Fig. A.2. The original function is the heavy line, and the linear function is the thinner straight line that is tangent to it. The arbitrarily selected point x_o is shown as a gray dot on the abscissa, and the value of the original function $g(x_o)$ is shown as a gray dot on the ordinate. The gray circle where the curve and line intersect is the place where the original function and its linear counterpart have the same value and slope, i.e., $\tilde{g}(x_o) = g(x_o)$ and $\tilde{g}'(x_o) = g'(x_o)$. The open circle on the x axis is the solution we seek. Since $g(x_o) \neq 0$, we know that x_o is *not* the solution we seek.

Newton's idea was to recognize that we could actually solve the equation that results from setting the linear function equal to zero, i.e.,

$$\tilde{g}(x) = g(x_o) + (x - x_o)\, g'(x_o) = 0 \tag{A.2}$$

The value of x found by solving this equation is the solid black dot on the abscissa in the figure. This value depends upon our choice of x_o. Taking the result of one iteration as the x_o for the next iteration establishes an algorithm

$$x_{i+1} = x_i - \frac{g\,(x_i)}{g'\,(x_i)} \tag{A.3}$$

that generates a sequence of values $\{x_0, x_1, x_2, \ldots, x_n\}$. This iterative algorithm is Newton's method.[1] The sequence converges to a solution to the original nonlinear equation, if all goes well. Newton's method is not guaranteed to converge, but when

[1] Newton did not present the algorithm as we have stated it. Joseph Raphson (1688–1715), an English mathematician, put Newton's method into the form that we recognize today. Many authors call the method the Newton-Raphson method to acknowledge this contribution.

it does, the convergence is fast in the neighborhood of the solution. For problems where convergence is an issue, other strategies can be used. To see how the algorithm works, let us apply Newton's method to a simple nonlinear equation.

Example A.1
Find the solution to the equation $x^2 - 3 = 0$.

Solution. The function is $g(x) = x^2 - 3$ with derivative $g'(x) = 2x$. The Newton iteration equation takes the form

$$x_{i+1} = x_i - \frac{g(x_i)}{g'(x_i)} = x_i - \frac{x_i^2 - 3}{2x_i}$$

From any starting value (except for $x_0 = 0$ in this case), we can compute a sequence of estimates. Table A.1 shows how the iteration goes, starting from the value $x_0 = 10$. We could probably guess that 10 is not anywhere near the solution we seek, but this choice shows how quickly Newton's method converges to the true solution (which is $x = \sqrt{3}$).

Code A.1 MATLAB code `Newton` that solves this example

```
%. Solve g(x) = x^2 - 3 = 0 by Newton's Method

%. Set parameters
   x = 10;                    % Initial guess of x
   tol = 1.e-8;               % Tolerance for convergence of Newton's method
   err = 1.0;                 % Initial error > tol

%. Newton iteration
   while err > tol
     g = x^2 - 3;
     A = 2*x;
     err = norm(g);
     fprintf('%15.9f%15.9f\n',x,err)
     x   = x - g/A;
   end
```

To run Newton's method, you need to specify the starting value x_0 and an error tolerance that tells the algorithm when to stop. For the example in Table A.1, the tolerance was $tol = 10^{-8}$. It may also be a good idea to limit the number of allowable iterations in case the sequence diverges.

Table A.1 *Results of Newton iteration on a simple example.* With a starting point of $x_0 = 10$, Newton's method solves $g(x) = x^2 - 3 = 0$ to within an error tolerance of $tol = 10^{-8}$ in 6 iterations

n	x_n	$g(x_n)$
0	10.000000000	97.000000000
1	5.150000000	23.522500000
2	2.866262136	5.215458632
3	1.956460732	0.827738595
4	1.744920939	0.044749084
5	1.732098271	0.000164421
6	1.732050808	0.000000002

For the simple function in this example, the convergence is monotonic, which means that the error gets smaller in each successive iteration. The absolute value of $g(x_n)$ is a good measure of error at iteration n because the aim is for $g(x)$ to be zero. Notice that the errors in the last three iterations are on the order 10^{-2}, 10^{-4}, and 10^{-9}, which is the hallmark of *quadratic convergence*, where the negative exponent on the error doubles with each successive iteration. Why didn't the method find the other solution? Newton's method can only go to one solution at a time. To find the other root, a different starting point must be specified. Any negative value will work in this case.

A.2 The Directional Derivative

The *directional derivative* of a function $\mathbf{g}(\mathbf{x})$ is defined as

$$\mathcal{D}\mathbf{g}(\mathbf{x}) \cdot \mathbf{h} = \frac{d}{d\varepsilon}\Big[\mathbf{g}(\mathbf{x} + \varepsilon\mathbf{h})\Big]_{\varepsilon=0} \tag{A.4}$$

The notation on the left side stands for the *derivative* of the function $\mathbf{g}$ *at* location $\mathbf{x}$ *in the direction* of the vector $\mathbf{h}$, which is constant and has the same character as $\mathbf{x}$. In fact, the value $\varepsilon\mathbf{h}$ is a *perturbation* of $\mathbf{x}$—a little nudge to test how $\mathbf{g}$ changes in the neighborhood of $\mathbf{x}$. This approach reduces the rate of change of $\mathbf{g}$ to be an ordinary scalar derivative with respect to ε, holding the vectors $\mathbf{x}$ and $\mathbf{h}$ constant. Once the differentiation step is complete, we set $\varepsilon = 0$, giving the rate of change as we just start to move in the direction $\mathbf{h}$.

The directional derivative gives the rate of change in a specific direction, but since we did not restrict $\mathbf{h}$ in any way, it can represent *any* direction. In that sense, the directional derivative provides general information about the rate of change of the function. The reason the directional derivative is of interest to us is because

$$\mathcal{D}\mathbf{g}(\mathbf{x}) \cdot \mathbf{h} = \left[\nabla \mathbf{g}(\mathbf{x}) \right] \mathbf{h} \tag{A.5}$$

In other words, if we can factor the **h** out of the final expression for the directional derivative, the matrix that multiplies it is the gradient of the function $\mathbf{g}(\mathbf{x})$. The components of the gradient matrix can be expressed as

$$\left[\nabla \mathbf{g} \right]_{ij} = \frac{\partial g_i}{\partial x_j} \tag{A.6}$$

where g_i is the ith component of the vector **g** and x_j is the jth component of the vector **x**.

Example A.2
Scalar function of a vector variable. Compute the directional derivative of the function $r(\mathbf{x}) = \sqrt{\mathbf{x} \cdot \mathbf{x}}$. Identify the gradient of the function.

Solution. This function simply computes the length of a vector **x**, which is the square root of the vector dotted with itself. The directional derivative of $r(\mathbf{x})$ is

$$\begin{aligned}
\mathcal{D}r(\mathbf{x}) \cdot \mathbf{h} &= \frac{d}{d\varepsilon}\Big[r(\mathbf{x} + \varepsilon\mathbf{h}) \Big]_{\varepsilon=0} \\
&= \frac{d}{d\varepsilon}\Big[\big((\mathbf{x} + \varepsilon\mathbf{h}) \cdot (\mathbf{x} + \varepsilon\mathbf{h}) \big)^{\frac{1}{2}} \Big]_{\varepsilon=0} \\
&= \Big[\tfrac{1}{2} \big((\mathbf{x} + \varepsilon\mathbf{h}) \cdot (\mathbf{x} + \varepsilon\mathbf{h}) \big)^{-\frac{1}{2}} \big(2(\mathbf{x} + \varepsilon\mathbf{h}) \cdot \mathbf{h} \big) \Big]_{\varepsilon=0} \\
&= \frac{\mathbf{x} \cdot \mathbf{h}}{\sqrt{\mathbf{x} \cdot \mathbf{x}}} = \frac{\mathbf{x}}{r(\mathbf{x})} \cdot \mathbf{h}
\end{aligned}$$

We have invoked the product rule and the chain rule in taking these derivatives with respect to ε. Note that the vector $\mathbf{x}/r$ is a unit vector pointing in the direction **x**. We can identify $\mathbf{x}/r$ as the gradient of the function $r(\mathbf{x})$. In this case, the natural factoring out of the vector **h** is through the dot product. The directional derivative produces a scalar rate of change. The gradient of a scalar function of a vector variable is a vector. The dot product of vectors is a scalar.

Example A.3
Vector function of a vector variable. Compute the directional derivative of the function $\mathbf{g}(\mathbf{x}) = r(\mathbf{x})\mathbf{x}$, where $r(\mathbf{x}) = \sqrt{\mathbf{x} \cdot \mathbf{x}}$ is the same function as the previous example. Identify the gradient of the function.

Solution. We can build the directional derivative in pieces. First, the directional derivative of the function $\mathbf{x}$ is

$$\mathcal{D}\mathbf{x} \cdot \mathbf{h} = \frac{d}{d\varepsilon}\Big[\mathbf{x} + \varepsilon\mathbf{h}\Big]_{\varepsilon=0} = \mathbf{h} \tag{A.7}$$

The directional derivative of $r(\mathbf{x})\mathbf{x}$ can be computed with the product rule of differentiation, which works the same for directional derivatives as it does for ordinary derivatives. Since we have already found the directional derivatives of each component, we can compute

$$\begin{aligned}\mathcal{D}\mathbf{g}(\mathbf{x}) \cdot \mathbf{h} &= \Big(\mathcal{D}r(\mathbf{x}) \cdot \mathbf{h}\Big)\mathbf{x} + r(\mathbf{x})\Big(\mathcal{D}\mathbf{x} \cdot \mathbf{h}\Big) \\ &= \Big(\frac{\mathbf{x}}{r} \cdot \mathbf{h}\Big)\mathbf{x} + r(\mathbf{x})\mathbf{h}\end{aligned}$$

Now, we need to factor out the vector $\mathbf{h}$ to identify the gradient matrix. First, note that in matrix notation the dot product is $\mathbf{a} \cdot \mathbf{b} = \mathbf{a}^T\mathbf{b}$. Also, note that the scalar that results from the dot product can be moved to the right side of $\mathbf{x}$ in the first term without violating the rules of matrix algebra. Thus, we can write

$$\mathbf{x}\Big(\frac{1}{r}\mathbf{x}^T\mathbf{h}\Big) + r\mathbf{h} = \Big[\frac{1}{r}\mathbf{x}\mathbf{x}^T + r\mathbf{I}\Big]\mathbf{h}$$

where $\mathbf{I}$ is the identity matrix of dimension equal to the dimension of $\mathbf{x}$ (which we will call n here). We put the identity matrix in because the identity times any vector returns the same vector. In order to group the two terms together, they must have the same matrix character. The term $\mathbf{x}\mathbf{x}^T$ is an $n \times n$ matrix, which can be added to an $n \times n$ identity matrix. Finally, we can identify the gradient matrix

$$\nabla\mathbf{g} = \frac{1}{r(\mathbf{x})}\mathbf{x}\mathbf{x}^T + r(\mathbf{x})\mathbf{I}$$

such that $\mathcal{D}\mathbf{g}(\mathbf{x}) \cdot \mathbf{h} = [\, \nabla\mathbf{g}\,]\mathbf{h}$.

A.3 Numerical Integration of Differential Equations

In our computational strategies for the axial bar and beam problems, we convert the boundary-value problem into an initial-value problem for the purpose of computing the distribution of the state variables along the axis. We can create these plots by integrating the differential equations numerically. This section goes over the theory behind the numerical solution of initial-value problems.

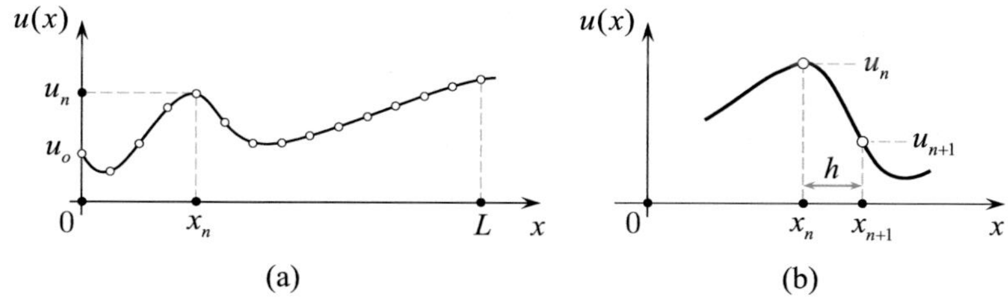

Fig. A.3 *Discrete solution.* A sequence of points in space that represent an underlying function. **a** The complete discrete series, showing the initial value u_o, the value u_n at point x_n, and the final point at L. **b** Close-up of two adjacent points, showing their respective values and the size of the step, h

In numerical integration we discretize the problem to find the response at certain points, and then connect the dots to visualize the function. If we are trying to approximate a function $u(x)$, then we will find a sequence of values

$$\{u_0, u_1, u_2, \ldots, u_n, \ldots, u_N\}$$

where u_n represents the value of the exact function $u(x_n)$ at the specific location x_n. In general, u_n will only be an approximation of the actual response function evaluated at that point. Figure A.3 shows a discrete representation of a function $u(x)$. While it is possible to develop methods with variable step sizes, in this book we will always take equally spaced points, so that $h = x_{n+1} - x_n$ is constant for all n. The final value is $x_N = L$.

We call the set of values $\{u_n, N_n\}$ for the axial bar problem the *state* at location x_n. For the beam problem, $\{V_n, M_n, \theta_n, w_n\}$ is the state. In the continuous setting, the state variables, e.g., $u(x)$ and $N(x)$ for the axial bar problem, are related to each other through differential relationships. In the discrete setting we view u_n and N_n as independent quantities, and we need to add a relationship between them that causes them to emulate the original differential relationship. These discrete relationships distinguish one numerical integrator from another. The key constraint is that only the values at the discrete points in the approximation can be used.

For the axial bar problem, the internal axial force is related to the displacement through the relationship $u' = N/EA$, which can be integrated between x_n and x_{n+1} to give

$$u(x_{n+1}) = u(x_n) + \frac{1}{EA}\int_{x_n}^{x_{n+1}} N(\xi)\, d\xi \tag{A.8}$$

This relationship is exact. The problem is that we do not know how the axial force varies between the discrete locations x_n and x_{n+1}. The situation is depicted in Fig. A.4. The integral we seek is the area under the curve. We know N_n and we seek N_{n+1}. Let us approximate the variation of the axial force between the two points as a linear function. To simplify the computations, let $\zeta = (x - x_n)/h$. Now, the axial force between the discrete points can be expressed as the linear function

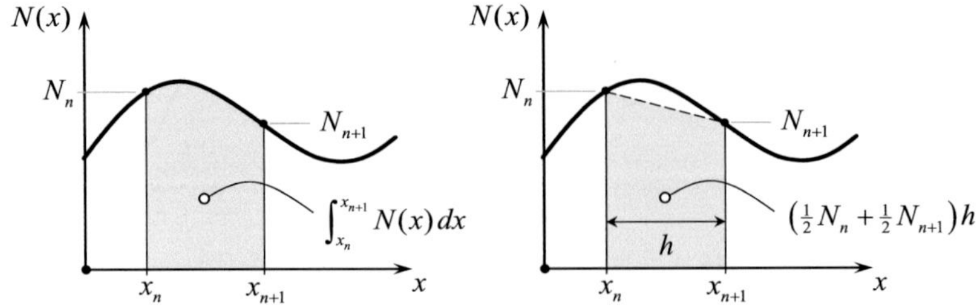

Fig. A.4 *The trapezoidal rule.* The idea behind the trapezoidal rule is to approximate the integral using the area of the trapezoid defined by connecting two adjacent points in the discrete series

$$N(\zeta) = (1-\zeta)\,N_n + \zeta\,N_{n+1} \tag{A.9}$$

Even though we still do not know the value of N_{n+1}, noting that $dx = h\,d\zeta$, we can execute the integral as

$$\begin{aligned}\int_{x_n}^{x_{n+1}} N(\xi)\,d\xi &= \int_0^1 \big[(1-\zeta)N_n + \zeta\,N_{n+1}\big]h\,d\zeta \\ &= \big[\tfrac{1}{2}N_n + \tfrac{1}{2}N_{n+1}\big]h\end{aligned}$$

The result is simply the area of a trapezoid defined by the linear interpolation of the two discrete values. This method is called the *trapezoidal rule* of integration. We can generalize this formula by asserting that

$$\int_{x_n}^{x_{n+1}} N(\xi)\,d\xi \approx \big[\beta N_n + (1-\beta)N_{n+1}\big]h$$

where $\beta \in [\,0,\,1\,]$ is a parameter. This parameterization creates a family of integrators that we call the *generalized trapezoidal rule*. One member of that family, $\beta = 0.5$, is the trapezoidal rule. If we take $\beta = 1$, then we are approximating the integral with the left Riemann sum, a method often called the *backward Euler* method. If we take $\beta = 0$, then we are approximating the integral with the right Riemann sum, a method often called the *forward Euler* method.

Applying the generalized trapezoidal rule to the axial bar problem gives the recursion

$$\begin{aligned}N_{n+1} &= N_n - h\big[\,\beta p_n + (1-\beta)p_{n+1}\,\big] \\ u_{n+1} &= u_n + h\big[\,\beta N_n + (1-\beta)N_{n+1}\,\big]/EA\end{aligned}$$

to numerically integrate the differential equations

$$N'(x) + p(x) = 0, \qquad u'(x) = \frac{N(x)}{EA}$$

In this algorithm, we step forward from known initial conditions $\{u_o, N_o\}$, computing the new state from the old state and then moving on to the next step. Since the axial bar problem is a boundary-value problem, we must first compute the initial values u_o and N_o, only one of which is known from the boundary conditions for any given problem.

From Fig. A.4, it is evident that the area of the trapezoid is not the same as the actual integral unless the variation of the real function is linear. But if the two points are taken close together, then the error is small. Hence, the step size h is crucial to the accuracy of the approximation.

A.4 Numerical Quadrature

In the study of the axial bar and beams, integrals over the domain of the system show up. For example, in the *method of initial parameters*, the computations of $\mathcal{I}_0, \mathcal{I}_1, \mathcal{I}_2$, and $\mathcal{I}_3$ are definite integrals over the domain of the bar. The process of executing these integrals is called *quadrature*.

The *definite integral* of the function $g(x)$ between the limits a and b is shown in Fig. A.5. The geometric significance of this integral is that it is the shaded area under the curve between those two points. By the fundamental theorem of calculus, we can evaluate this integral as

$$\int_a^b g(x)\,dx = G(b) - G(a) \tag{A.10}$$

The function $G(x)$ is the *antiderivative* of the function $g(x)$, i.e., $G'(x) = g(x)$, where the prime indicates differentiation with respect to x. When the antiderivative is simple to compute, that approach is ideal, because it is exact. However, if the function is complicated, or if you just want to write a general-purpose computer routine to do integrals, then numerical quadrature is a good alternative. Numerical quadrature rules can be put into the general form

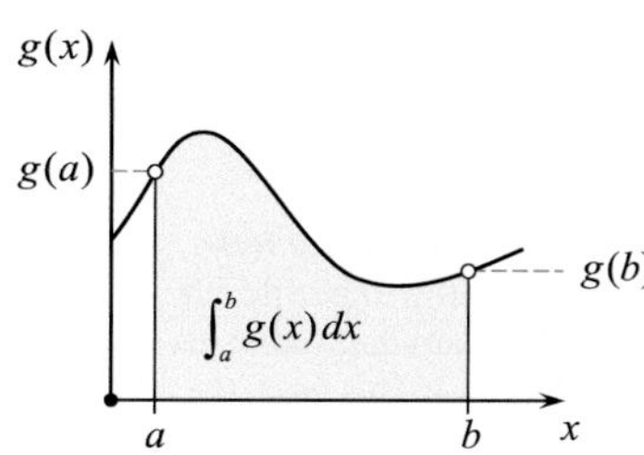

Fig. A.5 *Integral.* Geometrically, the integral of $g(x)$ from a to b is the area under the curve between those two points

$$\int_a^b g(x)\,dx = \sum_{i=1}^{N} g(x_i)w_i \tag{A.11}$$

where x_i are called the *stations* and w_i are called the *weights*. Numerical quadrature, then, reduces the problem of integration to the evaluation of the function at certain points, multiplying those values by their respective weights, and adding them up. The only distinction between one numerical quadrature rule and another is the specific values of the stations and weights.

Any integral from a to b can be transformed to an integral from 0 to 1 with the change of variable $\xi = (x-a)/h$, where $h = b - a$, i.e.,

$$\int_a^b g(x)\,dx = h\int_0^1 g(\xi)\,d\xi$$

The change of variable changes the limits of integration, the element of integration is $dx = h\,d\xi$, and the function $g(x)$ can be converted by substituting $x = a + \xi h$. For some applications it might be useful to transform the region $[\,a, b\,]$ to the interval $[-1, 1\,]$. The change of variable $\xi = -1 + 2(x-a)/h$ converts the integral to

$$\int_a^b g(x)\,dx = \tfrac{1}{2}h\int_{-1}^1 g(\xi)\,d\xi$$

The change of variable will simplify the specification of the integration stations for the numerical methods.

Trapezoidal Rule The simplest version of numerical quadrature is the trapezoidal rule. The idea is based upon the formula for area of a trapezoid, which is illustrated in Fig. A.6, and has the explicit expression

$$A_{trap} = \tfrac{1}{2}h\big[\,g(b) + g(a)\,\big] \tag{A.12}$$

where $h = b - a$ is the distance between the two points. The vertices of the trapezoid lie on the curve $g(x)$. Hence, the area formula only involves the evaluation of the function at those two points. It is evident that in regions where the curvature of the function is negative (as shown in the figure), the area of the trapezoid underestimates

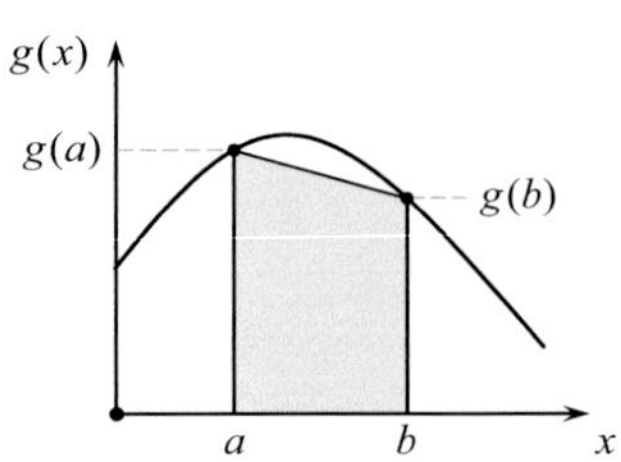

Fig. A.6 *Trapezoidal Rule.* The area of the trapezoid (the shaded region) approximates the integral. In this case, the shaded area of the trapezoid underestimates the area under the curve

the area under the curve. On the other hand, in regions where the curvature is positive, the opposite happens.

If we divide the region of integration into segments defined by *nodes* located at $\{x_0, \ldots, x_N\}$, then we can create a trapezoid between each pair of nodes. This gives $N+1$ nodes with N trapezoidal segments. Let us further assume that the nodes are equally spaced, i.e., all $x_{i+1} - x_i = h$ are the same. If the goal is to integrate from a to b, for example, then $h = (b-a)/N$. With this definition, we can write the composite rule as

$$\int_a^b g(x)\,dx = \tfrac{1}{2}h \sum_{i=0}^{N} \left[\, g(x_{i+1}) + g(x_i)\,\right] \tag{A.13}$$

This equation can be put into the standard form of Eq. A.11 if we identify the weights as

$$w_i = \tfrac{1}{2}h \begin{cases} 1 & i = 0 \text{ or } N \\ 2 & i \neq 0 \text{ or } N \end{cases} \tag{A.14}$$

with stations $x_i = a + ih$ for $i = 0, \ldots, N$. The reason that the interior points have twice the weight is because they participate in two adjacent trapezoids. This method is called the *composite trapezoidal rule*.

Simpson's Rule Polynomial functions are simple to integrate. If we fit a quadratic function through three adjacent points that lie on the given function, then we can approximate the area under the curve by integrating the quadratic function exactly. This approach is called *Simpson's rule*. The setup is shown in Fig. A.7.

Imagine that we want to integrate $g(x)$ from a to c. The point b is halfway between a and c, the distances between points being $h = b - a = c - b$. Rather than integrate $g(x)$, we will find a quadratic function $q(x)$, the gray curve in Fig. A.7, and integrate that instead. To simplify this calculation, let $\zeta = (x - b)/h$, and note that $dx = h\,d\zeta$. The point a is at $\zeta = -1$, b is at $\zeta = 0$, and c is at $\zeta = 1$. Now the approximate quadratic function can be expressed as

$$q(\zeta) = A + B\zeta + C\zeta^2 \tag{A.15}$$

where the values of A, B, and C are determined by making sure the quadratic function passes through the three points on the actual curve. To wit,

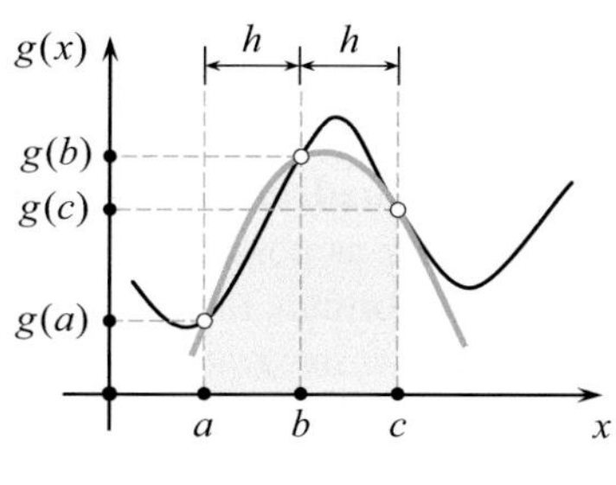

Fig. A.7 *Simpson's rule.* Gray curve is a quadratic function that is equal to the original function (black curve) at a, b, and c. Integrate the quadratic function (the shaded area) exactly to approximate the integral

$$\begin{aligned} q(-1) &= A - B + C = g(a) \\ q(0) &= A = g(b) \\ q(1) &= A + B + C = g(c) \end{aligned}$$

We can solve this system of equations to give

$$A = g(b), \quad B = \tfrac{1}{2}g(c) - \tfrac{1}{2}g(a), \quad C = \tfrac{1}{2}g(a) - g(b) + \tfrac{1}{2}g(c)$$

Now, integrate the quadratic function to give

$$\int_{-1}^{1} q(\zeta)h\, d\zeta = h \int_{-1}^{1} \left(A + B\zeta + C\zeta^2\right) d\zeta = 2h\left(A + \tfrac{1}{3}C\right)$$

Finally, substitute the values of A and C to give the value of the integral

$$\int_{a}^{c} q(x)\, dx = \tfrac{1}{3}h\left[\, g(a) + 4g(b) + g(c)\,\right] \tag{A.16}$$

This is the Simpson approximation of the integral of $g(x)$. It should be evident that we can chop any integration region up into multiple pieces and then treat each piece just like this one. When you do that, the c point from the first region becomes the a point for the next one, etc. As with the trapezoidal rule, adjacent interior points contribute twice. Therefore, the weights for the composite Simpson rule are

$$w_i = \tfrac{1}{3}h \begin{cases} 1 & i = 0 \text{ or } N \\ 4 & i = \text{odd interior} \\ 2 & i = \text{even interior} \end{cases} \tag{A.17}$$

with stations $x_i = a + ih$, where $h = (b - a)/N$, for $i = 0, \ldots, N$. Note that N must be an even number (because each Simpson segment involves three points). This method is called the *composite Simpson rule*. A simple way to generate the weights in MATLAB is to define M as the number of Simpson segments and then use the `repmat` function

```
w = [1, repmat([4,2],1,M), 4, 1]
```

to generate the values w = [1, 4, 2, 4, ..., 2, 4, 1]. Now, compute $N = \text{length}(\text{w})$ and the length of the Simpson segment as $2h = L/M$. Finally, complete the weights by multiplying w by $h/3$.

Gaussian Quadrature We want our numerical quadrature formulas to integrate as accurately as possible. One way to interpret that desire is to have the numerical quadrature formula integrate a certain set of simple functions exactly. For example, any numerical integrator should integrate the function $g(x) = 1$ exactly. If it cannot do that one, then there is little hope of integrating more complex functions. Since any function can be represented as a polynomial through Taylor's series, we might also think of the accuracy of quadrature as being how many of the terms in the series it can integrate exactly. The trapezoidal rule can integrate a linear function exactly. Simpson's rule can integrate a quadratic function exactly (recall, that is how we derived the rule). Gaussian quadrature is a strategy for integrating polynomials of higher order exactly.

Gauss-Legendre quadrature When we derived Simpson's rule, we assumed that the integration stations were fixed, and we sought weights that would integrate the quadratic function exactly. Gauss had the great idea that you could optimize the integrator by adjusting both the stations x_i and the weights w_i to find the integrator that could exactly integrate a polynomial of the highest order possible. Consider, for example, a two-point quadrature rule

$$\int_{-1}^{1} g(\xi)\, d\xi = w_1 g(\xi_1) + w_2 g(\xi_2) \tag{A.18}$$

with four parameters: two stations (ξ_1, ξ_2) and two weights (w_1, w_2). If we select the stations and weights to integrate the functions $\left\{1, \xi, \xi^2, \xi^3\right\}$ exactly, then we get four equations to solve for the four parameters. For the two-point rule, we find that

$$w_1 = 1, \qquad w_2 = 1, \qquad \xi_1 = -\tfrac{1}{\sqrt{3}}, \qquad \xi_2 = \tfrac{1}{\sqrt{3}}$$

This quadrature rule will integrate cubic functions exactly with only two integration stations.

Gauss-Legendre quadrature rules can be derived from the Legendre polynomials. The integration stations x_i for an n-point Gauss-Legendre rule lie at the zeros of the Legendre polynomial $P_n(x)$. The weights are then given by

$$w_i = \frac{2}{\left(1 - x_i^2\right)\left[P_n'(x_i)\right]^2}$$

The Legendre polynomials can be computed symbolically in MATLAB with the function `legendreP(n,x)`, where `x` is declared symbolic. An n-point Gauss-Legendre quadrature rule will integrate a polynomial of degree $2n-1$ exactly.

Gauss-Lobatto quadrature In some circumstances it is desirable to specify that two of the integration stations are at the ends of the integration region. Gaussian quadrature rules that include the end points are called *Gauss-Lobatto* quadrature. Since we are specifying the first and last station, that takes away two of our equations, and hence, an n-point Gauss-Lobatto quadrature rule will integrate a polynomial of degree $2n-3$ exactly.

The integration stations x_i for an n-point Gauss-Lobatto rule lie at the zeros of the Legendre polynomial $P'_{n-1}(x)$, plus the end points. The weights are then given by

$$w_i = \frac{2}{n(n-1)\left[P_{n-1}(x_i)\right]^2}$$

The weights of the end points are half of one minus the sum of the weights of the interior points. Again, the stations and weights can be computed symbolically in MATLAB.

Implementation Code A.2, `NumInt`, computes the weights and stations for a specified numerical quadrature rule. This function is used in various codes throughout the book.

Code A.2 MATLAB code `NumInt` finds the weights and stations for Gauss-Legendre, Gauss-Lobatto, composite Simpson, and composite trapezoidal quadrature rules

```
function [wt,st,nQpts] = NumInt(QType,n)
%. This function establishes the weights and stations for numerical
%  integration by various rules on segment x = [0,1]
%
%   QType = Specify quadrature rule
%       n = Number of integration points (nQpts going out)
%      wt = Weights
%      st = Stations
%------------------------------------------------------------------

   switch QType

     case 1   % Gauss-Legendre
       syms x
       P = legendreP(n,x);                    % nth order Legendre polynomial
       dP = diff(P,x);                        % Compute first derivative of P
       st = vpasolve(P==0);                   % Stations are roots of Pn
       a = vpa(subs(dP,x,st));                % Evaluate dP at stations
       wt = eval(1./((1-st.^2).*a.^2));       % Weights for integration
       st = eval((st + 1)/2);                 % Convert to region [0,1]
       clear x;                               % Release symbolic x
       nQpts = n;                             % Number of points in rule

     case 2   % Gauss-Lobatto
       syms x
       P = legendreP(n-1,x);                  % Legendre polynomial
       dP = diff(P,x);                        % Compute first derivative of P
       st = vpasolve(dP==0);                  % Stations are roots of dP
       a = vpa(subs(P,x,st));                 % Evaluate dP at stations
       wt = eval(1./(n*(n-1)*a.^2));          % Weights for integration
       st = eval((st + 1)/2);                 % Convert to region [0,1]
       st = [0; st; 1];                       % Add end points
       R = (1-sum(wt))/2;                     % Compute leftover weights
       wt = [R; wt; R];                       % Add weights for end points
       clear x;                               % Release symbolic x
       nQpts = n;                             % Number of points in rule

     case 3  % Composite Simpson
       m = ceil((n-3)/2);                     % Number of simpson segments
       wt = repmat([4;2],m,1);                % [4,2,4,2,4,2...]
```

```
        wt = [1;wt;4;1];                       % Simpson weights
        nQpts = length(wt);                    % No. Simpson Integration pts.
        wt = wt/(3*(nQpts-1));                 % Simpson weights
        st = (0:1/(nQpts-1):1)';               % Simpson stations

      case 4   % Composite Trapezoidal
        wt = 2*ones(n-2,1);                    % [2,2,2,...,2]
        wt = [1;wt;1];                         % Trapezoidal weights
        nQpts = length(wt);                    % No. Integration pts
        wt = wt/(2*(nQpts-1));                 % Trapezoidal weights
        st = (0:1/(nQpts-1):1)';               % Trapezoidal stations

    end

end
```

Note that this function assumes that the range of integration is [0, 1]. The routine works for any number of points. The number of integration points is specified in the input to the function, as is the type of rule.

Examples To demonstrate the performance of the quadrature rules, let us consider a couple of examples. The functions are shown in Fig. A.8. We will compute the integrals using quadrature rules with different numbers of integration stations, n.

The first case is a polynomial function $g(x) = -5x + 9x^3 + 10x^6$, shown in Fig. A.8 (the function on the left), integrated from $x = 0$ to $x = 1$. The results are given in Table A.2. Gauss-Legendre gets the exact result with $n = 4$ and Gauss-Lobatto does so with $n = 5$. This result is consistent with the observation that Gauss-Legendre will integrate a polynomial up to order $2n-1$ exactly, while Gauss-Lobatto will exactly integrate one of order $2n-3$. Hence, the fourth order Gauss-Lobatto rule can integrate a fifth order polynomial exactly, but not a sixth. Note that Simpson's rule is not exact, even up to 11 integration points. In fact, even with 59 integration points, Simpson's rule still has an error in the seventh digit. The trapezoidal rule is accurate only to three places with 51 integration points. Observe that Gauss-Lobatto and Simpson are the same for $n = 3$.

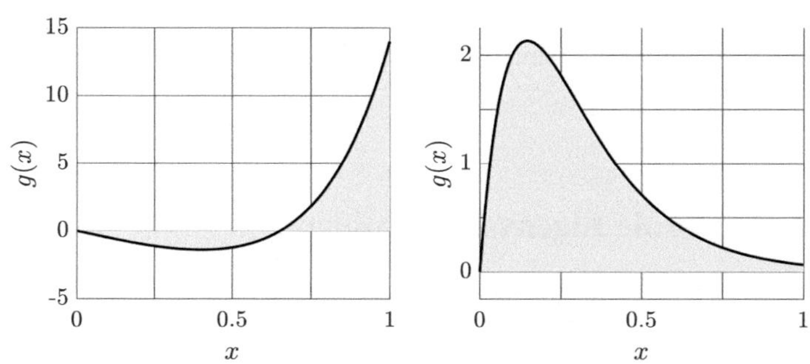

Fig. A.8 *Functions to compare numerical quadrature rules.* The function on the left, Example 1, is $g(x) = -5x + 9x^3 + 10x^6$. The results for this function are in Table A.2. The function on the right, Example 2, is $g(x) = 10\left(e^{-5x} - e^{-9x}\right)$. The results for this function are in Table A.3

Table A.2 *Example 1.* Integration with Gauss-Legendre, Gauss-Lobatto, Simpson, and Trapezoidal quadrature rules integrating the function $g(x) = -5x + 9x^3 + 10x^6$ between $x = 0$ and $x = 1$

n	G.-Legendre	n	G.-Lobatto	n	Simpson	n	Trapezoidal
2	0.9537037	3	1.5208333	3	1.5208333	11	1.2509050
3	1.9537037	4	1.1833333	5	1.2034505	21	1.1966860
4	1.1785714	5	1.1785714	7	1.1836134	31	1.1866249
5	1.1785714	6	1.1785714	9	1.1801809	41	1.1831020
6	1.1785714	7	1.1785714	11	1.1792333	51	1.1814712
Exact	1.1785714		1.1785714		1.1785714		1.1785714

Table A.3 *Example 2.* Integration with Gauss-Legendre, Gauss-Lobatto, Simpson, and Trapezoidal quadrature rules integrating the function $g(x) = 10\left(e^{-5x} - e^{-9x}\right)$ between $x = 0$ and $x = 1$

n	G.-Legendre	n	G.-Lobatto	n	Simpson	n	Trapezoidal
2	1.0845107	3	0.4841976	3	0.4841976	11	0.8427671
3	0.9211570	4	0.8110151	5	0.8019802	21	0.8672011
4	0.8797249	5	0.8701934	7	0.8559785	31	0.8718266
5	0.8757772	6	0.8752738	9	0.8685867	41	0.8734531
6	0.8755585	7	0.8755403	11	0.8725302	51	0.8742073
Exact	0.8755501		0.8755501		0.8755501		0.8755501

The second example is *not* a polynomial function, but rather it is an exponential rise followed by exponential decay, i.e., $g(x) = 10(e^{-5x} - e^{-9x})$, as shown in Fig. A.8 (the function on the right). Again, the function is integrated from $x = 0$ to $x = 1$. The results are given in Table A.3. None of the quadrature rules get the integral exactly right. Gauss-Legendre and Gauss-Lobatto are right to five digits with 6 and 7 stations, respectively. Simpson and trapezoidal rule get 2 digits with 11 and 51 stations, respectively. The superiority of the Gaussian quadrature rules is evident.

The Gaussian integration rules are particularly well-suited for integrating polynomials and perform well on many functions with a small number of integration stations.

A.5 The Algebraic Eigenvalue Problem

In mechanics, the *eigenvalue problem* shows up in several different contexts. As such, it is a fundamental building block for understanding the subject. In Chap. 2, the eigenvalue problem gave insight into the static determinacy and stability of truss structures. In Chaps. 4 and 5, the eigenvalue problem appears in the context of the principal values and principal directions of the strain and stress tensors. In Chap. 9 it shows up in the context of the computation of the principal value of the moment of inertia. In Chap. 12 it shows up in the context of stability of equilibrium. This appendix provides a review of the mathematics of the symmetric matrix *eigenvalue problem*, as well as a description of some approaches to solving eigenvalue problems.

The *standard eigenvalue problem* takes the form

$$\mathbf{A}\boldsymbol{\varphi} = \lambda\boldsymbol{\varphi} \tag{A.19}$$

where $\mathbf{A}$ is a *symmetric* $N \times N$ matrix. We assume that the matrix $\mathbf{A}$ is known and that we seek λ and $\boldsymbol{\varphi}$ as the solution to the eigenvalue problem. In essence, the eigenvalue problem seeks a vector $\boldsymbol{\varphi}$ with the property that multiplying by the matrix $\mathbf{A}$ is the same as scaling the length of the vector without changing its direction.

Characteristic Equation Equation A.19 can be written as a linear system of equations. To wit,

$$\left[\mathbf{A} - \lambda\,\mathbf{I}\right]\boldsymbol{\varphi} = \mathbf{0} \tag{A.20}$$

where $\mathbf{I}$ is the $N \times N$ *identity* matrix. The condition for a solution to the eigenvalue problem other than the trivial solution $\boldsymbol{\varphi} = \mathbf{0}$ is that the determinant of the coefficient matrix must be zero, i.e.,

$$\det\left[\mathbf{A} - \lambda\,\mathbf{I}\right] = 0 \tag{A.21}$$

This equation is called the *characteristic equation* associated with the eigenvalue problem. It is an Nth order polynomial equation in λ. Notice that this equation does not involve $\boldsymbol{\varphi}$. The zeros of the polynomial are called the *eigenvalues* of $\mathbf{A}$. For small systems, solving the characteristic equation is an effective approach to finding the eigenvalues.

As an example, consider the matrix

$$\mathbf{A} = \begin{bmatrix} 5 & 2 & 1 & 0 & 0 \\ 2 & 6 & 3 & 1 & 0 \\ 1 & 3 & 7 & 2 & 1 \\ 0 & 1 & 2 & 8 & 2 \\ 0 & 0 & 1 & 2 & 9 \end{bmatrix} \tag{A.22}$$

The plot of $\det[\,\mathbf{A} - \lambda\mathbf{I}\,]$ versus λ is shown in Fig. A.9. The roots of the characteristic equation are shown with open circles in the plot and have the values

$$\lambda = 2.891, \quad 4.388, \quad 6.233, \quad 9.060, \quad 12.428$$

These roots can be found by any root-finding algorithm for a scalar equation of a scalar unknown, e.g., Newton's method.

Once the eigenvalues λ_i for $i = 1, \ldots, N$ are known, the *eigenvectors* can be found by solving the N homogeneous linear systems of equations

$$\left[\mathbf{A} - \lambda_i\mathbf{I}\right]\boldsymbol{\varphi}_i = \mathbf{0}, \qquad i = 1, \ldots, N \tag{A.23}$$

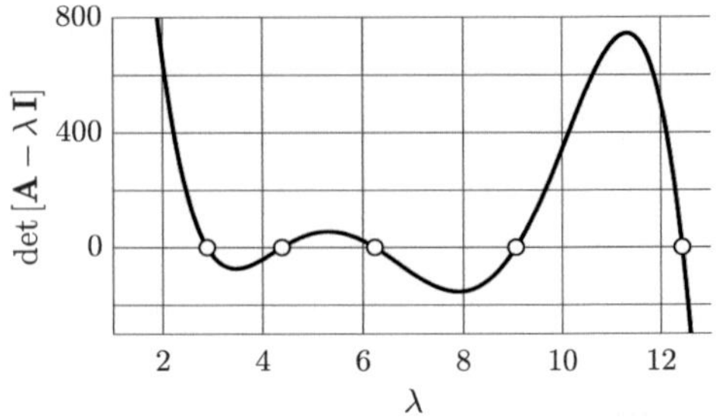

Fig. A.9 Plot of $\det[\mathbf{A} - \lambda\,\mathbf{I}]$ versus λ. The eigenvalues are the roots of the characteristic equation

The coefficient matrix for each of these equations is singular. That is how we got the λ_i values in the first place. Therefore, solving these equations must assume some knowledge of at least one of the components of $\boldsymbol{\varphi}_i$, and more than one if the eigenvalue is repeated. Notice that we could multiply $\boldsymbol{\varphi}_i$ by any scalar, including negative values, and the resulting vector would still satisfy the eigenvalue problem. For that reason, the eigenvectors are often called *mode shapes* because they represent a shape whose magnitude is indeterminate. It is common to normalize the eigenvectors to unit length, i.e., $\boldsymbol{\varphi}_i^T\boldsymbol{\varphi}_i = 1$, in which case they are called *orthonormal*.

Computing the Eigenvectors If an eigenvalue λ_i is known, e.g., by solving the characteristic equation, then we can find the corresponding eigenvector $\boldsymbol{\varphi}_i$ as follows. First, partition the problem as

$$[\mathbf{A} - \lambda_i\mathbf{I}]\hat{\boldsymbol{\varphi}}_i = \begin{bmatrix} b_i & \mathbf{b}_i^T \\ \mathbf{b}_i & \mathbf{B}_i \end{bmatrix}\begin{Bmatrix} 1 \\ \mathbf{x}_i \end{Bmatrix} = \begin{Bmatrix} 0 \\ \mathbf{0} \end{Bmatrix}$$

where b_i and $\mathbf{b}_i$ make up the first column, b_i and $\mathbf{b}_i^T$ the first row, and $\mathbf{B}_i$ the lower right $(N-1)\times(N-1)$ submatrix of $[\mathbf{A} - \lambda_i\mathbf{I}]$. Solve for $\mathbf{x}_i$ from $\mathbf{b}_i + \mathbf{B}_i\mathbf{x}_i = \mathbf{0}$, and construct the eigenvector $\boldsymbol{\varphi}_i$ as

$$\mathbf{x}_i = -\mathbf{B}_i^{-1}\mathbf{b}_i, \quad \rightarrow \quad \hat{\boldsymbol{\varphi}}_i = \begin{Bmatrix} 1 \\ \mathbf{x}_i \end{Bmatrix} \quad \rightarrow \quad \boldsymbol{\varphi}_i = \frac{\hat{\boldsymbol{\varphi}}_i}{\|\hat{\boldsymbol{\varphi}}_i\|}$$

Repeat this computation for $i = 1, \ldots, N$, as shown in Code A.3 (which includes the example matrix $\mathbf{A}$ in Eq. A.22 again).

Code A.3 MATLAB code `SolveEVP` computes the eigenvalues and eigenvectors of a matrix

```
%  Solve eigenvalue problem with characteristic equation approach
   clear; clc;

%. Input matrix
   A = [ 5, 2, 1, 0, 0 ;
         2, 6, 3, 1, 0 ;
         1, 3, 7, 2, 1 ;
         0, 1, 2, 8, 2 ;
         0, 0, 1, 2, 9 ];
```

```
%. Establish problem size, create identity matrix
   N = size(A,1);
   I = eye(N);

%. Find eigenvalues by solving characteristic equation
   syms z
   f=@(z) det(A-z*I);
   lambda = double(vpa(solve(f(z)==0)));

%. Find eigenvectors by solving linear system of equations
   PHI = zeros(N,N);
   for i=1:N
     B = A-lambda(i)*I;
     b = B(2:N,1);
     B = B(2:N,2:N);
     x = -B\b;
     phi = [1;x];
     PHI(:,i) = phi/norm(phi);
   end
```

This process assumes that the first component of the eigenvector is not zero. We assume a value of one for that element, solve the remaining $N-1$ equations for $\mathbf{x}_i$, and then normalize the vector to unit length. For this matrix, all five of the eigenvectors are amenable to this approach. If any eigenvector has a zero for that first component, then a different element of the eigenvector should be assumed as one with the appropriate change to the partitioning scheme. At least one of the components must be nonzero. If the element assumed to be one is actually supposed to be zero, then the matrix $\mathbf{B}_i$ will be singular. If an eigenvalue is repeated, then additional values in the eigenvector must be prescribed, again with corresponding changes to the partitioning scheme.

Orthogonality of Eigenvectors The eigenvectors are orthogonal if the eigenvalues are distinct. The proof is straightforward. Take two different eigenpairs $(\lambda_n, \boldsymbol{\varphi}_n)$ and $(\lambda_m, \boldsymbol{\varphi}_m)$. Each pair satisfies Eq. A.23. Premultiply the equation with index n by $\boldsymbol{\varphi}_m^T$ and the equation with index m by $\boldsymbol{\varphi}_n^T$ and subtract the first equation from the second one to get

$$\boldsymbol{\varphi}_n^T\left[\mathbf{A}-\lambda_m\mathbf{I}\right]\boldsymbol{\varphi}_m-\boldsymbol{\varphi}_m^T\left[\mathbf{A}-\lambda_n\mathbf{I}\right]\boldsymbol{\varphi}_n=0 \qquad \text{(A.24)}$$

The result equals zero because of Eq. A.23. The matrix $\mathbf{A}$ is symmetric, so

$$\boldsymbol{\varphi}_n^T\mathbf{A}\boldsymbol{\varphi}_m=\boldsymbol{\varphi}_m^T\mathbf{A}\boldsymbol{\varphi}_n$$

Of course, $\boldsymbol{\varphi}_n^T\boldsymbol{\varphi}_m=\boldsymbol{\varphi}_m^T\boldsymbol{\varphi}_n$, too. Therefore, Eq. A.24 reduces to

$$\left(\lambda_n-\lambda_m\right)\boldsymbol{\varphi}_n^T\boldsymbol{\varphi}_m=0$$

which proves orthogonality, i.e., $\boldsymbol{\varphi}_n^T\boldsymbol{\varphi}_m=0$, as long as the eigenvalues are not equal, i.e., $\lambda_n\neq\lambda_m$. If two or more eigenvalues *are* equal, then the associated eigenvectors form a subspace in which any linear combination of vectors in that subspace satisfies the eigenvalue problem. Say, for example, that $\lambda_n=\lambda_m=\lambda$ are repeated eigenvalues associated with the distinct eigenvectors $\boldsymbol{\varphi}_n$ and $\boldsymbol{\varphi}_m$. In other words, $\mathbf{A}\boldsymbol{\varphi}_n = \lambda\boldsymbol{\varphi}_n$ and $\mathbf{A}\boldsymbol{\varphi}_m = \lambda\boldsymbol{\varphi}_m$ for the same eigenvalue λ. Now, let $\boldsymbol{\varphi} = a\boldsymbol{\varphi}_n + b\boldsymbol{\varphi}_m$ and compute

$$\begin{aligned}\mathbf{A}\boldsymbol{\varphi} &= \mathbf{A}\left(a\boldsymbol{\varphi}_n + b\boldsymbol{\varphi}_m\right)\\ &= a\mathbf{A}\boldsymbol{\varphi}_n + b\mathbf{A}\boldsymbol{\varphi}_m\\ &= a\lambda\boldsymbol{\varphi}_n + b\lambda\boldsymbol{\varphi}_m\\ &= \lambda\left(a\boldsymbol{\varphi}_n + b\boldsymbol{\varphi}_m\right) = \lambda\boldsymbol{\varphi}\end{aligned}$$

Thus, the vector $\boldsymbol{\varphi}$, which is a linear combination of $\boldsymbol{\varphi}_n$ and $\boldsymbol{\varphi}_m$, satisfies the eigenvalue problem for any values of a and b. The vectors $\boldsymbol{\varphi}_n$ and $\boldsymbol{\varphi}_m$ might not be orthogonal, but it is always possible to select vectors from the subspace that are orthogonal or to select two vectors and orthogonalize them. Therefore, the eigenvectors form an orthonormal basis for N-dimensional space.

It is also straightforward to prove that the eigenvectors are $\mathbf{A}$-orthogonal. Premultiply Eq. A.23 (with index m) by $\boldsymbol{\varphi}_n$ to get

$$\boldsymbol{\varphi}_n^T\left[\mathbf{A} - \lambda_m\mathbf{I}\right]\boldsymbol{\varphi}_m = 0$$

Since $\boldsymbol{\varphi}_n^T\boldsymbol{\varphi}_m = 0$ if $\lambda_n \neq \lambda_m$, it must also be true that $\boldsymbol{\varphi}_n^T\mathbf{A}\boldsymbol{\varphi}_m = 0$. Also, note that for orthonormal eigenvectors, $\boldsymbol{\varphi}_n^T\mathbf{A}\boldsymbol{\varphi}_n = \lambda_n$.

The Spectral Representation The matrix $\mathbf{A}$ can be put into *spectral* form using the eigenvalues and eigenvectors. In particular, for orthonormal eigenvectors, we can write

$$\mathbf{A} = \sum_{n=1}^{N}\lambda_n\boldsymbol{\varphi}_n\boldsymbol{\varphi}_n^T$$

The proof is simple. The eigenvector $\boldsymbol{\varphi}_m$ should satisfy $\mathbf{A}\boldsymbol{\varphi}_m = \lambda_m\boldsymbol{\varphi}_m$. Use the spectral representation to compute

$$\mathbf{A}\boldsymbol{\varphi}_m = \sum_{n=1}^{N}\lambda_n\boldsymbol{\varphi}_n\boldsymbol{\varphi}_n^T\boldsymbol{\varphi}_m = \lambda_m\boldsymbol{\varphi}_m$$

Because the eigenvectors are orthogonal, only the mth term of the sum survives, and $\boldsymbol{\varphi}_m^T\boldsymbol{\varphi}_m = 1$. The inverse of $\mathbf{A}$ can also be represented in spectral form. For orthonormal eigenvectors, we can write

$$\mathbf{A}^{-1} = \sum_{n=1}^{N}\frac{\boldsymbol{\varphi}_n\boldsymbol{\varphi}_n^T}{\lambda_n}$$

The proof is similar to the previous one. First, note that $\mathbf{A}^{-1}\mathbf{A} = \mathbf{I}$ and that $\mathbf{I}\boldsymbol{\varphi}_m = \boldsymbol{\varphi}_m$. Now, compute

$$[\mathbf{A}^{-1}\mathbf{A}]\boldsymbol{\varphi}_m = \sum_{n=1}^{N} \frac{\boldsymbol{\varphi}_n \boldsymbol{\varphi}_n^T \mathbf{A} \boldsymbol{\varphi}_m}{\lambda_n} = \frac{\boldsymbol{\varphi}_m \lambda_m}{\lambda_m} = \boldsymbol{\varphi}_m$$

The second equality is true because the eigenvectors are orthogonal and, since the eigenvectors have unit length, $\boldsymbol{\varphi}_m^T \mathbf{A} \boldsymbol{\varphi}_m = \lambda_m$.

Index

A
Acceleration, 96, 450
Anticommutative, 6, 41
Antiderivative, 491
Applied loads, 247
Area, 315, 316
Associated flow rule, 445
Axial bar, 93, 356, 363
  `AxialBar` (code), 123
  `LoadFunction` (code), 125
Axial force, 97, 255
Axiom of locality, 222
Axis, 93

B
Basis, 10
Beam computations
  `BeamCode1` (main code), 369
  `BeamCode2` (main code), 377
  `DistrLoadFcn` (code), 371
  `MultispanBeam` (main code), 385
  `PointLoadFcn` (code), 376
Beam kinematics
  rotation-curvature, 251
  strain-displacement, 252
Bending moment, 255
Bernoulli
  Daniel, 268
  Jacob, 268
Bernoulli-Euler beam theory, 253, 268
Bifurcation diagram, 455
Body force, 191
Boundary conditions, 358
  axial bar, 98
  beam, 268, 269
  essential, 467
  natural, 467
  torsion, 404
Boundary-value problem, 358, 488
Bredt's formula, 430
Bubble functions, 477
Buckling
  load, 455, 463
  mode shape, 457
Bulk modulus, 225

C
Calculus of variations
  fundamental theorem, 467
Cantilever beam, 365
Cauchy
  Augustin-Louis, 192
  formula, 196, 232
  tetrahedron, 193
  wafer, 193
Cayley-Hamilton theorem, 31
Centroid, 55, 98, 247, 258, 315, 317, 331
Characteristic equation, 462
  buckling, 469
  plane strain, 172
  standard eigenvalue problem, 499
  strain, 28, 168
  stress, 205
Characteristic function, 28, 168
Commutative, 5, 6
Compression, 67, 99
Configuration
  deformed, 138

K. D. Hjelmstad, *Mechanics of Deformable Solids: An Introduction*,
https://doi.org/10.1007/978-3-031-89873-0

reference, 138
undeformed, 138
Constitutive model, 103, 221
Continuity conditions
axial bar, 116
beam, 270, 382
torsion, 418
Continuous medium, 139
Continuum mechanics, 139
Convergence
quadratic, 486
Coordinate geometry, 2
Coordinate system, 93
Couple, 58
Creep, 238
Critical state, 454
Cross product, 6, 14
Cross section, 93, 247
nonprismatic, 112
prismatic, 416
Curvature, 251

D
Definite integral, 491
Deflection curve, 282
Deformation
gradient, 140, 178, 182
homogeneous, 137, 139
map, 138, 176
nonhomogeneous, 139, 176
Deformed configuration, 176
Density, 104, 191, 228
Derivative
chain rule, 249, 290, 453, 487
directional, 177, 486
divergence, 212
partial, 212
product rule, 338, 487
Determinant, 23
Differential equation
separable, 268
Dilatation, 174
Directional derivative, 177, 486
Displacement, 138, 181
axial bar, 95
beam, 249
transverse, 247
Divergence, 212
Divergence theorem, 336, 430
Dot product, 5, 14

E
Effective stiffness
axial bar, 105
beam, 259
torsion, 406
Eigenfunction, 463
Eigenpair, 29
Eigenvalue, 27, 456, 499
Eigenvalue problem, 27, 168, 205, 474
matrix, 499
`SolveEVP` (code), 500
stability test, 456
standard, 499
Eigenvector, 27, 457, 499
Elasticity
anisotropic, 237
nonlinear, 237
Elastoplasticity, 237
End-state variables
axial bar, 121
beam, 381
Energy
conservation of, 451
criterion for stability, 455, 467
Equations, condition
overdetermined, 78
singular, 78
underdetermined, 78
Equilibrium
axial bar, 100
beam (force), 257
beam (moment), 257
torsion, 400, 401
Euler, Leonhard, 40, 268
Euler load, 463
Extreme fiber, 300

F
Factor of safety, 454
Failure mode, 228
Finite element method, 482
First yield, 446
Fixed support, 47, 98
Force, 96
Frames, 49, 81
Free body diagram, 39, 40, 42
Free end, 98
Functional, 465

G
Gaussian quadrature, 494

Gauss-Legendre, 495
Generalized trapezoidal rule, 367, 368, 371
Gradient, 178, 487
Green deformation tensor, 178

H
Head-to-tail rule, 3
Hermitian functions, 477
Hooke's law, 222
Hydrostatics, 39, 59

I
Incompressible, 228
Index notation, 336
Inelasticity, 237
Initial parameters
method of, 358, 361, 491
Initial-value problem, 358, 367, 488
Inner product, 19
Integration
by parts, 356, 363
strip method, 321
techniques of, 316
Interior support conditions
beam, 383
Internal forces, 96
Internal hinge, 48, 271, 277, 383
Internal pin joint, 48
Invariants, 23
plane strain, 172
plane stress, 210
primary, 24
principal, 28
strain, 168
stress, 205
Isotropic, 222

K
Kinematic, 404
Kinematic hypothesis
axial bar, 95
beam, 248, 250
torsion, 397
Kinetic, 404

L
Lagrange, Joseph-Louis, 250
Lagrangian strain, 150
tensor, 150
Lamé
Gabriel, 210
parameters, 222
stress ellipsoid, 210
Law of cosines, 5
Leibniz rule, 357
Linear differential equation, 266
Linear function, 484
Linearization, 248, 265
Linearized strain, 182
Local equilibrium equation, 100

M
Machines, 49, 81
Material constants, 221
Matrix, 18
inner product, 19
orthogonal, 12, 24
outer product, 19
Maximum shear angle, 160
Mechanical advantage, 51
Mensuration formulas, 316
Method of joints, 68
Method of sections, 70
Microstrain, 164
Mode shape, 463
Mohr circle, 158, 333
strain, 162, 163
stress, 202
Moment-area method, 362, 365
Moment arm, 41
Moment of area
first, 263, 315, 345
second, 259, 315, 318
Moment of force, 41
Moment of inertia, 259, 315
principal, 333
polar, 407
tensor, 332
Motion, 137
rigid body, 142
Multispan beam, 380, 389

N
Neutral axis, 261
Newton-Raphson, 484
`Newton` (code), 485
Newton, Sir Isaac, 40
Newton's laws of motion, 40
Nodes, 66
Nonhomogeneous deformation, 176
Nonlinear algebraic equations, 483

Nonprismatic, 94, 416
Nonprismatic beam, 291
Normal flow rule, 445
Normal strain, 179
Normal stress, 260
Numerical integration
  backward Euler method, 490
  forward Euler method, 490
  generalized trapezoidal rule, 490
  Simpson's rule, 370, 388
Numerical quadrature
  composite rule, 493
  `NumInt` (code), 496
  stations, 492
  weights, 492

O
Origin, 41
Orthogonal, 9, 10
  matrix, 12, 24
Orthonormal, 10, 24, 500
Outer product, 19

P
Parallel axis theorem, 318, 332
Parallelepiped, 11
Parallelogram rule, 3
Patch loading, 383
Perturbation, 486
Piecewise functions, 114, 417
Pinned support, 47
Planar problems, 22
Plane sections, 250
Plane strain, 155, 172, 230
Plane stress, 199, 210, 229, 230, 233
Plasticity, 237
Plastic strain, 442
Point load, 51, 127, 373
Poisson
  Denis Siméon, 223
  effect, 223, 226
  ratio, 224, 228, 406
Polar moment of area, 407
Position vector, 248
Pressure, 191, 225
Principal
  angle, 158, 202, 333
  directions, 27, 206
  invariants, 34
  strain, 159
  stress, 202
  values, 26, 28, 160, 169, 206
Principia, 40
Prismatic, 94, 396
Product of inertia, 333
Progressive yielding, 445
Projection, 16
Properties of cross sections
  `ComputeProperties` (code), 342
  `CutsForQ` (code), 347
  `InputCrossSection` (code), 343
  `XSectionProp` (Q version), 348
  `XSectionProp` (simple code), 343
Pulley, 49
Pure shear, 405
Pythagorean
  identity, 157
  theorem, 2, 153

Q
Quadratic convergence, 486
Quadrature, 491
  composite Simpson rule, 494
  composite trapezoidal rule, 493
  Gauss-Legendre, 495
  Gauss-Lobatto, 495

R
Radius of curvature, 251
Rank, 78
Raphson, Joseph, 484
Rayleigh, Lord, 472
Reaction force, 42
Reference configuration, 94, 176
Relaxation, 238
Restoring forces, 450
Resultant, 54
Right-hand rule, 6, 255
Rigid body, 248
  modes, 79
Rigid body motion, 142
Ritz function, 473
Ritz method
  `AssembleEqns` (code), 480
  `CantileverRitz` (code), 476
  `RitzBuckling` (main code), 479
  `RitzFcns` (code), 478
Ritz, Walther, 473
Roller support, 47
Rotation, 22
  beam, 249
  torsion, 397

S

Sandwich construction, 106
Second moment of area
  polar, 407
Second-order theory, 460
Section modulus, 300
Self-equilibrating, 79
Shear
  angle, 398
  deformation, 141, 155, 180
  flow, 429
  force, 255
  modulus, 223, 405
Shooting method, 120
Sign convention, 259
Simpson's rule, 493
Spectral representation, 33, 502
Stability
  definition, 451
Static determinacy
  degree of, 287
  determinate, 109
  indeterminate, 65, 109, 287, 413, 418
Static equilibrium, 40, 42, 450
Statics, 39
Step size, 489
Stiffness, 103, 228
Strain, 96, 137, 146, 222
  axial bar, 96
  `Eigen` (code), 170
  engineering, 95
  gauge, 165
  gauge rosette, 165
  linearized, 182
  normal, 141
  resultant, 251
  shear, 141, 151
Strain-displacement relationship
  axial bar, 96
  beam, 251, 266
  linear elasticity, 182
Strain-rotation relationship, 398
Strength, 228, 441, 447, 450
Stress, 191, 222
  axial, 97
  deviatoric, 443
  `Mohr3D` (code), 209
  normal, 198
  shear, 198
Stress formula
  axial bar, 106
  beam, normal stress, 261
  beam, shear stress, 263
  torsion, 409
Stress resultant, 57
  axial bar, 97
  beam, 254
  torque, 399
Stress tensor, 195
  symmetry, 216
Stretch, 146, 177
Structures, 39, 65
Support conditions, 46
Symmetric tensor, 20

T

Tangential deviation, 364
Taylor series, 483
Tension, 67, 99
Tensor, 1, 18, 191
  components, 19
  invariants, 24, 157
  moment of inertia, 333
  product, 333
  projection, 17
  second-order tensor, 20
  symmetric, 20
  trace, 157, 168
Thin-walled section, 425
  closed, 428
  open, 425
Timoshenko
  beam theory, 253
  Stephen P., 253
Torque, 397
Torque-rotation relationship, 406
Torsion, 396
Torsion constant
  closed thin-walled section, 432
  open thin-walled section, 426
  rectangular section, 425
  solid section, 424
  thin rectangle, 426
Trace, 23, 24
Traction, 192, 254, 399
Transformation equations
  strain, 157
  stress, 201
Trapezoidal rule, 492
Triaxial test, 227
Triple scalar product, 6, 340
Trivial solution, 28
Truss, 66
  `CheckCondition` (code), 80
  degree of instability, 79

degree of static indeterminacy, 79
`InputExample` (code), 77
joints, 66
method of joints, 68
method of sections, 70
`TrussAnalysis` (code), 75
Twist, 396
Typical particle, 51, 274, 357
beam theory, 248
construction, 39

U
Ultimate strength
beam, 450
torsion, 447
Undeformed configuration, 94
Unit vectors, 9
Universal ball joint, 47

V
Vector, 1
addition, 3
components, 11
head-to-tail rule, 3, 11
multiplication, 4
outer product, 7
parallelogram rule, 3
subtraction, 3, 4
tensor product, 7
unit, 9
Vector addition
associative law, 7
commutative law, 7
distributive law, 8
Vector multiplication
cross product, 6
dot product, 5
multiplication by scalar, 4
triple scalar product, 6
Velocity, 450
Viscoelasticity, 238
Viscoplasticity, 238
Von Mises, Richard, 443

Y
Yield function, 443
Drucker-Prager, 443
Mohr-Coulomb, 443
von Mises, 443
Yielding, 228
Yield surface, 443
Young's modulus, 103, 104, 224, 228, 247, 258, 406

Z
Zero vector, 3

MIX
Papier aus verantwortungsvollen Quellen
Paper from responsible sources
FSC® C105338

If you have any concerns about our products, you can contact us on
ProductSafety@springernature.com

In case Publisher is established outside the EU, the EU authorized representative is:
Springer Nature Customer Service Center GmbH
Europaplatz 3, 69115 Heidelberg, Germany

Printed by Libri Plureos GmbH
in Hamburg, Germany